AF556686

PLANT GENETIC ENGINEERING

Volume 1
Applications and Limitations

The Editors

Dr Rana P. Singh is a Reader in the Department of Biosciences, M. D. University, Rohtak, India. He is teaching courses in Plant Biology and Biotechnology to graduate and postgraduate students since 1986. He earned his first class M. Sc. in Botany from DDU Gorakhpur Univ., Gorakhapur and Ph.D. in Life Sciences (Plant Biochemistry) from Devi Ahilya University, Indore, India. He has more than 70 research publications including original research papers, reviews and book chapters in reputed International research journals and books. He has edited 5 books published by International publishers. He is fellow and member of various academic societies. Dr Singh has been awarded R. D. Asana Medal from Indian Soc. Plant Physiol., New Delhi and JEB Young Scientist award of The Academy of Environmental Biology, India for significant research contributions. He is recipient of INSA, New Delhi and UNESCO, Paris Biotech. Fellowships and visited twice at University of Guelph, Canada for advance research. He has visited Yunnan Agri. Univ., Kunming, China and hosted three Senegalese Scientists for research collaborations. He has guided many students for Ph.D., M. Phil. and M. Sc. degrees and completed many projects in areas of his interest. His group has contributed significant work in nitrogen relation to plants. Dr Singh is Editor-in-chief of an International research journal, Physiology and Molecular Biology of Plants and on editorial board of three other journals. His current research interests are metabolic engineering for nutrient use efficiency, stress tolerance and nutritional improvement in plants.

Dr Pawan K. Jaiwal is a Reader in the Department of Biosciences, M. D. University, Rohtak, India. He is teaching Molecular Biology, rDNA technology and Plant Biotechnology courses to graduate and postgraduate students since 1986. He has throughout first division academic career with Ph.D. in Botany (Plant Physiology) from Kurukshetra University, Kurukshetra, India. He has more than 70 research publications including original research papers, reviews and book chapters in reputed International research journals and books. He has edited three books published by International publishers. Dr Jaiwal has been awarded DST Young Scientist Project, INSA Visiting Fellowship, New Delhi and DBT Overseas Associateship (Long Term) for advance research at laboratory of Prof Ingo Potrykus, ETH, Switzerland. Dr Jaiwal is a member of several academic bodies and on the Editorial Board of four International research journals. His has guided many students for Ph.D., M. Phil. and M. Sc. degrees. He has completed several research projects on Biotechnology for improvement of important grain legumes. His group has developed gene transfer protocols in *Vigna radiata* and *V. mungo* for the first time. His current research interests are metabolic engineering for resistance to abiotic and biotic stresses, improvement of nutrient use efficiency and nutritional quality of grain legumes.

Other Volumes in this series:

Volume 2 **Improvement of Food Crops :** *Eds.* Pawan K. Jaiwal & Rana P. Singh

Volume 3 **Improvement of Commercial Plants - I :** *Eds.* Rana P. Singh & Pawan K. Jaiwal

Volume 4 **Improvement of Commercial Plants - II :** *Eds.* Pawan K. Jaiwal & Rana P. Singh

Volume 5 **Improvement of Vegetables :** *Eds.* Rana P. Singh & Pawan K. Jaiwal

Volume 6 **Improvement of Fruits :** *Eds.* Pawan K. Jaiwal & Rana P. Singh

PLANT GENETIC ENGINEERING

Volume 1

Applications and Limitations

Editors

Rana P. Singh **Pawan K. Jaiwal**

Department of Biosciences
M. D. University,
Rohtak-124001 India

2003

SCI TECH PUBLISHING LLC, U.S.A.

ISBN : 1-930813-02-3
SERIES ISBN : 1-930813-17-1

First Published 2003

Published by
SCI TECH PUBLISHING LLC.
9207, Country Creek Drive
Houston, Texas - 77036 (U.S.A.)
Tele. : 713-541-94-- Fax : 713-541-9401
E-mail : sci@hotmail.com

Printed by
RAINBOW PROCESSORS & PRINTERS
New Delhi-110059 India

PREFACE

The world population is expected to increase from about 6.5 billion at present to 10 billion by 2050. Most of this increase will be in the developing countries. The burgeoning population and detoriation of the natural resources have challenged the planners, scientists and farmers to meet the food security and nutritional quality for all. Improvement of crops using hybrid technology was a major breakthrough which could witnessed several high yielding crop varieties. However, many limitations have been realized for this technology e.g. limited gene pool and sexual incompatibility with wild relatives. The extensive use of water, fertilizers and agrochemicals etc. for the cultivation of these crops has resulted in serious environmental degradation and health hazards. The gene technology developed during the last 15 years have improved crops in more precise and faster ways as compared to the traditional Mendelian methods. Designer crops based on novel genetic combinations have been created by moving genes across the sexual barriers.

Susceptibility of the crop plants to various biotic and abiotic factors in the various agroclimatic conditions has been a major concern for more crop productivity per unit area. Development of the transgenics for resistance/tolerance of these factors have been achieved in many crops by genetic engineering. Certain novel molecular techniques have been identified to cause great impact on gene manipulation. RNA editing in plant organelles is one of those which has been advanced considerably during the recent years (Chapter 1). The strategies, current status, state of art, applications and limitations of genetic engineering approaches for the development of transgenic plants resistant to insect pests, viral and bacterial pathogens (Chapters 2-4) and abiotic stress tolerance (Chapters 5-8) have been discussed by the researchers actively working in these areas.

Depletion of soil nutrients over the years has also resulted in stagnation or even decrease in crop productivity in many agroclimatic zones. Genetic transformation of micro-organisms enriching soil nutrients and plants with improved nutrient use efficiency has been achieved. Chapter 9 deals with genetic manipulations of diazotrophic bacteria to enhance the potential of biological nitrogen fixation in plants. Such approaches have great significance for sustainable agriculture. Nutritional quality improvement of plant produces has become a focus of attention in the recent years using genetic engineering techniques. The gene technology has opened potentials for production of valuable commercial products from plants which have made possible the establishment of wide range biotechnology industries throughout the world. As a result the area under cultivation of transgenic plants has increased upto 45 million hectares in 2001 in less than a decade. Manipulation of carotenoids, lignin and production of antibodies, edible vaccines and secondary metabolites of pharmaceutical interest has been discussed in chapters 10-13. Asthetic values are often related to flowers and the global market for floricultural products are expanding rapidly. It is possible now to modify flower colour in desired manner by genetic engineering (Chapter 14).

The possible environmental and health concerns of GMO crops have generated vital public and scientific debate throughout the world. One such concern relates with the gene(s) that code for antibiotic and herbicide resistance which are co-transferred into the host genome with desirable gene(s). The product of these genes may not be necessarily harmful, albeit their presence in transgenic plants increases the chances of their escape through pollen or seed dispersal to the wild and weed relatives of the crops. Antibiotic resistance genes may get passed into pathogenic micro-organisms in the gastro-intestinal tract or soil, making them resistant to the treatment with such antibiotics. Similarly the resultant gene products in the transgenic plants may cause unknown health hazards to the consumers (Chapter 16). A case to case examination of transgenes product and the subsequent flow of transgene to the neighbouring crops as well as horizontal flow should be carefully evaluated in accordance with the desired regulations before reaching to a conclusion on their commercialization. The Intellectual Property Rights under TRIPs is another debated issue which is complex in context of agricultural practices and socio-economic conditions in many countries. Chapter 17 discusses this aspect in the context of Indian agriculture. Bioinformatics has recently emerged as an important area for designing gene manipulation in genomic and post genomic era. Chapter 18 deals this aspect of knowledge.

The response to our invitations for the contribution in this series has been overwhelming. We are grateful to our contributors for their kind response and patience during the production of all these six volumes simultaneously. Meeting the dealine of a publication of this magnitude puts a heavy toll on editors in partitioning the time between family, routine professional activities and publication work. We are thankful to our family members and research students who might have suffered on various occassions due to our deep involvement in this multi volume series. bringingout this series to a reality.

Rohtak, India
25 March, 2002

Rana P Singh
Pawan K Jaiwal

CONTENTS

Contents of other volumes in the series

PLANT GENETIC ENGINEERING

SERIES ISBN : 1-930813-17-1

ISBN : 1-930813-03-1

Volume 2 : **Improvement of Food Crops**
Editors : Pawan K. Jaiwal & Rana P. Singh

ISBN: 1-930813-04-X

Volume 3 : **Improvement of Commercial Plants - I**
Editors : Rana P. Singh & Pawan K. Jaiwal

ISBN : 1-930813-04-X

Volume 4 : **Improvement of Commercial Plants-II**
Editors : Pawan K. Jaiwal & Rana P. Singh

ISBN: 1-930813-14-7

Volume 5 : **Improvement of Vegetables**

Editors : Rana P. Singh & Pawan K. Jaiwal

ISBN : 1-930813-15-5

Volume 6 : **Improvement of Fruits**
Editors : Pawan K. Jaiwal & Rana P. Singh

Chapter 1

RNA EDITING IN PLANT ORGANELLES

Valérie Blanc[1], Jean-Claude Farré, Dominique Bégu and Alejandro Araya*

Laboratoire de Replication et Expression des Génomes Eucaryotes et Rétroviraux (REGER), UMR 5097, Centre National de la Recherche Scientifique and Université Victor Segalen-Bordeaux 2, 146 rue Léo Saignat, 33076 Bordeaux Cedex, France.

Summary

Organellar RNAs are subject to post-transcriptional C-to-U conversions at highly specific sites by a deamination mechanism. This RNA modification involves most plant mitochondrial mRNAs and a restricted number of chloroplast transcripts. The C-to-U changes alter the coding capacity of mRNAs without modification of the reading frame and lead to the synthesis of proteins with an amino acid sequence different to that predicted from the gene. Proteins produced from edited mRNAs are more similar to those from organisms where this process is absent. Some tRNAs and intron regions can also be subjected to RNA editing. Adaptive strategies in plant organelles allow the RNA editing process to be implicated in control of gene expression such as the production of functional proteins or the modulation of RNA maturation process. It has also been proposed as a correction step. The hypothesis that RNA editing leads to functional proteins was tested by constructing transgenic plants expressing an unedited mitochondrial protein. These proteins were characterized by a male-sterile phenotype. Some examples of natural cytoplasm male-sterile plants correlate with defect in RNA editing and are interesting models to study this process.

[1]Present address : Washington University School of Medicine, Division of Gastroenterology, 660 South Euclid Avenue, Campus Box 8124, Saint Louis, Missouri 63110-1093, U.S.A.

*Corresponding author : E-mail : Alexandre.Araya@reger.u-bordeaux2.fr

Keywords : Chloroplast, mitochondria, RNA editing, precursors, intron, splicing, mRNA, cytoplasmic male sterility (CMS).

1. INTRODUCTION

The concept of RNA editing was first used to describe the insertion of uridine residues within the coding region of mitochondrial mRNAs from kinetoplastid protozoa (Benne *et al.*, 1986). Then, the definition of editing was extended to any co- or post-transcriptional process that changes the primary sequence of an RNA from that encoded by the corresponding gene. The editing process includes changes resulting in nucleotide insertion or deletion and also in base conversion reactions (Blanc *et al.*, 1996; Bass, 1997; Estevez and Simpson, 1999). RNA editing has been described in the nuclear compartment of eukaryotic cells, in the genome of some viruses and in the mitochondria and chloroplasts of land plants (Table 1).

RNA editing in plant organelles is characterized by C-to-U and a few U-to-C transitions. Evidence for RNA editing is found in all vascular plant organelles (seed plants, ferns and fern allies) and in all bryophytes group except the subclass of thalloid liverworts (Marchantiidae) (Malek *et al.*, 1996; Freyer *et al.*, 1997). Recently, the complete analysis of *Arabidopsis thaliana* mtDNA and their transcripts reveals 456 C-to-U editing events (Giege and Brennicke, 1999). In wheat mitochondria 1200 editing sites have been estimated (Bonen, 1991; Schuster and Brennicke, 1994). U-to-C transitions are rare in higher plant (Schuster *et al.*, 1990) and fern mitochondria (Malek *et al.*, 1997) but are frequent in hornworts (Steinhauser *et al.*, 1999). In chloroplasts, only some mRNAs undergo editing in coding regions in comparison to the high number of C-to-U transitions observed in mitochondria. Similar to the situation described in mitochondria, a high frequency of U-to-C transitions is observed in hornwort chloroplasts (Yoshinaga *et al.*, 1996). RNA editing is also observed in non-photosynthetic plastids (Hirose *et al.*, 1994).

2. RNA EDITING IN CODING REGIONS

RNA editing events occur mostly in the coding region of the mRNAs from plant organelles, altering the identity of the encoded amino acid. The editing process changes the genetic message to produce proteins which are more similar to the functional homologous from mammals and yeast organelles where RNA editing does not occur. In general, RNA

Table 1. Different types of RNA editing. More details, which are not described in the text, can be found in the reviews from (Araya *et al.*, 1994; Gray, 1996; Chan *et al.*, 1997; Smith *et al.*, 1997; Stuart *et al.*, 1997). This table has been adapted from Smith *et al.* (Smith *et al.*, 1997).

	Compartment	Organism	Mechanism
I. Insertion-Deletion (Reading frame changes)			
U insertion/deletion	Kinetoplastids	*Trypanosoma, Leishmania, Crithidia*	Cleavage/ligation
C insertion (rarely U, AA CU, GU, GC insertions)	Mitochondria. mRNAs, rRNAs, tRNAS	*Physarum polycephalum*	Unknown
G insertion	Unknown	Paramyxoviruses	Co-transcriptional, viral replicase
A insertion	Unknown	Ebola viruses	Co-transcriptional
3' mRNA poly A synthesis	Mitochondria. mRNAs	Vertebrates	Poly A addition after processing cleavage
II. Base modifications or replacement (no reading frame changes)			
C-to-U	Mitochondria, mRNAs, tRNAs	Land plants,	C deamination
C-to-U	Chloroplasts, mRNAs	Land plants	C deamination?
C-to-U	Mitochondria	*Physarum polycephalum*	Unknown

Table 1. Continued

	Compartment	Organism	Mechanism
C-to-U	Nuclei ApoB mRNA	Mammals	C deamination
U-to-C	Mitochondria, Chloroplasts, mRNA	Land plants.	Unknown
A-to-I	Nuclei, Glutamic and serotonin receptor subunits mRNAs	Vertebrates	A deamination (ds-RNA)
A-to-I	Nuclei? Antigenomic RNA	Hepatitis Virus	A deamination (ds-RNA)
U-to-A	Nuclei α-galactosidase mRNA	Humans	
C-to-A, A-to-G U-to-A, U-to-G	Mitochondria, tRNAs	*Acanthamoeba castellani*	

editing increases the hydrophobicity of the encoded mitochondrial proteins. Comparison of a given gene between various plants shows the conservation of editing sites amongst many species (Zanlungo *et al.*, 1993). Most modifications in mRNAs occur in C residues located in the first or the second position of the codon thus generating changes in the encoded protein. It is interesting to note that some codon families can be found edited either at first, second or at both position and thus are potentially capable to generate several different codons, depending on the base found at the third position of the codon. For example the codon CCN (Pro) may generate the codons *U*CN (Ser), C*U*N (Leu), *UU*R (Leu) and *UU*Y (Phe). The codon family CGN (Arg) may be edited as *U*GY (Cys), *U*GG (Trp) or *U*GA (Stop). Nevertheless, some changes appear to be without effect because they are produced at the third position of the codon which do not alter the amino acid identity (silent editing). In such case, the importance of editing events does not appear obvious. Probably, the "silent" editing sites may represent an adaptation of the mitochondria to the individual tRNA abundance or are required for other purposes.

In some cases, the RNA editing event delimitates the reading frame in the messenger by introducing new translation signals in the mRNA such as initiation or stop codons (Fig. 1) (Bégu *et al.*, 1990; Hoch *et al.*, 1991; Quiñones *et al.*, 1995). In hornwort and in some ferns, U-to-C RNA editing reestablishes the open reading frame by the removal of several stop codons (Malek *et al.*, 1996; Steinhauser *et al.*, 1999).

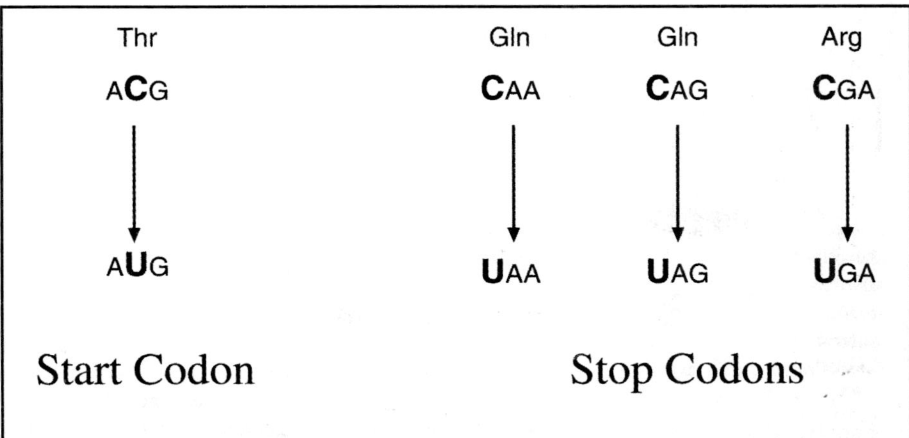

Figure 1 : Scheme of base changes involved in the generation of initiation and stop codons by RNA editing in plant organelles.

3. EDITING IN INTRONS

Higher plant mitochondria are characterized by the presence of *cis*- and *trans*-splicing introns (Binder *et al.*, 1996). With one exception, all of them are group II introns presenting a well-conversed secondary structure (Michel *et al.*, 1989). At least 25 editing sites have been described in different intronic sequences in plant organellar transcripts (Fig. 2).

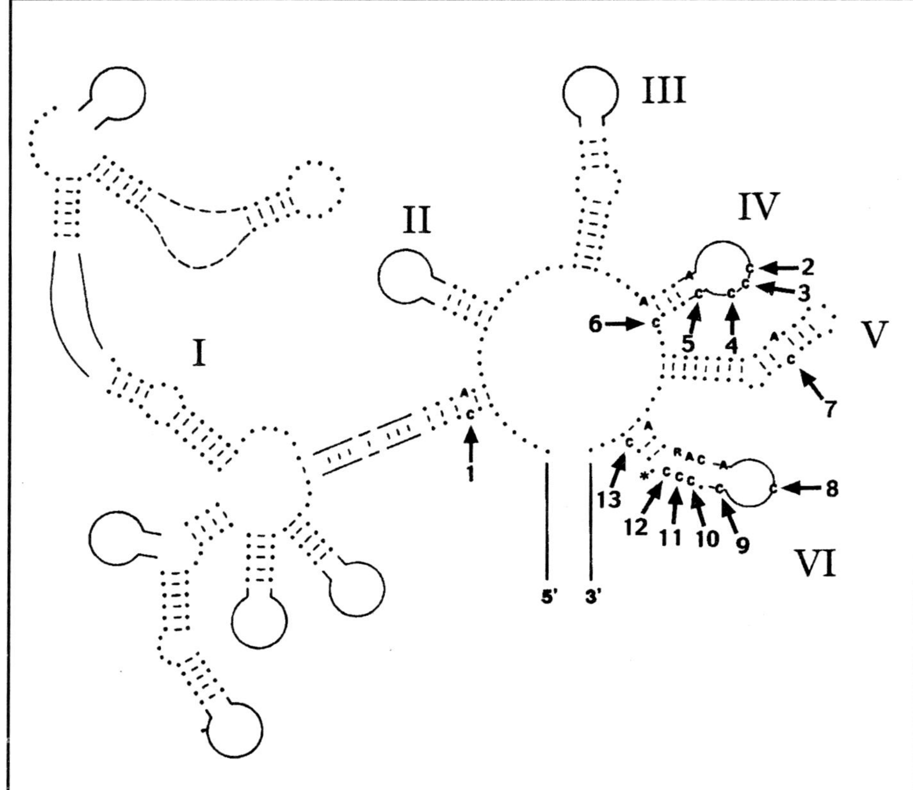

Figure 2 : Scheme of the canonical structure of group II introns proposed by Michel *et al.* (1989). Arrows indicate the residues edited in different introns from several plant mitochondria : 1, *cis*-intron of *S. tuberosum rsp*10 (Zanlungo *et al.*, 1995) and intron 2 from Oenothera *nad2* (Binder *et al.*, 1992); 2 and 3, intron 4 from Oenothera *nad5* (Knoop *et al.*, 1991); 4 and 5, intron 2 from Oenothera *nad2* (Binder *et al.*, 1992); 6, intron 1 from Oenothera *nad2* (Lippok *et al.*, 1994); 7 and 8, introns 3 and 4 from wheat *nad7* (Carrillo and Bonen, 1997); 9, intron 1 from *A. thaliana nad5* (Giege and Brennicke, 1999); intron 1 of *nad5* from *A. thaliana* (Giege and Brennicke, 1999) and Oenothera (Knoop *et al.*, 1991).

The RNA editing sites in intron sequences described so far in different species are shown in Figure 2. At the difference of coding regions, the number of editing sites is restricted to one or two sites in each intron. A majority of edited residues are found in highly structured regions located in domains I, IV, V and VI. Most of them revert A-C mispairing to a correct base paring A-U. Because these editing events improve the conformation of the intron, they have been considered necessary for efficient splicing (Wissinger *et al.*, 1991; Zanlungo *et al.*, 1995). Particularly interesting is the situation of domain VI where many editing sites are found in the vicinity of an A residue in a bulge, which is involved in forming the lariat structure during splicing (Michel *et al.*, 1989). The canonical structure of domain VI, suitable for splicing, is predicted when the C residue is edited (Farré and Araya, 1999). We analyze the editing status of the C residue at position 11 (Fig. 2) in the intermediate (lariat) molecules generated by splicing of exons *nad1*d-e and *nad5* II-III. All the molecules engaged in splicing were edited. In contrast, the unspliced precursors showed partial editing at the respective sites (personal communication). By grafting domain VI from *trans*-intron 3 of Oenothera *nad1* in yeast *al5c* intorn, Börner *et al.* (1995). Found that the editing in the stem (site 12 in Fig. 2) was necessary for splicing of the chimeric intron. These results clearly indicate that editing process may be involved in the control of the mRNA maturation events. However, editing in domain V in introns 3 and 4, or in domain VI of intron 4 of wheat *nad7* (sites 7 and 8 in Fig. 2 respectively), shown no evident correlation between editing and maturation status (Carrillo and Bonen, 1997). Interesting is the situation of site 1 where the C-to-U editing site is located in domain I of *Solanum tuberosum rps10* intron (Zanlungo *et al.*, 1995). This position is crucial for splicing as inferred from mutation studies in a yeast mitochondial intron model (Schmelzer and Schweyen, 1986). Some editing sites are located in regions that have not an apparent double-stranded structure such as sites, 2, 3 and 4 in domain IV and site 8 in domain VI (Fig. 2). To understand the role of RNA editing at these sites, more information concerning the actual structure and the possible 3-D interactions of these regions are required.

Some mitochondrial tRNAs also undergo editing process that correct mispairing in the acceptor stem (Maréchal-Drouard *et al.*, 1993). The precise role of editing in the "adaptor" function of tRNA in protein synthesis remains to be determined. However, tRNA editing seems to be necessary for precursor maturation (Marchfelder *et al.*, 1996; Maréchal-Drouard *et al.*, 1996).

4. RNA EDITING FUNCTION

As is discussed above, the emergence of RNA editing in land plants provided the organelles with an additional step during gene expression. This process can fulfill diverse function as to adapt the genetic message to synthesize better adapted polypeptides in a kind of "correction" process. In other cases, partial edited RNAs may generate variant proteins from one message (Lu and Hanson, 1996; Lu *et al.*, 1996; Phreaner *et al.*, 1996). However, the last is not a general situation because, in some case, only the "edited" form of ATP6 protein accumulates in mitochondria despite the presence of partially edited *atp6* transcripts (Lu and Hanson, 1994). In maize mitochondria only edited translation products are detected (Williams *et al.*, 1998). The different situations observed might result from an evolutionary process whereby the cell takes advantage of the emergence of protein variants to fulfill new functions. By contrast, generation of protein diversity by RNA editing may be negatively selected when this concerns proteins that are crucial for mitochondrial functions as ATPase subunits 6 or 9. The overall extent of transcript editing may vary in different genotypes and is affected by nuclear genotype (Lu and Hanson, 1992; Wilson and Hanson, 1996) indicating that nuclear factors may be involved in the editing process.

In non-coding sequences, RNA editing seems to be used as a signal to perform subsequent maturation steps, or to correct mismatch pairing in highly structured RNAs. However, this does not seems to be a general situation for all transcripts (Schock *et al.*, 1998).

5. MECHANISM OF RNA EDITING

The finding of partially edited RNAs *in vivo* indicates that RNA editing is a post-transcriptional process. Whereas RNA editing in trypanosome mitochondria proceeds with a 3' to 5' polarity within the editing domain, in plant organelles the conversions occur in a non-oriented way and independently at each editing site (Schuster *et al.*, 1990).

Different biochemical processes may account for a C-to-U transition: (i) cleavage of the RNA; nucleotide exchange and religation of the RNA, (ii) base exchange by transglycosylation and (iii) deamination of the base. The last process is involved in C-to-U RNA editing of the apolipoprotein B (apoB) mRNA (Navaratnam *et al.*, 1995). Several independent approaches, *in vitro* (Blanc *et al.*, 1995; Yu and Schuster, 1995) and *in organello* (Rajasekhar and Mulligan, 1993), showed that the sugar-phosphate backbone remains intact during the conversion reaction ruling out an enzyme cascade mechanism. Furthermore, performing the editing reaction on a double-labeled RNA with ^{3}H on the base and ^{32}P on the α-phosphate of the C residues, it was found that the C-to-U transition occurs by a deamination reaction and discards a transglycosylation pathway. The mechanism of U-to-C change remains unknown. So far, no cytosine deaminase has been described as being able to perform the reverse reaction. It is possible that a different enzyme, catalyzing either a transamination or a transglycosylation reaction, may be responsible for the U-to-C changes.

6. SPECIFICITY OF RNA EDITING

A major concern in RNA editing studies is to understand how the editing machinery recognizes specific cytosine residues. The comparison of the sequences flanking editing sites does not show any apparent conserved motif at the primary structure level (Giege and Brennicke, 1999), suggesting requirement of individual *cis*-acting sequence elements determining each editing site.

The recombinant ability of plant mitochondria DNA in generating chimeric genes that contain fragments of normal functional genes (Hanson and Folkerts, 1992) give us a clue to this problem. In chimeric transcripts, the sequences derived from normal mitochondrial genes are edited at the same positions as found in the functional mRNAs. This supports the idea that only the neighboring sequences are involved in editing specificity. It has been suggested that the 5' sequence, upstream of editing sites, is sufficient for recognition (Kubo and Kadowaki, 1997; Williams *et al.*, 1998). It should be noted that as little as 39 nucleotides are sufficient to direct the editing machinery towards the editing sites of the *atp9* sequences found in the chimeric *C-atp6* gene of maize mitochondria (Kumar and Levings III CS, 1993). Similar observations were obtained in chloroplasts using transformation techniques to introduce exogenous sequences containing editing sites (Chaudhuri and Maliga, 1996; Bock *et al.*, 1996). Scanning point mutagenesis experiments in the tobacco *ndhB* transcripts shown that only few nucleotide in 5' and 3' of an editing site have an effect on RNA editing (Bock *et al.*, 1997). Recent studies indicate that a precisely defined distance between the editing site and upstream *cis*-element is a crucial parameter for selection of the correct cytosine to be edited (Hermann and Bock, 1999). The role of 5' sequences is well illustrated in the case of *ndhH-D* transcripts where the editing at site III is performed only in spliced transcripts but not in the unspliced precursor (Del Campo *et al.*, 2000). This notion of distance between editing and recognition sites recalls the C-to-U editing of apoB mRNA. In this case, the editing site is located 4 to 6 bases 5' of this sequence (Backus and Smith, 1991). It is also possible that particular secondary structures play a role in editing site recognition, similar to those directing A-to-I or C-to-U conversions in mammals (Higucki *et al.*, 1993; Anant and Davidson, 2000).

There are compelling evidences on the participation of nucleus-encoded *trans*-acting factors in the editing reactions in chloroplasts (Chaudhuri *et al.*, 1995; Bock and Koop, 1997). A spinach plastid DNA containing a gene with an editing site remains unedited after being introduced into tobacco plastid genome (Bock *et al.*, 1994). In contrast, the same site is edited when the tobacco plastids are associated with a spinach nuclear background. Furthermore, the introduction of a transgene containing additional RNA copies in chloroplast reduces the editing of the corresponding transcript suggesting that the recognition factors for some editing sites are depletable (Chaudhuri *et al.*, 1995; Bock *et al.*, 1996). The molecular identity of the trans-acting factors remains unknown. The participation of small RNA molecules such as the guide RNAs (gRNA) necessary for editing in kinetoplastids, have been postulated. If such gRNAs exist, they will undergo a compartment-specific import considering that a mitochondrial sequence is not edited in transgenic plastids (Sutton *et al.*, 1995) and that a promiscuous chloroplast sequences is not edited in rice mitochondria. (Zeltz *et al.*, 1996).

Land plant mitochondria has large genomes that may code for putative gRNA molecules (Bock *et al.*, 1996; Zeltz *et al.*, 1996). Some RNAs with antisense properties were identified for some but not for all editing sites. These putative gRNAs sequences have been looked for unsuccessfully by several approaches. In the case of editing in non-coding, highly structured regions in introns or in some tRNAs, a kind of *cis*-acting guide can be apparent but, as discussed above, this is not a general rule. As long as their general relevance for all the editing sites, in both organelles and in all plants, will not be established their significance in RNA editing will remain doubtful.

It has been suggested that protein factor could be involved. If it is the case, RNA editing should be dependent on non-organellar factors because this process is independent from organellar protein synthesis (Zeltz *et al.*, 1993).

7. EVOLUTIONARY ORIGIN OF RNA EDITING

Considering the distribution of RNA editing in plant mitochondria, this process appears to originate from an early ancestor of the land plants. The evidence for this timing is the presence of editing in organelles of all land plants including mosses, liverworts and hornworts and, in contrast, the absence of RNA editing in green algae (Malek *et al.*, 1996; Steinhauser *et al.*, 1999).

The experimental evidences indicating the absence of the RNA editing process in mitochondria of Marchantiidae (Oda *et al.*, 1992) is intriguing. One can suppose that marchantiidae has lost the editing capacity after they diverged from other bryophytes, unless a reduced frequency of editing remains undetected in that group.

Covello and Gray (1993) have proposed an interesting model to explain the emergence of RNA editing. Initially, the editing activity is an extant of a pre-existing enzyme (deaminase) that gained the capacity to bind and act on polynucleotides sequences. A mutation introduced into the genome was rendered neutral by the editing activity that repairs the default on the RNA. As the number of editing events increases the editing activity becomes essential in the gene expression pathway and is maintained by natural selection. Considering this theory, if the number of editing events is low or non-existing like in Marchantia, the editing activity appears dispensable and potentially may disappear.

8. RNA EDITING AND CMS

In flowering plants, male sterility is due to abnormal development of the male gametophyte resulting in the inability to produce functional pollen. In the case of cytoplasmic male sterility (CMS), the abnormalities observed in tapetal cells are originated by dysfunction of the mitochondria that is essential for pollen development. In several cases, this tissue specificity may result from interactions between the mitochondrial DNA and nuclear genes that regulate mitochondrial gene expression (Schnable and Wise, 1998; Vedel *et al.*, 1994; Wen and Chase, 1999).

Rearrangements in the mitochondrial genome that create novel chimeric genes and the synthesis of new proteins are correlated with CMS phenotype (Dewey *et al.*, 1986). In the case of Texas CMS maize, the production of a 13 kDa protein (URF13) encoded by the chimeric gene *T-urf13* results in mitochondria abnormalities observed in soporophytic tissues, particularly the tapetal cells of the anther (Levings and Siedow, 1992).

The multiplicity of the genes involved in CMS plants reported so far suggests that mitochondrial impairment may be the common consequence of the action of different chimeric genes products. If this is true, then a mitochondrial dysfunction could also be produced by other means. Considering that the RNA editing process is crucial for mitochondria gene expression, an RNA editing dysfunction may also generates a male-sterile phenotype. We have reported the production of male-sterile tobacco plants by expression of the unedited mitochondrial ATPase subunit 9 gene (*u-atp9*), fused to a mitochondrial transit sequence (Hernould *et al.*, 1993). In transgenic tobacco, the expression of *u-atp9* causes a mitochondrial dysfunction that affects normal anther development, specifically in the tapetal cell layer, and reduces pollen formation (Hernould *et al.*, 1998). The expression of the unedited gene was suppressed by antisense strategy resulting in the recovery of male fertility (Zabaleta *et al.*, 1996). These results indicate first that the editing process ensures the synthesis of functional mitochondrial proteins and, second that RNA editing process may be involved as one of the causes of mitochondrial dysfunction leading to male sterility.

Several authors have searched for a correlation between male sterility and RNA editing in plants with variable results (Handa and Nakajima, 1992; Iwabuchi *et al.*, 1993; Stahl *et al.*, 1994). A more carefully analysis have been performed in *Sorghum bicolor*. In a male-sterile line of *S. biocolor*, the editing of one *atp6* mRNA was strongly reduced in anthers. Restoration of fertility in F_1 and F_2 lines correlated with an increase in editing levels of *atp6* transcripts (Howad and Kempken, 1997; Howad *et al.*, 1999). These data suggest that RNA editing is necessary to produce functional ATP6 and that the loss of *atp6* mRNA editing contributes to induce CMS in this plant. This situation is reminiscent of some human mitochondrial pathologies where point mutations in some genes result in mitochondria dysfunction and tissue-specific diseases (Kempken *et al.*, 1998).

Interestingly, in *S. bicolor* highly edited transcripts were efficiently processed indicating that a more complex events can be involved in the emergence of the CMS phenotype and confirm the idea that RNA editing may influence the transcript processing efficiency (Howad *et al.*; Pring *et al.*, 1999). The *S. bicolor* is an interesting model since it shows tissue- and transcript-specific modulation of the RNA editing process.

9. CONCLUSION AND FUTURE PROSPECTS

It is intriguing that a process that can be lost during evolution by a natural gene mutation process such as is observed in liverworts, is maintained in most plant lineage. Probably this situation reflects the fact that RNA editing plays a major role in land plants organellar gene expression. Several aspects of the gene expression pathways seem to be affected by editing such as RNA processing and the synthesis of functional proteins. Some other putative functions remain to be discovered such as the role of editing at silent positions and the function of variant proteins generated from partially edited mRNAs. While knowledge on chloroplast RNA editing is being unveiled thanks to the transgenic technology and the ability of chloroplast to undergo homologous recombination, mitochondrial editing process lacks of efficient experimental approaches. The design of plant mitochondrial transformation systems should be now considered as an important key to open the black box of mitochondrial RNA editing. The main challenges are focused on the identification of the factors responsible for site-specific editing reactions. The correlation observed between RNA editing events, male sterility and nuclear restorer genes will allow to gain insight on the factors involved in editing site recognition in plant mitochondria but also open the way to control male fertility and to design new biotechnological tools to produce male-sterile lines.

REFERENCES

Anant S and Davidson NO (2000). An AU-rich sequence element (UUUN[A/U]U) downstream of the edited C in apolipoprotein B mRNA is a high-affinity binding site for Apobec-1: binding of Apobec-1 to this motif in the 3' untranslated region of c-myc increases mRNA stability. *Mol. Cell. Biol.,* **20** : 1982-1992.

Araya A, Bégu D and Litvak S (1994). RNA editing in plants. *Physiol. Plant.,* **91** : 543-550.

Backus JW and Smith HC (1991). Apolipoprotein B mRNA sequences 3' of the editing site are necessary and sufficient for editing and editosome assembly. *Nucleic Acids Res.,* **19** : 6781-6786.

Bass BL (1997). RNA editing and hypermutation by adenosine deamination. *Trends Biochem. Sci.,* **22** : 157-162.

Bégu D, Graves PV, Domec C, Arselin G, Litvak S and Araya A (1990). RNA editing of wheat mitochondrial ATP synthase subunit 9: direct protein and cDNA sequencing. *Plant Cell,* **2** : 1283-1290.

Benne R, Van den Burg J, Brakenhoff JP, Sloof P, Van Boom JH and Tromp MC (1986). Major transcript of the frameshifted *coxII* gene from trypanosome mitochondria contains four nucleotides that are not encoded in the DNA. *Cell,* **46** : 819-826.

Binder S, Marchfelder A and Brennicke A (1996). Regulation of gene expression in plant mitochondria. *Plant Mol. Biol.,* **32** : 303-314.

Binder S, Marchfelder A, Brennicke A and Wissinger B (1992). RNA editing in trans-splicing intron sequences of *nad2* mRNAs in *Oenothera* mitochondria. *J. Biol. Chem.,* **267** : 7615-7623.

Blanc V, Jordana X, Litvak S and Araya A (1996). Control of gene expression by base deamination: the case of RNA editing in wheat mitrochondria. *Biochimie,* **87** : 511-517.

Blanc V, Litvak S and Araya A (1995). RNA editing in wheat mitochondria proceeds by a deamination mechanism. *FEBS Lett.,* **373** : 56-60.

Bock R, Kossel H and Maliga P (1994). Introduction of a heterologous editing site into the tobacco plastid genome: the lack of RNA editing leads to a mutant phenotype. *EMBO J.,* **13** : 4623-8.

Bock R, Hermann M and Kössel H (1996). *In vivo* dissection of *cis*-acting determinants for plastid RNA editing. *EMBO J.,* **15** : 5052-5059.

Bock R and Koop HU (1997). Extraplastidic site-specific factors mediate RNA editing in chloroplasts. *EMBO J.,* **16** : 3282-3288.

Bock R, Hermann M and Fuchs M (1997). Identification of critical nucleotide positions for plastid RNA editing site recognition. *RNA,* **3** : 1194-1200.

Bonen L (1991). "The mitochondrial genome: so simple yet so complex." *Curr. Opin. Genet. Dev.,* **1** : 515-522.

Börner GV, Mörl M, Wissinger B, Brennicke A and Schmelzer C (1995). RNA editing of a group II intron in Oenothera as a prerequisite for splicing. *Mol. Gen. Genet.,* **246** : 739-744.

Carrillo C and Bonen L (1997). RNA editing status of *nad7* intron domains in wheat mitochondria. *Nucleic Acids Res.,* **25** : 403-409.

Chaudhuri S, Carrer H and Maliga P (1995). Site-specific factor involved in the editing of the *psbL* mRNA in tobacco plastids. *EMBO J.,* **14** : 2951-2957.

Chaudhuri S and Maliga P (1996). Sequences directing C to U editing of the plastid psbL mRNA are located within a 22 nucleotide segment spanning the editing site. *EMBO J.,* **15** : 5958-5964.

Covello PS and Gray MW (1993). On the evolution of RNA editing. *Trends Genet.* **9** : 265-268.

Del Campo EA, Sabater B and Martin M (2000). Transcripts of the *ndhH-D* operon of barley plastids: possible role of unedited site III in splicing of the ndhA intron. *Nucleic Acids Res.,* **28** : 1092-1098.

Dewey RE, Levings CS and Timothy DH (1986). "Novel recombinations in the maize mitochondrial genome produce a unique transcriptional unit in the Texas male-sterile cytoplasm. *Cell,* **44** : 439-449.

Estevez AM and Simpson L (1999). Uridine insertion/deletion RNA editing in trypanosome mitochondria-a review. *Gene,* **240** : 247-260.

Farre JC and Araya A (1999). 'The *mat-r* open reading frame is transcribed from a non-canonical promoter and contains an internal promoter to co-transcribe exons *nad1e* and *nad5III* in wheat mitochondria. *Plant Mol. Biol.,* **40** : 959-967.

Freyer R, Kiefer-Meyer MC and Kossel H (1997). Occurrence of plastid RNA editing in all major lineages of land plants. *Proc. Natl. Acad. Sci. USA.,* **94** : 6285-6290.

Giege P and Brennicke A (1999). RNA editing in *Arabidopsis* mitochondria effects 441 C to U changes in ORFs. *Proc. Natl. Acad. Sci. USA.,* **96** : 15324-15329.

Gray MW (1996). RNA editing in plant organelles: a fertile field. *Proc. Natl. Acad. Sci. USA*. **93** : 8157-8159.

Handa H and Nakajima K (1992). RNA editing of atp6 transcripts from male-sterile and normal cytoplasms of rapeseed (*Brassica napus* L.)" *FEBS Lett.,* **310** : 111-114.

Hanson MR and Folkerts O (1992). Structure and function of the higher plant mitochondrial genome. *Int. Rev. Cytol.,* **141** : 129-172.

Hermann M and Bock R (1999). Transfer of plastid RNA-editing activity to novel sites suggests a critical role for spacing in editing-site recognition. *Proc. Natl. Acad. Sci. USA.,* **96** : 4856-4861.

Hernould M, Suharsono S, Zabaleta E, Carde JP, Litvak S, Araya A and Mouras A (1998). "Impairment of tapetum and mitochondria in engineered male-sterile tobacco plants. *Plant Mol. Biol.,* **36** : 499-508.

Hernould M, Suharsono S, Litvak S, Araya A and Mouras A (1993). Male-sterility induction in transgenic tobacco plants with an unedited *atp9* mitochondrial gene from wheat. *Proc. Natl. Acad. Sci. USA.,* **90** : 2370-2374.

Higuchi M, Single FN, Köhler M, Sommer B, Sprengel R and Seeburg PH (1993). RNA editing of AMPA receptor subunit GluR-B: a base-paired intron-exon structure determines position and efficiency. *Cell,* **75** : 13661-1370.

Hirose T, Wakasugi T, Sugiura M and Kossel H (1994). RNA editing of tobacco *petB* mRNAs occurs both in chloroplasts and non-photosynthetic proplastids. *Plant Mol. Biol.,* **26** : 509-13.

Hoch B, Maier RM, Appel K, Igloi GL and Kissel H (1991). Editing of a chloroplast mRNA by creation of an initiation codon. *Nature,* **353** : 178-180.

Howad W and Kempken F (1997). Cell type-specific loss of *atp6* RNA editing in cytoplasmic male sterile *Sorghum bicolor. Proc. Natl. Acad. Sci. USA.,* **94** : 11090-11095.

Howad W, Tang HV, Pring DR and Kempken F (1999). Nuclear genes from Tx CMS maintainer lines are unable to maintain atp6 RNA editing in any anther cell-type in the *Sorghum bicolor* A3 cytoplasm. *Curr. Genet.,* **36** : 62-8.

Iwabuchi M, Kyozuka J and Shimamoto K (1993). Processing followed by complete editing of an altered mitochondrial atp6 RNA restores fertility of cytoplasmic male sterile rice. *EMBO J.,* **12** : 1437-1446.

Kempken F, Howard W and Pring DR (1998). Mutations at specific *atp6* codons which cause human mitochondrial diseases also lead to male sterility in a plant. *FEBS Lett.,* **441** : 159-160.

Knoop V, Schuster W, Wissinger B and Brennicke A (1991). Trans-splicing integrates an exon of 22 nucleotides into the *nad5* mRNA in higher plant mitochondria. *EMBO J.,* **10** : 3483-93.

Kubo N and Kadowadi K (1997). Involvement of 5' flanking sequence for specifying RNA editing sites in plant mitochondria. *FEBS Lett.,* **413** : 40-44.

Kumar R and Levings CS (1993). RNA editing of a chimeric maize mitochondria gene transcript is sequence specific. *Curr. Genet.,* **23** : 154-9.

Levings CSD and Siedow JN (1992). Molecular basis of diseases susceptibility in the Texas cytoplasm of maize. *Plant Mol. Biol.,* **19** : 135-47.

Lippok B, Brennicke A and Wissinger B (1994). Differential RNA editing in closely related introns in *Oenothera* mitochondria. *Mol. Gen. Genet.,* **243** : 39-46.

Lu B and Hanson MR (1992). A single nuclear gene specifies the abundance and extent of RNA editing of a plant mitochondrial transcript. *Nucleic Acids Res.,* **20** : 5699-703.

Lu B and Hanson MR (1994). A single homogeneous form of ATP6 protein accumulates in petunia mitochondria despite the presence of differentially edited atp6 transcripts. *Plant Cell,* **6** : 1955-68.

Lu B and Hanson MR (1996). Fully edited and partially edited *nad9* transcripts differ in size and both are associated with polysomes in potato mitochondria. *Nucleic Acids Res.,* **24** : 1369-1374.

Lu B, Wilson RK, Phreaner CG, Mulligan RM and Hanson MR (1996). Protein polymorphism generated by differential RNA editing of a plant mitochondrial *rps12* gene. *Mol. Cell. Bol.,* **16** : 1543-1549.

Malek O, Brennicke A and Knoop V (1997). Evolution of trans-splicing plant mitochondrial introns in pre-Permian times. *Proc. Natl. Acad. Sci. USA.,* **94** : 553-558.

Malek O, Lätting K, Hiesel R, Brennicke A and Knoop V (1996). RNA editing in bryophytes and a molecular phylogeny of land plants. *EMBO J.,* **15** : 1403-1411.

Marchfelder A, Brennicke A and Binder S (1996). RNA editing is required for efficient excision of tRNA(Phe) from precursors in plant mitochondria. *J. Biol. Chem.,* **271** : 1898-1903.

Maréchal-Drouard L, Kumar R, Remacle C and Small I (1996). RNA editing of larch mitochondrial tRNA (His) precursors is a prerequisite for processing. *Nucleic Acids Res.,* **24** : 3229-3234.

Maréchal-Drouard L, Ramamonjisoa D, Cosset A, Weil JH and Dietrich A (1993). Editing corrects mispairing in the acceptor stem of bean and potato mitochondrial phenylalanine transfer RNAs. *Nucleic Acids Res.,* **21** : 333-342.

Michel F, Umesono K and Ozeki H (1989). Comparative and functional anatomy of group II catalytic introns-a review. *Gene,* **82** : 5-30.

Navarathnam N, Bhattacharya S, Fujino T, Patel D, Jarmaz AL and Scott J (1995). Evolutionary origins of apoB mRNA editing: catalysis by a cytidine deaminase that has acquired a novel RNA-binding motif at its active site. *Cell.,* **81** : 187-195.

Oda K, Yamato K, Ohta E, Nakamura Y, Takemura M, Nozato N, Akashi K, Kanegae T, Ogura Y and Kohchi T (1992). Gene organization deduced from the complete sequence of liverwort Marchantia polymorpha mitochondrial DNA. A primitive form of plant mitochondrial genome. *J. Mol. Biol.* **223** : 1-7.

Phreaner CG, Williams MA and Mulligan RM (1996). Incomplete editing of rps12 transcripts results in the synthesis of poymorphic polypeptides in plant mitochondria. *Plant Cell,* **8** : 107-17.

Pring DR, Tang HV, Howad W and Kempken F (1999). A unique two-gene gametophytic male sterility system in sorghum involving a possible role of RNA editing in fertility restoration. *J. Hered.,* **90** : 386-93.

Quiñones V, Zanlungo S, Holuigue L, Litvak S and Jordana X (1995). The *coxl* initiation codon is created by RNA editing in potato mitochondria. *Plant Physiol.,* **108** : 1327-1328.

Rajasekhar VK and Mulligan MR (1993). RNA editing in plant mitochondria: a phosphate is retained during C to U conversion in mRNAs. *Plant Cell,* **5** : 1843-1852.

Schmelzer C and Schweyen RJ (1986). Self-splicing of group II introns *in vitro* : mapping of the branch point and mutational inhibition of lariat formation. *Cell,* **46** : 557-565.

Schnable PS and Wise RP (1998). The molecular basis of cytoplasmic male sterility and fertility restoration. *Trends Plant Sci.,* **3** : 175-180.

Schock I, Marechal-Drouard L, Marchfelder A and Binder S (1998). Processing of plant mitochondrial tRNAGly and tRNASer(GCU) is independent of RNA editing. *Mol. Gen. Genet.,* **257** : 554-560.

Schuster W and Brennicke A (1994). The plant mitochondrial genome: structure, information content, RNA editing and gene transfer. *Annu. Rev. Plant Physiol. Plant Mol. Biol.,* **45** : 61-78.

Schuster W, Wissinger B, Unseld M and Brennicke A (1990). Transcripts of the NADH-dehydrogenase subunit 3 gene are differentially edited in *Oenothera* mitochondria. *EMBO J.,* **9** : 263-269.

Smith HC, Gott JM and Hanson MR (1997). A guide to RNA editing. *RNA,* **3** : 1105-1123.

Stahl R, Sun S, L'Homme Y, Ketela T and Brown GG (1994). RNA editing of transcripts of a chimeric mitochondrial gene associated with cytoplasmic male-sterility in *Brassica. Nucleic Acids Res.,* **22** : 2109-2113.

Steinhauser S, Beckert S, Capesius I, Malek O and Knoop V (1999) Plant mitochondrial RNA editing. *J. Mol. Evol.* **48** : 303-312.

Stuart K, Allen TE, Heidmann S and Sewiert SD (1997). RNA editing in kinetoplastid protozoa. *Microbiol. Mol. Biol. Rev.,* **61** : 105-120.

Sutton CA, Zoubenko OV, Hanson MR and Maliga P (1995). A plant mitochondrial sequence transcribed in transgenic tobacco chloroplasts is not edited. *Mol. Cell. Biol.,* **15** : 1377-1381.

Vedel F, Pla M, Vitart V, Gutierres S, Chêtrit P and De Paepe R (1994). Molecular basis of nuclear and cytoplasmic male sterility in higher plants. *Plant Physiol. Bochem.,* **32** : 601-18.

Wen L and Chase CD (1999). Mitochondrial gene expression in developing male gametophytes of male-fertile and S male-sterile maize. *Sex. Plant Reprod.,* **11** : 323-330.

Williams MA, Kutcher BM and Mulligan RM (1998). Editing site recognition in plant mitochondria: the importance of 5'flanking sequences. *Plant Mol. Biol.,* **36** : 229-237.

Williams MA, Tallakson WA, Phreaner CG and Mulligan RM (1998). Editing and translation of ribosomal protein S13 transcripts: unedited translation products are not detectable in maize mitochondria. *Curr. Genet.,* **34** : 221-226.

Wilson RK and Hanson MR (1996). Preferential RNA editing at specific sites within transcripts of two plant mitochondrial genes does not depend on transcriptional context or nuclear genotype. *Curr. Genet.,* **30** : 502-508.

Wissinger B, Schuster W and Brennicke A (1999). Trans splicing in *Oenothera* mitochondria: *nad1* mRNAs are edited in exon and trans-splicing group II intron sequendes. *Cell,* **65** : 473-482.

Yoshinaga K, Iinuma H, Masuzawa T and Uedal K (1996). Extensive RNA editing of U to C in addition to C to U substitution in the *rbcL* transcripts of hornwort chloroplasts and the origin of RNA editing in green plants. *Nucleic Acids Res.* **24** : 1008-1014.

Yu W and Schuster W (1995). Evidence for a site-specific cytidine deamination reaction involved in C to U RNA editing of plant mitochondria. *J. Biol. Chem.,* **270** : 18227-18233.

Zabaleta E, Mouras A, Hernould M, Shharsono S and Araya A (1996). Transgenic male-sterile plant induced by an unedited *atp9* gene is restored to fertility by inhibiting its expression with antisense RNA. *Proc. Natl. Acad. Sci. USA.*, **93** : 11259-11263.

Zanlungo S, Bégu D, Quinones V, Araya A and Jordana X (1993). RNA editing of apocytochrome b (cob) transcripts in mitochondria from two genera of plants. *Curr. Genet.*, **24** : 344-348.

Zanlungo S, Quiñones V, Moenne A, Holuigue L and Jordana X (1995). Splicing and editing of *rsp10* transcripts in potato mitochondria. *Curr. Genet.*, **27** : 565-571.

Zeltz P, Hess WR, Neckermann K, Börner T and Kössel H (1993). Editing of the chloroplast *rpoB* transcript is independent of chloroplast translation and shows different patterns in barley and maize. *EMBO J.*, **12** : 4291-4296.

Zeltz P, Kadowaki K, Kubo N, Maier RM, Hirai A and Kössel H (1996). A promiscuous chloroplast DNA fragment is transcribed in plant mitochondria but the encoded RNA is not edited. *Plant Mol. Biol.*, **31** : 647-656.

Chapter 2

GENETIC ENGINEERING FOR INSECT RESISTANCE

Lise Jouanin★, Christophe Ripoll and Céline Deraison

Boilogie cellulaire, INRA, 78000 VERSAILLES Cedex, France

Summary

Plant genetic engineering offers new possibilities for plant improvement. Phytophagous insects belonging to the Lepidoptera, the Coleoptera and the Homoptera orders are major pests for the most important crops worldwide. Different approaches have been considered in order to obtain insect-resistant crops including the expression of entomotoxic proteins. The main strategy used until now consists in the expression of endotoxins (Cry) originating from Bacillus thuringiensis (Bt). Many Bt-expressing plants have been obtained and some are already commercialized. In order to enlarge the spectra of activity against insects and to co-express different toxins in transgenic crops, screenings for new entomotoxic proteins of plant, bacterial, and insect origins, have been performed and some genes encoding these toxins have already been introduced into crops and tested against selected insect pests. Current work area is presented in this chapter. In addition, the management strategies for deploying insect-resistant crops as well as the potential risks for humans and non-target insects of these plants are discussed.

Keywords : Transgenic plants; entomotoxin; *B. thuringiensis*; resistance management; non-target insects; risk assessment.

★Corresponding author : E-mail : jouanin@versailles.inra.fr

1. INTRODUCTION

Pest control is largely accomplished by the use of chemical pesticides; however, losses in the major crops remains important (Oerke, 1994; Fig. 1). In addition, major problems related to the use of these products have been reported, the most important being detrimental impacts on the environment, such as pollution of land and water tables, toxicity towards non-target organisms and accumulation in food chains. Therefore, it is necessary to develop more friendly pest-protection procedures. The use of other types of pest controls such as breeding for resistant varieties, modified agricultural practices, biological products, must be developed. It this context, transgenic plants represent a very promising technology. The first transgenic plants were obtained in 1983 (Horsch *et al.*, 1983) and reports of the first applications to insect resistance were published in 1987 (Barton *et al.*, 1987; Fischhoff *et al.*, 1987; Hilder *et al.*, 1987; Vaeck *et al.*, 1987). Many field trials have been performed in different

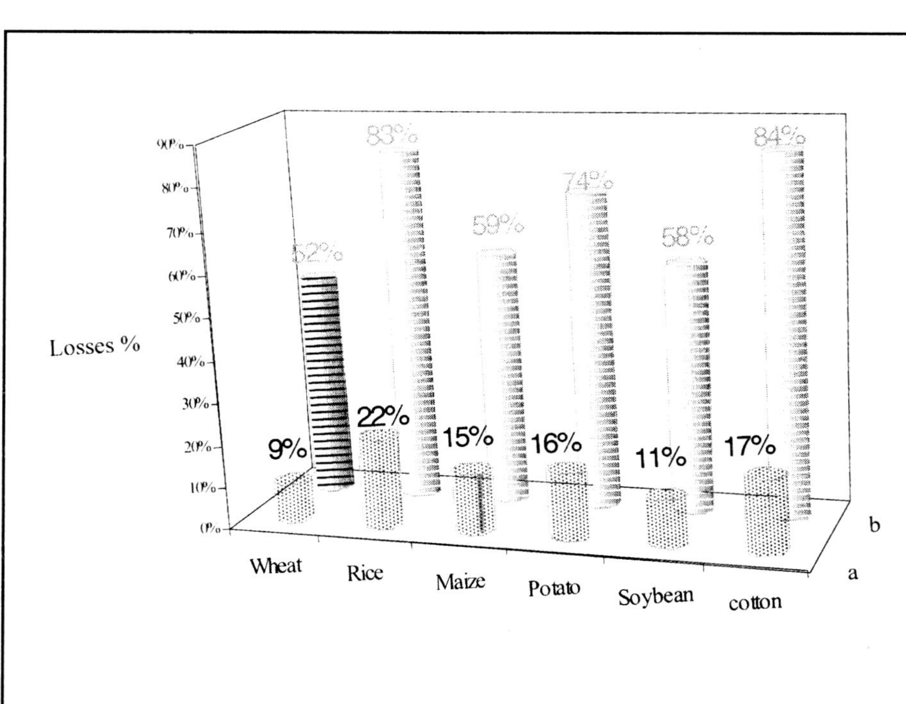

Figure 1 : Woldwide crop losses (according to Oerke *et al.*, 1994).
A : Crop losses caused by insects
B : Predicted crop losses caused by insects in absence of pesticides.

countries during the following years, and in 1996 lepidopteran-insect-resistant cotton was commercialized and cultivated in the USA (Estruch *et al.*, 1997).

The expression of an insecticidal protein in plants presents many advantages over the spray of chemicals. The "toxin", confined in the plant, is active at the early stage of the insect attack reducing the damages and affecting only phytophagous insects. It can be effective against insects feeding inside the plant (borers) and on parts of the plant difficult to reach (roots). The protection is more durable since the toxin is confined in the plant. The culture costs are reduced (but the seeds are more expensive) and the environment is more protected. Before introduction and expression in a transgenic plant, the gene(s) encoding the insecticidal protein must be identified. To confer the resistance trait, the "toxin" must be active after ingestion since it is present into the insect food. This consideration has, up until now, excluded the use of neurotoxins. Insecticidal proteins can be of diverse origins and the most well-known are derived from bacteria or plants. Until now, the insects which have been the main targets of this approach have been Lepidoptera which cause large losses in cultivated crops. While, the expression of endotoxins originating from the bacteria, *Bacillus thuringiensis,* has been the most successful strategy for obtaining resistant plants, many other strategies are also under evaluation. The aim of this paper is to summarize the data obtained and to discuss the potential problems posed by the use of this new method. The reader is also referred to other recent reviews (Estruch *et al.*, 1997; Jouanin *et al.*, 1998; Schuler *et al.*, 1998; Hilder and Boulter, 1999).

2. STRATEGIES FOR CREATING INSECT RESISTANT PLANTS

Insects are one of the major cause of losses in the cultivated crops. Depending upon the plant considered, different classes of insects are the main pests (Table 1). The first part of this chapter will list the genes encoding known insecticidal proteins and the second part will describe their introduction into transgenic plants.

2.1. Genes encoding insecticidal proteins

2.1.1. *Bacillus thuringiensis* δ-endotoxins

Bacillus thuringiensis is a gram-positive bacteria that synthesizes insecticidal *Cry*stalline inclusions during sporulation. The *Cry*stalline

Table 1. Main insect pests on different important crops

Plant	Common name	Insect (Latin name)	Order
Tobacco	Tobacco budworm	*Manduca sexta*	Lepidoptera
Maize	Western corn rootworm	*Diabrotica virgifera virgifera*	Coleoptera
	Northern corn rootworm	*Diabrotica longicornis barberi*	Coleoptera
	European corn borer	*Ostrinia nubilalis*	Lepidoptera
	Fall armyworm	*Spodoptera frugiperda*	Lepidoptera
Cotton	Tobacco budworm	*Heliothis virescens*	Lepidoptera
	Bollworm	*Helicoverpa zea*	Lepidoptera
	American budworm	*Heliothis armigera*	Lepidoptera
	Egyptian cotton worm	*Spodoptera littoralis*	Lepidoptera
	Cotton boll weevil	*Anthonomus grandis*	Coleoptera
Rice	Striped stem borer	*Chilo suppressalis*	Lepidoptera
	Yellow rice borer	*Scirpophaga incertulas*	Lepidoptera
	Brown planthopper	*Nilaparvata lugens*	Homoptera
	Green leafhopper	*Nephotetix virescens*	Homoptera
Wheat	Aphid	*Rhopalosiphum padi*	Homoptera
	Aphid	*Sitobion avenae*	Homoptera

Table 1. Continued

Plant	Common name	Insect (Latin name)	Order
Oilseed rape/Canola	Diamondback moth	*Plutella xylostella*	Lepidoptera
	Stem flea beetle	*Psylliodes chrysocephala*	Coleoptera
	Pod weevil	*Ceuthorhynchus assimilis*	Coleoptera
	Cabbage aphid	*Brevicoryne brassicae*	Homoptera
Potato	Colorado potato beetle	*Leptinotarsa decemlineata*	Coleoptera
	Potato tuber moth	*Phthorimaea operculella*	Lepidoptera
		Aulacorthum solani	Homoptera
		Myzus persicae	Homoptera

structure of the inclusion is made up of protoxin subunits called δ-endotoxins (*Cry* for *Cry*stal proteins). Most *Bt* strains produce several *Cry* proteins, each possessing a specific host range. The narrow host range of each individual toxin makes this group of insecticidal proteins very attractive with respect to both efficiency and environmental safety. Many different genes have being isolated and sequenced. The classification of the *Cry* proteins has been modified recently (Crickmore *et al.*, 1998). The nomenclature was initially based on insecticidal activity (Höfte and Whiteley, 1989) but many exceptions to this systematic arrangement became apparent making the nomenclature system inconsistent. A new nomenclature, based on hierarchical clustering using amino-acid sequence identity, was proposed using Roman numerals in place of Arabic numerals (See *Bt* toxin nomenclature at: http://epunix.biols.susx.ac.uk/Home/Neil_Crickmore/Bt/insdex.html). A large number of the isolated and characterized genes encode toxins active against Lepidoptera (*Cry*1A, *Cry*1B, *Cry*1C, *Cry*2, *Cry*9..) although some are toxic for Coleoptera (*Cry* 3). Most of these proteins, even in the *Cry*1 subfamily, have a distinctive insecticidal spectrum. The size of most of these *Cry* proteins is about 130 kD and they are produced in an inactive form (Fig. 2). After ingestion, the alkaline environment of the insect midgut causes the *Cry*stals to dissolve and release their protoxins (several protoxins can be included in the same *Cry*stal). The protoxin then needs to be cleaved by gut proteases to give a 65-70 kD truncated form which is the active toxin. The toxin binds to specific receptors on

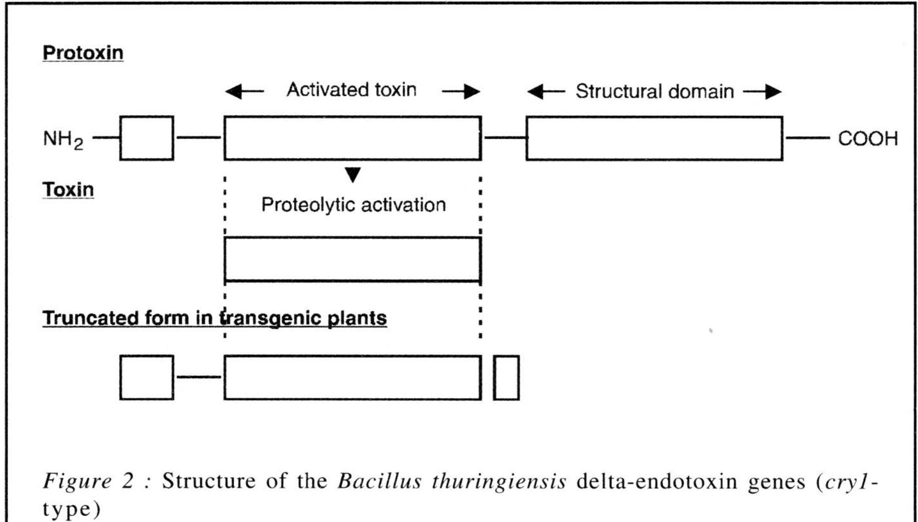

Figure 2 : Structure of the *Bacillus thuringiensis* delta-endotoxin genes (*cry1*-type)

the cell membranes and forms pores that destroy the epithelial cells by colloid osmotic lysis (Knowles and Dow, 1993) resulting in the death of the insect. Specificity is, to a large extent, determined by a toxin-receptor interaction (van Rie *et al.*, 1990) although solubility of the *Cry*stal and protease activation also play a role (de Maagd *et al.*, 1999).

B. thuringiensis was first used as a bioinsecticide against different lepidopteran pests (Lambert and Perferoen, 1992), however, due to the low field-persistance the use of *Bt* sprays is relatively limited. The fact that *Bt* toxins are safe for non-target organisms and for mammals together with their high and rapid toxicity towards target insects, as well as the availability of a large number of genes possessing different specificities, make these toxins very interesting for introduction into plants.

2.1.2. Plant genes

2..1.2.1. Plant proteinase inhibitors : Plant proteinase inhibitors (PI) are small proteins which participate in the natural defence of plants against herbivory (Ryan *et al.*, 1990) Hydrolysis of dietary proteins in insects can involve different types of digestive proteinases: serine-, cysteine-, aspartic- and metallo-proteinases, and different proteinases predominate in the gut according to the insect order. Serine proteinases are dominant in Lepidopteran larvae (Christeller *et al.*, 1992) and cysteine proteinases in Coleopteran larvae (Murdock *et al.*, 1987). Many different plant serine PI have been characterized and cloned; they can be classified according to their sequence homology (Boulter, 1993). The most known are the Bowman-Birk, the Kunitz and the potato PI. Fewer plant cysteine PI have been characterized and cloned.

The action mode of serine- and cysteine-PI at the molecular level is known (Bode and Huber, 1992). They are competitive inhibitors and form non-covalent complexes with proteases. The antimetabolic action of these PI against insects is not fully understood : direct inhibition of digestive enzymes or enzyme hypersecretion (to overcome the inhibition) inducing depletion in essential amino-acids (Reeck *et al.*, 1997).

2.1.2.2. Plant lectins : Lectins are proteins containing at least one non-catalytic domain which binds reversibly to a specific mono- or oligosaccharide (Peumans and Van Damme, 1995). They can be subdivided into four categories: the merolectins which contain only one carbohydrate binding domain, the hololectins which present at least two such domains, the chimerolectins are a fusion of a carbohydrate-binding domain with an unrelated domain, and the superlectins contain at least two carbohydrate-binding domain with different sugar specificities. Lectins have been isolated from many plant tissues such as seeds, storage and vegetative tissues of dicots and monocots. On the basis of molecular and structural analysis, the plant lectins can be classified in seven families, legume lectins, chitin-binding lectins monocot mannose-binding lectins, type 2 RIP and related lectins, jacalin-related lectins, amarathin lectins and cucurbitaceae phloem lectins (Van Damme *et al.*, 1998). The role of lectins in the plant is not well characterised actually, but they are supposed to be involved in different physiological processes such as storage proteins, sugar transport, cell-to-cell recognition, interaction with microorganisms and defense against pest and pathogens. The first report of an insecticidial role was made by Jansen with the lectin of *Phaseolus vulgaris* PHA against the bruchid beetle (Jansen *et al.*, 1976). For the past few years, lectins from a wide variety of sources have been tested for their entomotoxic properties in intensive screening programmes. These experiments have shown that lectines belonging to different families and with different sugar specificities exert interesting effects on different genus of insects. The effects were characterized by growth delay of insects, a decrease of fecundity and sometimes mortality (Czapla, 1997). The mechanism of lectin action on insects is not well determined. It involves binding to the midgut of the insect (Sauvion, 1995; Powell *et al.*, 1998; Zhu *et al.*, 1998; Habidi *et al.*, 1998; Habidi *et al.*, 2000). This binding could lead to severe or slight damages to the midgut epithelium depending on the insect species and lectins used. In some cases, the lectin crosses the epithelium and is found in the haemolymph (Powell *et al.*, 1998). All the informations available until now show differences according to the lectin-insect interaction studied, and do not allow the determination of a specific mode of action. It seems that lectins could induce different effects such as disorganisation of the midgut cells, antifeedant effects by preventing absorption of nutrients or by blocking glycoproteins involved in digestion or transport (Czapla, 1997).

2.1.2.3. α-amylase inhibitors (α-AI) : The commun bean, *Phaseolus vulgaris*, contains a family of related seed proteins (PHA-E and L, arcelin and α-amylase inhibitors). PHA-E and –L are classical lectins with strong agglutinin activity while α-AI can complex insect α-amylases and is supposed to play a role in plant defense. α-AI inhibits the α-amylase present in the midgut of coleopteran pests of stored-products (*Callosobruchus, Bruchus...*) (Shade *et al.*, 1994). Genes encoding α-AI have also been isolated from cereals.

2.1.3 Other toxins of bacterial origin

In order to identify new insecticidal proteins, large screenings of bacterial extracts have been initiated in different laboratories (Estruch *et al.*, 1997). These programs have allowed the identification of new gene candidates for creating insect-resistant crops.

2.1.3.1. Vegetative Bt proteins : In the search for novel insecticidal proteins, it was discovered that supernatants from exponential cultures of *B. thuringiensis* contained toxins active against Lepidoptera such as *Argotis ipsilon* (black cutworm, BCW). Two of these toxins, VIP (for vegetative insecticidal proteins), toxic for lepidopteran larvae have been isolated (Estruch *et al.*, 1996). The toxins VIP3A(a) and (b) do not belong to the δ-endotoxin family but are toxic in the same range. The symptomatology developed upon VIP3A-ingestion resembles that of δ-endotoxins with a specific binding to gut epithelium cells (Yu *et al.*, 1997).

2.1.3.2. Vegetative Bacillus cereus toxins : Estruch *et al.* (1996) have also reported the isolation of proteins possessing insecticidal properties from supernatants of *Bacillus cereus* isolates. The toxicity for the Northern- and Southern-corn rootworms (*Diabrotica longicornis barbieri* and *Diabrotica virgifera virgifera*) is associated with a binary system called VIP1 and VIP2. The genes encoding this binary system are good candidates for expression in plants and especially in corn.

2.1.3.3. Streptomyces cholesterol oxidase : *Streptomyces* cultures secrete a cholesterol oxidase (COX) active against the boll weevil (*Anthonomus grandis*), a major cotton pest worldwide. This protein is active within the same range as Bt toxins (Purcell *et al*,. 1993) and has been expressed in tobacco protoplasts (Corbinet *et al.*, 1994). However, until now, no report of its successful introduction into cotton has been reported.

2.1.3.3. Photorhadus luminescens toxins : *Photorhadus luminescens* is a gram-negative bacterium, mutualistic with entomophagous nematodes. It is released in the hemocoel of infected insects and participates in the lethality of the insects. The toxins produced by the bacteria are normally delivered directly to the hemocoel but they are also active after ingestion (Bowell and Ensign, 1998). During the culture of *P. luminescens*, several toxins are secreted into the growth media during the stationary phase of culture (Guo *et al.*, 1999). The most abundant, named Toxin complex a (Tca), is highly active against

M. sexta with a mode of action similar to that observed with *B. thuringiensis* toxins (Blackburn *et al.*, 1998).

2.1.4 Use of genes of insect origin

In the search for new toxin genes, several studies have raised the possibility of interfering with insect physiology using proteinase inhibitors or chitinase of insect origin. For example, one serine PI isolated from the hemolymph of *M. sexta* induces negative effects on insect development (Thomas *et al.*, 1994, 1995a, 1996b). In a plant, the constitutive expression of a *M. sexta* chitinase gene, which is normally expressed during insect moulting, interferes with digestion since chitin is present in insects, not only as an exoskeletal material, but also in the peritrophic membrane (Kramer *et al.*, 1997).

2.2 Insect-resistant transgenic plants

The availability of genes encoding toxins active against insects after ingestion has allowed the development of programs for testing the possibility of using the transgenic strategy to create insect-resistant crops.

2.2.1 Lepidopteran-resistance plants

The main phytophagous insects are the larvae of Lepidoptera (about 44% of chemical treatments are directed against these insects). *B. thuringiensis* strains active against this class of insects were the first to have been discovered and *Bt* is, therefore the most well-known strategy used to create crops resistant to these insects, however, many other approaches are at the experimental stage.

2.1.1.1 ***Bt plants :*** The first published reports of the introduction of *Cry*1A genes into plants were published in 1987 (Barton *et al.*, 1987; Fishhhof *et al.*, 1987; Vaeck *et al.*, 1987) where transformed plants were tobacco and tomato, which can be considered as model plants. *Bt* genes have been transferred to a number of other crops such as cotton, maize, rice.. (Table 2; reviewed in Mazier *et al.*, 1997; Peferoen, 1997a). In early experiments, both full-length (encoding the protoxin) and truncated (encoding the Nterminal part of the protein) *Cry* genes (Fig. 2) were introduced into plants via *Agrobacterium tumefaciens* transformation. Only plants expressing the truncated genes presented some protection against the insect larvae. Field trials performed on these first generation *Bt*-plants demonstrated a low protection level in natural conditions (Peferoen, 1997a). Several attempts were made to increase the level of

Table 2. Some examples of Lepidopteran-resistant *Bt* transgenic plants.

Plant	*Cry* gene	Type	Target insects	Reference
Tobacco	*Cry1Aa*	WT	*Manduca sexta*	Barton *et al.*, 1987
	Cry1Ab	WT	*M. sexta*	Vaeck *et al.*, 1987
	Cry1Ab & c	PM	*M. sexta*	Perlak *et al.*, 1991
	Cry1C	S	*Spodoptera littoralis*	Strizhov *et al.*, 1996
Chloroplasts	*Cry1Ac*	WT	*H.virescens, H. zea, S. exigua*	McBride *et al.*, 1995
	Cry2Aa2	WT	*H. virescens, H. zea, S. exigua*	Kota *et al.*, 1999
Tomato	*Cry1Ab*	WT	*H. virescens*	Fischoff *et al.*, 1987
Canola	*Cry1Ac*	S	*Plutella xylostella, H. zea*	Stewart *et al.*, 1996
Rice	*Cry1Ab*	S	*Chilo suppressalis*	Fujimoto *et al.*, 1993
	Cry1Ab	S	*Scirpophaga incertulas*	Nayak *et al.*, 1997
	Cry2A	S	*S. incertulas, C. medinalis*	Ghareyazie *et al.*, 1998; Cheng *et al.*, 1998
Potato	*CryV*	S	*Phthorimaea operculella*	Douches *et al.*, 1998
Alfalfa	*Cry1C*	S	*S. liitoralis*	Strizhov *et al.*, 1996
Cotton	*Cry1Ab & c*	S	*H. zea H. virescens*	Perlak *et al.*, 1990
Maize	*Cry1Ab*	S	*Ostrinia nubilabis*	Koziel *et al.*, 1993
	Cry9C	S	*O. nubilabis*	Pefereon, 1997

WT = native gene; P = partially-modified gene, S = synthetic gene expression: the use of improved promoters (CaMV 35S with a duplicated enhancer sequence) and the use of untranslated virus leader sequences. However, the best improvement was observed by using partial or entirely synthetic genes. *B. thuringiensis* genes are gram+ bacterial genes and possess a high A/T-content (60-70%) compared with plant genes (40-50%). As a consequence *Cry*-gene codon-usage is inefficient in plants, A/T-rich regions may contain transcription termination (polyadenylation) sites, Cryptic mRNA splice sites and mRNA instability motifs. In order to avoid this problem, the nucleotide sequences were modified without changing the amino-acid sequences. The effects of gene modification were first investigated by Perlak *et al.* (1991) for the *Cry1Ab* and *Cry1Ac* genes. An important increase in the amount of Cry protein was observed and field trials of these *Bt*-cotton demonstrated that the plants were totally protected against important lepidopteran pests (Wilson *et al*, 1992). Different synthetic *Cry* genes (*Cry1Aa, b, c, Cry1C, Cry9C*) have been synthetized (Perlak *et al.*, 1991; Fujimato *et al.*, 1993; Koziel *et al.*, 1993; Sardana *et al.*, 1996; Strizhov *et al.*, 1996) and many reports of the successful introduction of these genes into various plants have been published (see Table-3, for some examples), together with the results of field trials (Peferoen, 1997b).

In order to increase the level of expression of the native *Bt* gene, McBride *et al.* (1995) expressed the *Cry1Ab* gene, and Kota *et al.* (1999) the *Cry2Aa2* gene, in chloroplasts by homologous recombination. The large number of chloroplasts in a cell leads to a very high level of toxin production (3-5% of soluble proteins) in tobacco. Nevertheless, chloroplast transformation is far from being routinely achieved and this technology needs to be adapted to crops.

2.2.1.1 Plants expressing serine PI : Serine-like proteinases are dominant in lepidopteran larvae (Christeller *et al.*, 1992). It has been shown that different serine PI are able to inactivate lepidopteran proteases and to cause deleterious effects on development and growth when incorporated into artificial diets (reviewed in Boulter, 1993; Reeck, 1997). The first constitutive expression of a PI in a plant was reported by Hilder *et al.* in 1997, who showed that a trypsin/trysin inhibitor derived from cowpea (*Vigna unguiculata*). CpTI, confers resistance against *Heliothis virescens* when expressed in tobacco. Many reports (Table-3 and reviewed in Reeck *et al.*, 1997; Jouanin *et al.*, 1998; Schuler *et al.*, 1998) detail the production of transgenic plants expressing PI of various origins and their anti-feeding effects on different lepidopteran larvae. However, to be effective, the level of PI expression must be high

Table 3. Some examples of transgenic plants expressing serine PI targeted against Lepidoptera

Plant	PI gene	PI type	Target insects	Reference
Tobacco	Cowpea PI (CpTI)	Bowman Birk (T/T)	*M. sexta*	Hilder *et al.*, 1987
	Potato PI (PPI-II)	Potato (T/C)	*M. sexta*	Johnson *et al.*, 1989
	Tomato PI (TI-I)	Tomato I (C)	*M. sexta*	Johnson *et al.*, 1989
	Mustard PI (MTI-2)	Mustard I (T)	*S. littoralis*	De Leo *et al.*, 1998
	Nicotiana alata PI, (NaPI)	*N. alata* I (T/C)	*Heliothis armigera*	Chiarity *et al.*, 1999
	Soybean PI (SBTI)	*Kunitz*	*S. litura*	McManus *et al.*, 1999
Rice	CpTI	Bowman Bird (T/T)	*Seramia inferens*	Duan *et al.*, 1996
	PPI-II	Potato I (T/C)	*S. inferens, Chilo suppressalis*	Xu *et al.*, 1996
Potato	CpTI	Bowman Birk (T/T)	*Lacanobia oleracea*	Gatehouse *et al.*, 1997
Wheat	Barley PI (Cme)	Cereal I (T)	*Sitotroga cerealealla*	Altpeter *et al.*, 1999
Pea	*N. alata* PI (NaPI)	*N. alata* I (T/C)	*H. armigera*	Charity *et al.*, 1999

T = Typsin; C – Chymotrypsin

(De Leo *et al.*, 1998). In addition, insects can rapidly adapt to the ingestion of PI by overexpressing existing proteases or inducing the production of new types, less-sensitive to the introduced PI (reviewed in Jongsma and Bolter, 1997). Lepidopteran larvae possess a complex pool of serine proteases and the PI strategy needs to be improved in order to obtain durable protection. This could be achieved by co-expressing PI of different types and/or improving the affinity of intoduced PI for the target insect proteases (Jongsma *et al.*, 1996; Michaud, 1997). Until now, even if increased mortality and reduced growth of lepidopteran larvae have been observed after ingestion of serine PI-expressing plants, these effects have not been deemed sufficiently convincing to permit the commercialization of these plants.

2.2.1.3. Other lepidopteran-resistant plants : Several other strategies have been tested to obtain lepidopteran-resistant plants. Bioassays using artificial diets containing lectins have shown the deleterious effects of these proteins on several Lepidoptera, and transgenic plants expressing selected lectins have been obtained; however, very few promising results have been published. The expression of the snowdrop lectin (GNA, *Galanthus nivalis*; Fitches *et al.*, 1997) or of the concanavalin A (ConA, *Canavalia ensifirmis*; Gatehouse *et al.*, 1999) in potato results mainly in a larval weight reduction for the tomato moth (*Lacanobia oleracea*). These lectins can, therefore, be considered as anti-feedant rather than insecticidal for Lepidoptera.

There is no report of the successful use of plant chitinase in effectively controlling insects (Kramer *et al.*, 1997). In contrast, the constitutive expression of a *M. sexta* (tobacco hornworm) gene encoding a chitinase in tobacco induces a significant reduction of growth in tobacco budworm (*H. virescens*) larvae whereas no difference was observed in tobacco hornworm (*M. sexta*) (Ding *et al.*, 1998). A synergistic effect was observed when this insect chitinase was used with sub-lethal doses of *Bt* toxin, and detrimental effects were observed in the case of *M. sexta* (Ding *et al.*, 1998). To date, while no report of transgenic plant expressing the recently identified bacterial toxins (VIP3 from *B. thuringiensis*, Tca from *P. luminescens*) has been published, Estruch *et al.* (1997) have nevertheless described the use of these genes to create the second generation of insecticidal plants.

2.2.2 Coleopteran-resistant plants

Coleoptera are important insect pests for crops (about 12% of the total expenditure of chemical insecticides). While different strategies have been developed to fight against this insect order, the results obtained are less spectacular that with Lepidoptera and few reports of successful strategies against coleopteran insects are available (Table-4). This insect order seems to be composed of more variable species than the lepidopteran order and any given toxin is active on only a few of them.

2.2.2.1. Plants expressing Bt δ-endotoxins : Among the *Bt* δ-endotoxin genes cloned, several genes (*Cry*3A, B) encode toxins active against Coleoptera such as the colorado potato beetle (CPB, *Leptinotarsa decemlineata*). Synthetic *Cry*3A genes have been designed and successfully introduced into potatoes (Sutton *et al.*, 1992; Adang *et al.*, 1993; Perlak *et al.*, 1993). The *Cry*3A gene has been introduced into aspen and shown to be effective against *Chrysomela tremulae* (Genissel *et al.*, personnel com.). However, the activity spectra of coleopteran *Bt*-toxins is restricted to a limited number of insects from this order and is not toxic for several very important insect pests such as the Southern- or Northern-corn rootworms and the boll weevil.

2.2.2.2. Plants expressing cysteine protease inhibitors : Studies of the protease content of the gut of different Coleoptera have shown the presence of cysteine proteases, which can sometimes represent the major type (Murdock *et al.*, 1987). Few cysteine PI of plant origin have been cloned. The cDNA of OC-I, a rice cysteine PI, has been constitutely expressed in different plants. When expressed to a level of 1% of the soluble proteins in poplar, it induces an increase in mortality; however, this lethal effect is mainly observed at the end of the larval stages (Leplé *et al.*, 1995). A significant growth reduction in colorado potato beetle larvae was observed when OC-I was expressed in potatoes (Lecardonnel *et al.*, 1998). OC-I expression in oilseed rape does not confer resistance against several Coleoptera feeding on this plant (reviewed in Jouanin *et al.*, 2000). As already observed with Lepidoptera, the lack of effects can be linked to a number of factors: the need for a high expression level (which was not obtained in oilseed rape), overexpression of cysteine proteases, compensation by serine proteases (Bonadé-Bottino *et al.*, 1999) and degradation of the introduced PI by insensitive proteases (Girard *et al.*, 1998). The digestive complex of coleopteran insects involves proteases of different types (serine, cysteine, aspartyl) and it may be difficult to obtain a durable protection using PI for this insect order even if PI of several types (serine and cysteine for example) are expressed simultaneously.

Table 4. Coleopteran-resistant transgenic plants

Plant	Gene	Target Insect	References
Potato	Cry3A (S)	*Leptinotarsa decemlineata*	Adang *et al.*, 1993
	Cry3A (S)	*L. decemlineata*	Perlak *et al.*, 1993
	Cysteine PI, OCI	*L. decemlineata*	Lecardonnel *et al.*, 1999
Poplar	Cysteine PI, OCI	*Chrysomela tremula*	Leple *et al.*, 1995
	Cry3A (S)	*C. tremulae*	Gennisel *et al.*, pers com
Pea	Bean α-amylase I	*Callosobruchus maculatus*	Sahe *et al.*, 1994
	Bean α-AI	*Bruchus pisorum*	Schroeder *et al.*, 1995
Azuki bean	Bean α-AI	*Callosobruchus chinensis*	Ishimoto *et al.*, 1996a
Eggplant	Cry3B (WT, P, S)	*L. decemlineata*	Iannacone *et al.*, 1997

WT – native gene : P = partially-modified gene, S = synthetic gene

2.2.2.3. Plant expressing α-amylase inhibitors : Inhibitors of α-amylases (α-AI) have been used to obtain an improved protection against Coleoptera. The common bean α-AI is supposed to play a role in plant defense against insects and has been expressed in pea and in Azuki bean, where its expression confers resistance to the bruchid beetles, *Callosobruchus maculatus* and *C. chinensis* (Shade *et al.*, 1994; Ishimoto *et al.*, 1996a). As well as being active against these pests of stored grain, Schroeder *et al.* (1995) demonstrated that the expression of this gene in pea is also bale to confer resistance to *Bruchus pisorum*, a bruchid feeding on crops in the field. However, other bruchids can nevertheless feed on plants expressing the bean α-AI because they possess a serine protease able to cleave the α-AI (Ishimoto *et al.*, 1996b). It is, therefore, difficult to evaluate the long-term interest of α-AI in transgenic plants. A recent study (Norton *et al.*, 2000) demonstrates a complete protection under field conditions of transgenic peas expressing the α-AI-1 against the pea weevil. The expression in pea of another α-AI (α-AI-2, which shares 78% amino-acid sequence identity and presents a different specificity) is partially effective in this protection.

2.2.2.4. Plants expressing other bacterial toxins : As mentioned for lepidopteran-resistant plants, many new candidate toxins are under evaluation. Guo *et al.* (1999) have identified a *Photorhabdus luminescens* toxin active against the Southern-corn rootworm, an important coleopteran pest of corn. Genes encoding several of these toxins have been cloned and represent good candidates for expression in plants (Bowen *et al.*, 1998). Other bacterial toxins active against Coleoptera have been identified in *B. cereus* (VIP1 and 2) and in *Streptomyces* (COX). Although no report of VIP expression in plants has yet been published, work is underway (Estruch *et al.*, 1997) COX has been expressed in tobacco protoplasts (Corbin *et al.*, 1994), but, its successful introduction into cotton in order to fight against the boll weevil has not yet been reported.

2.2.3 Homoptera-resistant plants

Homoptera are sap-sucking insects. This order includes the aphids, whiteflies, plant hoppers and leafhoppers. Many of these species are serious pests of agricultural and horticultural crops and of ornamental plants. Damages result from direct feeding but also because these insects are major vectors of viruses. About 27% of the total expenditure on chemical pesticides is concerned with their elimination. Very few toxins effective against this insect order have been identified, mainly because artificial diets to test potential candidates are difficult to compose and the digestive physiology of these insects is not well known. Nevertheless, some positive results have been obtained in transgenic plants (Table-5).

2.2.3.1. Plants expressing lectins : Recently, protein screening in artificial diets has shown that some lectins are toxic against aphids (Rahbé *et al.*, 1993). These lectins are mainly mannose-binding lectins belonging to the monocot mannose-binding lectins or to legume lectins showing glucose/mannose specificity. The most effective lectins found in those tests were the snowdrop lectins (GNA, *Galanthus nivalis* agglutinin) and the jackbean lectin (ConA, *Canavalia ensiformis* agglutinin). The GNA coding-sequence has been inserted into tobacco, potato, wheat and rice plants under the control of the CAMV 35S constitutive promoter or the phloem-specific promoter of the rice sucrose synthase gene *RSS-1* (Hilder *et al.*, 1995; Down *et al.*, 1996; Gatehouse *et al.*, 1996; Stoger *et al.*, 1999). The expression of the lectin in plants leads to a reduction of fecundity for three aphid species : *Myzus persicae* on transgenic tobacco and potato, *Aulacorthum solani* on transgenic potato and *Sitobion avenae* on transgenic wheat. When expressed in rice, GNA reduced the survival of green leafhopper (*Nephotettix virescens*) and brown planthopper (*Nilaparvata lugens*) (Rao *et al.*, 1998). ConA transgenic potatoes were also obtained, these plants cause a reduced development of *Myzus persicae* (Gatehouse *et al.*, 1999). Until now, no field trials of plants expressing lectins have been reported.

2.2.3.2. Other plants expressing toxins active against sap-sucking insects : Sap-sucking insects are generally not considered to need proteins to fulfil their nutritional requirements due to the abundance of free amino-acids in their food. However, some aphids can be sensitive to serine proteinase inhibitors (Rahbé et Febvay, 1993). Recently, Lee *et al.* (1999) reported that the expression of the soybean trypsin inhibitor (SKTI) confers resistance against the brown planthopper. PI are present in different kingdoms and have been isolated from insect hemolymph. Thomas *et al.* (1995a and b) observed that the expression of a serine PI derived from *M. sexta* in tobacco and cotton reduces reproduction in *Bemisia tabacci,* a whitefly. Expression of the cysteine PI, OC-I, in oilseed rape also causes a reduced fecondity in *M. persicae* (Jouanin *et al.*, 2000). However, as already mentioned for other insect orders, the PI strategy needs to be improved to assure a sufficient protection necessary for commercial exploitation.

The expression of the tryptophan decarboxylase gene from

Table 5. Homopteran-resistant transgenic plants

Plant	Gene	Target insect	References
Tobacco	*Galanthis nivalis* lectin, GNA	*M. persicae*	Hilder *et al.*, 1995
	Manduca sexta PI	*Bemesia tabacci*	Thomas *et al.*, 1995a
	Catharanthus tryptophan decarboxylase, TDC		Thomas *et al.*, 1995c
Alfalfa	*M. sexta* PI	*Thrips*	Thomas *et al.*, 1994
Potato	*G. nivalis* lectin, GNA	*M. persicae*	Gatehouse *et al.*, 1996
	GNA	*Aulacorthum solani*	Down *et al.*, 1996
	C. ensiformis lectin, ConA	*M. persicae*	Gatehouse *et al.*, 1999
Rice	*G. nivalis* lectin, GNA	*Nilaparvata lugens*	Rao *et al.*, 1998
	G. nivalis lectin, GNA	*V. virescens*	Foissac *et al.*, 2000
	Soybean Kunitz trypsin I, SKTI	*N. lugens*	Lee *et al.*, 1999
Wheat	*G. nivalis* lectin, GNA	*Sitobion avenue*	Stoger *et al.*, 1999
Cotton	*M. sexta* PI	*B. tabacci*	Thomas *et al.*, 1995b
Oilseed rape	Oryzacystatin, OC-I	*M. persicae*	Jouanin *et al.*, 2000

Catharanthus roseus in tobacco enables the plant to synthesize tryptamine and tryptamine-based alkaloids which cause a decrease in whitefly emergence (Thomas *et al.*, 1995c). The mechanism by which tryptamine may interfere with insects is unknown.

2.3 Field trials and commercialization of insect-resistant transgenic crops

Many experimental field trials have been performed to test the performances of insect-resistant transgenic plants in natural conditions in different countries and they will not be described here. To our knowledge, only *Bt*-transgenic plants are commercialized, it is why we will concentrate on these crops (http://www.aphis.usda.gov/bbep/bp/for USA).

The first *Bt*-cotton field trial was reported in 1992 (Wilson *et al.*, 1992) and since 1996 only one *Bt*-cotton (Bolgard™, Monsanto) has been released. This plant expresses the *Cry*1Ac protein which protects it against tobacco budworm (*Heliothis virescens*); cotton bollworn (*Helicoverpa zea*) and pink bollworm (*Pectinophora gossypiella*). In 1999, 27% of the total acreage of cotton was planted with *Bt*-cotton in the USA. Approval for release has been granted in other countries such as Australia, China and South Africa. A first evaluation of the interest of planting such modified cotton is available (http:/www.econ.ag.gov/whatsnew/issues/biotech/) and has demonstrated a reduction in the use of chemicals.

Similarly, *Bt*-Maize have been developed to fight against the European corn borer (ECB; *Ostrinia nubilabis*), and the first report of a field trial was published by Koziel *et al.* (1993). The commercialized *Bt* varieties originate from five different transformation events which vary according to which gene is expressed *(CrylAb, CrylAc* and *Cry 9C),* and the promoter associated to the coding sequence (which affects the quantity and location of the *Cry* protein). These transformation events were used by several companies to develop hybrid varieties. In 1999, 30% of the cultivated area in the USA consisted in transgenic varieties. Some of these *Bt*-hybrids are also permitted for plantation in Argentina, Canada and within the European community. The evaluation of use of these *Bt*-crops demonstrates the need for fewer insecticide treatments for target pests together with significantly higher yields for some regions (http:/www.bio.org/food&ag/ncfap.htm).

In 1995, *Bt*-potato (NewLeaf™, Monsanto) became the first *Bt*-crop commercialized. Approval for release has been granted in countries such as Canada, Mexico and Georgia (de Maagd *et al.*, 1999). Until now, *Bt*-potatoes are not cultivated on large surfaces (4% acreage in 1998 in the USA). This small percentage reflects the need of potato growers to control other insect pests (aphids and other foliar feeders) and the reduction in insecticidal application was minor. In 1999, a new transgenic potato resistant to CPB and to PLRV (the main pathogenic virus of potato) was commercialized and this combined resistance should be of benefit to the environment since it will reduce the use of insecticides aimed against CPB and aphids (the main vectors of viruses).

3. GESTION OF INSECT RESISTANT PLANTS

The repeated and unmanaged use of chemicals has led to the rapid evolution of resistant insect populations. Resistance is defined as "a genetically based decrease in susceptibility of a population to an insecticide" (Tabashnik, 1994). Field uses of *B. thuringiensis*-based biopesticide products have led, in the case of one insect *Plutella xylostella*, to the occurrence of resistant insect population in Hawaii (Tabashnik *et al.*, 1990) and in other areas (reviewed in Frutos *et al.*, 1999). The important increase in the cultivation of transgenic insect-resistant crops could led to the same problem. Most of the introduced genes work as monogenic traits and could, therefore, be easily overcome. Until now, only crops expressing *Cry* genes have been allowed to be planted in fields and no cases of insect resistance have yet been reported. However, there is no doubt that the potential for resistance is present (Gould *et al.*, 1997). In addition, under laboratory conditions many strains of *Cry*-resistant insects have been selected (reviewed in Frutos *et al.*, 1999). As a result, the problems of insect-resistance has been taken into account from the beginning, whenever large plantations of insect resistant crops are planned (Riebe, 1999).

3.1. Mechanisms of insect resistance to entomotoxins

3.1.1. Mechanisms of resistance to *Cry* proteins

Different mechanisms of resistance which are related to the mode of action of the *Cry* toxins can be observed. When ingested, the toxin needs to be activated by serine-type midgut proteases (Milne and Kaplan, 1993). A resistance mechanism could, therefore, be due to a reduction in protoxin activation as a result of modifications in the midgut protease profile (Oppert *et al.*, 1994; 1997). Degradation of the toxin in the midgut could also be involved (Forcada *et al.*, 1996). After activation, the toxin penetrates the peritrophic membrane that separates epithelium cells from the lumen and recognizes specific binding sites at the surface of the brush border membrane (Hoffman *et al.*, 1988a and b; van Rie *et al.*, 1989 and 1990). Although binding is an essential step of toxicity, the affinity of the toxin for the receptor is not always correlated with toxicity. A single toxin can recognize several different receptors in the same or in different insects (Masson *et al.*, 1995; Fiuza *et al.*, 1996). Despite high amino-acid sequence identity, the affinities of toxins depend on the target-insect binding-site, for example, *Cry*Aa shares a binding site with *Cry*1Ab and *Cry*1Ac in *M. sexta* (Hoffman *et al.*, 1995; van Rie *et al.*, 1989) and *L. dispar* (Wolfersberger, 1990) but not in *O. nubilalis* where only *Cry*1Aa and *Cry*1Ac compete for binding sites. Many other examples have shown that the binding domains recognized by a particular toxin vary from one insect species to another (reviewed in Frutos *et al.*, 1999). The most frequently encountered mechanism of resistance among Lepidoptera is the modification of the receptor site. This phenomenon has been studied in a large range of insects and toxins, and it has been concluded that resistance is due to either a change in the level of receptor affinity for a toxin, or to a lower number of receptor sites (Ferre *et al.*, 1991; van Rie *et al.*, 1990; Masson *et al.*, 1995).

The last mode of resistance corresponds to an as yet unknown mechanism. Gould *et al.* (1992) reported that a *H. virescens* strain selected for its resistance to *Cry*1Ac was also resistant to *Cry*1Ab and *Cry*2Aa. Similar observations have been made for *S. exigua* and *Cry*1C (Moar *et al.*, 1995) and *P. interpunctella* and *Cry*1Ac (Mohamed *et al.*, 1996). In the former case, the *S. exigua* resistant-strain was also resistant to *Cry*1Ab, *Cry*2Aa, *Cry*9C and *Cry*1E-*Cry*1C fusion protein and in the latter case, the *P. interpunctella* strain was resistant to *Cry*1Ab; *Cry*1Ac, *Cry*1B, *Cry*1C, *Cry*2Aa. This broad resistance mechanism is not due to a decrease in receptor affinity or in the number of binding sites but is considered to occur at the post-binding level.

3.1.2. Mechanism of resistance to other entomotoxins

Since only *Cry*-expressing plants are currently on the market, less work has been devoted to the emergence of insect resistance towards other types of insect-resistant transgenic plants. However, many reports have related adaptation problems with proteinase inhibitors (Jongsma *et al.*, 1997; Jouanin *et al.*, 2000). Depending upon the insect, different ways for the digestive proteases to circumvent the action of PI have been demonstrated. These mechanisms include the synthesis of proteases insensitive to the ingested PI – found both in Lepidoptera (Broadway, 1995; Jongsma *et al.*, 1995; Wu *et al.*, 1997) and Coleoptera (Bolter and Jongsma, 1995; Bonadé-Bottino *et al.*, 1999), and the degradation of PI by sensitive proteases – observed in Lepidoptera (Giri *et al.*, 1998) and Coleoptera (Girard *et al.*, 1998a).

For other strategies of plant resistance based upon the use of lectins, VIP and COX, no studies of insect resistance have been reported. In addition, the receptors of these entomotoxic proteins in the target insects are, at the moment, not well characterized.

3.2. Management of insect-resistant transgenic plants in the fields

Transgenic crops will be cultivated over large surfaces. The occurrence of resistant strains in the target insect must be considered, since it is known that resistance alleles exist in the insect populations. The initial allelic frequency of resistance genes in the population is assumed to range from 10^{-6} to 10^{-2} (Gould *et al.*, 1997). Recessivity of resistance alleles is a frequently encountered situation, but partial recessivity and dominant alleles could also be observed (reviewed in Frutos *et al.*, 1999). Partially recessive and recessive alleles are generally associated with modification of binding sites whereas dominant ones seem to be associated with other mechanisms such as broad-spectrum resistance which can be controlled by one locus or very few loci (Lui and Tabashnik, 1997a; Gould *et al.*, 1997). Resistance management strategies are oriented toward a reduction of selection (reviewed in Tabashnik, 1989; Peferoen, 1997b; MeGaughey *et al.*, 1998). These strategies are of different types: tissue- or time-specific expression of toxins, transfer of multiple toxins with different modes of action, low doses associated with natural enemies, high doses plus refuge, and culture techniques.

3.2.1. Use of tissue or time-specific promoters

In most cases, the toxin is expressed under the control of constitutive promoters such as the CaMV 35S promoter and its derivatives, or monocot ubiquitin or actin promoters. Tissue- and time-specific promoters can be used to limit toxin production to the tissues fed upon by the pest, or to periods when the pest attacks the plant (Table-6). For example, in order to fight against seed-attacking insects, the promoter of the bean phytohaemgglutinin have been associated to the α-Amylase inhibitor (Ishimoto *et al.*, 1996a). The rice sucrose synthase promoter which confers phloem-specific expression has been used to create plants resistant to sap-sucking insects such as aphids and planthoppers (Shi *et al.*, 1994; Rao *et al.*, 1998).

Use of promoters allowing toxin expression only after insect feeding has also been considered. Duan *et al.* (1996) obtained Lepidoptera-resistant transgenic rices expressing a potato PI under the control of its own promoter. Induction of expression by chemicals (salicylic acid) has also been observed using the tobacco promoter of the pathogenesis-related protein 1a (Williams *et al.*, 1992). However, when using these strategies, the production of the toxin must be rapid and high, and for chemical induction, the impact on the environment must be considered (Roush, 1997). In our opinion, the main interest of this approach is to restrict the 'toxin' expression to tissues where or when the insects feed and to avoid the presence of the toxin in the part of the plant to be used (for example in animal and human nutrition). In the light of these considerations, the questions of the control of toxin production by different promoters is more related to risk assessment than to management.

3.2.2. Gene combination

The use of multiple resistance genes or gene-pyramiding requires the incorporation into the plant genome of genes encoding two or more entomotoxins each possessing different modes of action. This could be achieved though the use of several *Cry* genes after demonstration that the target receptors of the Cry proteins are different. Carprio (1998) considered five resistance management strategies: sequential introduction of two toxins, yearly alternation of two toxins, mosaic distribution of two toxins, and a full-rate or a half-rate combination of two toxins. These strategies were simulated by spraying and evaluation of the results demonstrated that the full-rate combination is the best strategy to delay resistance. However, when using *Cry* genes, it is important to be sure that no cross-resistance is possible, and the presence of a refuge is also required.

Until now, few reports have detailed the introduction of two toxins. However, Douches *et al.* (1998) transformed a potato genotype possessing natural resistance against insects with a *CryV* gene so as to produce a combined resistance. Co-transformation with two *Cry* genes (*Cry1Ab* and *Cry1Ac*) has also been performed in rice (Cheng *et al.*, 1998) via *Agrobacterium*. Maqbool and Christou (1999) introduced the *CryIAc, Cry2A* and the lectin GNA genes *via* particle bombardement into rice

Table 6. Indicible or tissue-specific promoters used for creating insect-resistant plants

Promoter	Expression site	Associated gene	Plant	Reference
Bean phytohaemagglutinin L (PHA-L)	Seeds	Bean α-amylse I (α-AI)	Pea,	Shade *et al.*, 1994, Shroeder *et al.*, 1995
			Azuki bean	Ishimoto *et al.*, 1996a
Maize phosphoenolpyruvate-carboxylase (PEPC)	Green tissues	Cry1Ab	Maize	Koziel *et al.*, 1993
Rice sucrose synthase (RSS1)	Phloem	Lectin GNA	Tobacco	Shi *et al.*, 1994
		GNA	Rice	Rao *et al.*, 1998, Foissac *et al.*, 2000
		GNA	Wheat	Stroger *et al.*, 1999
Potato proteinase inhibitor (pin2)	Wounded-inducible	Potato PI-II	Rice	Duan *et al.*, 1996
Tobacco pathogenesis-related protein 1a (PR-1a)	Chemically induced	Cry1Ab	Tobacco	Williams *et al.*, 1996
Arabidopsis Ribulose-1, 5 biphosphate carboxylase small subunit (RBCS)	Green tissues	*N. alata* PI, (NaPI)	Tobacco	Charity *et al.*, 1999

and observed cosegregation of the transgenes. It is certain that more and more work will be devoted to the study of the co-expression of different genes. It is for this reason that it is important for the future to identify new toxins since, for some insects, the actual choice of genes available for transfer is limited.

3.2.3. Low doses in association with natural enemies

This strategy is an extension of the partial resistance developed for field management of varieties bred for resistance to pests (Bottrell *et al.*, 1996). It relies on the negative impact of the "resistance" gene on growth and development of the pest larvae making it more susceptible to attacks by natural enemies. This strategy could be difficult to use with *Cry*-expressing plants because of the high induced mortality. It could be used with an insecticidal protein less effective than the *Cry* proteins such as proteinase inhibitors or lectins. However, it will be important to determine the potential negative effects of the transgene expression on natural enemies (Schuler *et al.*, 1999a). Nevertheless, the low-dose strategy will most likely ensure developing resistance as quickly as high doses (Roush, 1997).

3.2.4. High dose and refuge

The high dose strategy is considered to be the most efficient and promising way of managing resistance, if used in conjunction with refuges (Roush, 1996). Refuges are areas planted with non-transgenic plants where the pest population can survive and act as a reservoir of wild-type susceptible alleles. The success of this strategy depends upon the initial frequency of allele resistance. Several ways of implanting such areas can be envisaged: seed mixtures, internal or external zones. The size of the refuge zone is another subject of debates. According to the model, the resistant homozygote individual insect which is able to survive on the transgenic crop, will mate with an adult from the important population present in the refuge, the resulting heterozygotes will be sensitive to the toxin expressed in the transgenic plant. However, several assumptions must be made so that the model will work: adult movements, recessive inheritance, random mating of adults and the simultaneous development of sensitive and resistant adults. A recent study (Liu *et al.*, 1999) has demonstrated that the resistance could have a cost and that the larval development of resistant insects was delayed. However, this observation has been made under laboratory conditions. A recent study (Shelton *et al.*, 2000) performed under field conditions indicates that a separate refuge could be more effective in conserving susceptible larvae than a mixed refuge. This work was performed with *Cry*1Ac-expressing brocoli and the diamondback moth. However, the decision to implement the high dose/refuge strategy will most likely be a case by case situation based on local conditions and on pest biology and mobility (Frutos *et al.*, 2000). There is no clear evidence of the minimal refuge size and estimates range from 10% to 50%, depending upon the crop, the pest and the simulation used program (Lui and Tabashnik, 1997b). The actual recommendations for insect-resistant plants is a 20% refuge (50% for cotton) that can be sprayed with chemicals or a 4% refuge that remains unsprayed.

3.2.5. Culture techniques

Mixtures, rotation or mosaic of transgenic plants can be envisaged. The use of these strategies depend upon the availability of different types of transgenic plants all resistant to the same insect- which is far from being the case- and the instability or cost of the resistance. Most of the studies in the area have demonstrated that these strategies would not be efficient in delaying resistance (McGaughey and Whalon, 1992; Frutos *et al.*, 2000).

4. RISK ASSESSMENT OF INSECT RESISTANT PLANTS

When using the transgenic plants or derived products, it is important to determine the entomotoxin toxicity towards other organisms. Three categories need to be considered: humans, animals and non-target insects. In this chapter, we will only detail the risks which are specific to insect-resistant transgenic plants, and not those relevant to all transgenic plants and which are more related to the plant biology itself (impact on biodiversity by crossing with wild relatives, pollen dispersion..). Another point which will not be discussed here concerns the potential risks associated with the marker genes (coding for antibiotic- or herbicide-resistance) generally used to select the transformed cells at the first stages of the transformation procedure. Some studies have demonstrated the innocuity of the proteins encoded by such marker genes (Flavell *et al.*, 1992). In addition, different strategies are now available which avoid or eliminate these marker genes in the plants available commercially.

4.1. Risk for humans and animals

4.1.1. Direct toxicity

The potential risks must be considered according to the final use of the transgenic crop. It is not the same in the case of cotton (industrial use), of maize (use of derived products), and in the case of a vegetable used for human food, either eaten raw or after cooking. However, even if eaten raw, the ingested proteins are very rapidly degraded by the digestive enzymes and, in most cases, loose their activity and properties. The term "toxin" is negatively perceived by the public even if most of these entomotoxins are specific for insects. This is particularly true in the case of *Cry* toxins which act on specific receptors present in the gut of several classes of insects (but not all).

Bacillus thuringiensis sprays have been used for a long time and different studies have demonstrated its innocuity for humans and mammals. In the case of proteins of plant origin, most of them are already present in vegetables and fruits. However, in some cases (proteinase inhibitors and α-amylase-inhibitors), they are considered as anti-nutritional and vegetables containing them in large amounts much be cooked before consumption (potatoes, bean...). Some lectins are toxic for mammals (ConA)) while others (GNA) are considered to be safe (Czapla, 1997). While, a recent work of Ewens and Pustzi (1999) suggests that GNA ingestion can be harmful in rats, this study has been the subject of a large controversy and further experiments are necessary before conclusion can be drawn. For some new entomotoxins (COX, VIP...) toxicity studies have not yet been published but they will be undertaken before commercialization of crops expressing the corresponding genes. In such studies, long-term tests at low doses may be more appropriate than short-term tests at high doses, since the former pattern is more related to the normal consumation of transgenic crops in food.

4.1.2. Risks of allergy

Another criteria for the safety of transgenic plants that must be considered concerns their potential allergenicity. The allergy could be due to the introduced protein or to side effects related to transformation or to modification of the metabolism of the transgenic plant. For the first point, it is possible to determine the potential allerginicity of a protein before its transfer (Käppeli and Auberson, 1998). The second point is more difficult to evaluate, however, the food industry is aware of this possibility. It is interesting to note that such a problem is also of concern in traditional breeding programs.

4.2. Risk for non-target insects

The main interest of insect-resistance is the confinement of the entomotoxin inside the plant and therefore, only insects feeding on the plant will be exposed to the toxin. However, secondary pests, predators, or parasites of pests could ingest or come into contact with the toxin. Natural enemies form part of the Integrated Pest Management (IPM) approach and, therefore, it is imperative to investigate possible adverse effects upon natural biological agents (Schuler *et al.*, 1999a). Apart from *Cry*-expressing crops, most of the studies on non-target insects have been performed under laboratory conditions and must be considered as the "worst-case-scenario" (Poppy, 1999). A compilation of such studies on *Bt*-maize is available on the Novartis web site (www.cp.novartis.com).

4.2.1. Secondary pests or non-pest insects

Even if the main target of a toxin is an insect which causes considerable damages to the crop, very often other insects can feed on the plant. If they are sensitive to the expressed toxin, they will also be affected and this represents an advantage for the crop protection. However, some insects could be affected in a non-deliberate way. The typical example of this is the monarch butterfly (*Danaus plexippus*), a mythic butterfly of North America. Losey *et al.* (1999), observed a higher mortality rate in butterfly larvae fed milkweed coated with *Bt*-maize pollen as compared to larvae fed leaves coated with non-transformed maize pollen or with leaves free of pollen. This study has been performed under artificial laboratory conditions which do not reflect most of the characteristics of the monarch way of life (Hodgson, 1999). A consortium of biotechnology and pesticide companies has funded different studies to quantify the risk of *Bt*-maize for monarchs and, even if some complementary studies will be performed in 2000, the risk seems to be very weak (http://www.fooddialogue.com/monarch/index.html).

4.2.2. Predators

Predators are insects which feed on phytophagous insects. They can be found in most insect orders. Typical examples are the 2-spot ladybird (*Adalia bipunctata*) and the green lacewing (*Chrysoperla carnea*). These insects fed on aphids and different studies have been performed to determine the impact of feeding on aphids reared on insect-resistant plants (Reviewed in Schuler *et al.*, 1999). In most cases, no negative impact have been observed and even if some deleterious effects have been observed in the case of green lacewing and *Bt*-maize (Hilbeck *et al.*, 1998), the effects are reduced compared to that observed with chemical sprays. Another study (Pilcher *et al.*, 1997) performed on three predators (*Coleomegilla maculata* De Geer, *Orius insidiosus* Say and *Chrysoperla carnea* Stephens) feeding on *Bt*-corn pollen under field conditions did not revealed any effects on the predator populations.

Studies have also been performed on the impact of feeding aphids reared on *GNA*-expressing potatoes to 2-spot ladybirds (Birch *et al.*, 1999; Down *et al.*, 2000). The small negative impact observed for some development criteria seems to be more due to indirect effects (reduced aphid availability) than to direct adverse effects.

The digestive proteases of the two-spotted stinkbug (*Perillus bioculatus*), a predator of colorado potato beetle, (Ashouri *et al.*, 1998) and of the 2-spot ladybird (Walter *et al.*, 1998) are of the cysteine-type. In the former case, deleterious effects on fecondity were observed when fed on prey raised on *OC-I*-expressing plants (Ashouri *et al.*, 1998). It is obviously of importance to determine the impact of *OC-I*-expressing plants on ladybirds feeding aphids reared on *OC-I*-expressing plants. Similar studies must be performed when using new toxins especially those intended to fight against aphids.

4.2.3. Parasitoids

Parasitoids are insects that complete their larval development in a single host insect (either externally or internally), and which can result in the host insect death. Compared to predators they are relatively host-specific. While few studies dealt with those insects (reviewed in Schuler *et al.*, 1999a). Schuler *et al.* (1999b) work on *Bt*-oilseed rape, *P. xylostella*, and on a parasitic wasp, *Cotesia plutella*, demonstrated that *Bt* does not affect the parasitic insect directly. The adult wasp does not emerge from the parasitized insect because of the latter's early death and a *P. xylostella Bt*-resistant strain is a suitable host for the wasp. In these tritrophic systems, parasitoids find their host by sensing the chemicals released by the attacked plant. The slight damage caused by an herbivore on a *Bt*-resistant plant does not allow wasp attraction (Poppy, 1999).

The protease activity of the wasp, *Eulophus pennicornis*, has been shown to be of the serine type (Down *et al.*, 1999) and therefore, use of serine PI could have a deleterious effect on this wasp. The lectin GNA expressed in potatoes had no effect on this wasp feeding on *Lacanobia oleracea* (Bell *et al.*, 1999). As for the predators, it is sure that a reduction in the host population could affect the parasitoid abundance, however, this effect is already observed with the use of chemicals and biopesticides.

4.2.4. Pollinating insects

Pollinating insects such as honeybees and bumblebees play a major role in seed production and fruit set of many crops. As a result, it is crucial to evaluate the impact of insect-resistant plants on these insects. They fed exclusively on pollen and nectar and there are therefore two possibilities to avoid risk: non-expression of the toxin in the tissues they fed on, or innocuity of the toxin for them. In the former case, since the CaMV 35S promoter is often used and shown to be inactive in pollen from plants such as cotton, maize and oilseed rape (Schuler *et al.*, 1998; Jouanin *et al.*, 2000) the risk is zero even if the toxin is toxic for bees. Nectar is composed mainly of sugar and of very few proteins and the risk is limited because no accumulation of toxins will occur. With other promoters, studies must be performed to determine the accumulation of toxins in these tissues. The risk of toxicity of the expressed protein must be considered case by case. Many studies have been performed on *B. thuringiensis* endotoxins used as sprays or expressed in transgenic plants and have demonstrated the non-toxicity for pollinating insects (Arpaia, 1996). Bees relie on serine proteases for digestion and it has been determined that only ingestion of serine PI at doses much higher than the expected level of expression in transgenic plants have a negative impact on bees (Burgess *et al.*, 1996; Girard *et al.*, 1998; Jouanin *et al.*, 2000).

5. CONCLUSIONS AND FUTURE PROSPECTS

To increase the yield and reduce the use of chemicals in modern agriculture, it is important to develop new approaches and genetic engineering is one of them. This strategy has opened new ways of obtaining crops resistant to their main insect pests. Expression of bacterial *Bacillus thuringiensis Cry* endotoxins is the most advanced of these strategies. *Bt*-expressing maize, cotton and potatoes are already commercialized in some countries. However, until now, they have been mainly grown in industrial countries and it is of importance that these plants extend to developing countries (Toenniessen, 1995; Serageldin, 1999).

To avoid problems with the emergence of resistant population of insects it is important to cultivate these crops under resistance-management conditions. In the long term, and with the aim of extending the range of insect pests, it is important to increase the number of genes which can be expressed in plants. Many studies, currently at the laboratory stage, are performed with this objective in mind. Another factor which will affect the future of these crops is the public acceptance of products derived from transgenic plants (Boulter, 1995; Ruibal-Mendieta *et al.*, 1998; Gaskell *et al.*, 1999). Better consumer information is necessary to allow a well-informed decision based on the comparison of the potential benefits of using transgenic plants as against the continued reliance on chemical insecticides.

ACKNOWLEDGEMENTS

This research was partly supported by the European Union (FAIR program CT 98-4239). The authors thank Simon Hawkins for critical reading of the manuscript.

REFERENCES

Adang MJ, Brody MS, Cardineau G, Eagan N, Rousch RT, Shewmaker CK, Jones A, Oakes JV and McBride KE (1993) The reconstruction and expression of a *Bacillus thuringiensis Cry*IIIA gene in protoplasts and potatoes plants. *Plant Mol. Biol.,* **21** : 1131-1145.

Ashouri A, Overney S, Michaud D and Cloutier C (1998) Fitness and feeding are affected in the two-spotted stinkbug, *Perillus bioculatus,* by the cysteine proteinase inhibitor, Oryzacystatin I. *Arch. Insect Biochem. Physiol.,* **38** : 74-83.

Altpeter F, Diaz I, McAuslane H, Gaddour K, Carbonero P and Vasil IK (1999) Increased insect resistance in transgenic wheat stably expressing trypsin inhibitor Cme. *Mol. Breed.,* **5** : 53-63.

Arpaia S (1996) Ecological impact of Bt-transgenic plants : 1. Assessing possible effects of *Cry*IIIB toxin on honey bee (*Apis mellifera*) colonies. *J. Genet. Breed.,* **50** : 315-319.

Barton KA, Whiteley HR and Yang NS (1987) *Bacillus thuringiensis* δ-endotoxin expressed in transgenic *Nicotiana tabacum* provides resistance to Lepidopteran insects. *Plant Physiol.,* **85** : 1103-1109.

Blackburn M, Golubeva E, Bowen D, Ffrench-Constant R (1998) A novel insecticidal toxin from *Photorhabdus luminescens*, toxin complex a (Tca), and its histopathological effects on the midgut of *Manduca sexta. Appl. Environ. Microbiol.,* **64** : 3036-3041.

Bell HA, Fitches EC, Down RE, Marris GC, Edwards JP, Gatehouse JA and Gatehouse AMR (1999) The effect of snowdrop lectin (GNA) delivered *via* artificial diet and transgenic plants on *Eulophus pennicornis* (Hymenoptera: Eulophidae), a parasitoid of the tomato moth *Lacanobia oleracea* (Lepidoptera : Noctuidae). *J. Insect Physiol.,* **45** : 983-991.

Birch ANE, Geoghegan IE, Majerus MEN, McNicol JW, Hackett CA, Gatehouse AMR and Gatehouse JA (1999) Tri-trophic interactions involving pest aphids, predatory 2-spot ladybirds and transgenic potatoes expressing snowdrop lectin for aphid resistance. *Mol. Breed.,* **5** : 75-83.

Bode W and Huber R (1992) Natural protein proteinase inhibitors and their interactions with proteinases. *Eur. J. Biochem.,* **204** : 433-451.

Bonade-Bottino M, Lerin J, Zaccomer B, Jouanin L (1999) Physiological adaptation explains the insensitivity of *Baris coerulescens* to transgenic oilseed rape expressing oryzacystatin. *Insect Biochem. Mol. Biol.,* **29** : 131-138.

Bolter CJ and Jongsma MA (1995) Colorado potato beetles (*Leptinotarsa decemlineata*) adapts to proteinase inhibitor induced in potato leaves by methyl jasmonate. *J. Insect Physiol.,* **41** : 1071-1078.

Bottrell DG, Barbosa P and Gould F (1997) Manipulating natural enemies by plant variety selection and modification : a realistic strategy? *Annu. Rev. Entomol.,* **43** : 343-367.

Boulter D (1993) Insect pest control by copying nature using genetically engineered crops. *Phytochem.,* **34** : 1453-1466.

Boulter D (1995) Plant Biotechnology: Facts and public perception. *Phytochemistry,* **40** : 1-9.

Bowen DJ and Ensign JC (1998) Purification and characterization of a high molecular-weight insecticidal protein complex produced by the entomophathogenic bacterium *Photorhabdus* luminescens. *Appli. Environ. Microbiol.,* **59** : 1828-1837.

Bowen D, Rocheleau TA, Blackburn M, Andreev O, Golubeve E, Bhartia R, Ffrench-Constant RH (1998) Insecticidal toxins from the bacterium *Photorhabdus luminescens. Science,* **280** : 2129-2132.

Broadway RM (1995) Are insects resistant to proteinase inhibitors? *J. Insect Physiol.,* **41** : 107-116.

Burgess EPJ, Lalone LA and Christeller JT (1996) Effect of two proteinase inhibitors on the digestive enzymes and survival of honey bees. *J. Insect Physiol.,* **42** : 823-828.

Caprio MA (1998) Evaluating resistance management strategies for multiple toxins in the presence of external refuges. *J. Econ. Entomol.,* **91** : 1021-1031.

Charity JA, Anderson MA, Bittisnich DJ, Whitecross M and Higgins TJV (1999) Transgenic tobacco and peas expressing a proteinase inhibitor from *Nicotiana alata* have increased insect resistance. *Mol. Breed.,* **5** : 357-367.

Chen L, Marmey P, Taylor NJ, Brizard JP, Espinoza, C, D'Cruz P, Huet, H, Zhang, S, de Kochko A, Beacy RN and Fauquet CM (1998) Expression and inheritance of multiple transgenes in rice plants. *Nature Biotech.,* **16** : 1060-1064.

Christeller JT, Laing WA, Markwick NP and Burgess EPJ (1992) Midgut protease activities in 12 phytophagous lepidopteran larvae – Dietery and protease inhibitor interactions. *Insect Biochem. Mol. Biol.,* **22** : 735-746.

Corbin DR, Greenplate JT, Wong EY and Purcell JP (1994) Cloning of an insecticidal clolesterol oxydase gene and its expression in bacteria and in plant protoplasts. *Appl. Environ. Microbiol.,* **60** : 4239-4244.

Crickmore N, Zeigler DR, Feitelson J, Schnepf E, Van Rie J, Lereclus D, Baum J and Dean DH (1998) Revision of the nomenclature for the *Bacillus thuringiensis* pesticidal *Cry*stal proteins. *Microbiol. Mol. Boil. Rev.,* **62** : 807-813.

Czapla H (1997) Plant lectins as insect control proteins in transgenic plants. In: *Advance in insect control: the role of transgenic plants* (Eds Carozzi N and Koziel M) Taylor and Francis, London pp: 123-138.

De Leo F, Bonade-Bottino M, Ceci LR, Gallerani R and Jouanin L (1998) Opposite effects on *Spodoptera littoralis* larvae of low and high expression level of a trypsin proteinase inhibitor in transgenic plants. *Plant Physiol.*, **118** : 997-1004.

De Maagd, Bosch D and Stiekema W (1999) *Bacillus thuringiensis* toxin-mediated insect resistance in plants. *Trends Plant Sci.,* **4** : 9-13.

Ding X, Gopaladrishnan B, Johnson LB, White FF, Wang X and Muthukrishnan S (1998) Insect resistance in transgenic tobacco expressing a insect chitinase gene. *Transgenic Res.*, **7** : 77-84.

Douches DS, Westedt AL, Zarda K and Schoeter (1998) Potato transformation to combine natural and engineered resistance for controlling tuber moth. *HortSci.,* **33** : 1053-1056.

Down RE, Gatehouse AMR, Hamilton WDO and Gatehouse JA (1996) Snowdrop lectin inhibits development and decreases fecondity of the glasshouse potato aphid (*Aulacorthum solani*) when administrated *in vitro* and *via* transgenic plants in laboratory and glasshouse trials. *J. Insect Physiol.,* **42** : 1035-1045.

Down RE, Ford L, Mosson HJ, Fitches EC, Gatehouse AMR (1999) Protease activity in the larval stage of parasitoid wasp, *Eulophus pennicornis* (nees) (Hymonoptera : Eulophidae); effects of protease inhibitors. *Parasitology*, **119** : 157-166.

Down RE, Ford, L, Woodhouse SD, Raemeakers RJM, Leitch B, Gatehouse JA and Gatehouse AMR (2000) Snowdrop lectin (GNA) has no acute toxic effects on the beneficial insect predator, the 2-spot ladybird (*Adalia bipunctata* L.). *J. Insect Physiol.,* **46** : 379-391.

Duan X, Li X, Xue Q, Abo-El-Saad M, Xu D and Wu R (1996) Transgenic rice plants harboring an introduced potato proteinase inhibitor II gene are insect resistant. *Nature Biotech.*, **14** : 494-498.

Estruch JJ, Warren G, Mullins MA, Nye GJ, Craing JA and Koziel MG (1996) Vip3A, a novel *Bacillus thuringiensis* vegetative insecticidal protein with a wide specturm of activities against *lepidopteran* insects. *Proc. Natl. Acad. Sci. USA,* **93** : 5389-5394.

Estruch JJ, Carozzi NB, Desai N, Duch NB, Warren GW and Koziel MG (1997) Transgenic plants: an emerging approach to pest control. *Nature Biotech.,* **15** : 137-141.

Ewen SWB and Pusztai (1999) Effect of diets containing genetically modified potatoes expressing *Galanthus nivalis* lectin on rat small intestine. *The Lancet,* **354** : 1353-1354.

Ferre J, Real MD, van Rie J, Janbsens S and Peferoen M (1991) Resistance to *Bacillus thuringiensis* biopesticide in a field population on *Plutella xylostella* is due to the change in midgut membrane receptor. *Proc. Natl. Acad. Sci. USA,* **88** : 5119-5123.

Fischhoff DA, Bowdish KS, Perlak FJ, Marrone P, McCormick S, Niedermeyer J, Dean D, Kusano-Kretzmer K, Mayer E, Rochester D, Rogers S and Fraley R (1987) Insect tolerant transgenic tomato plants. *Bio/Technol.*, **5** : 805-813.

Fitches E, Gatehouse AMR and Gatehouse JA (1997) Effects of snowdrop lectin (GNA) delivered *via* artificial diet and transgenic plants on the development of tomato moth (*Lacanobia oleracea*) larvae in laboratory and glasshouse trials. *J. Insect Physiol.,* **44** : 1213-1224.

Fiuza LM, Nielsen-Leroux C, Goze E, Frutos R and Charles JF (1996) Binding of *Bacillus thuringiensis Cry*1 toxins to the midgut brush border membranes vesicles of *Chilo suppressalis* (Lepidoptera: Pyralidae): evidence of shared binding sites. *Appli. Environ. Microbiol.,* **62** : 1544-1549.

Flavell B, Dart E, Fuchs RL and Fraley RT (1992) Selectable marker genes: Safe for plants. *Bio/Technol.,* **10** : 141-144.

Foissac X, Thi Loc N, Christou P, Gatehouse AMR and Gatehouse JA (2000) Resistance to green leafhopper (*Nephotettix virescens*) and brown planthopper (*Nilaparavata lugens*) in transgenic rice expressing snowdrop lectin (*Galanthus nivalis* agglutinin; GNA). *J. Insect Physiol.,* **46** : 573-583.

Forcada C, Alcacer E, Garcera MD and Martinez R (1996) Differences in the midgut proteolytic activities of two *Heliothis virescens* strains, one susceptible and one resistant to *Bacillus thuringiensis* toxins. *Arch. Insect Biochem. Physiol.,* **31** : 257-272.

Frutos R, Rang C and Royer M (1999) Managing insect resistant to plant producing *Bacillus thuringiensis* toxins. *Critical Red. Biotech.,* **19** : 227-276.

Fujimoto H, Itoh K, Yamamoto M, Kyozuka J and Shimamoto K (1993) Insect resistant rice generated by introduction of a modified δ-endotoxin gene of *Bacillus thuringiensis. Bio/Technol.,* **11** : 1151-1155.

Gaskell G, Bauer MW, Durant J, Allum NC (1999) Worlds apart? The reception of genetically modified food in Europe and the US. *Science*, **285** : 384-387.

Gatehouse AMR, Down RE, Powell KS, Sauvion N, Rahbe Y, Newell CA, Merryweather A, Hamilton WDO and Gatehouse J (1996) Transgenic potato plants with enhanced resistance to the peach-potato aphid *Myzus persicae. Entomol. Exp. Appl.,* **79** : 295-307.

Gatehouse AMR, Davison G, Newell CA, Merryweather A, Hamilton WDO, Burgess EJ, Gilbert RJC and Gatehouse JA (1997) Transgenic potato plants with enhanced resistance to the tomato moth, *Lacanobia oleracea. Mol. Breed.,* **3** : 49-63.

Gatehouse AMR, Davison GM, Stewart JN, Gatehouse LN, Kumar A, Geoghegan IE, Birch ANE and Gatehouse J (1999) Concavalin A inhibits development of tomato moth (*Lacanobia oleracea*) and peach-potato aphid (*Myzus persicae*) when expressed in transgenic potato plants. *Mol. Breed.,* **5** : 153-165.

Gjareuazoe B, Alinia F, Menguito CA, Rubia LG, de Palma JM, Liwanag EA, Cohen MB, Khusch GS and Bennett J (1997) Enhanced resistance to two stem borers in an aromatic rice containing a synthetic *Cry*IA(b) gene. *Mol. Breed.,* **3** : 410-414.

Girard C, Le Metayer M, Bonade-Bottino M, Pham Delegue MH and Jouanin L (1998a) High level of resistance to proteinase may be conferred by proteolytic cleavage in beetle larvae. *Insect Biochem. Mol. Biol.,* **28** : 229-237.

Girard C, Picard-Nizou AL, Zaccomer B, Grallien E, Jouanin L and Pham-Delegue MH (1998b) Effects of proteinase inhibitors ingestion on survival, learning abilities and digestive proteases of the honeybee. *Transgenic Res.,* **7** : 239-246.

Giri AP, Harsulkar AM, Desphande VV, Sainani MN, Gupta VS and Ranjekar PK (1998) Chickpea defensive proteinase inhibitors can be inactived by podborer gut proteinases. *Plant Physiol.,* **116** : 393-401.

Gould F, Martinez-Ramirez A, Anderson A, Ferre J, Silva FJ and Moar WJ (1992) Broad-specturm resistance to *Bacillus thuringiensis* toxins in *Heliothis virescens. Proc. Natl. Acad. Sci. USA.*, **89** : 7986-7990.

Gould F, Anderson A, Jones A, Sumerford D, Heckel DG, Lopez J, Micinshi S, Leonard R, Laster M (1997) Initial frequency of alleles for resistance to *Bacillus thuringiensis* in field populations of *Heliothis virescens. Proc. Natl. Acad. Sci. USA.*, **94** : 3519-3523.

Guo L, Fatig III RO, Orr GL, Schafer BW, Strickland JA, Sukhapinda K, Woodsworth AT and Petell JK (1999) *Photorhabdus luminescens* W-14 insecticidal activity consists of a least two similar but distinct proteins. *J. Biol. Chem.*, **274** : 9836-9842.

Habidi J, Backus EA and Czapla TH (1998). Subcellular effects and localization of binding sites of phytohemagglutinin in the potato leafhopper, *Empoasca fabae* (Insecta: Homoptera: Cicadellidae). *Cell Tissue Res.* **294** : 561-574.

Habidi J, Backus EA and Huesing JE (2000). Effect of phytohemagglutinin (PHA) on the midgut epithelial cells and localization of its binding sites in western tarnished plant bug, *Lygus hesperus* Knight. *J. Insect Physiol.*, **46** : 611-619.

Hilbeck A, Baumgartner M, Fried PM and Bigler F (1998) Effect of transgenic *Bacillus thuringiensis* corn-fed prey on mortality and development time of immature *Chrysoperla carnae* (Neuroptera: Chrysopidae). *Environ. Entomol.*, **27** : 480-487.

Hilder VA, Gatehouse AMR, Sheerman SE, Barker RF and Boulter D (1987) A novel mechanism of insect resistance engineered into tobacco. *Nature*, **333** : 160-163.

Hilder VA, Powell KS, Gatehouse AMR, Gatehouse JA, Gatehouse LN, Shi Y, Hamilton WDO, Merryweather A, Newell CA, Timans JC, Peumans WJ, van Damme E and Boulter D (1995) Expression of snowdrop lectin in transgenic tobacco plants results in added protection against aphids. *Transgenic Res.*, **4** : 18-25.

Hilder VA and Boulter (1999) Genetic engineering of crop plants for insect resistant – a critical review. *Crop Protection*, **18** : 177-191.

Hodgson J (1999) Monarch *Bt*-corn paper questionned. *Nature Biotech.*, **17** : 627.

Hoffmann C, Luthy P, Hutter R and Pliska V (1988a) Binding of the delta-endotoxin from *Bacillus thuringiensis* to brush-border membranes vesicles of the cabbage butterfly (*Pieris brassicae*). *Eur. J. Biochem.*, **173** : 85-91.

Hoffmann C, Vanderbruggen H, Höfte H, van Rie J, Jansens S and van Mellaert H (1988b) Specificity of *Bacillus thuringiensis* δ-endotoxins is correlated with the presence of high-affinity binding sites in the brush-border membrane of target insect midguts. *Proc. Natl. Acad. Sci. USA.*, **85** : 7844-7848.

Hofte H and Whiteney HR (1989) Insecticidal *Cry*stal proteins of *Bacillus thuringiensis. Microbiol. Rev.*, **54** : 2010-2017.

Horsch RB, Fry JE, Hoffman NL, Eichholtz D, Rogers SG and Fraley RT (1985) A simple and general method for transferring genes into plants. *Science*, **227** : 1229-1231.

Iannacone R, Grieco P and Cellini F (1997) Specific sequence modifications of a *Cry*3B endotoxin gene result in high levels of expression and insect resistance. *Plant Mol. Biol.*, **34** : 485-496.

Ishimoto M, Sato T Chrispeels MJ and Mitamura K (1996a) Bruchid resistance of transgenic azuki bean expressing seed α-amylase inhibitor in the commun bean. *Entomol. Exp. Appl.*, **79** : 309-315.

Ishimoto M and Chrispeels MJ (1996b) Protective mechanism of the Mexican bean weevil against high levels of α-amylase inhibitor in the commun bean. *Plant Physiol.,* **111** : 393-401.

Jansen DH and Juster HB (1976). Insecticidal action of the phytohemagglutinin in black beans on a bruchid beetle. *Science* **192** : 795-796.

Johnson KA, Narvaez J, An G and Ryan CA (1989) Expression of proteinase inhibitors I and II in transgenic tobacco plants : Effect on natural defense against *Manduca sexta* larvae. *Proc. Natl. Acad. Sci. USA.,* **86** : 9871-7875.

Jongsma MA, Bakker PL, Peters J, Bosch D and Stiekema WJ (1995) Adaptation of *Spodoptera exigua* larvae to plant proteinase inhibitors by induction of gut proteinase activity insensitive to inhibitors. *Proc. Natl. Acad. Sci. USA,* **92** : 8041-8045.

Jongsma MA, Stiekema WJ and Bosch D (1996) Combatting inhibitor-insensitive proteases of insect pests. *Trends Biotech.,* **14** : 331-333.

Jongsma MA and Bolter C (1997) The adaptation of insects to plant protease inhibitors. *J. Insect Physiol.,* **43** : 885-895.

Jouanin, L, Bonade-Bottino M, Girard C, Morrot G, and Giband M (1998) Transgenic plants for insect resistance. *Plant Sci.,* **131** : 1-11.

Jouanin, L, Bonadé-Bottino M, Girard C, Lerin J and Pham Délégue MH (2000) Expression of protease inhibitors in rapeseed. In *Recombinant Protease Inhibitors in Plants.* (Ed. Michaud D). Academic Press, New York, USA pp 182-194.

Käppeli O and Auberson L (1998) How safe is safe enough in plant genetic engineering. *Trends Plant Sci.,* **3** : 276-281.

Knowles BH and Dow JAT (1993) The *Cry*stal delta-endotoxins of *Bacillus thuringiensis* – models for their mechanism of action in the insect gut. *BioEssays,* **15** : 469-476.

Kota M, Daniell H, Varma S, Garczynski SF, Gould F and Moar WJ (1999) Overexpression of the *cacillus thuringiensis* (Bt) *Cry*2Aa2 protein in chloroplasts confers resistance to plants against susceptible and *Bt*-resistant insects. *Proc. Natl. Acad. Sci. USA,* **96** : 1840-1845.

Koziel MG, Beland GL, Bowman C, Carozzi N, Crenshaw R, Crossland L, Dwason J, Desai N, Hill M, Kadwell S, Launis K, Lewis K, Maddox D, NcPherson K, Neghji HR, Nerlin E, Rhodes R, Warren GW, Wright H, Eoola SV (1993) Field performance of elite transgenic maize plants expressing an insecticidal protein derived from *Bacillus thuringiensis. Bio/Technol.* **11** : 194-200.

Kramer KJ, Muthukrishnan S, Johnson L and White F (1997) Chitinases for insect control. In: *Advance in Insect Control: The Role of Transgenic Plants* (Eds Carozzi N and Koziel M) Taylor and Francis, London pp 185-193.

Lambert B and Peferoen M (1992) Insecticidal promise of *Bacillus thuringiensis.* Facts and mysteries about a successful biopesticide. *BioScience,* **42** : 112-122.

Lecardonnel A, Chauvin L, Jouanin L, Beaujean A, Prevost G and Sangwan-Norreel B (1999) Effects of the rice cystatin I expression in transgenic potato on colorado potato beetle larvae. *Plant Sci.,* **140** : 71-79.

Lee SI, Lee SH, Koo JC, Chun HJ, Lim CO, Mun JH, Song YH and Cho MJ (1999) Soybean Kunitz trypsin inhibitor (SKTI) confers resistance to the brown planthopper (*Nilaparvata lugens* Stal) in transgenic rice. *Mol. Breed.,* **5** : 1-9.

Leple JC, Bonade-Bottino M, Augustin S, Pilate G, Dumanois-Le Tan V, Delplanque A and Jouanin L (1995) Toxicity to *Chrysomela tremulae* (Coleoptera: Chrysomelidae) of transgenic poplars expressing a cysteine proteinase inhibitor. *Mol. Breed.*, **1** : 319-328.

Liu YB and Tabashnik BE (1997a) Inheritance of resistance to the *Bacillus thuringiensis Cry*IC in the diamondback moth. *Appl. Environ. Microbiol.*, **63** : 2218-2223.

Liu YB and Tabashnik BE (1997b) Experimental evidence that refuges delay insect adaptation of *Bacillus thuringiensis*. *Proc. R. Soc. Lond.*, **264** : 605-610.

Liu YB Tabashnik BE, Dennehy TJ, Patin AL and Bartlett AC (1996) Development time and resistance to *Bt* crops. *Nature*, **400** : 519.

Losey JE, Rayor LS and Carter ME (1999) Transgenic pollen harms monarch larvae. *Nature*, **399** : 214.

Maqbool S, Husnain T, Masson L and Christou P (1998) Effective control of yellow stem borer and rice leaf folder in transgenic rice indica varieties Basmati 370 and M7 using the novel δ-endotoxin *Cry*2A *Bacillus thuringiensis* gene. *Mol. Breed.*, **4** : 501-507.

Maqbool SB and Christou P (1999) Multiple traits of agronomic importance in transgenic indica rice platns: analysis of transgene integration patterns, expression levels and stability. *Mol. Breed.*, **5** : 471-480.

Masson L, Juo K, Mazza A, Brousseau R and Adang M (1995) The *Cry*1Ac receptro purified from *Manduca sexta* displays multiple specificities. *J. Biol. Chem.*, **270** : 20309-20315.

Mazier M, Pannetier C, Tourneur J, Jouanin L and Giband M (1997) The expression of *Bacillus thuringiensis* toxin genes in plant cells. *Biotech. Annu. Rev.*, **3** : 313-347.

McBride KE, Svab Z, Schaaf DJ, Hogan PS and Maliga P (1995) Amplification of a chimeric *Bacillus* gene in chloroplasts leads to an extraordinary level of an insecticidal protein in tobacco. *Biotechnology*, **13** : 362-365.

McGaughey WH Whalon ME (1992) Managing insect resistance to *Bacillus thuringiensis* toxins. *Science*, **258** : 1451-1455.

McGuaghey WH, Gould F and Gelernter W (1998) *Bt* resistance management. *Nature Biotech.*, **16** : 144-146.

McManus MT, Burgess EPJ, Philip B, Watson LM, Laing WA, Voisey CR and White DWR (1999) Expression of a soybean (Kunitz) inhibitor in transgenic tobacco : Effects on larval development of *Spodoptera litura*. *Transgenic Res.*, **8** : 383-395.

Michaud D (1997) Avoiding protease-mediated resistance to herbivorous pests. *Trends Biotech.*, **15** : 4-6.

Milne RE and Kaplan H (1993) Purification and characterization of a trypsin-like digestive enzymes from spruce budworn (*Choristoneura fumiferansa*) responsible for the activation of δ-endotoxins form *Bacillus thuringiensis*. *Insect Biochem. Mol. Biol.*, **25** : 1101-1114.

Moar WJ, Pusztai-Carey M, van Faassen H, Bosch D, Frutos R, Rang C, Luo K and Adang MJ (1995) Development of *Bacillus thuringiensis Cry*IC resistance by *Spodoptera exigua* (Hubner) (Lepidoptera: Noctuidae). *Appl. Environ. Microbiol.*, **61** : 2086-2092.

Mohamed SI, Johnson DE and Aronson AI (1996) Altered binding of the *Cry*1Ac toxin to larval membranes but not to the toxin-binding protein in *Plodia interpunctella* selected for resistance to different *Bacillus thuringiensis* isolates. *Appl. Environ. Microbiol.,* **62** : 4168-5173.

Morton RL, Schroeder HE, Bateman KS, Chrispeels MJ, Armstrong E and Higgins TJV (2000) Bean a-amylase inhibitor 1 in transgenic peas (*Pisum sativum*) provides complete protection from pea weevil (*Bruchus pisorum*) under field conditions. *Proc. Natl. Acad. Sci. USA,* **97** : 3820-3825.

Murdock LL, Brookhart G, Dunn PE, Foard DE, Kitch L, Shade RE, Shukle RH and Wolfson JL (1987) Cysteine digestive proteinases in coleoptera. *Comp. Biochem. Physiol.,* **87B** : 783-787.

Nayak P, Basu D, Das S, Basu A, Ghosh D, Ramakrishman NB, Ghosh M and Sen S (1997) Transgenic elite indica rice plants expressing *Cry*IAc δ-endotoxin of *Bacillus thuringiensis* are resistant against yellow stem borer (*Scirpophaga incertitus*). *Proc. Natl. Acad. Sci. USA,* **94** : 2111-2116.

Oerke EC (1994) Estimated crop losses due to pathogens, animals pests and weeds. In : *Crop Production and Crop Protection : Estimated Losses in Major Food and Cash Crops.* (Eds. Oerke EC, Dehne HW, Schonbech F and Weber A) Elsevier Amsterdam. pp : 72-88.

Oppert B, Kramer KJ, Johnson D, McIntosh SC and McGaughey WH (1994) Altered protoxin activation by midgut from a *Bacillus thuringiensis* resistance strain of *Plodia interpunctella. Biochem. Res. Commun.,* **198** : 940-947.

Oppert B, Mramer KJ, Berman RW, Johnson D and McGaughey WH (1997) Proteinase-mediated insect resistance to *Bacillus thuringiensis* toxins. *J. Biol. Chem.,* **272** : 23473-23476.

Peferoen M (1997a) Insect control with transgenic plants expressing *Bacillus thuringiensis Cry*stal proteins. In: *Advance in Insect Control: The Role of Transgenic Plants* (Eds Carozzi N and Koziel M) Taylor and Franics, London pp: 21-48.

Peferoen M (1997b) Progress and prospects for field use of *B. thuringiensis* genes in crops. *Trends Biotech.,* **15** : 173-177.

Perlak FJ, Deaton RW, Armstrong TO, Fuchs RL, Sims SR, Greeneplate JT and Fishhoff DA (1990) Insect resistant cotton plants. *Bio/Technol.*, **8** : 939-943.

Perlak FJ, Fuchs DA, Dean RL, McPherson SL and Fishhoff DA (1991) Modification of the coding sequence enhances plant expression of insect control genes. *Proc. Natl. Acad. Sci. USA.*, **88** : 3324-3328.

Perlak FJ, Stone TB, Muskopf YM, Peterson LJ, Parker GB, McPherson SA, Wyman J, Love S, Reed G, Beaver D and Fishhoff DA (1993) Genetically improved potatoes: protection from damage by Colorado potato beetles. *Plant Mol. Biol.,* **22** : 313-321.

Peumans WJ and Van Damme EJM (1995) Lectins as plant defense proteins. *Plant Physiol.*, **109** : 347-352.

Plicher CD, Obrycki JJ, Rice ME and Lewis LC (1997) Preimaginal development, survival, and field abundance of insect predators an transgenic *Bacillus thuringiensis* corn. *Environ. Entomol.*, **26** : 446-454.

Poppy G (2000) GM crops: environmental risks and non-targets effects. *Trends Plants Sci.*, **5** : 4-6.

Powell KS, Spence J, Bharathi M, Gatehouse JA and Gatehouse AMR (1998) Immunohistochmecial and developmental studies to elucidate the mechanism of action of the snowdrop lectin on the rice brown planthopper, *Nilaparvata lugens* (Stal). *J. Insect Physiol.*, **44** : 529-539.

Purcell JP, Greenplate JT, Jennings MG, Gyerse JS, Pershing JC, Sims SR, Prinsen MJ, Corbin DR, Tran MH, Sammons RD and Stonard RJ (1993) Cholesterol oxydase: a potent insecticidal protein active against boll weevil larvae. *Biochem. Biophys. Res. Commun.*, **196** : 1406-1413.

Rahbé Y and Febway G (1993) Protein toxicity to aphids : an *in vitro* test on *Acyrthosiphon pisum. Entomol. Exp. Appl.*, **67** : 149-150.

Rao KV, Rathore KS, Hodges TK, Fu X, Stoger E, Sudhakar D, Williams S, Christou P, Bharathi M, Bown DP, Powell KS, Spence J, Gatehouse AMR, Gatehouse JA (1998) Expression of snowdrop lectin (GNA) in transgenic rice plants confers resistance to rice brown planthopper. *Plant J.*, **15** : 469-477.

Reeck GR, Kramer KJ, Baker JE, Kanost R, Fabrick JA and Brehnke CA (1997) Proteinase inhibitors and resistance of transgenic plants to insects. In: *Advance in Insect Control: the Role of Transgenic Plants* (Eds Carozzi N and Koziel M) Taylor and Francis, London pp: 157-183.

Riebe JF (1999) The development and implementation of strategies to prevent resistance to *Bt* expressing crops. *Can. J. Plant Pathol.,* **21** : 101-105.

Roush RT (1996) Can we slow adaptation by pest to insect transgenic crops? In *Biotechnology and Integrated Pest Management.* (Ed. Persley GL). CAB International, Cambridge. pp: 242-263.

Roush RT (1997) Managing resistance to transgenic crops. In: *Advance in Insect Control : The Role of Transgenic Plants.* (Eds Carozzi N and Koziel M) Taylor and Frangis, London UK. pp: 271-294.

Ruibal-Mendieta NL and Lints FA (1998) Novel and transgenic food crops: Overview of scientific versus public perception. *Transgenic Res.,* **7** : 379-386.

Ryan CA (1990) Proteinase inhibitors in plant: genes for improving defenses against insects and pathogens. *Annu. Rev. Phytopathol.*, **28** : 839-943.

Sardana R, Dukiandjiev S, Giband M, Cheng X, Cowan K, Sauder C and Altosaar I (1996) Construction and rapid testing of synthetic and modified toxin gene sequences *Cry1*A (b & c) by expression in maize endosperm cutlure. *Plant Cell Rep.,* **15** : 677-681.

Sauvion N (1995). Mode d'action de deux lectins a mannose sur le puceron de pois, *Acyrtosiphum pisum* (Harris). Potential d'utilisation des lectines végétables dans une création de plantes transgéniques résistantes aux pucerons. *Ph.D. thesis,* INSA, laboratoire de biologie appliquee, Villeurbane, France.

Schuler TH, Popply GM, Kerry BR and Denholm I (1998) Insect-resistant transgenic plants. *Trends Biotech.*, **16** : 147-196.

Schuler TH, Popply GM, Kerry BR and Denholm I (1999a) Potential side effects of insect-resistant transgenic plants on arthropod natural enemies. *Trends Biotech.*, **16** : 210-216.

Schuler TH, Potting RP, Denholm I and Poppy GM (1999b) Parasitoid behaviour and *Bt* plants. *Nature*, **400** : 825.

Serageldin I (1999) Biotechnology and food security in the 21st century. *Science*, **285** : 387-389.

Shade RE, Schroeder HE, Pueyo JJ, Tabe LM, Murdock LL, Higgins TJV and Chrispeels MJ (1994) Transgenic pea seeds expressing the α-amylase inhibitor of the common bean are resistant to bruchid beetles. *Bio/Technol.,* **12** : 793-796.

Shelton AM, Tang JD, Roush RT, Metz TD and Earle ED (2000) Field tests on managing resistance to *Bt*-engineered plants. *Nature Biotech.,* **18** : 339-342.

Shi Y, Wang MB, Powell KS, Van Damme E, Hilder VA, Gatehouse AMR, Boulter D, Gatehouse JA (1994) Use of the rice sucrose synthase-1 promoter to direct phloem-specific expression of beta-glucuronidase and snowdrop lectin in transgenic tobacco plants. *J. Exp. Bot.,* **45** : 623-631.

Schroeder HE, Gollasch S, Moore A, Tabe LM, Graing S, Hardie D, Chrispeels MJ, Spencer D and Higgins TJV (1995) Bean α-amylase inhibitor confers resistance to the pea weevil (*Bruchus pisorum*) in transgenic pea (*Pisun sativum* L). *Plant Physiol.,* **111** : 393-401.

Stewrt CN, Adang MJ, All JN, Raymer PL, Ramachandran S and Parrott WA (1996) Insect control and dosage effects in transgenic canola containing a synthetic *Bacillus thuringiensis Cry*IAc gene. *Plant Physiol.*, **112** : 15012-15017.

Stoger E, Willians S, Christou P, Down R and Gatehouse J (1999) Expression of the insecticidal lectin from snowdrop (*Galanthis nivalis* agglutinin; GNA) in transgenic wheat plant: effects on predation by the grain aphid *Sitobion avenae. Mol. Breed.*, **5** : 63-73.

Strizov M, Keller M, Mathur J, Koncz-Kalman Z, Bosch D, Prudovsky E, Schell J, Sneh B, Moncz C and Zilberstein A (1996) A synthetic *Cry*IC gene, encoding a *Bacillus thuringiensis* δ-endotoxin, confers *Spodoptera* resistance in alfalfa and tobacco. *Proc. Natl. Acad. Sci. USA,* **93** : 15012-15017.

Sutton DW, Havstad PK and Kemp JD (1992) Synthetic *Cry*IIIA gene from *Bacillus thuringiensis* improved for high expresion in plants. *Transgenic Res.*, **1** : 228-236.

Tabashnik BE (1989) Managing resistance with multiple pesticide tactics: theory, evidence and recommendations. *J. Econ. Entomol.*, **82** : 1263-1269.

Tabashnik BE, Cushing NL, Finson N and Johnson MW (1990) Field development of resistance to *Bacillus thuringiensis* in diamondback moth (Lepidoptera: Pyrallidae). *J. Econ. Entomol.,* **83** : 1671-1676.

Tabashnik BE, (1994) Evolution of resistance to *Bacillus thuringiensis. Annu. Rev. Entomol.,* **39** : 47-79.

Thomas JC, Wasmann CC, Echt C, Dunn RL and Bohnert HJ (1994) Introduction and expression of an insect proteinase inhibitor in alfalfa. *Plant Cell Rep.,* **14** : 31-36.

Thomas JC, Adams DG, Kepenne VD, Wasmann CC, Brown JK, Manost MR and Bohnert HJ (1995a) *Manduca sexta* encoded protease inhibitors expressed in *Nicotiana tabacum* provide protection against insects. *Plant Physiol. Biochem.,* **33** : 611-614.

Thomas JC, Adams DG, Kepenne VD, Wasmann CC, Brown JK, Kanost MR and Bohnert HJ (1999b) Protease inhibitors of *Manduca sexta* expressed in cotton. *Plant Cell Rep.,* **14** : 758-762.

Thomas JC, Adams DG, Nesler CL, Brown JK and Bohnert HJ (1995c) Tryptophan decarboxylase, tryptamine, and reproduction of the whitefly. *Plant Physiol.,* **109** : 717-720.

Toenniessen GH (1995) Plant biotechnology and developing countries. *Trends Biotech.,* **13** : 404-409.

Vaeck M, Reynaert A, Hofte H, Jansens S, De Beuckeleer M, Dean C, Zabeau M, van Montagu M and Leemans J (1987) Transgenic plants protected from insect attach. *Nature*, **328** : 33-37.

Van Damme EJM, Peumans WJ, Barre A and Rougé P (1998) Plant lectins: a composite of severla distinct families of structually and evolutionary related proteins with diverse biological roles. *Crit. Revu. Plant Sci.* **17** : 575-692.

Van Rie J, Jansens S, Höfte H, Degheele D and van Mellaert H (1989) Specificity of *Bacillus thuringiensis* delta-endotoxins. Importance of specific receptors on the brush-border membrane of the midgut of targets insects. *Eur. J. Biochem.*, **186** : 239-247.

Van Rie J, McGaughey WH, Johnson DE, Barnett BD and van Mellaert H (1990) Mechanism of resistance to the microbial insecticide *Bacillus thuringiensis. Science.*, **247** : 72-74.

Walter AJ, Ford L, Majerus MEN, Geoghegan IE, Birch ANE, Gatehouse JA and Gatehouse AMR (1998) Characterisation of the proteolytic activity in the larval midgut of two-spot ladybird (*Adalia punctata* L.) and its sensitivity to proteinase inhibitors. *Insect Biochem. Mol. Biol.*, **28** : 173-180.

Williams S, Friedrich L, Dincher S, Carozzi N, Kessmann H, Ward E, Ryals J (1993) Chemical regulation of *Bacillus thuringiensis* delta-endotoxin expression in transgenic plants. *Bio/Techol.*, **7** : 194-200.

Wilson WD, Lfint HM, Deaton RW, Fischhoff DA, Perlak FJ, Armstrong TA, Fuchs RL, Parks NJ, Stapp BR (1992) Resistance of cotton lines containing a *Bacillus thuringiensis* toxin to pink bolloworn (Lepidoptera: Gelechiidae) and the insects. *J. Econom. Entomol.*, **85** : 1516-1521.

Wolfersberger MG (1990) The toxicity of two *Bacillus thuringiensis* δ-endotoxins to gypsy moth larvae is inversely related to the affinity of binding sites in midgut brush border membranes for the toxins. *Experientia*, **46** : 475-477.

Wu Y, Llewellyn D, Matthews A and Dennis ES (1997) Adaptation of *Helicoverpa armigera* (Lepidoptera: Noctuidae) to a proteinase inhibitor expressed in transgenic tobacco. *Mol. Breed.*, **3** : 371-380.

Xu D, Xue Q, McElroy D, Mawal Y, Hilder VA and Wu R (1996) Constitutive expression of a cowpea trypsin inhibitor gene, CpTI, in transgenic rice confers resistance to two major rice pests. *Mol. Breed.*, **2** : 167-173.

Yu GG, Mullins MA, Warren GW, Koziel MG and Estruch JJ (1997) The *Bacillus thuringiensis* vegetative insecticidal protein Vip3A lyses midgut epithelium cells of susceptible insects. *Appli. Environ. Microbiol.*, **63** : 532-536.

Zhu-Salzman K, Shade RE, Koiwa H, Salzman RA, Narasimhan M, Bressan IA, Hasegawa PM and Murdock LL (1998) Carbohydrate binding and resistance to proteolysis control insecticidial activity of *Grifinia simplicifolia* lectin II. *Proc. Natl. Acad. Sci.*, **95** : 15123-15128.

Chapter 3

GENETICALLY ENGINEERED RESISTANCE TO PLANT VIRUSES

Kirthi Narayanaswamy and HS Savithri*

Department of Biochemistry, Indian Institute of Science, Bangalore - 560 012, India

Summary

The term genetically engineered resistance to plant viruses refers to the generation of transgenic plants that can be resistant to viral diseases. A number of foreign genes capable of conferring resistance to viral infection can be introduced into plants using transformation techniques. Resistance to plant viruses can be conferred either by expressing part of a viral genome or virally associated sequences. Resistance can also be conferred by other non viral genes such as antibodies and antiviral proteins. The former is termed as the pathogen derived resistance (PDR). Strategies for PDR are divided into those that require the production of proteins and those that require only the accumulation of the viral nucleic acid sequences. PDR can be mediated by expression of (i) coat protein which can interfere with the viral disassembly process (ii) mutant movement protein which can sequester the host factors (iii) complete or partial replicase protein which can interfere with viral replication. Also viral resistance can be conferred by modified transgenes that encode untranslatable RNAs such as DI RNAs and antisense RNAs. In this case the transgenic RNA would interact with viral RNA to account for homology dependent virus resistance and also nuclear derived RNA which accounts for homology dependent gene silencing. Resistance due to non viral genes is mediated by genes encoding the antibodies for a viral protein, or expression of ribosomal inactivating protein like the pokeweed antiviral protein, or use of

*Corresponding author : E-mail : bchss@biochem.iisc.ernet.in

antisense RNA for key enzymes in the plant. Yet another antiviral protein that has been used to confer resistance in plants is protease inhibitors. Especially among those viruses which use the strategy of producing a polyprotein which is processed to give the individual proteins, the production of a protease inhibitor in the plant limits the viral life cycle. Also satellite RNA, can mediate resistance. These transgene mRNA molecules are assumed to act as parasitic RNAs and result in protection of the transgenic plant to an infection of a helper virus. Also increase in salisylic acid in plants can confer resistance to viruses by producing an enhanced hypersensitive response. Thus engineering resistance in plants is a very useful and innovative method to develop viral resistant plants. However these trangenic plants expressing viral pathogen derived sequences have been considered as sites for hyper evolution of viruses through recombination between a mild or defective viral genome. Hence a proper evolution of the transgenic lines would be essential before they can be commercialized.

Keywords : Plant viruses, Engineered resistance, pathogen derived resistance (PDR), Coatprotein mediated resistance (CPMR), Movement protein mediated resistance (MPMR) replicase mediated resistance (Rep-MR).

1. INTRODUCTION

Plant viruses are among the smallest pathogens that cause severe losses in numerous crops worldwide. They are simple in their organization; composed of a single type of nucleic acid (RNA or DNA) with protective coats made up of protein subunits and in some cases surrounded by a lipid envelope. The small size of the viral genomes severely limits their information content. Viruses have therefore evolved suitably to maximize the use of their genomes and redirect host cell machinery towards their multiplication.

Plant viruses face special problems initiating an infection. The outer surfaces of plants are composed of protective layers of waxes and pectin. Furthermore, each cell is surrounded by a thick wall of cellulose overlying the cytoplasmic membrane. To date, no plant virus is known to use a specific cellular receptor of the type that animal and bacterial viruses use to attach to cells. Rather, plant viruses rely on a mechanical injury to the cell wall to directly introduce a virus particle into a cell. This is achieved either by the vector associated with transmission of the virus or simply by mechanical damage to cells. After replication in an initial cell, the lack of receptors poses special problems for plant viruses in recruiting new cells to the infection. The problem these viruses face in reinfection and recruitment of new cells is the same as they face initially - how to cross the barrier of the plant cell wall. Plant cell walls necessarily contain channels called plasmodesmata which allow plant cells to communicate with each other and to pass metabolites between them. However, these channels are too small to allow the passage of virus particles or genomic nucleic acids.

Many (if not most) plant viruses have evolved specialized movement proteins which modify the plasmodesmata. One of the best known examples of this is the 30k protein of tobacco mosaic virus (TMV). This protein is expressed from a sub-genomic mRNA and its function is to modify plasmodesmata causing genomic RNA coated with 30k protein to be transported from the infected cell to neighbouring cells (Grdzelishvili *et al.*, 2000). Other viruses, such as cowpea mosaic virus (CPMV - Comovirus family) have a similar strategy but employ a different molecular mechanism. In CPMV, the 58/48k proteins form tubular structures allowing the passage of intact virus particles to pass from one cell to another (Kasteel *et al.*, 1997).

Typically, virus infections of plants might result in effects such as growth retardation, distortion, mosaic patterning on the leaves, yellowing, wilting, etc. These macroscopic symptoms result from *necrosis* of cells, caused by direct damage due to virus replication *hypoplasia*, *i.e.* localized retarded growth frequently leading to mosaicism (the appearance of thinner, yellow areas on the leaves) *hyperplasia*, which is excessive cell division or the growth of abnormally large cells, resulting in the production of swollen or distorted areas of the plant.

2. METHODS FOR DEVELOPING RESISTANCE

Conventional breeding programs to develop resistant plant varieties are effective, but expensive and often in addition to the desired character some of the undesirable characters may be transferred. Also, the resistance can be circumvented by virus variation. Alternative and additional approaches for protecting plants from infection with viruses include, control of vectors that transmit them by use of insecticides, use of virus free propagating material and cultural practices such as sanitation, crop rotation, programmed sowing, etc. With the development of plant transformation techniques, astonishing progresses have been made to introduce foreign genes into plants for virus resistance. Since Hamilton (1980, 1985) first suggested that virus resistance could be obtained by introducing viral genes into a plant, the possibility of viral genes functioning as resistance genes has been demonstrated in many virus-plant systems. The number of plant species amenable to transformation and regeneration has increased greatly in recent years and it will likely to be possible to produce fertile transformants of most, if not all, significant crops in the near future.

There are a number of ways in which resistance could potentially be achieved by using plant transformation and regeneration; several of these methods have been explored in considerable detail. For example, plant transformation might allow the natural virus resistance genes present in one plant species to be introduced into another species. However, the problems in identifying and characterizing such resistance genes limit the use of this approach. In General, there are two large categories of genetically engineered resistance to plant viruses. The first one is the so called "pathogen-derived resistance". The second category is resistance conferred by other genes including antibodies and antiviral proteins.

Genetically engineered resistance by expressing part of a viral genome or virally associated sequences in transgenic plants has its roots in empirically observed "cross protection". In cross protection phenomenon, inoculation of a host plant with a milder or symptomless strain can protect the plant from infection by more severe strains of the same or very closely related viruses. From these empirical observations on cross protection arises the *pathogen-derived resistance (PDR)* concept which states that certain key gene products of a pathogen present in the plant in a dysfunctional form, in excess, or at an inappropriate stage during viral replication cycle could disrupt infection by the invading pathogen. The first illustration that PDR is indeed a viable way of producing virus-resistant plants was provided by the experiments of Powell-Abel *et al.* (1986), who demonstrated that tobacco plants transgenic for, and expressing, the TMV coat protein were resistant to infection with the virus. This discovery opened a new field of research in plant sciences, and many papers have been published in the past decade illustrating the utility of PDR.

3. LIFE CYCLE OF A VIRUS : STRATEGIES THAT CAN BLOCK VIRUS MULTIPLICATION

Plant viruses differ considerably in their morphology and in the genetic material used to encode the virus genes. These genomes include single- or double stranded DNA, double stranded RNA, or single-stranded RNA of either plus- or minus-sense RNA. They are either monopartite (with one genome) or multipartite (with divided genomes). They replicate and propagate by various methods but the common feature in the life cycle of all these is the transcription of mRNAs for translation of structural and nonstructural proteins that are required to fulfill the viral life cycle (Scholthof *et al*., 1993). Despite differences in their replication strategies, all plant viruses have broadly similar steps in their life cycles (Fig I). Upon gaining entry into the host, via fungi, insects, mites, nematodes, etc., through necrotic lesions, or through injury caused by abrasives used during mechanical inoculations, the virus particles partially disassemble, to expose the viral DNA or RNA to the cellular milieu (Verduin, 1992). DNA viruses generally enter the nucleus and utilize host enzymes to produce mRNAs for translation while viruses having mRNA as genetic material undergo translation to produce the virus-specific proteins necessary for propagating themselves.

A crucial process in the infection by most plus-sense RNA viruses is the production of replicase protein(s) that, along with the cellular machinery, produce progeny by replicating the parental genome. The progeny viral RNA moves from cell-to-cell via the plasmodesmata, as a complex with viral protein (movement protein) that aids in the movement or, for optimum long-distance movement, in association with a functionally active coat protein (Citovsky and Zambryski, 1991). It is theoretically possible to block the viral life cycle at any of these stages *i.e.* uncoating, translation, replication, and/or movement (Fig 1). The aim of generating transgenic plants resistant to virus infection is to express a portion of the viral genome, with or without expression of an encoded protein, that will inhibit some of the above mentioned steps in the multiplication cycle.

Initially, all attempts to use PDR to produce virus-resistant plants followed the lead of Powell-Abel *et al.* (1986) by making plants transgenic for viral coat protein genes. In a large number of cases, transgenic lines showed resistance to different extents when challenged with the virus from which the gene was derived (Beachy *et al*., 1990; Fitchen and Beachy, 1993; Grumet, 1994). Subsequently, successful lines of resistant transgenic plants were also obtained by using the viral replicase gene instead of coat protein. This was, as in coat protein-mediated protection, first demonstrated with TMV (Golemboski *et al*., 1990). Further, it was shown that satellite RNAs, antisense and sense defective RNAs and

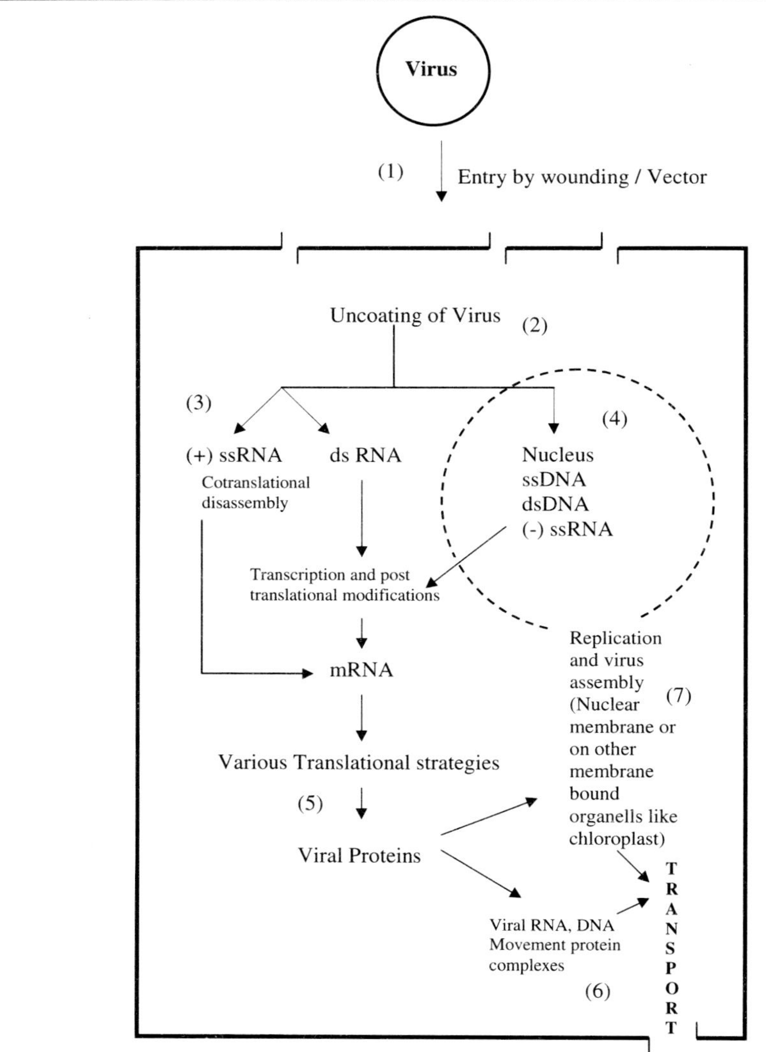

Figure 1 : Schematic representation of viral life cycle. The virus enters the plant by either a vector or through wounds on the plant surface (1). After entry it is uncoated and the respective nucleic acid is released. The (+) ss RNA can also undergo translational disassembly (2). The (+) ssRNA remains in the cytosol as does ds RNA (3), while the DNA viruses and (-) ss RNA viruses need to enter the nucleus (4) to complete their life cycle. The various viral proteins are produced within the plant cell by using the plant cellular machinery (5). The movement protein may form complex with nucleic acid and enable in transport of the virus locally and systemically (6). Finally replication and assembly of viruses gives intact viruses which can reinfect fresh plants (7).

defective-interfering sequences also give rise to resistant phenotypes (Gerlach *et al.*, 1987; Harrison *et al.*, 1987; Jacquemond *et al.*, 1988; Kim *et al.*, 1995). Other portions of the plant viruses such as NIa protease of potyviruses (Maiti *et al.*, 1993; Vardi *et al.*, 1993), the movement proteins of tobamoviruses, bromoviruses and potexviruses (Beck *et al.*, 1994; Lapidot *et al.*, 1993; Malyshenko *et al.*, 1993) and the 3' noncoding region of a tymovirus (Zaccomer *et al.*, 1993) have also been used to generate transgenic plants. Thus in all probability, any part of a plant viral genome can potentially give rise to PDR. However, expression of not all portions of the viral genome can confer resistance. For example, transgenic lines expressing tobacco vein mottle virus (TVMV) protease are resistant to infection by TVMV, while those expressing cylindrical inclusion protein were found to be more susceptible. (Maiti *et al.*, 1993)

4. PDR APPROACH

Strategies for PDR are divided into those that require the production of proteins and those that require only the accumulation of viral nucleic acid sequences. According to Beachy (1997), the former confer resistance to a broader range of virus strains and viruses, whereas the latter provide very high levels of resistance to a specific virus strain.

4.1. Coat protein-mediated resistance

Powell-Abel *et al.* (1986) first reported that constitutive expression of TMV coat protein (CP) gene in transgenic *N. tabacum* plants confers protection against challenge by TMV, a phenomenon termed CP-mediated resistance (CPMR). In these experiments, transgenic plants were more resistant to TMV virions than to TMV RNA inocula. Based on this observation, it was suggested that CPMR against TMV was through the inhibition of the virion disassembly in the initially infected cells (Register and Beachy, 1988; Powell-Abel *et al.*, 1990). Similarly, transgenic plants producing low levels of AlMV CP were found to be resistant only to AlMV virion inocula and plants expressing a mutant AlMV CP with the N-terminal Ser residue changed to Gly were susceptible to infection with wild type viral particles or RNAs but were resistant to mutant virions (Taschner *et al.*, 1994). However, the CPMR against potato virus X (PVX) was found to be effective against both RNA and virion inocula. It was suggested that an interaction of transgenic CP with the PVX origin of assembly could lead to resistance (Hemenway *et al.*, 1988). The origin of assembly was mapped to the 5' region of the viral genome (Sit *et al.*, 1994) and it was suggested that an interaction of CP would have a potential to suppress translation of the viral RNA-dependent RNA polymerase (RdRp) that is encoded by the 5' most ORF. It is also possible that, transgenic expression of PVX CP inhibited cell-to-cell movement of PVX, for which CP is an essential factor (Chapman *et al.*, 1992). The PVY coat protein mediated resistance is shown to have been carried on up to the T4 progenitor lines. Most of the T4 plants accumulated extremely low levels of CP protein and steady state mRNA and exhibited almost complete resistance to PVY (Hans *et al.*, 1999) Also R1 progenies of transgenic *Pisum sativum* L were shown to harbor the transgene (coat protein of Pea enation mosaic virus) by PCR and also trangenics R2, R3, R4 plants displayed delayed or transient PEMV multiplication and attenuated symptoms as compared to control inoculated individuals (Chowrira *et al.*, 1998). Transgenic tobacco and tomato plants expressing the coat protein of Physalis mottle tymovirus were shown to be ressistant to challenge inoculation by the virus (Ranjith kumar *et al.*, 1999; Srividhya *et al.*, 2000).

Several hypotheses have been proposed to explain the mechanisms of CPMR (Fitchen and Beachy, 1993; Lomonossoff, 1995). It was proposed from the data obtained with transgenic plants and protoplasts that express the TMV CP that, the transgenic CP prevents the uncoating of the challenge virus (Osbourn *et al.*, 1989; Register and Beachy, 1988; Wu *et al.*, 1990). The model suggests that either endogenous CP assembles to form virus-like particles (VLPs) that block the site for uncoating the challenge virus or CP inhibits virus disassembly by shifting the disassembly-assembly reaction in favor of assembly, thereby preventing virus infection in the inoculated cells (Register and Nelson, 1992). It was proposed earlier (Clark *et al.*, 1995a; Register and Beachy, 1988) that when the challenge virus undergoes disassembly in transgenic plants the CP in the host re-encapsidates the RNA, thereby blocking ribosome binding and translation of viral RNA by driving the uncoating reaction towards assembly rather than disassembly. Reassembly requires that the constitutively expressed CP be assembly competent to produce virus like particles (VLPs) and that the CP is sufficiently similar to the challenge virus CP to establish strong CP-virion interactions that reduce infection. This hypothesis implies an eventual exchange of the CP subunits of the challenge virus with the subunits of the CP produced by the transgenic plant (Clark *et al.*, 1995a; 1995b). It was shown that specific interactions between the transgenic CP and the CP of the challenge virus as well as an interference with the virus movement are necessary components of

CPMR against TMV (Bendahmane *et al.*, 1997). However, It was also shown in the case of TMV and AlMV that, the assembly into virus-like particles is not a prerequisite for CP mediated protection by using assembly defective mutants (Clark *et al.*, 1995a; Yusibov and Loesch-Fries, 1995).

Culver *et al.* (1995) generated site-specific mutants (E50Q and D77N) of TMV coat protein in the full length cDNA clone. When these mutant transcripts were inoculated into *N. tabacum* cv. Xanthi, both the mutants infected tobacco but were unable to spread and cause systemic infection. Purified virions from these infected plants were much more stable than wildtype virus, thereby confirming that negative charges on these side chains play a crucial role in the disassembly of TMV. It was suggested that these mutants might confer better resistance to virus infection. Transgenic plants expressing a mutant of TMV CP which could form stable VLPs were shown to confer better resistance (Bendahmane *et al.*, 1997). In another study, mutations in the coat protein gene of plum pox virus was shown to suppress particle assembly, heterologous encapsidation and complementation in transgenic plants of *Nicotiana benthamiana* (Varrelmann and Maiss, 2000).

Recently a series of constructs carrying deletions and frame shifts of *Cymbidium Ringspot Tombusvirus (CymRSV)* showed that an 860nt long RNA sequence in the *CymRSV* CP coding region (between nucleotide 2666 and 3526) is an elicitor of a very rapid HR-like response of *Datura stramonium* which limits the virus spread. This finding provides the first evidence that an untranslatable RNA can trigger an HR like resistance response in virus infected plants (Szittya, and Burgyan, 2001). However, it is likely that the mechanisms of CPMR are not the same for all viruses and could involve any interaction of the CP required for the virus infection cycle.

The role of viral CP in virus movement has been investigated using tobacco etch virus expressing β-glucuronidase (TEV-GUS) (Dolja *et al.*, 1994). The expression of GUS gene permits visualization of virus translocation. Mutations affecting subunit interactions also debilitated virus movement. But these defects could be rescued in transgenic plants expressing wildtype TEV CP. These results suggested that potyviral CP possesses distinct separable activities required for virion assembly, cell-to-cell movement and long distance transport (Dolja *et al.*, 1994).

Graham *et al.* (1997), by using GUS fusion constructs driven by *Agrobacterium rhizogenes* RolC and maize Sh (Shrunken; sucrose synthase-1) promoters, reported that RolC promoter was specific for phloem tissue, bundle sheath cells and vascular parenchyma but not for xylem or non-vascular tissues. The expression was exclusively confined to phloem tissues. Potato leafroll luteovirus (PLRV) replicates only in phloem tissues and hence virus-resistant lines could be obtained when RolC was used to drive the expression of PLRV CP gene (Graham *et al.*, 1997). Thus transgenic plants in which individual viral functions are expressed separate from the other viral genes, are a useful tool to study viral processes. CPMR was the first strategy used to develop transgenic viral-resistant plants and is still the most popular mode of obtaining resistance.

4.2. Movement protein-mediated resistance

The cell-to-cell movement of virions involves many steps (Carrington *et al.*, 1996). In most plus sense RNA viruses, replication of viral RNA takes place in the cytoplasm and this is followed by transfer of newly synthesized genomes from the site of replication to intracellular transport system. Interaction between the movement proteins (MPs), replication proteins or nascent genomes may initiate this transport process and provide specificity for movement of viral RNA. This is followed by transport of viral genomes to plasmodesmata and transit through plasmodesmata. The MPs form complexes with viral genome which can range in size. TMV MP/nucleic acid complex for example, is 1.5 to 2.0 nm in diameter (Citovsky *et al.*, 1992) whereas red clover necrotic mosaic dianthovirus (RCNMV) MP/nucleic acid complex retains considerable secondary structure (Fujiwara *et al.*, 1993). The interaction of MP/RNA complex with plasmodesmata alters the gating properties and directly facilitates the transfer of these large molecules. Long distance or phloem dependent movement requires that the virus be able to enter and exit bundle sheath cells, phloem parenchyma and companion cells, and sieve elements. This long distance transport involves viral and host functions that are distinct from those involved in movement through mesophyll cells. Most viruses require CP for long distance movement.

The effectiveness of movement protein mediated resistance (MPMR) was illustrated by the transgenic expression of viral movement proteins, which conferred resistance only when the transgene specified a dysfunctional MP (Lapidot *et al.*, 1993; Malyshenko *et al.*, 1993). Transgenic expression of functional MP either had no effect on virus infection or increased susceptibility. MP produced in transgenic plants could enable movement defective mutants of TMV to move to adjacent cells (Deom *et al.*, 1990; Holt and Beachy, 1991).

It was suggested that the transgenic expression of a dysfunctional MP would lead to competition for plasmodesmatal binding sites between the mutant MP and the wildtype MP of the inoculated virus leading to the blockage of cell-to-cell movement of the virus (Lapidot *et al.*, 1993). An interesting and important aspect of MPMR is the broad spectrum efficacy of the resistance mechanism (Deom *et al.*, 1992). The mutant MP of TMV, mediates resistance not only to tobamoviruses but also to viruses of the potex-, cucumo-, and tobraviral groups (Cooper *et al.*, 1995). Similarly, transgenic expression of the brome mosaic virus MP in a nonhost plant conferred resistance to TMV (Malyshenko *et al.*, 1993). Transgenic plants expressing Potato virus X ORF 2 protein (p24) are shown to be resistant to Tobacco mosaic virus and Ob Tobamoviruses (Ares *et al.*, 1998). Transgenic potato plants expressing mutant alleles of PLRV ORF 4, the gene for the movement protein pr17 were generated for broad range protection against virus infection. When tested for protection against infection by PLRV, all transgenic lines showed a significant reduction of antigen. Potato lines accumulating N or C terminally extended PLRV pr 17 mutant proteins were resistant to infection by the unrelated potato viruses PVY and PVX (Tacke *et al.*, 1996). These examples of broad spectrum resistance indicate that MPs of several different viruses may interact with the same plasmodesmatal components (Carrington *et al.*, 1996).

In potex-, carla-, hordei-, and furovirus groups, three MPs are encoded by a series of overlapping reading frames referred to as triple gene block (TGB) (Petty and Jackson, 1990; Beck *et al.*, 1991). Transgenic expression of mutant TGB protein confers resistance against a narrower range of viruses than does the TMV MPMR (Beck *et al.*, 1994). This indicates that the TGB proteins may not interact with the plasmodesmata in the same way as does the TMV MP.

Transgenic plants could also be very helpful to study viral movement processes. It has been shown for various viruses that MPs produced in transgenic plants are functional and can transport movement-deficient mutants of the virus (Wolf *et al.*, 1991; Cooper *et al.*, 1996). Increase in the plasmodesmatal permeability could be studied using plants expressing MPs (Vacquero *et al.*, 1994; Citovsky, 1993). Plasmodesmata between mesophyll cells of transgenic plants expressing MP was shown to result in ~10-fold increase in diameter than that of plasmodesmata of control plants (Wolf *et al.*, 1989). In view of the broad spectrum of resistance conferred by dysfunctional MPs, this may prove to be a better strategy than CPMR.

4.3. Replicase-mediated resistance

Genes that encode complete or partial replicase proteins can confer near immunity to infection that is generally limited to virus strain from which the gene sequence was obtained. Replicase-mediated resistance (Rep-MR) to TMV was first reported in plants transformed with a sequence encoding a 54 kD fragment of replicase (Golemboski *et al.*, 1990). The protein fragment was not detected in these transgenic plants suggesting that certain examples of Rep-MR was RNA- rather than protein-mediated (Baulcombe, 1996). However, in the case of PVY (Audy *et al.*, 1994) and AlMV (Brederode *et al.*, 1995) it was observed that the resistance mechanism was protein based rather than RNA based because the resistance phenotype was influenced by mutations affecting the primary structure of the replicase protein encoded by the transgene. Only the mutant protein but not the wild-type replicase conferred resistance to infection. A similar approach provided resistance to tomato yellow leaf curl virus (TYLCV), a geminivirus (Noris *et al.*, 1996). The presence of an open reading frame and apparently, production of protein was also required for Rep-MR in the case of TMV and CMV (Carr and Zaitlin, 1991; Zaitlin *et al.*, 1994). A truncated mutant of replicase derived from a cucumber mosaic virus (CMV) subgroup I virus conferred high levels of resistance in tobacco plants to all subgroup I CMV strains, but not to subgroup II strains or other viruses (Zaitlin *et al.*, 1994). Recently it was shown that nonviral expression of a 50 kD TMV helicase fragment (p50) is sufficient to induce the N gene mediated hypersensitive response in tobacco (Erickson *et al.*, 1999) Transgenic pea lines carrying the replicase (*NIb*) gene of pea seed borne mosaic poty virus (PsbMV) were generated and used in experiments to determine the effectiveness of induced resistance upon infection by heterologous isolates. Three pea lines showed inducible resistance in which an initial infection by the homologous isolate was followed by a highly resistant state (Jones *et al.*, 1998). In many plant RNA viruses domains 1,2 and 3 are conserved in replicase proteins and genetically engineered transgenic plants with the domain 1 sequence of tobacco mosaic virus 126 kD protein gene are completely resistant to viral infection (Song *et al.*, 1999).

The mechanisms effecting Rep-MR are not clearly understood, although it was shown that plants exhibiting Rep-MR can strongly repress replication. In many cases, these transgenic plants were resistant to high levels of challenge inoculum. It is proposed that protein produced by the transgene interferes in some manner with the function of replicase produced by the virus, perhaps by binding to host factors or to viral proteins that regulate replication and virus gene expression (Beachy, 1997). The expression of intact replicase component could lead to defective assembly of the multimeric replication complex due to presence of an excess amount of one of the components. On the other hand, the resistance occurring from the expression of defective (truncated) replicase proteins, might be caused by modified interactions of the truncated proteins with host factors essential for replication cycle. Such a host factor could be common for the replication processes of various unrelated viruses. If such a host factor is identified, transgenic plants harboring wildtype or mutant form of it might provide a more general strategy for obtaining resistance.

The use of transgenic plants expressing replicase protein in studying the viral replication processes was demonstrated in the case of AlMV. The transgenically produced replicase proteins were capable of forming active replication complexes and these transgenic plants allowed replication of AlMV-RNA segments not capable of replication themselves (Van Dun *et al.*, 1988; Taschner *et al.*, 1991). It was also shown that transgenic tobacco plants expressing replicase genes P1 and P2 of AlMV could be infected with RNA 3 of the tripartite AlMV genome (Van der Vassan *et al.*, 1996). Further, with the use of these transgenic plants it was shown that a deletion mutant of RNA 3, in which 5' terminal 22 nucleotides were deleted was as infectious as the wildtype, indicating that the 5' terminal sequence of AlMV RNA 3 is dispensable for replication. Thus, by generating transgenic plants expressing different forms of replicase proteins the process of viral replication could be studied.

4.4. Nucleic acid mediated resistance

With increasing number of reports on the use of viral genes for PDR, deviations from the original PDR concept became more frequent. A consistent lack of correlation between expression level of the transgenic protein and levels of resistance was reported (Gielen *et al.*, 1991; Golemboski *et al.*, 1990; Lawson *et al.*, 1990; Stark and Beachy, 1989; Van der Wilk *et al*., 1991). In some cases the protein product was not detectable, suggesting that the expression of the protein was not essential for resistance. Further, the subsequent findings demonstrated that resistance was conferred by modified viral transgenes that encoded untranslatable RNAs (De Haan *et al.*, 1992, Lindbo and Dougherty, 1992; Van der Vlugt *et al.*, 1992). The first clue on the molecular basis of RNA-mediated resistance was shown by Lindbo *et al.* (1993). They observed that the new shoots that developed after TEV infection, in transgenic plants expressing TEV CP sequences, remained virus free and were resistant to subsequent inoculations. Recovered plants could not be infected with TEV but were susceptible to infection with a closely related virus, PVY. The occurrence of recovery coincided with a substantial drop in cytoplasmic transgenic RNA levels. But the nuclear RNA levels showed no appreciable difference between unchallenged and recovered tissue. It was concluded that a post-transcriptional, cytoplasmic activity was responsible for the reduction of transgenic RNA levels and consequently the same activity may be responsible for virus resistance. This proposal was substantiated by the findings that gene silencing with nonviral transgenes is due to a post transcriptional mechanism (Ingelbrecht *et al.*, 1994; de Carvalho Niebel *et al.*, 1995). This post transcriptional mechanism operates at the RNA level and would therefore have the potential to suppress the accumulation of viral RNA that shares sequence identity with the silenced transgene (English *et al.*, 1996). Such a mechanism has been referred to as homology-dependent resistance to reflect the relationship with homology-dependent gene silencing (Mueller *et al.*, 1995; Ratcliff *et al.*, 1997).

As per this model, the transgenic RNA would interact with viral RNA to account for homology-dependent virus resistance and with nucleus-derived RNA to account for homology-dependent gene silencing (Mueller *et al.*, 1995). In principle, the interaction leading to suppression of viral RNA could involve base pairing of the sense RNA transcript of the transgene and the negative strand of the viral RNA, which is produced as an intermediate in the replication cycle of most viral RNAs. However, such a mechanism cannot account for the decreased levels of cytoplasmic transgene transcripts. It has therefore been proposed that as the levels of transgenic RNA increase above a certain threshold level in the nucleus, methylation of the transgenic loci would occur (Ingelbrecht *et al.*, 1994; Wassenegger *et al.*, 1994) (Fig. 2). Extensive methylations of the transgene may cause aberrations in the transcribed messenger RNAs (aberrant RNA), that subsequently trigger a resident RNA-dependent RNA polymerase (Schiebel *et al.*, 1993a, 1993b) present in the cytoplasm to synthesize (short) antisense RNA molecules. This antisense RNA would have the potential to base pair with both transgenic and viral RNAs (Fig. 2).

There are several ways in which the formation of duplex RNA could influence accumulation of host and viral RNAs to cause virus resistance and gene silencing (Fig. 2). For example, the base pair region may render the duplex RNA susceptible to degradation by RNases specific for double-stranded RNA (Nicholson, 1996). The base pair region could also arrest translation and consequently have an indirect effect on RNA stability (Green, 1993).These effects could cause reduced accumulation of both nucleus- and virus- derived RNAs (Fig. 2). Furthermore, if the base pair region is required in *cis* for replication of viral RNA, the formation of a duplex with an antisense RNA could inhibit virus replication.

5. RISK ISSUES CONCERNED WITH PDR

Transgenic plants expressing viral pathogen-derived sequences have

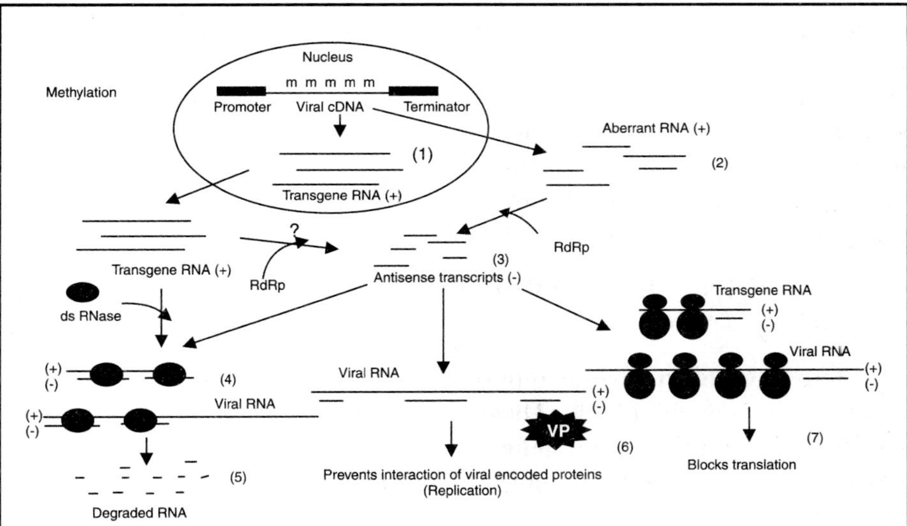

Figure 2 : Model of homology-dependent resistance and gene silencing. Initially the viral DNA is transcribed and the transgene transcript (1) inhibits viral replication by base pairing with the negative strand of the viral RNA. As the concentration of the transgene transcript increases above a certain level in the nucleus, it leads to methylation of the transgene which results in aberration in the transcribed RNA. (2) (Aberrant RNA). This in turn activates a host RNA dependent RNA polymerase which synthesizes short antisense transcript (3). These transcripts can base pair with transgene (gene silencing (4)) or with viral RNA (homology dependent resistance (5) and make them susceptible for degradation by dsRNase. (5) the base pairing can also prevent binding of other Society of Plant Physiology and Biochemistry proteins (6) and inhibit translation (7). © Soc. Plant Physiol. Biochem.

been considered sites for hyperevolution of viruses through recombination between a mild or defective viral genome (DNA or RNA) and the transgene or its transcript. It has been reported that pathogenic variants can develop from transgenically expressed satRNAs (Palukaitis and Roossinck, 1996). Transencapsidation or heteroencapsidation of viral RNAs by transgenically expressed CP has been reported (Osbourn *et al.*, 1990; Holt and Beachy, 1991). The CP of plum pox potyvirus has been shown to confer aphid transmissibility on a nontransmissible isolate of zucchini yellow mosaic virus. This effect would apply only for a single acquisition-transmission cycle. In an attempt to detect recombinant viral RNA, transgenic plants harboring a 3kb TMV transgene was inoculated with a TMV vector with wild type and mutant coat protein separately. Although, the recombinant RNA between TMV RNA and host mRNA did not accumulate to detectable levels under non-selective conditions, it did accumulate in presence of selective pressure. However, even in the later case an encapsidated recombinant viral population did not develop (Adair and Kearney, 2000). The risk of generating more severe strains exists not only in considering PDR but also in nature. For example, a gemini virus isolate associated with severe cassava mosaic disease in Uganda was shown to have arisen by interspecific recombination of East African cassava mosaic virus (EACMV) and African cassava mosaic virus (ACMV) (Zhou *et al.*, 1997).

6. ALTERNATE APPROACHES TO ENGINEER RESISTANCE

There have been several reports regarding the use of novel genes to create virus-resistant plants. These include genes encoding for antibodies specific for a viral protein, genes encoding for antiviral proteins such as pokeweed protein, and host genes that may be required for normal viral replication cycle.

Exogenous application of pokeweed antiviral protein (PAP), a ribosome-inhibiting protein found in the cell walls of *Phytolacca americana* (pokeweed), protects heterologous plants from viral infection. Lodge *et al.* (1993) isolated a cDNA clone for PAP and introduced it into tobacco and potato plants. Transgenic plants that expressed either PAP or a double mutant derivative of PAP showed resistance to infection by different viruses. Resistance was effective against both mechanical and aphid transmission. Analysis of the vacuum infiltrate of leaves expressing PAP showed that it is enriched in the intercellular fluid. Analysis of resistance in transgenic plants suggests that PAP confers viral resistance by inhibiting an early event in infection. Previous methods for creating virus-resistant plants have been specific for a particular virus or closely related viruses. To protect plants against more than one virus, multiple genes must be introduced and expressed in a single transgenic line. The advantage of using PAP in transgenic plants for virus resistance is resistance to a broad spectrum of plant viruses by expression of a single gene. The disadvantage of using PAP could be that it affects the host protein synthesis also and this in turn might affect the growth and production of the transgenic plant.

Yet another ribosomal inactivating protein is Dianthin, isolated from *Dianthus caryopyllus*. Dianthin has been expressed from the African cassava mosaic virus (ACMV) virion sense promoter that is transactivated by the product of viral gene AC2. This avoids the need for constitutive expression of dianthin, facilitating the regeneration of phenotypically normal plants, and ensures that the transgene expression is localized to virus infected cells. When challenged with ACMV, transgenic plants produce atypical necrotic lesions on inoculated leaves, indicative of dianthin expression, viral DNA accumulation is significantly reduced in these tissues and plants exhibit attenuated systemic symptoms from which they recover (Hong *et al.*, 1996).

S-Adenosylhomocysteine hydrolase (SAHH) is a key enzyme in transmethylation reactions that use S-adenosylmethionine as the methyl donor. Because of the importance of SAHH in a number of S-adenosylmethionine-dependent transmethylation reactions, particularly the 5' capping of mRNA during viral replication, SAHH has been considered as a target of potential antiviral agents against animal viruses. To test the possibility of engineering a broad type of resistance to plant viruses, Masuta *et al.* (1995) expressed the antisense RNA for tobacco SAHH in transgenic tobacco plants. As expected, transgenic plants constitutively expressing an antisense SAHH gene showed resistance to infection by various plant viruses. Among those plants, about half exhibited some level of morphological change (typically stunting). Analysis of the physiological change in those plants showed that they contained excess levels of cytokinin. Because cytokinin has been found to induce acquired resistance, there is also a strong possibility that the observed resistance was induced by cytokinin

Other types of resistance genes apparently do not require expression of functional protein products. These are the so-called RNA-based resistance genes. Satellite RNA (satRNAs), defective-interfering (DI)

RNA, and antisense RNA belong obviously to this category. Resistance to some viruses could also be achieved by the production of such RNAs in transgenic plants (Gerlach *et al.*, 1987; Harrison *et al.*, 1987; Jacquemond *et al.*, 1988; Kim *et al.*, 1995). Both satRNAs and DI RNAs depend on helper virus for their replication; DI RNAs are truncated forms of the helper virus, while satRNAs are not related by sequence similarity. These transgene mRNA molecules are assumed to act as parasitic RNAs and result in protection of the transgenic plant to an infection of the helper virus.

RNA-mediated virus resistance has been observed in transgenic plants at varying frequencies, suggesting that a nuclear requirement or other pre-condition must be met. Goodwin *et al.* (1996) conducted a study to characterize genetically transgenes that confer a highly resistant state to infection by tobacco etch virus (TEV). Transgenic tobacco line 2RC-6.13, expressing an untranslatable mRNA containing the TEV coat protein open reading frame, had three distinct transgene integration events that segregated as two linkage groups. A genetic series of plants that contained zero, one, two, or all three transgene inserts in both homozygous and heterozygous conditions was produced and examined. Genetic and biochemical data suggested that RNA-mediated virus resistance is a multigenic trait in line 2RC-6.13; three or more transgenes were necessary to establish the highly resistant state, One or two transgene copies resulted in an inducible form of resistance (*i.e.,* recovery). Transcription rates and steady state RNA levels of the transgene-derived transcript present in different members of the genetic series supported a post-transcriptional RNA degradation process as the underlying mechanism for transgene transcript reduction and virus resistance, This degradation process appeared to initiate via cleavage of specific sites within the target RNA sequence, as determined by RNA gel blot and primer extension analyses of transgene-derived mRNA from various transgenic plant lines.

Antisense RNA mediated resistance has not been working well for RNA viruses. Antisense inhibition of plant nuclear gene expression is well documented. Thus this strategy of engineered virus resistance may have a better potential with DNA viruses that transcribe their genome into mRNA in the nuclei.

Ribozyme targeting specific viral RNA sequences, mild satellite RNA species, and DI have not been proven to be very effective although some of the transgenic plant lines do exhibit a certain level of resistance. It was shown that two enzymes of an interferon induced mammalian antiviral 2-5A system, human RNase L and human 2-5A synthetase, when expressed in transgenic plants could provide resistance to TMV, AlMV or TEV (Mitra *et al.*, 1996).

As the processing mechanism of all known potyviruses involves the activity of cysteine proteinases, the constitutive expression of rice cysteine proteinase inhibitor could induce resistance against two important potyviruses, tobacco etch virus (TEV) and potato virus Y (PVY), in transgenic tobacco plants. Also there was a direct correlation between the level of oryzacystatin message, inhibition of papain a cysteine proteinase and resistance to TEV and PVY in all lines tested (Gutierrez-Campos *et al.*, 1999).

Salisylic acid plays a crucial role in triggering SAR. When two bacterial genes coding for enzymes that convert Chorismate into salicylic acid was targeted into chloroplasts, the transgenic plants showed 500 to 1000 fold increased accumulation of salisylic acid and salicylic acid glucoside as compared to control plants. This expression also did not affect the plant phenotype (Verberne *et al.*, 2000). Hence this could be another means of engineering resistance in plants at a broard spectrum level. Multiple virus resistance is also possible through the simple strategy of linking viral gene segements to a silencer DNA such as GFP. It has been demonstrated that transgenic plants with GFP gene fused to segments (> 60 nt) of N gene sequence of Tomato spotted wilt virus in transgenic plants shows RNA mediated resistance in tospoviruses (Jan *et al.*, 2000).

7. CONCLUSIONS AND FUTURE PROSPECTS

From the foregoing discussion it is clear that genetically engineered resistance to plant viruses is an important alternative approach in disease management and has potential for wide spread application.

ACKNOWLEDGEMENTS

This review chapter is an updated version of the review article "Mechanism and application of engineered resistance to plant viruses" by C.T. Ranjith-Kumar and H.S. Savithri (1999), *J. Plant Biol.* **26** : 97-100. We thank the department of Biotechnology, New Delhi, India for financial support.

REFERENCES

Adair TL and Kearney CM (2000) Recombination between a 3-kilobase tobacco mosaic virus transgene and a homologous viral construct in the restoration of viral and non viral genes. *Arch. Virol.*, **145** : 9-17.

Ares X, Calamante G, Cabral S, Lodge J, Hemenway P, Beachy RN and Mentaberry A (1998) Transgenic plants expressing potato virus X ORF2 protein (p24) are resistant to tobacco mosaic virus and Ob tobamoviruses *J. Virol.,* **72** : 731-738.

Audy P, Palukaitis P, Slack SA and Zaitlin M (1994) Replicase-mediated resistance to potato virus Y in transgenic plants. *Mol. Plant-Microbe Inter.,* **7** : 15-22.

Baulcombe DC (1996) Mechanism of pathogen derived resistance to viruses in transgenic plants. *Plant Cell,* **8** : 1833-1844.

Beachy RN (1997) Mechanism and applications of pathogen-derived resistance in transgenic plants. *Curr Op Biotechnol* **8** : 215-220.

Beachy RN, Loesch-Fries S and Tumer NE (1990) Coat protein mediated-resistance against virus infection. *Annu. Rev. Phytopathol.,* **28** : 451-474.

Beck DL, Guilford PJ, Voot DM, Andersen MT and Forster RLS (1991) Triple gene block proteins of white clover mosaic potexvirus are required for transport. *Virology,* **183** : 695-702.

Beck DL, Van Dolleweerd CJ, Laugh TJ, Balmori E, Voot DM, Andersen MT O'Brien IEW and Forster RLS (1994) Distruption of virus movement confers broad spectrum resistance against systemic infection by plant viruses with a triple gene block. *Proc. Nat. Acad. Sci. USA* **91** : 10310-10314.

Bendahmane M, Fitchen JN, Zhang G and Beachy RN (1997) Studies of coat protein mediated resistance to tobacco mosaic tobamovirus: Corelation between assembly of mutant coat protein and resistance. *J. Virology,* **71** : 7942-7950.

Brederode FT, Taschner PEM, Posthumus E and Bol JF (1995) Replicase-mediated resistance to alfalfa mosaic virus. *Virology,* **207** : 467-474.

Carr JP and Zaitlin M (1991) Resistance in transgenic tobacco plant expressing a non-structural gene sequence of tobacco mosaic virus is a consequence of markedly reduced virus replication. *Mol. Plant-Microbe Inter.,* **4** : 579-585.

Carrington JC, Kasschau KD, Mahajan SK and Schaad MC (1996) Cell to cell and long distance transport of viruses in plants. *Plant Cell* **8** : 1669-1681.

Chapman S, Hills GJ, Watts J and Baulcombe DC (1992) Mutational analysis of the coat protein gene in potato virus X: Effect of virion morphology and viral pathogenicity. *Virology,* **191** : 223-230.

Chowrira GM, Cavileer TD, Gupta SK, Lurquin PF, Berger PH (1998) Coat protein-mediated resistance to pea enation mosaic virus in transgenic *Pisum sativum* L. *Transgenic Res.,* **7** : 265-71

Citovsky V (1993) How do plant virus nucleic acids move through intercellular connections? *Plant Physiol.,* **102** : 1071-1076.

Citovsky V and Zambryski P (1991) How do plant virus nucleic acids move through intercellular connections? *Bioessays,* **13** : 373-379.

Citovsky V, Wong ML, Shaw AL, Prasad BMV and Zambryski P (1992) Visualization and characterization of tobacco mosaic virus movement protein binding to single stranded nucleic acid. *Plant Cell,* **4** : 397-411.

Clark WG, Fitchen JH and Beachy RN (1995b) Studies of coat protein-mediated resistance to TMV. *Virology,* **208** : 485-491.

Clark WG, Fitchen JH, Nejidat A, Deom CM and Beachy RN (1995a) Studies of coat protein-mediated resistance to tobacco mosaic virus TMV II, challenge by a mutant with altered virion surface does not overcome resistance conferred by TMV coat protein. *J. Gen. Virol.,* **76** : 2613-2617.

Cooper B, Lapidot M, Heick JA, Dodds JA and Beachy RN (1995) A defective movement protein of TMV in transgenic plants confer resistance to multiple viruses whereas the functional analog increases susceptibility. *Virology,* **206** : 307-313.

Culver JN, Dawson WO, Plonk K and Stubbs G (1995) Site directed mutagenesis confirms the involvement of carboxylate groups in the disassembly of tobacco mosaic virus. *Virology,* **206** : 724-730.

De Carvalho Niebel F, Frendo P, Van Montagu M and Cornelissen M (1995) Post transcriptional co-suppression of α-1,3-glucanase gene does not affect accumulation of transgene nuclear mRNA. *Plant Cell,* **7** : 347-358.

Deom CM, Schubert KR, Wolf S, Holt CA, Lucas WJ and Beachy RN (1990) Plant virus movement proteins. *Proc. Natl. Acad. Sci. USA,* **87** : 3284-3288.

Deom CM, Lapido M and Beachy RN (1992) Molecular characterization and biological function of the movement protein of TMV in transgenic plants. *Cell,* **69** : 221-224.

Dolja VV, Haideman R, Robertson NL, Dougherty WG and Carrington JC (1994) Distinct functions of capsid protein in assembly and movement of TEV in plants. *EMBO J.,* **13** : 1482-1491.

English JJ, Mueller E and Baulcombe DC (1996) Suppression of virus accumulation in transgenic plants exhibiting silencing of nuclear genes. *Plant Cell,* **8** : 179-188.

Erickson FL, Holzberg S, Calderon-Urrea A, Handley V, Axtell M, Corr C and Baker B (1999) The helicase domain of the TMV replicase proteins induces the N-mediated defence response in tobacco. *Plant J.,* **18** : 67-75.

Fitchen JH and Beachy RN (1993) Genetically engineered protection against viruses in transgenic plants. *Annu. Rev. Microbial.,* **47** : 739-763.

Fujiwara T, Giesman-Cookmeyer D, Ding B, Lommel SA and Lucas WJ (1993) Cell-to-cell trafficking of macromolecules through plasmodesmata potentiated by the red clover necrotic mosaic virus movement protein. *Plant Cell,* **5** : 1783-1794.

Gerlach WL, Llewellyn D and Haseloff J (1987) Construction of a disease resistance gene using the satellite RNA of tobacco ringspot virus. *Nature,* **328** : 802-806.

Gielen JJL, De Haan P, Kool AJ, Peters D, Van Griensven MQJM and Goldbach RW (1991) Engineered resistance to tomato spotted wilt virus, a negative stranded RNA virus. *Bio/Technol.,* **9** : 363-367.

Golemboski DB, Lomonossoff GP and Zaitlin M (1990) Plant transformed with tobacco mosaic virus non structural gene sequence are resistant to the virus. *Proc. Natl. Acad. Sci. USA,* **87** : 6311-6315.

Goodwin J, Chapman K, Swaney S, Parks TD, Wernsman EA and Dougherty WG (1996) Genetic and biochemical dissection of transgenic RNA-mediated virus resistance. *Plant Cell,* **8** : 95-105

Graham MW, Craig S and Waterhouse PM (1997) Expression patterns of vascular specific promoters RolC and Sh in transgenic potatoes and their use in engineered PLRV resistant plant. *Plant Mol. Biol.,* **33** : 729-735.

Green PG (1993) Control of mRNA stability in higher plants. *Plant Physiol.,* **102** : 1065-1070.

Grdzelishvili VZ, Chapman SN, Dawson WO and Lewandowski DJ (2000) Mapping of the Tobacco mosaic virus movement protein and coat protein subgenomic RNA promoters *in vivo. Virology,* **275** : 177-192.

Grumet R (1994) Development of virus-resistant plants via genetic engineering; *Plant Breed Rev.,* **12** : 47-79.

Gutierrez-Campos R, Torres-Acosta JA, Saucedo-Arias LJ and Gomez-Lim MA (1999) The use of cysteine proteinase inhibitors to engineer resistance against potyviruses in transgenic tobacco plants. *Nature Biotech.,* **17** : 1223-1226.

Hackland AF, Rybicki EP and Thomson JA (1994) Coat protein mediated resistance in transgenic plants. *Arch. Virol.,* **139** : 1-22.

Hamilton RI (1980) Defense triggered by previous invaders. In : *Viruses in Plant Disease: An Advanced Treatise.* (Eds, JG Horsefall and EB Cowling) Academic Press, New York. **5** : 279-303.

Hamilton RI (1985) Using plant viruses for disease control. *Hort. Sciecne,* **22** : 848-852.

Han SJ, Cho HS, You JS, Nam YW, Park EK, Shin JS, Park YI, Park WM, Paek KH (1999) Gene silencing-mediated resistance in transgenic tobacco plants carrying potato virus Y coat protein gene. *Mol. Cells,* **9** : 376-83.

Harrison BD, Mayo MA and Baulcombe DC (1987) Virus resistance in transgenic plants that express cucumber mosaic virus satellite RNA. *Nature,* **328** : 799-802.

Hemenway CL, Fang RX, Kaniewski WK, Chua NH and Tumer NE (1988) Analysis of the mechanism of protection in transgenic plants expressing the potato virus X coat protein or its antisense RNA. *EMBO J.,* **7** : 1273-1280.

Holt CA and Beachy RN (1991) *In vivo* complementation of infectious transcripts from mutant tobacco mosaic cDNAs in transgenic plants. *Virology,* **181** : 109-117.

Hong Y, Saunders K, Hartley MR and Stanley J (1996) Resistance to geminivirus infection by virus-induced expression of dianthin in transgenic plants. *Virology,* **220** : 119-127.

Ingelbrecht I, Van Houdt H, Van Montagu M and Depicker A (1994) Post-transcriptional silencing of reporter transgenes in tobacco correlates with DNA methylation. *Proc. Natl. Acad. Sci. USA,* **91** : 10502-10506.

Jacquemond M, Anselem J and Tepfer M (1988) Gene coding for a monomeric form of cucumber mosaic virus satellite RNA confer tolerance to CMV. *Mol. Plant-Microbe Inter.,* **1** : 311-316.

Jan FJ, Fagoaga C, Pang SZ and Gonsalves D (2000) A minimum length of N gene sequence in transgenic plants is required for RNA-mediated tospovirus resistance. *J. Gen. Virol.,* **81** : 235-42.

Jones AL, Johansen IE, Bean SJ, Bach I and Maule AJ (1998) Specificity of resistance to Pea seed borne mosaic potyvirus in transgenic peas expressing the viral replicase (NIb) gene. *J. Gen. Virol.,* **79** : 3129-3137

Kasteel DT, Wellink J, Goldbach RW and Van Lent JW (1997) Isolation and characterization of tubular structures of cowpea mosaic virus. *J. Gen. Virol.,* **78** : 3167-3170

Kim SJ, Paek KH and Kim BD (1995) Delay in diseases development in transgenic petunia plants expressing CMVt-17N Satellite RNA. *J. Am. Soc. Hortic. Sci.,* **120** : 353-359.

Lapidot M, Gafny R, Ding B, Wolf S, Lucas WJ and Beachy RN (1993) A dysfunctional movement protein of tobacco mosaic virus that partially modifies the plasmodesmata and limits virus spread in transgenic plants. *Plant J.,* **4** : 959-970.

Lawson C, Kaniewski W, Haley L, Rozman R, Newell C, Sanders P and Tumer NE (1990) Engineering resistance to mixed virus infection in a commercial potato cultivar: Resistance to potato virus X and potato virus Y in transgenic Russet Burbank. *Bio/Technology,* **8** : 127-134.

Lindbo JA and Dougherty WG (1992) Pathogen-derived resistance to potyvirus: Immune and resistant phenotypes in transgenic tobacco expressing altered forms of a potyvirus coat protein nucleotide sequence. *Mol. Plant-Microbe Inter.,* **5** : 144-153.

Lindbo JA, Silva-Rosales L, Proebsting WM and Dougherty WG (1993) Induction of a highly specific antiviral state in transgenic plants : implication for regulation of gene and viral resistance. *Plant Cell,* **5** : 1749-1759.

Lodge JK, Kaniewski WM and Tumer NE (1993) Broad spectrum virus resistnce in transgenic plants expression poke weed antiviral protein. *Proc. Natl. Acad. Sci. USA,* **90** : 7089-7093.

Lomonossoff GP (1995) Pathogen-derived resistance to plant viruses. *Annu. Rev. Phytopathol.,* **33** : 323-343.

Maiti IB, Murphy JF, Shaw JG and Hunt AG (1993) Plants that express a potyvirus proteinase gene are resistant to viral infection. *Proc. Natl. Acad. Sci. USA,* **90** : 6110-6114.

Malyshenko SI, Kondakova OA, Nazarova JV, Kaplan IB, Taliansky ME and Atabekov JG (1993) Reduction of tobacco mosaic virus accumulation in transgenic plants producing non-functional viral transport protein. *J. Gen. Virol.,* **74** : 1149-1156.

Masuta C, Tanaka H, Uchara K, Kuwata S, Koiwai A and Noma M (1995) Broad resistance to plant viruses in transgenic plants conferred by antisense inhibition of a host gene essential in S-adenosyl methionine dependent transmethylation reactions. *Proc. Natl. Acad. Sci. USA,* **92** : 6117-6121.

Mitra A, Higgins DW, Langenberg WG, Nie H, Sengupta DN and Silverman RH, (1996) A mammalian 2-5A system functions as an antiviral pathway in transgenic plants. *Proc. Natl. Acad. Sci. USA,* **93** : 6780-6785.

Mueller E, Gilbert J, Davenport G, Brigneti G and Baulcombe DC (1995) Homology-dependent resistance: transgenic virus resistance in plants related to homology-dependent gene silencing. *Plant J.,* **7** : 1001-1013.

Nicholson AW (1996) Structure, reactivity and biology of double-stranded RNA. *Prog. Nucleic Acid Res. Mol. Biol.,* **52** : 1-65.

Noris E, Accotto GP, Tavazza R, Brunetti A, Crespi S and Tavazza M (1996) Resistance to tomato yellow leaf curl geminivirus in *N. benthamania* plant transformed with a truncated viral *Cl* gene. *Virology,* **224** : 130-138

Osbourn JK, Sarkar S and Wilson TMA (1990) Complementation of coat protein-defective TMV mutants in transgenic tobacco plants expressing TMV coat protein. *Virology,* **179** : 921-925.

Osbourn JK, Watts JW, Beachy RN and Wilson TMA (1989) Evidence that nucleocapsid disassembly and a later step in viral replication are inhibited in transgenic tobacco protoplast expressing TMV coat protein. *Virology,* **172** : 370-373.

Palukaitis P and Roossinck MJ (1996) Spontaneous change of a benine satellite RNA cucumber mosaic virus to a pathogenic variant. *Nature Biotechnol.,* **14** : 1264-1268.

Petty ITD and Jackson AO (1990) Mutational analysis of barley stripe mosaic virus RNA B. *Virology,* **179** : 712-718.

Powell-Abel PA, Nelson RS, De B, Hoffmann N, Rogers SG, Fraley RT and Beachy RN (1986) Delay of disease development in transgenic plants that express the tobacco mosaic virus coat protein gene. *Science,* **232** : 738-743.

Powell-Abel PA, Sanders PR, Tumer N, Fraley RT and Beachy RN (1990) Protection against tobacco mosaic virus in transgenic plants that express tobacco mosaic virus antisense RNA. *Virology,* **175** : 370-373.

Ranjith Kumar CT, Manoharan M, Krishna Prasad S, Shobha Cherian, Umashankar M, Lakshmi Sita G and Savithri HS (1999) Engineering resistance against Physalis mottle tymovirus by expression of the coat protein and 3' non coding region. *Curr. Sci.,* **77** : 1542-1548.

Ratcliff F, Harrison BD and Baulcombe DC (1997) A similarity between viral defence and gene silencing in plants. *Science,* **276** : 1558-1560.

Register III JC and Beachy RN (1988) Resistance to TMV in transgenic plants results from interference with an early event in infection. *Virology,* **166** : 524-532.

Register III JC and Nelson RS (1992) Early events in the plant virus interaction relationship with genetically engineered protection and gene resistance. *Sem. Virol.,* **3** : 441-451.

Schiebel W, Haas B, Marinkovic S, Klanner A and Sanger HL (1993a) RNA-directed RNA polymerase from tomato leaves. I. Purification and physical properties. *J. Biol. Chem.,* **268** : 11851-11857.

Schiebel W, Haas B, Marinkovic S, Klanner A and Sanger HL (1993b) RNA-directed RNA polymerase from tomato leaves. II. Catalytic *in vitro* properties. *J. Biol. Chem.,* **268** : 11858-11867.

Scholthof KBG, Scholthof HB and Jackson AO (1993) Control of plant virus disease by pathogen –derived resistance in transgenic plants. *Plant Physiol.,* **102** : 7-12.

Sit TL, Leclerc D and Abouhaidar MG (1994) The minimal 5' sequence for *in vitro* initiation of papaya mosaic potexvirus assembly. *Virology,* **199** : 238-242.

Song EK, Koh HK, Kim JK and Lee SY (1999) Genetically engineered transgenic plants with the domain 1 sequence of tobacco mosaic virus 126 kD protein gene are completely resistant to viral infection. *Mol. cells,* **9** : 569-575

Sree Vidya CS, Manoharan M, Ranjith Kumar CT, Savithri HS and Lakshmi Sita G (2000) *Agrobacterium*-mediated transformation of tomato (*Lycopersicon esculentum var* Pusa Ruby) with coat protein gene of *Physalis* mottle tymovirus. *J. Plant. Physiol.,* **156** : 106-110.

Stark DM and Beachy RN (1989) Protection against potyvirus in transgenic plants: evidence for broad spectrum resistance. *Bio/Technology,* **7** : 1257-1262.

Szittya G and Burgyan J (2001) Cymbidium Ringspot Tombusvirus Coat Protein Coding Sequence Acts as an Avirulent RNA. *J. Virol.,* **75** : 2411-2420

Tacke E, Salamini F and Rohde W (1996) Genetic engineering of potato for broad-spectrum protection against virus infection. *Nature Biotechnol.,* **14** : 1597-601.

Taschner PEM, Van Der Kuyl AC, Neelman L and Bol J (1991) Replication of an incomplete alfalfa mosaic virus genome in plants transformed with viral replicase gene. *Virology,* **181** : 445-450

Taschner PEM, Van Marie G, Brederode FT, Tumer NE and Bol JF (1994) Plants transformed with a mutant alfalfa mosaic virus coat protein resistance to mutant but not to wild type virus. *Virology,* **203** : 269-276.

Vacquero C, Turner AP, Demangeat G, Sanz A, Serra MT, Roberts K and Garcia-Luque I (1994) The 3a protein from CMV increases the gating capacity of plasmodesmata in transgenic tobacco plants. *J. Gen. Virol.,* **75** : 3193-3197.

Van der Vassan EAG, Neeleman L and Bol JF (1996) The 5' terminal sequence of alfalfa mosaic virus RNA 3 is dispensable for replication and contains a determinant of symptom formation. *Virology,* **221** : 271-280.

Van der Vlugt RAA, Ruiter RK and Goldbach R (1992) Evidence for sense RNA mediated resistance to PVY^{N} in tobacco plants transformed with the viral coat protein cistron. *Plant Mol. Biol.,* **20** : 631-639.

Van der Wilk F, Willink DPL, Huisman MJ, Huttinga H and Goldbach R (1991) Expression of the potato leaf roll luteovirus coat protein gene in transgenic potato plants inhibits viral infection. *Plant Mol. Biol.,* **17** : 431-439.

Van Dun CMP, Van Vlotel-Doting L and Bol J (1988) Expression of alfalfa virus cDNA 1 and 2 in transgenic tobacco plants. *Virology,* **163** : 572-578.

Vardi E, Sela I, Edelbaum O, Livnch O, Kuznetsova L and Stram Y (1993) Plants transformed with a cistron of a potato virus Y protease (NIa) are resistant to virus infection. *Proc. Natl. Acad. Sci. USA,* **90** : 7513-7517.

Varrelmann M and Maiss E (2000) Mutations in the coat protein gene of plum pox virus suppress particle assembly, heterologous encapsidation and complementation in transgenic plants of *Nicotiana benthamiana. J. Gen. Virol.,* **81** : 567-76.

Verberne MC, Verpoorte R, Bol JF, Mercado-Blanco J and Linthorst HJ (2000) Over production of salicylic acid in plants by bacterial transgenes enhances pathogen resistance. *Nature Biotechnol.,* **18** : 779-783

Verduin BJM (1992) Early interactions between virus and plants. *Sem. Virol.,* **3** : 423-431.

Wassengger M, Heimes S, Riedel L and Sanger H (1994) RNA directed *de novo* methylation of genomic sequence in plants. *Cell,* **76** : 567-576.

Wolf S, Deom CM, Beachy RN and Lucas WJ (1989) Movement protein of TMV modified plasmodesmata size exclusion limit. *Science,* **246** : 377-379.

Wolf S, Deom CM, Beachy RN and Lucas WJ (1991) Plasmodesmata function is probed using transgenic tobacco that express viral movement protein. *Plant Cell,* **3** : 593-604.

Wu X, Beachy RN, Wilson TMA and Shaw JG (1990) Inhibition of uncoating of TMV particles in protoplast from transgenic plant tobacco plants that express the viral coat protein gene. *Virology,* **179** : 893-895.

Yusobov Y and Loesch-Fries LS (1995) High affinity RNA binding domains of alfalfa mosaic virus coat protein are not required for coat protein mediated resistance. *Proc. Natl. Acad. Sci. USA,* **92** : 8980-8984.

Zaccomer B, Cellier F, Boyer JC, Haenni AL and Tepfer M (1993) Transgenic plant that express genes including the 3' untranslated region of to turnip yellow mosaic virus (TYMV) genome are partially protected against TYMV infection. *Gene,* **136** : 87-94.

Zaitlin M, Anderson JM, Perry KL, Zhang L and Palukaitis P (1994) Specificity of replicase –mediated resistance to cucumber mosaic virus. *Virology,* **201** : 200-205.

Zhou X, Liu Y, Calvert L, Munoz C, Otim-Nape W, Robinson DJ and Harrison BD (1997) Evidence that DNA-A of a geminivirus associated with severe cassava mosaic disease in Uganda has arisen by interspecific recombination. *J. Gen. Virol.,* **78** : 2101-2111.

Chapter 4

GENETIC ENGINEERING FOR BACTERIAL RESISTANCE

Ram C Yadav[1★], Neelam R Yadav[1] and Rajiv Mishra[2]

[1]*Department of Biotechnology and Molecular Biology, CCS Haryana Agricultural University, Hisar - 125 004 India*
[2]*Department of Agronomy and Range Sciences, University of California, Davis CA 95616 USA*

Summary

Various molecular approaches are being used for enhancing resistance to bacterial diseases viz., inhibiting bacterial pathogenecity, production of antibacterial proteins of non plant origin, inducing artificially programmed cell death and enhancing natural plant defenses. The most important class of genes that has been used by breeders for disease control are the plant resistance (R) genes. Many R genes have been cloned from different plant species which confer resistance to many types of pathogens. It is a major breakthrough in the elucidation of the molecular basis of disease resistance to a wide range of pathogens. It is remarkable that despite their specificity, all the R genes identified to date encode polypeptides that share similar structural motifs, allowing them to be classified into five broad categories. Among these, three classes are more prevalent which consist of nucleotide binding domain (NBD), Leucine rich repeats (LRR) and a serine/threonine protein kinase (PK) which are considered components of a signal transduction pathway. Engineering R genes into desirable crop species will result in the first launches of bacterial resistant crops within 4-5 years.

Keywords : Avirulence, resistance, hypersensitive response,

★Corresponding author : E-mail : rcyadav@hau.nic.in

systemic acquired resistance, transgenic, genetically engineered plant

Abbreviations : Avr : Avirulence, R : Resistance, HR : Hypersensitive response, LRR : Leucine rich repeat, NBS : Nucleotide binding site, SAR : Systemic acquired resistance, PK : Protein Kinase, LZ : Leucine Zipper, TM : Transmembrane, TIR : Toll/interleukin-1-receptor

1. INTRODUCTION

Plant breeders have made significant contributions in developing high yielding varieties in the last 50 years which led to the green revolution. But field crops are subjected to biotic and abiotic stresses that limit their productivity. These adverse conditions impair normal development of the crop and result in lower yields and poor food quality. The burgeoning population will put considerable pressure on agriculture since the gap between demand and supply of food commodities is continuously widening. To bridge the gap, efforts are underway to increase the yields by developing high yielding varieties and hybrids resistant to insect/ pests and diseases. One of the alternatives to protect the crop from yield losses due to biotic and abiotic stresses can be through the introduction and development of transgenic plants using resistance (R) genes from varied sources. Bacterial diseases are one of the main biotic stresses which result in heavy yield losses globally.

Application of protective agrochemicals is not enough to control bacterial diseases. Secondly, application of chemicals is subject to increasing restrictions because of their harmful impact on the environment and human health. Resistant varieties have been developed through conventional breeding using the natural sources of resistance against bacterial diseases. Sometimes it is difficult and time consuming to transfer gene of interest from wild species into cultivated cultivars. With the advent of recombinant DNA technology and advances in genetic transformation techniques has enabled to develop transgenic plants resistant to diseases (Fig. 1.) Düring (1996) and Morgues *et al.* (1998) have reviewed the strategies to improve plant resistance for bacterial diseases through genetic engineering in plants. We present here the various approaches for engineering bacterial blight resistance in plants.

2. EXPRESSION OF ANTIBACTERIAL PROTEINS

Antibacterial proteins are one of the important components of many groups of animals *viz.*, amphibians, mammals and particularly arthropods. These proteins probably work in synergy and have a bactericidal action on a large range of bacteria. The genes encoding antibacterial proteins have been cloned and expressed in plants to confer resistance to bacterial diseases.

2.1. Lysozymes

Lysozymes are defined as enzymes with a specific hydrolytic activity against the bacterial cell wall peptidoglycan, commonly known as murein. Lysozyme activity has been reported in a variety of organisms *viz.*, bacteriophages, mammals and recently in plants also. Most of the known plant lysozymes are basic proteins and display chitinase activity. Expression of heterologous lysozymes in transgenic plants has been reported as a mechanism to enhance resistance against pathogenic bacteria. Different lysozyme genes *viz.*, egg white, T4 Bacteriophage and human lysozyme have been expressed in plants (Nakajima *et al.*, 1997).

2.2. Expression of other antibacterial peptides

The expression of a lactoferrin gene from human in tobacco delayed the symptoms by 5 to 25 days caused by *R. solanacearum* (Mourgues *et al.*, 1998). The resistance has been due to formation of smaller peptide in truncation of loctoferrin. Another gene tachyplesin isolated from horse shoe crab has also been expressed in potatoes, resulting in a reduced amount of tuber rot caused by *E. carotovora*. Attacin, a 20 kD protein and Magainin from frog skin has been tested for enhancing resistance in plants. Genes encoding anti-microbial cysteine rich thionin have been transferred in toabcco. The high level expression of α-thionin driven by CaMV 35S promoter reduced the disease symptoms by *Pseudomonas syringae* (Carmona *et al.*, 1993).

The efficiency of different strategies may be improved by the expression of many lytic peptide genes having synergistic effects. It is a very hot area for the investigation of antibacterial proteins in a variety of organisms and it will lead to the characterization of many molecules.

2.3. Expression of thionins isolated from wheat and barley in transgenic tomato, potato and tobacco

Carmona *et al.* (1993) reported the effective use of cereal thionins in transgenic plants of tobacco using the *Pseudomonas syringae* model system for enhanced resistance. Wheat and barley thionins possess an antibacterial and antifungal activity (Carmona *et al.*, 1993). These are synthesized from larger precursors protein. Wheat and barley α-thionins genes containing their native signal peptide coding sequences have been introduced into the transgenic tobacco. Constitutive expression of the barley α-thionin led to a significant decrease in the growth of *Pseudomonas syringae* pv. syringae. However, the wheat α-thionin gene introduced into tobacco did not lead to significant effects.

2.4. Lytic peptides

Lytic peptides which are small proteins interact with and which form pores in the bacterial membranes. Cecropins insert into the outer membrane of bacterial cells as well as in eukaryotic cell membranes thereby disrupting their membrane structure and killing the protoplasts. Nordeen *et al.* (1992) has provided information on the required levels of cecropin B to kill many species of phytopathogenic bacteria. Transgenic potato and tobacco plants have been produced with cecropins (isolated from giant silk moth) and their synthetic analogs Shiva-1 and SB-37 (Dastefano-Beltran *et al.*, 1993). Progenies of tobacco plants expressing the Shiva-1 gene when inoculated with *Ralstonia solanacearum* (causal agent of bacterial wilt) showed delayed symptoms and reduced mortality. Another analog (MB 39) was introduced and expressed in tobacco and progenies of two plants showed no necrosis when infiltrated with *Pseudomonas syringae* pv. *tabaci* (Huang *et al.*, 1997). Recently, Osusky *et al.* (2000) reported a strategy for engineering transgenic plants with broad-spectrum resistance to bacterial as well as fungal phytopathogens. They expressed a synthetic gene in two potato cultivars encoding a N-terminus-modified, cecropin-melittin cationic peptide chimera (Msr A1), with broad spectrum antimicrobial activity. Tubers retained their resistance to infections challenge for more than a year and did not appear to be harmful when fed to mice. This approach of engineering cationic peptides into plants will help in developing disease resistant varieties in different crop species.

2.5. Resistance by toxin inactivation

The bacterial resistance developed through genetic engineering was reported by Anzai *et al.* (1989) for the fist time in tobacco. Wildfire disease of tobacco is caused by *Pseudomonas syringae* pv. *tabaci* which produces tabtoxin, a dipeptide toxin containing an uncommon βlactum amino acid (and either serine or threonine) which causes the chlorotic symptoms. The tabotoxin resistance gene ttr encoding an inactivating acetylating enzyme from the same bacterium was exposed at high levels in transgenic tobacco and successfully enhanced resistance to this bacterium.

3. TOOLS FOR ENGINEERING BACTERIAL RESISTANCE IN PLANTS

The basic tools and procedures for recombinant DNA research are almost the same in all plants which include DNA isolation protocols, enzymes for DNA synthesis, restriction, insertion and ligation. Vectors or plasmids for cloning and propagating in *E. coli* and different analytical tools *viz.*, gel electrophoresis, blotting and hybridization, PCR and DNA sequencing (Ausubel *et al.*, 1997; Gelvin and Schilperoort, 1998). One of the key steps for genetic engineering is genetic transformation of plants where the exogenous DNA is introduced and expressed in the cells and developed into transgenic plants. Currently, there are many techniques which can be used but mainly there are two methods (1) *Agrobacterium tumefaciens* which introduces specific DNA segments into plant genome (Fig. 1) and (2) by biolistic or micro-projectile bombardment of plant cells and tissues with DNA coated tungsten or gold particles (Fig. 1). These two methods have been used extensively along with plant tissue culture methods for selection of antibiotic resistance and regeneration of intact plants from the transformed cells or tissues.

The introduced DNA sequences, often referred as the transgene, integrates randomly in plant genome and its expression depends on the site of insertion. It is always desirable to generate and screen a few dozens of independent transformants to identify few lines which display the desired sequences and its expression.

Genetic transformation can be exploited for crop improvement programs. One or two genes can be introduced directly without backcrossing and introgression. Genes from any source (plant or animal), closely or distantly related organisms can be utilized. A few resistance genes have been cloned from wild relatives and have been used for engineering bacterial blight resistance in different plant species (Table 1).

Usually, the genes are put under the control of a constitutive promoter like CaMV 35S. The plant growth and development are occasionally affected. In such cases, the use of chemically inducible promoters might

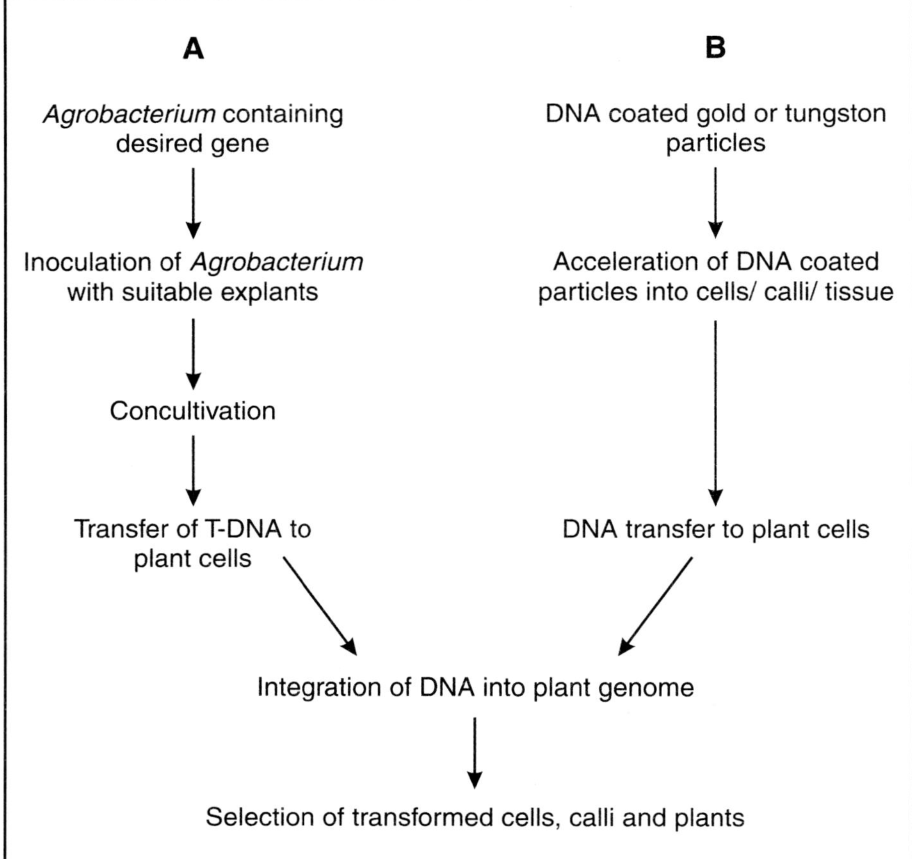

Figure 1 : Gene trasnfer methods in plants cells. (A) Indirect gene transfer mediated by *Agrobacterium*. It is a natural genetic engineer which transfers T-DNA from the Ti plasmid into plant cells. The gene of interest (*e.g.* For Bacterial blight resistance) is cloned in a Ti plasmid or a binary vector and used for transformation. (B) Direct gene transfer methods where DNA is delivered directly to the nucleus of the plant cells through different techniques such as particle bombardment, electroporation and PEG-mediated.

prove beneficial. Farmers could spray the fields with an inducer that would activate synthesis of the products when pathogen problems are detected. Recently, a promising inducible system based on glucocorticoid –mediated gene activation has been described. A chimeric transcription factor, designated GVG consisting of the DNA binding domain of the yeast transcription factor GAL4, the transactivating domain of the herpes viral protein VP16 and the receptor domain of the rat glucocorticoid receptor (GR) activates the gene expression in transgenic plants after treatment with glucocorticoids (Aoyama and Chua, 1997).

4. EXPLOITING PLANT DISEASE RESISTANCE GENES (R GENES)

Disease resistance genes enable the plant to resist attack from specific races of viruses, bacteria, fungi and nematodes. Plants contain a number of defense arsenals which get activated after microbial infection are governed by directly or indirectly by a gene for gene interaction between the products of a plant resistance (R) gene and a pathogen avirulence (*avr*) gene (Baker *et al.*, 1997). When the plant R gene or the pathogen (*avr*) gene is lacking these defenses are not induced and the pathogen multiplies rapidly and colonizes the plant. The gene-for-gene specificity has been demonstrated by the tomato-*Pseudomonas syringae* pv. *tomato* system by a direct interaction between the products of the R gene Pto and the corresponding *avr* gene. The majority of (R) genes cloned so far show conserved sequences and amino acid domains irrespective of whether they confer resistance to bacterial, fungal, viral or nematode pathogens. R genes could be grouped into five major classes based on their structural features (Baker *et al.*, 1997). These can be brought under three groups containing three structural domains viz., nucleotide binding site or domain (NBS/NBD), leucine rich repeat (LRR) and a protein kinase (PK) in various combinations that form the components of a signal transduction pathway. The various structural features are schematically represented in Figure 2. The susceptible varieties can be transformed with the cloned disease resistance genes and noval resistant genotypes can be generated rapidly as compared to conventional breeding. Further, this process eliminates the unwanted genes or genetic 'drag' of negative characters. This can be defined as 'precision breeding' where only desired genes can be introduced precisely in a cultivar deficient in those characters through genetic engineering. Disease resistant varieties developed by the plant breeders in few crops and plant species with high degree of disease resistance serve as donors for the resistance genes. Experiments are underway to introduce the cloned resistance genes for bacterial blight resistance in crops. In fact, the R genes have already been introduced in different crops like *Xa21* gene in rice (Zhang *et al.*, 1998; Tu *et al.*, 2000) and disease resistant homologues (I1ag *et al.*, 2000). The genes cloned and sequenced so far for bacterial resistance are given in the Table 1.

Table 1. Cloned plant resistance genes to bacterial pathogens

Resistance gene	Plant	Pathogen	Structure	References
Pto	Tomato	*Pseudomonas syringae* pv. tomato	PK	Martin et al., 1993
Xa1	Rice	*Xanthomonas oryzae* pv. *oryzae*	NBS-LRR	Yoshimura et al., 1998
Xa21	Rice	*Xanthomonas oryzae* pv. *oryzae*	LRR-TM-PK	Song et al., 1995
RPM1	*Arabidopsis*	*P. syringae* pv. *meculicola*	LZ-NBS-LRR	Grant et al., 1995
RPS2	*Arabidopsis*	*P. syringae* pv. *tomato*	LZ-NBS-LRR	Bent et al., 1994
RPS4	*Arabidopsis*	*P. syringae* pv. *pisi*	TIR-NBS-LRR	Arts et al., 1998
RPS5	*Arabidopsis*	*P. syringae* pv. *phaseolicola*	LZ-NBS-LRR	Warren et al., 1998

Resistance gene encoded proteins: LRR-Leucine rich repeats, LZ-Leucine zipper, NBS-Nucleotide binding site, PK-Protein Kinase, TIR-Toll/interleukin-1 receptor, TM-Transmembrane

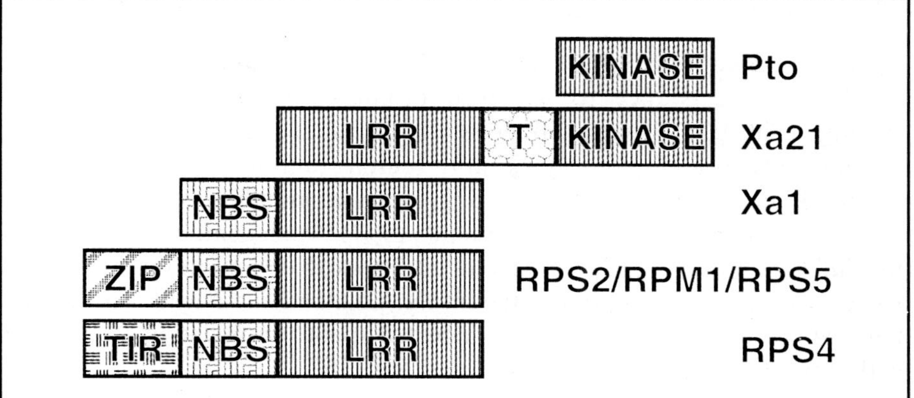

Figure 2 : Structural featurs of R genes that give resistance to bacterial pathogens. TIR-Toll/Interleukin-Receptor, NBS-Nucleotide binding domain or site, LRR-Leucine rich repeats, ZIP-Leucine Zipper, T-Transmembrane

4.1. *Pto* Gene

The *Pto* gene for bacterial speck resistance in tomato encodes serine/ threonine protein kinase (PK) with no obvious receptor-like LRR domain (Martin *et al.*, 1993). It has been reported recently that *Pto* is the receptor for the corresponding ligand encoded by the bacterial *avr* gene *Avr-Pto*. It has been reported that *Pto* interacts specifically with *AvrPto* protein in the yeast two-hybrid system and when disrupted by mutations lead to loss of recognition in the plant cell (Scofield *et al.*, 1996; Tang *et al.*, 1996). Recent results indicate that a threonine at position 204 plays a key role in the *AvrPto* interaction (Frederick *et al.*, 1998). This is interesting to note that threonine at this position is conserved in a large number of protein kinases and phosphorylation of this residue might lead to a conformational change of the kinase which allows *Pto-AvrPto* binding (Martin, 1999). How the *Pto-AvrPto* interaction stimulates disease resistance came from the recent report that over-expression of *Pto* activates a variety of defense responses against the pathogens (Tang *et al.*, 1999). Using the yeast two-hybrid system, Zhou *et al.* (1995) identified a second serine/threonine kinase, *Pto*-Interacting 1(*Pti1*) that physically interacts with *Pto.* Cross phosphorylation assays reveled that Pto phosphorylates *Pti1* and *Pti1* does not phophorylate *Pto. Pti1* transgene expression in tobacco plant enhanced the hypresensitive response to *P. syrinage* pv. *tabaci* carrying the *avrPto* gene.

4.2. The Extra-cellular LRR class

The extra-cellular LRR class of R genes includes the *Xa21* gene which gives resistance to bacterial blight (*Xanthomonas oryzae* pv. *oryzae*). The *Xa21* gene encodes an extracellular LRR domain, a transmembrane (a membrane spanning region) region and an intracellular protein kinase domain. The *Xa21* gene is a member of a multigene family clustered at the 21 locus (Ronald, 1997). Transgenic plants carrying the *Xa21* gene were highly resistant to 29 isolates and susceptible to 3 isolates from India, Philippines, Indonesia, Thailand and Nepal. This shows that a single gene is responsible for broad spectrum resistance observed in the donor line IRBB21 (Wang *et al.*, 1996).

Receptor kinase like proteins are involved in cellular signaling and peptide binding to the extracellular domain of Receptor tyrosine kinase (RTK) induces receptor homo-dimerization and stimulates auto-phosphorylation in all biological systems. *Xa21* LRR domain is extra-cellular and a polypeptide produced by the pathogen binds to it. This specific binding may lead to activation of the *Xa21* kinase with subsequent phosphorylation on specific serine threonine residues. These phosphorylated residues may serve as binding sites for proteins which can stimulate downstream responses. This process may lead to phosphorylation of transcription factor and these factors after phosphorylation can move into the nucleus (Ronald, 1997) from the cytoplasm and experiments are underway to know the link between the phosphorylated *Xa21* kinase and interacting proteins and *Xa21* mediated downstream plant defense responses.

Xa1, a second resistance gene from rice which gives resistance to *Xanthomonas oryzae* pv. *oryzae* (*Xoo*) strains carrying the *avrXa1* has been cloned (Yoshimura *et al.*, 1996) and found that it has NBS and LRR domains (different from *Xa21*). This shows that two unrelated classes of proteins are used in rice to perceive the same pathogen species (but different *avr* genes). The NBS-LRR proteins are predicted to be intracellular whereas LRR domain is predicted to be extra-cellular. The LRR domains of *Xa1* and *Xa21* are therefore, in different sub-cellular locations. Experiments are underway to clone *avrXa1* and *avrXa21* from *Xoo* which will help in deciphering the R-*avr* ligand relationships.

4.3. NBS-LRR Genes

The majority of the cloned R genes encode proteins with a nucleotide binding site (NBS) that is probably involved in signal transduction and at carboxy terminal a leucine rich repeat (LRR) domain (Gassmann *et al.*, 1999; Tai *et al.*, 1999) that is thought to be involved in ligand binding and pathogen recognition. The NBS-LRR class of genes is abundant in plant species. There are two major sub-classes, one with an amino-terminal Toll interleukin-1-receptor homology region (TIR) and thought to be involved in signal transduction (Ellis and Jones, 1998). RPS4 gene from *Arabidopsis thaliana* which gives resistance to *Pseudomonas syringae* pv. *pisi* falls in this category. The second subclass is without the TIR region and RPM1, RPS2 and RPS5 genes which give resistance to *P. syringae* pv *meculicola*, *P. syringae* pv. *tomato* and *P. syringae* pv. *phaseolicola* have leucine zipper (LZ) in addition to NBS-LRR domains. Leucine zippers are the members of this broader class of genes.

5. GENE FOR GENE INTERACTIONS

A gene for gene interactions as proposed by Flor (1971) states that dominant resistant genes interact with pathogen avirulance genes resulting in a hypersensititve response (HR). Gene for gene interactions have been interpreted in terms of receptor ligand model in which the products of resistance genes are receptors which detect the pathogen avirulence genes either directly (protein product) or indirectly (an enzyme product). The resistance receptor is envisaged to have two basic properties namely pathogen recognition and the ability to signal to downstream genes. It has been speculated that every resistance gene in the host could code for a receptor molecule that would directly or indirectly trigger the HR after interaction with corresponding avirulence gene product. HR is one of the most frequently occurring defense responses in plants against all pathogens *viz.* viruses, bacteria, fungi and nematodes. HR involves rapid death of a few cells around the pathogen penetration and observed as a necrotic lesion. The dead cells play an important role in resistance by restricting the growth of the bacteria or any pathogen (Fig. 3).

6. HOW DO *AVR* GENES FUNCTION

Avr genes determine the inability of a given bacterial strain to infect a plant carrying the corresponding R gene. More than thirty bacterial avr genes in the pathovars of *Pseudomonas syringae* and *Xanthomonas oryzae* have been cloned (Leach and White, 1996). Avr function requires bacterial hrp genes which are localized in 20-25 kilobases gene clusters and encodes a type III protein secretion system that appears to be capable of delivering Avr proteins across the walls and plasma

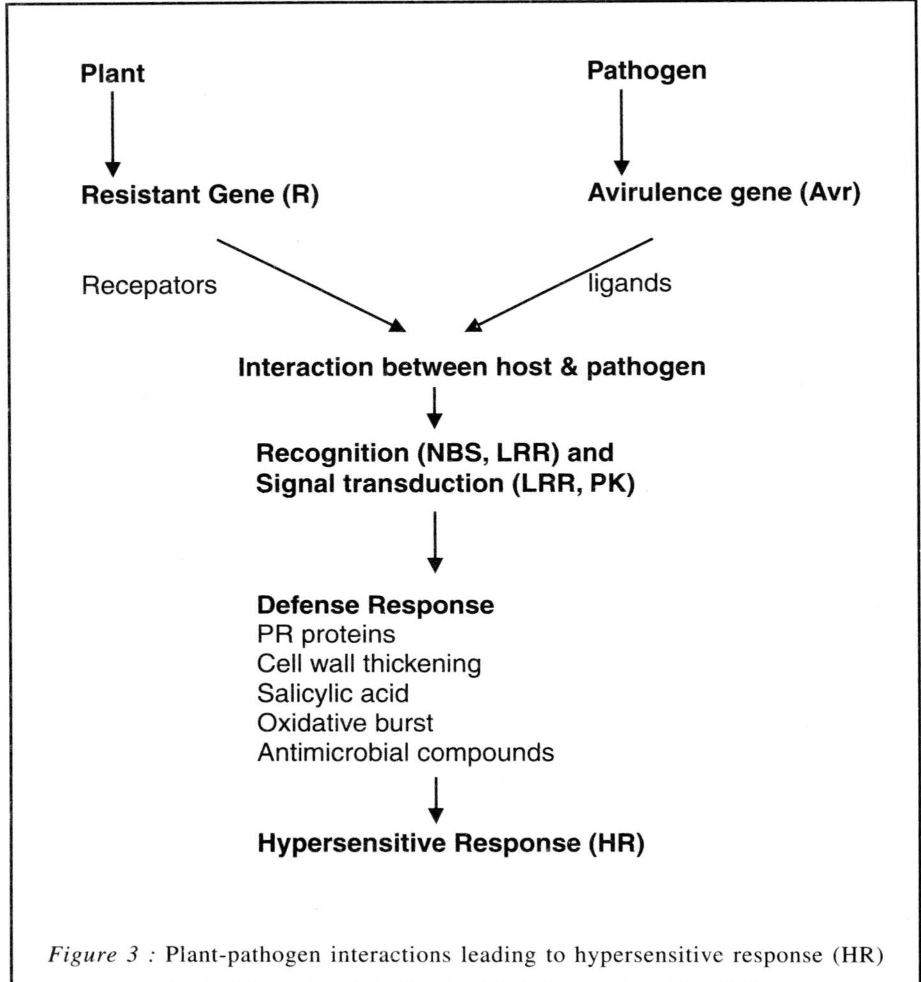

Figure 3 : Plant-pathogen interactions leading to hypersensitive response (HR)

membranes of living plant cells (Alfano and Collmer, 1997; Bonas and Ackerveken, 1999). Bacterial type III protein secretion systems are characterized by an ability to inject effector proteins into host cells. Nine of the hrp genes are universal components of type III secretion system and these have been renamed as hrc (HR and conserved). The hrc proteins enable protein movement across the bacterial outer and inner membranes independently of the general protein export pathway (Collmer, 1998). The genes encoding type III secretion are usually clustered. The hrp genes are expresssed when bacteria are inoculated onto plants. The hrp regulatory systems in plant pathogenic bacteria can be divided into two groups. In the Ist group of *Erwinia* and

Pseudomonas, hrp operons are activated by HrpL, a sigma factor and in second group of *Xanthomonas* (HrpX) and *Ralstonia* (HrpB) is activated by a homolog of AraC (Lindgren, 1997).

7. SYSTEMIC ACQUIRED RESISTANCE

Plants respond in a variety of ways to pathogenic microorganisms (Lamb, 1994). When the pathogen carries a specific avirulence (avr) gene and the plant host contains a corresponding resistance (R) gene, a defense response occurs at the site of infection that results in inhibition of pathogen growth. In addition to the defense response, a secondary defense response can be triggered that renders uninfected parts of the plant resistant to a variety of normally virulent pathogens (Ryals *et al.*, 1996). This is called systemic acquired response (SAR). SAR is an inducible defense mechanism that plays a central role in disease resistance (Delany *et al.*, 1994). SAR is induced by most pathogens that cause tissue necrosis and characterized by a long lasting, systemic resistance against a broad spectrum of pathogens (Ryals *et al.*, 1994). It has been proposed that SAR could serve as the basis for noval disease control strategies which include genetically engineered plants with enhanced disease resistance.

Genetic screens performed in *Arabidopsis* have revealed that NPR1 (for non-expressor of PR genes) which clearly functions downstream of SA (Dong, 1998). NPR1 gene has been cloned by two groups independently (Cao *et al.*, 1997; Ryals *et al.*, 1997). It is known that NPR1 (also known as NIM1 and SAI1) is a key regulator of SA-mediated SAR in *Arabidopsis* (Cao *et al.*, 1994; Delaney *et al.*, 1995). NPR1 gene encodes a noval protein with ankyrin-repeats (Cao *et al.*, 1997) which may be involved in protein-protein interaction. NPR1 also has a nuclear localization signal (NLS) is also important for inducing PR gene expression. Over-expression of NPR1 in *Arabidopsis* leads to enhanced disease resistance in a dose dependant manner to strains of *P. syringae* and *P. parasitica* that are normally virulent on wild type *Arabidopsis* (Cao *et al.*, 1998). NPR1 also participates in the jasmonate and ethylene regulated, SA-independent induced systemic resistance (ISR) (Dong, 1998). Over-expression of NPR1 gene in transgenic rice has also conferred resistance to *Xanthomonas oryzae* pv. *oryzae* (Chern *et al.*, 2001). Therefore, NPR1 gene represents an ideal target for engineering broad specturm disease resistance in crop plants.

8. ARITIFICALLY INDUCED PROGRAMMED CELL DEATH

The host-pathogen interaction leads to disruption in the cellular homeostasis often leading to cell death in both compatible and incompatible reactions (Gilchrist, 1998). This can take place by one of the mechanisms i.e. necrosis or apoptosis. Necrosis occurs in response to serious mechanical or toxic damage to the cell. It normally begins with an impairment of a cell's ability to maintain homeostasis, which in turn leads to an influx of water and extracellular ions. The entire cell swells and ruptures leading to cell lysis. Apoptosis or programmed cell death occurs in response to less severe damage to the cell and it is genetically controlled when the cell plays an important role in its own demise. The dead or dying cell is separated from its neighbouring cells with loss of specialized structures such as microvilli and desmosomes. Artificially programmed cell death as a mechanism of resistance could be achieved by expressing any toxic peptide as a response to pathogen infection. A number of genes could be used such as those encoding toxins, ribonucleases and other proteins whose enzymatic products are toxic to plant cells. One of the key factors for the success of this type of strategy is the choice of pathogen inducible promoter. A two component system was developed in which barnase, a gene from *Bacillus amyloliquefaciens* which encodes a cytotoxic protein with RNAs activity was placed under the control of the potato *prp1-1* promoter and its inactivator (barstar) was constitutively expressed in all the tissues (Strittmatter *et al.*, 1995). The promoter gets activated after the pathogen attack. A number of transgenic lines of potato expressing both the genes showed severe necrosis on leaves after inoculation with *Phytophthora infestans*. Similar strategy could be applied against diseases caused by bacteria if specific promoters are available.

9. CONCLUSIONS AND FUTURE PROSPECTS

The strategies outlined above have great potential for host-pathogen interactions. The antibacterial approaches like expression of cecropin, lysozyme, glucose oxidase and thionins etc can be applied easily to some of the important species for controlling bacterial diseases. Plant R genes have been used successfully in classical breeding programs for crop protection. The cloned R genes provide additional tools for genetic engineering of improved cultivars. The transgenic plants carrying *Xa21* (*japonica* as well as *indica* background) have the same resistance spectrum as the wild species donor (*O. longistaminata*) and the introgression line IRBB21 (*indica* backgroud) showing that the gene functions in diverse rice genotypes (Ronald, 1997). The R genes have the potential to be exploited in different plant species and there are successful cases of intra-specific transfer in Solanaceae.

Further, the recent data indicates that R genes from monocots and dicots are highly conserved, suggesting thereby that they share common functional domains (Song *et al.*, 1995). We may soon be able to design synthetic R genes on the basis of our existing knowledge of structure-function relationships and ligand-receptor interactions between pathogen *Avr* proteins and plant R gene proteins. Recently, He *et al.* (2000) demonstrated that the *Xa21* receptor serine oblique threonine kinase domain which recognizes the bacterial blight pathogen (*Xanthomonas oryzae* pv. *oryzae*) and signals the plant's defense response can be swapped with another receptor extracellular leucine rich repeat (LRR) and transmembrane domains of the *Arabidopsis* receptor (BRI 1). The chimeric receptor initiates plant defense responses in rice cells when treated with brassinosteroids. The new protein recognizes another molecule. Scientists may be able to engineer broad spectrum resistance to other pathogenes by altering the receptor domain to recognize them.

In addition, functional genomics approach has a great promise to accelerate the pace of discovery in disease control. In pharmaceuticals, these technologies have led to the identification of new drug targets. In agriculture, these technologies would have a dramatic impact on accelerating the pace of discovery. It should be straight forward to apply these technologies to small molecule discovery by identifying new validated targets for disease control.

REFERENCES

Alfano JR and Collmer A (1997) The type III (Hrp) secretion pathway of plant pathogenic bacteria: trafficking harpins, avr proteins and death. *J. Bacteriol.,* **179** : 5655-5662.

Allefs SJHM, de Jong ER, Florick DEA, Hoogendoom C and Steikema WJ (1996) *Erwinia* soft rot resistance of potato cultivars expressing antimicrobial peptide tachyplesin 1. *Mol. Breed.,* **2** : 97-105.

Aoyama T and Chua NH (1997) A glucocorticoid-mediated transcriptional induction system in transgenic plants. *Plant J.,* **11** : 605-612.

Anzai H, Yoneyama K and Yamaguchi I (1989) Transgenic tobacco resistant to a bacterial disease by detoxification of a pathogenic toxin. *Mol. Gen. Genet.,* **219** : 492-494.

Arts N, Metz M, Holub E, Staskawicz BJ, Daniels MJ and Parker JE (1998) Different requirements for EDS1 and NDR1 by disease resistance genes define at least two R-gene mediated signaling pathways in *Arabidposis. Proc. Natl. Acad. Sci. USA.,* **95** : 10306-10311.

Ausubel FM, Brent R, Kingston RE, Moore DD, Seidman JG, Smith JA and Struhl K (1997) Current protocols in molecular biology Wiley, New York.

Baker B, Zambryski P, Staskawicz B and Dinesh Kumar SP (1997) Signalling plant-microbe interactions. *Science,* **276** : 726-733.

Bent AF, Kunkel BN, Dahlbeck D, Brown KL, Schmidt R, Giraudat J, Laung J and Staskawicz BJ (1994) RPS2 of *Arabidposis thaliana*: a leucine-rich repeat class of plant disease resistance genes. *Science,* **265** : 1856-1860.

Bonas U and Ackerveken GVD (1999) Gene for gene interactions: bacterial avirulence proteins specifiy plant disease resistance. *Curr. Opion. Microbiol.* **2** : 94-98.

Cao H, Bowling SA, Gordon AS and Dong X (1994) Characterization of an *Arabidopsis* mutant that is non-responsive to inducers of systemic acquired resistance. *Plant Cell,* **6** : 1589-1592.

Cao H, Glazebrook J, Clarke JD, Volko S and Dong X (1997) The *Arabidopsis* NPR1 gene that controls systemic acquired resistance encodes a noval protein containing ankyrin repeats. *Cell,* **88** : 57-63.

Cao H, Li X and Dong X (1998) Development of broad spectrum disease resistance by overexpression of an essential regulatory gene in systemic acquired resistance *Proc. Natl. Acad. Sci. USA,* **95** : 6531-6536.

Chern MS, Fitzgerald HA, Yadav RC, Canlas PE, Dong X and Ronald PC (2001) Evidence for a resistance signaling pathway in rice similar to the NPR1-mediated signaling pathway in *Arabidopsis. Plant J.,* **27** : 101-113..

Carmona MJ, Molina A, Fernandez JA, Lo'pez-Fand JJ and Garcia-Almedo F (1993) Expression of α-thionin gene from barley in tobacco confers enhanced resistance to bacterial pathogens. *Plant. J.,* **3** : 457-462.

Collmer A (1998) Determination of pathogenicity and avirulence in plant pathogenic bacteria. *Curr. Opin. Plant Biol.,* **1** : 329-335.

Dastefano-Beltran LP, Nagpala PG, Cetiner MS, Denny T and Jaynes JM (1993) In: *Biotechnology in plant disease control.,* Wiley –Liss. pp 173-189.

Delany TP, Uknes S, Vernooij B, Friedrich L, Weymann K, Negrotto D, Gaffney T, Gut Rella M, Kessmann H, Ward E and Ryals J (1994) A central role of salicylic acid in plant disease resistance. *Science,* **266** : 1247-1250.

Delaney TP, Friedrich L and Ryals JA (1995) *Arabidopsis* signal transduction mutant defective in chemically and biologically induced disease resistance. *Proc. Natl. Acad. Sci. USA,* **92** : 6602-6606.

Dong X (1998) S.A.J.A., Ethylene, and disease resistance in plants. *Curr. Opin. Plant Biol,* **1** : 316-323.

Düring K (1996) Genetic engineering for resistance to bacteria in transgenic plants by introduction of foreign genes. *Mol. Breed.,* **2** : 297-305.

Düring K, Porsch P, Fladung M and Lörz H (1993) Transgenic potato plants resistant to the phytopathogenic bacterium *Erwinia carotovora. Plant J.,* **3** : 587-598.

Ellis J and Jones D (1998) Structure and function of proteins controlling strains specific pathogen resistance in plants. *Curr. Opin. Plant Bio.,* **1** : 288-293.

Flor HH (1971) Current status of the gene for gene concept. *Annu. Rev. Phytopath.,* **9** : 275-296.

Frederick RD, Thilmony RL, Sessa G, Martin GB (1998) Recognition specificity for the bacterial avirulence protein *AvrPto* is determined by Thr-204 in the activation loop of the tomato Pto kinase. *Mol Cell,* **2** : 241-245.

Gassmann W, Hinsch ME and Staskawicz BJ (1999) The *Arabidopsis* RPS4 bacterial resistance gene is a member of theTIR- NBS-LRR family of disease resistance genes. *Plant J.,* **20** : 265-277.

Gelvin SB and Schilerproot RA (1998) *Plant Molecular Biology Manual,* 2nd ed, Kluwer, Dordrecht, The Netherlands.

Gilchrist DG (1998) Programmed cell death in plant disease: The purpose and promise of cellular suicide. *Annu. Rev. Phytopath.,* **36** : 393-414.

Grant MR, Godiard L, Straube E, Ashfield T, Lewald J, Sattler A, Innes RW and Dangl JL (1995) Structure of the *Arabidopsis* RPM1 gene enabling dual specificity diasese resistance. *Science,* **269** : 843-846.

He Z, Wang ZY, Li J, Zhu Q, Lamb C, Ronaed PC and Chory J (2000) Perception of Brassinosteroids by the extracellular domain of the Receptor kinase BRI1. *Science,* **288** : 2360-2363.

Huang Y, Nordeen RO, Di M, Owens LD and McBeath JH (1997) Expression of an engineered cecropin gene cassette in transgenic tobacco plants confers disease resistance to *Pseudomonas syringae* pv *tabaci. Phytopathology,* **87** : 494-499.

Ilag LL, Yadav RC, Huang N, Ronald PC and Ausubel FM (2000) Isolation and characterization of disease resistance gene homologues from rice cultivar IR 64. *Gene,* **255** : 245-255.

Lamb CJ (1994) Plant disease resistance genes in signal perception and transduction. *Cell,* **76** : 419-422.

Leach JG and White FF (1996) Bacterial avirulence genes. *Annu. Rev. Phytopath.,* **34** : 153-179.

Lindgren PB (1997) The role of hrp genes during plant-bacterial interactions. *Annu. Rev. Phytopath.,* **35** : 129-152.

Martin GB (1999) Functional analysis of plant disease genes and their downstream effectors. *Curr. Opin. Plant Biol.,* **2** : 272-279.

Martin GB, Brommonschenkel SH, Chunwongse J, Frary A, Ganal MW, Spivey R, Wu T, Earle ED, Tanksley SD (1993) Map-based cloning of a protein kinase gene conferring disease resistance in tomato. *Science,* **262** : 1432-1436.

Morgues F, Brisset M and Chevreau E (1998) Strategies to improve plant resistance to bacterial diseaes through genetic engineering. *TIBTECH,* **16** : 203-209.

Nakajima H, Muranaka T, Ishige F, Akatsu K and Oeda K (1997) Fungal and bacterial resistance in transgenic plants expressing human lysozyme. *Plant Cell Rep.,* **16** : 674-679.

Nordeen RO, Sinden SL, Jaynes JM and Owens LD (1992) Activity of cecropin sb37 against protoplasts from several plant species and their bacterial pathogens. *Plant Sci.,* **82** : 101-107.

Osusky M, Zhou G, Osuska L, Hancock RE, Kay WW and Misra S (2000) Transgenic plants expressing cationic peptide chimeras exhibit broad-spectrum resistance to phytopathogens. *Nature Biotech.,* **18** : 1162-1166.

Ryals J, Ukens S and Ward E (1994) Systemic acquired resistance. *Plant Physiol.,* **104** : 1109-1112.

Ryals J, Neuenschwander UH, Willitis MG, Molina A, Stainer HY and Hunt MD (1996) Systemic acquired resistance. *Plant Cell,* **8** : 1809-1819.

Ryals J, Waymann K, Lawton K, Friedrich L, Ellis D, Stainer HY, Johnson J, Delany TP, Jesse T, Vos P and Uknes S (1997) The *Arabidopsis* NIM1 protein shows homology to the mammalian transcription factor inhibitor 1 kappa B. *Plant Cell.,* **9** : 425-439.

Ronald PC (1997) The molecular basis of disease resistance in rice. *Plant Mol. Biol.,* **35** : 179-186.

Scofield SR, Tobias CM, Rathjen JP Chang JH, Lavelle DT, Michelmore RW and Staskawicz BJ (1996) Molecular basis of gene-for-gene specificity in bacterial speck disease of tomato. *Science,* **272** : 2063-2065.

Song WY, Wang GL, Chen LL, Kim HS, Pi LY, Holsten T, Gardner J, Wang B, Zhai WX, Zhu LH, Fauquet C and Ronald P (1995) A receptor kinase like protein encoded by the rice disease resistance gene *Xa21. Science,* **270** : 1804-1806.

Staskawicz B (1999) Expression of Bs2 pepper gene confers resistance to bacterial spot disease in tomato. *Proc. Natl. Acad. Sci. USA,* **96** : 14153-14158.

Strittmatter J, Janssens J, Opsomer C and Botterman J (1995) Inhibition of fungal disease development in plants by engineering controlled cell death. *Bio/Technol.* **13** : 1085-1089.

Tai TH, Dahlbeck D, Clark ET, Gajiwala P, Pasion R and Whalon MC (1999) Expression of Bs2 peper gene confers resistance in tomato. *Proc. Natl. Acad. Sci. USA,* **96** : 14153-14158.

Tai TH, Dahlbeck D, Clark ET, Gajiwala P, Pasion R, Whalon MC, Stall RE, Tu J, Datta K, Khush GS, Zhang Q and Datta SK (2000) Field performance of Xa21 transgenic indica rice (*Oryza sativa* L.), IR72. *Theor. Appl. Genet.,* **101** : 15-20.

Tang X, Frederick RD, Zhou J, Halterman DA, Jai Y and Martin GB (1996) Initiation of plant disease resistance by physical interaction of AvrPto and Pto kinase. *Science,* **274** : 2060-2063.

Tang X, Xie M, Kim YJ, Zhou J, Klessing DF and Martin GB (1999) Overexpression of Pto activates defense responses and confers broad resistance. *Plant Cell.,* **11** : 15-30.

Tu J, Datta K, Khush GS, Zhang Q and Datta SK (2000) Field performance of Xa21 transgenic indica rice (*Oryza sativa* L.), IR72. *Theor. Appl. Genet.* **101** : 15-20.

Wang GL, Song WY, Ruan DL, Sideris S and Ronald PC (1996) *Mol. Plant-Microbe interact.,* **9** : 850-855.

Warren RF, Henk A, Mowery P, Holub E and Innes RW (1998) A mutation within the leucine rich repeat domain of the *Arabidopsis* disease resistance gene RPS5 partially suppresses multiple bacterial and downy mildew resistance genes. *Plant Cell,* **10** : 1439-1452.

Yoshimura S, Yamanouchi Y, Katayose Y, Toki S, Wang ZX, Kono I, Kurata N, Yano M, Iwata N and Sasaki T (1998) Expression of Xa1, a bacterial blight-resistance gene in rice, is induced by bacterial inoculation. *Proc Natl. Acad. Sci. USA,* **95** : 1663-1668.

Zhang S, Song WY, Chen L, Ruan D, Taylor N, Ronald P, Beachy R and Fauquet C (1998) Transgenic elite indica rice varieties, resistant to *Xanthomonas oryzae* pv *oryzae. Mol. Breed.,* **4** : 551-558.

Zhou J, Loh YT, Bressan RA and Martin GB (1995) The tomato gene Pti encodes a Serine/Threonine kinase that is phosphorylated by Pto and is involved in the hypersensitive response. *Cell,* **83** : 925-935.

Chapter 5

ABIOTIC STRESS INDUCED PROTEINS AND SIGNAL TRANSDUCTION IN PLANTS

Niharika Shankar*[1] and HS Srivastava

Department of Plant Science, MJP Rohilkhand University Bareilly -243006, India

Summary

Plant's defensive responses to abiotic stresses include induction of enzymes and metabolites involved in the scavenging of reactive oxygen species and of some specific defense proteins. Some of the proteins or peptides which provide protection against stresses are dehydrins, heat shock proteins, calcium binding proteins, phytochelatins, germin like proteins etc. These proteins defend cell against stress usually by protecting the structural and functional integrity of macromolecules and membranes. The stresses induce the synthesis of defense proteins generally at transcription level . But the signal transduction from stress to gene involves several secondary messenger or signal molecules. The signal molecules are common to several stresses and the sequence of participation of some well known signal molecules is stress → H_2O_2 → abscisic acid/ethylene/ polyamines → calcium → salicylic acid → protein kinases → gene in that order. The order or sequence for abscisic acid , ethylene and polyamines can not be discerned, but they probably either act synergisticaly or alternatively.

Keywords : Antioxidative defense system, defense proteins, heat shock proteins, phytochelatins, signal molecules, stress responsive genes

---

*Corresponding author : E-mail : niharikashankar@rediffmail.com

[1]Present address : 10B/7, Madan Mohan Malviya Marg, Lucknow - 226 001, India

1. INTRODUCTION

During their life time, plants have to deal with various biotic and abiotic stresses, because being almost immobile, they are unable to escape the extreme environmental conditions which prevail sometimes. Most abiotic stresses which affect morpho-physiological changes in plants act primarily through the production of reactive oxygen species (ROS), also called oxidative free radicals. ROS are also produced during normal metabolism of plants as in photosynthesis and in respiration. Three important ROS, superoxide anion, hydroxy free radical and H_2O_2, are highly toxic and cause changes in DNA, proteins and lipids and in membrane organisation. Therefore, there is a continuous need to scavenge these ROS so that the cell's integrity is maintained. Plants generally have adequate biochemical machinery, which comprises of antioxidants and their enzymes, to scavenge these molecular species. Antioxidants include low molecular weight compounds such as ascorbate, glutathione, polyphenols, α-tocopherol, flavonones, anthocyanins etc. and the enzymes include superoxide dismutase, peroxidases, ascorbate reductases and so on. The ROS also induce or activate certain protective or defense metabolites so that the minimum damage is done due to the stress. Thus, two types of protective responses are evoked at molecular level: 1. The increase in antioxidant enzymes and metabolites and 2. Induction of genes producing some specific defense related proteins (Srivastava, 1999). The hypersensitive response (HR) in fungal infected plants, which is characterized by localized cell collapse may also be considered as a defense mechanism because it prevents further spread of the fungus. The infection also induces the production of antimicrobial compounds such as phytoalexins and pathogenesis related proteins (Lamb and Dixon, 1991). Some signal molecules such as jasmonic acid, ethylene and salicylic acid are involved in the production of pathogenesis related proteins and in HR (Baker *et al.*, 1997). Many signal moecules of HR responses are same as those involved in the abiotic stresses. Genetically modified plants, which have an over-expression of antioxidant enzymes or some defense related proteins have better chance of survival under the stressed environmental conditions than non modified plants. For example, transgenic alfalfa (*Medicago sativa*) overexpressing Mn-superoxide dismutase have improved survival at 2 °C (McKersie *et al.*, 1999). Superoxide dismutase is an essential and perhaps the most important component of the ROS scavenging mechanism, because it dismutates two superoxide radicals to produce hydrogen peroxide and water. While bulk protein synthesis is inhibited, many specific proteins are induced during stressed conditions.

For example, a large number of genes encoding defense proteins are known to be induced during wounding in *Arabidopsis* (Raymond *et al.*, 2000). Induction of genes related to defense proteins and the signal transduction mechanism from stress to gene is described in this article.

2. STRESS INDUCED PROTEINS

2.1. Dehydrins

Dehydrins are glycine -rich, hydrophilic, 15 to 129 kD heat stable proteins and are induced in responses to environmental stresses which result in cellular dehydration, such as drought and cold (Close, 1997). Their presence has been reported in several species of monocots and dicots and two families of gymnosperms, Pinaceae and Ginkgoaceae (Close *et al.*, 1993). It is prosed that dehydrins and compatible solutes act synergistically to stabilize macromolecules such as proteins and nucleic acids and thereby stabilizing the protoplasm (Close, 1996). Their induction in response to other stresses such as wounding has also been demonstrated (Luo *et al.*, 1992; Danyluk *et al.*, 1994; Rouse *et al.*, 1996; Richard *et al.*, 2000). They belong to LEA (late -embryogenesis abundant) D11 family proteins and remain soluble at temperatures approaching 100 °C. They prevent damage of the embryo due to desiccation. These proteins accumulate in normal conditions also in dehydrating plant tissue, such as in seeds that are becoming mature.

The dehydrin genes have been cloned from conifers. Three D F65 cDNA genes have been cloned from *Pseudotsuga menzeisii* (Jarvis *et al.*, 1996). A cDNA clone encoding a dehydrin gene has been isolated from a cDNA library prepared from *Picea glauca* needle mRNAs (Richard *et al.*, 2000). The cDNA designated *PgDhn1,* is 1159 nucleotide long and has an open reading frame of 735 bp with a deduced amino sequence of 245 residues. The mRNA corresponding to PgDhn1 cDNA were induced by wounding and by jasmonic acid and methyl jasmonate treatments. Abscisic acid or etephon treatment of seedlings also induced gene transcript in the needles (Richard *et al.*, 2000).

2.2. Heat Shock Proteins

Heat shock proteins (HSPs) belong to a group of proteins termed molecular chaperones (Ellis and Vander-Vries, 1991). They render thermal tolerance in plants by assisting in the refolding of partially folded or denatured proteins under elevated temperature (Forreiter and Nover, 1998). They also assist in resolubilizing aggregated denatured proteins

(Parsell *et al.*, 1994). When plant species that are adopted to temperate environments, such as soybean, pea, maize and wheat are grown at elevated temperature, 20 to 40 different types of HSPs are induced (Vierling, 1991). The induction of HSPs in response to elevated temperature has been demonstrated in many other species also such as in *Phaseolus lunatus* (Keeler *et al.*, 2000) and *Beta vulgaris* suspension culture cells (Trofimova *et al.*, 1999). HSPs are also induced by other kinds of stresses such as drought, flooding, salinity and heavy metals (see Sabehat *et al.*, 1998). It is possible that specific stress induces a specific type of HSP.

Genes encoding major HSPs have been isolated from various species (Vierling, 1991). The expression of these genes is regulated primarily at transcription level. The thermo-inducibility of these genes is attributed to cis-regulatory prmotor elements (HSEs) located in the TATA box proximal to 5'flanking regions of the heat shock genes (Schoffl *et al.*, 1998). The activation of HSE is apparently through many activation proteins (HSFs). In eukaryotes, this activation protein is coverted from a monomer to a trimeric form, which is then become capable of high affinity binding to HSE and causing trascriptional induction.

Several types of experimental findings indicate the role of HSPs in thermal/cold tolerance. In plants, both high molecular mass HSPs (60-110 kD) and low molecular mass HSPs (10-15kD) are synthesized in response to high temperature. The accumulation specially of HSP70 is also induced by low temperature. The accumulation of HSP104 and HSP109 has been detected during acclimation of rice plant to low temperature (Pareek *et al.*, 1995). An *Arabidopsis* mutant (TU8), which has reduced thermostability, the cytosolic level of HSP70 is quite low (Ludwig-Muller *et al.*, 2000). At temperature above 25 ^{o}C, the mutants have a decreased growth and die after 2 weeks, whereas wild plants are unaffected. Further, transgenic *Arabidopsis* plants expressing less than usual amount of HSP_{101} have a reduced thermal tolerance, although they grow at normal rates at normal temperature (Queitsch *et al.*, 2000).

2.3. Anti-freeze proteins

Anti-freeze proteins (AFP) which provide tolerance towards freezing are produced during low temperature acclimation (Duman and Olsen, 1993; Griffith and Evart, 1995). They inhibit ice crystal formation in the cells (Evart *et al.*, 1999). In winter rye (*Secale cereale*) six AFPs ranging from 15 to 35 kD are formed during exposure to cold temperature (Griffith *et al.*, 1992; Hon *et al.*, 1994, 1995). These AFPs are similar to pathogen related proteins and have properties of glucanase, chitinase and thuamatin like proteins (Hon *et al.*, 1995) Genes encoding proteins similar to two chitinases have been cloned from winter rye and wheat. They are induced by cold and a lesser exent by drought but not by ABA (Yeh *et al.*, 2000)

2.4. Calcium binding proteins

Cytosolic level of calcium (Ca^{2+}) generally increases either due to uptake from the medium and/or due to release from the intracellular storage compartments in response to biotic stresses. Cytosolic Ca^{2+} is believed to play a fundamental role in the regulation of a variety of cellular processes leading to a specific response to each stress.

Calcium binding proteins calmodulin, calmodulin like proteins, centrin and calcineurin B like proteins bind with calcium in the signalling of abiotic stress induced genes. These proteins have Mr in the range of 14 to 22 kD, and with a conserved EF-hand motif. They are able to sense changes in Ca^{2+} concentration of less than 1 μM. Upon binding to Ca^{2+}, the proteins change conformation and interact with other protein molecules to regulate their activity. Many of these proteins are induced in response to the environmental stresses. Induction of calmodulin in response to physical and chemical stimuli (Botella and Arteca, 1994) and to salinity (Jang *et al.*, 1998) has been reported in several plants. An elicitor derived from *Typhula ishikariensis* is known to induce a 14 kD Ca^{2+} binding protein (centrin like) in wheat culture cells (Takezawa, 2000).

2.5. Phytochelatins

Phytochelatins are low molecular weight metal binding polypeptides present in all kinds of plants including monocots, dicots, gymnosperms, algae, diatoms and fungi. They are made up of only three amino acids glutamate, glycine and cysteine with empirical formulae (γ-Glu-Cys)n.Gly where n is generally 2 to 5. They are synthesized from glutathione by the enzyme phytochelatin synthase (γ-Glu-Cys dipeptidyl transferase ; E.C. 2.3.2.15) (Cobbett, 2000). The regulation of phytochelatin synthesis is primarily occur at the level of phytocheltin synthase which is a 95 kD tetramer with a K_m of 6.7 mM for glutathione. The enzyme is activated by heavy metals such as Cd, Ag, Bi, Pb, Zn, Cu, Hg and Au cations but induction is not known. The gene for phytochelatin synthase has been isolated from *Saccharomyces cerevisiae* (Clemens *et al.*, 1999) and *Arabidopsis* (Vatamanuk *et al.*, 1999; Ha *et al.*, 1999).

Many heavy metals induce the synthesis of phytochelatins in plants (Rauser, 1995; Leopold *et al.*, 1999). The synthesis is also induced in transgenic mustard which have overexpression of enzyme glutamyl cystein synthetase (Zhu *et al.*, 1999), the enzyme involved in the synthesis of glutathione, the precursor of phytochelatin. The mustard plants showing over expression of this enzyme show increased tolerance towards heavy metals and they also accumulated higher levels of Cd (Zhu *et al.*, 1999).

2.6. Germin like proteins

Germin like proteins (GLPs) have been found in several angiosperms including barley, rice, sugar cane, pea, *Pharbitis nil, Arabidopsis* etc. Germin was first reported to be associated with the germination of wheat seeds and other cereals (Lane, 1991). The wheat germin has been studied extensively; it is found in the apoplast and cytoplasm of the germinating embryo cells and it has oxalate oxidase activity (E.C. 1.2.3.4) an activity which produces H_2O_2. It is now known that germins and germin like proteins are part of superfamily of proteins together with seed storage globulins and sucrose binding proteins (Baumlein *et al.*, 1995; Braun *et al.*, 1996) Expression of *GLP* genes has been shown to be upregulated by salt stress in barley and *Mesembryathemum crystallinum* (Hurkman and Tanaka, 1995; Michalowski and Bohnert, 1992). In tobacco also the germin gene is expressed in some abiotic stresses such as Cd, Cu and Co and also by polyamines, wounding and TMV infection (Berna and Bernier, 1999).

2.7. Enzymes of proline biosynthesis

Proline acts as a osmoprotectant in plants and micro-organisms under drought and salt stress. A *Nicotiana plumbaginifolia* mutant having increased production of proline has enhanced tolerance to salt stress (Sumaryati *et al.*, 1992). Proline accumulates under stress through stimulation of its *de novo* synthesis together with the repression of its catabolism (Delauney and Verma, 1993; Peng *et al.*, 1996; Verbruggen *et al.*, 1996). It has also been demonstrated that salt and water stress induce the transcription of genes for Δ_1-pyrroline-5-carboxylate synthetase (P5CS) and Δ_1-pyrroline-5-carboxylase reductase (P5CR), the enzymes critical in the biosynthesis of proline (Hu *et al.*, 1992; Vebrugen *et al.*,

1993; Savoure *et al.*, 1995; Yoshiba *et al.*, 1995). The tobacco mutant tolerant to salt stress accumulates high amount of proline due to lack of feed back inhibition of the enzyme pyrroline-5- carboxylase synthetase by proline (Rosens *et al.*, 1999).

2.8. Antioxidant enzymes

Plants produce antioxidants and enzymes of their metabolism to scavenge the ROS, which are induced by the stresses. These include carotenoids, tocopherol, ascorbate and glutathione and the enzymes super oxide dismutase (SOD, E.C. 1.15.1.1), catalase (E.C. 1.11.1.6), and peroxidases.

Glutathione peroxidases (GPX, E.C. 1.11.1.9) are a family of multiple isozymes catalysing the reduction of H_2O_2, organic hydroperoxides and lipid hydroperoxides using glutathione as a reducing agent . This enzyme is considered one of the key enzymes involved in scavenging oxygen radicals in animals. The enzyme is important in plants also and is induced in response to mechanical stimulation in tomato plants (Depege *et al.*, 2000). Two m-RNA transcripts of this enzyme, namely GPLle-1 and GPXle-2 , increase within 1-6 h in response to rubbing of the stem internode.

2.9. Enzymes of secondary metabolites

Enzymes involved in the synthesis of certain secondary metabolites, which usually acts as structural components of barriers in the biotic and abiotic stresses are also induced by stresses. Lignin, for example, is a major structural component of plant secondary cell walls and is involved in protection against mechanical and microbial damage. Lignification of plant tissues often occurs in response to wounding and fungal infection. One of the enzymes of lignin biosynthesis, caffeoyl-coenzyme A-3-O methyl transferase (CCOAOMT), is known to be induced in response to wounding and fugal infection in poplar (Chen *et al.*, 2000). Another enzyme of lignin biosynthesis cinnamyl-alcohol dehydrogenase (CAD, E.C. 1.1.1.195) is also known to be induced by mechanical wounding in lucerne (Brill *et al.*, 1999)

Many other phenolics are also known as protectants of abiotic and biotic stresses. In Scots pine (*Pinus sylvestris*), induction of stilbenes is response to ultra-violet, ozone and fungal invasion has been reported (Schoppner and Kindl, 1979; Roseman *et al.*, 1991; Lieutier *et al.*, 1996). The imporant enzymes in the biosynthesis of methoxy stilbenes stsrting from phenyl alanine are, phenylalanine ammonia lyase (PAL), stilbene synthase (STS) and pinosylvin methyl transferase (PMT). Induction of *STS* and cinnamyl alcoholal dehyrogenase gene in response to ozone has been reported in Scots pine (Zinser *et al.*, 1998). Recently Chiron *et al.* (2000) have sequenced the s-adenosyl-L-methionine: pinosylvin -O-methyl transferase (*PMT*) gene from Scots pine and they have demonstrated that *STS* and *PMT* genes are induced by ozone. A dose dependent transient induction of these genes was observed with wounding also.

3. SIGNAL TRANSDUCTION IN GENE INDUCTION

Quite a few types of molecules participate in the transmission of the signal from stress to the target gene(s). Several investigations have demonstrated the existence of cross tolerance in plants. Exposure to one stress at moderate level provides protection against the other stress (Sabehat *et al.*, 1998). This indicates a common pathway of signal transduction in different stresses. Some of the molecules known to act as messenger molecules in the transmission of signals are hyrogen peroxide, salicylic acid, abscisic acid, ethylene, jasmonic acid, polyamines, nitric oxide , calcium, and so on. The evidences for the involvement of these molecules are two folds: (i) The internal concentration of the signal molecules increases during stresses and (ii) External application of these chemicals generally protects plants against the toxic effects of the environmental stresses.

Hydrogen peroxide is the most stable ROS and is capable of rapid diffusion across the cell membrane. Besides, being a product of oxidative stress, it is also produced in many enzymatic and non-enzymatic reactions. It is produced rapidly in response to the environmental stresses such as O_3 (Willekens *et al.*, 1995), and chilling (Prasad *et al.*, 1994). It also leads to altered cytosolic Ca^{2+} level and an increase in the expression of cold responsive (Prasad *et al.*, 1995) and antioxidant genes (Polidoros and Scandalios, 1999). Therefore, H_2O_2 may act as an early signal in the cellular defense systems against stress (Fig. 1).

An increase in ABA or ethylene also seems to be an early component of stress signal (Fig. 1). In bean for example, treatment of seedlings with 75 mM NaCl causes about 3 fold increase in ABA content within two hours (Montero *et al.*, 1997). Polyamines may be closely linked with ethylene in the signal transduction pathway (Fig.1), because the synthesis of both is closely linked, both shares-adenosyl methionine as an intermediate in their synthesis. This possibility is further strengthened

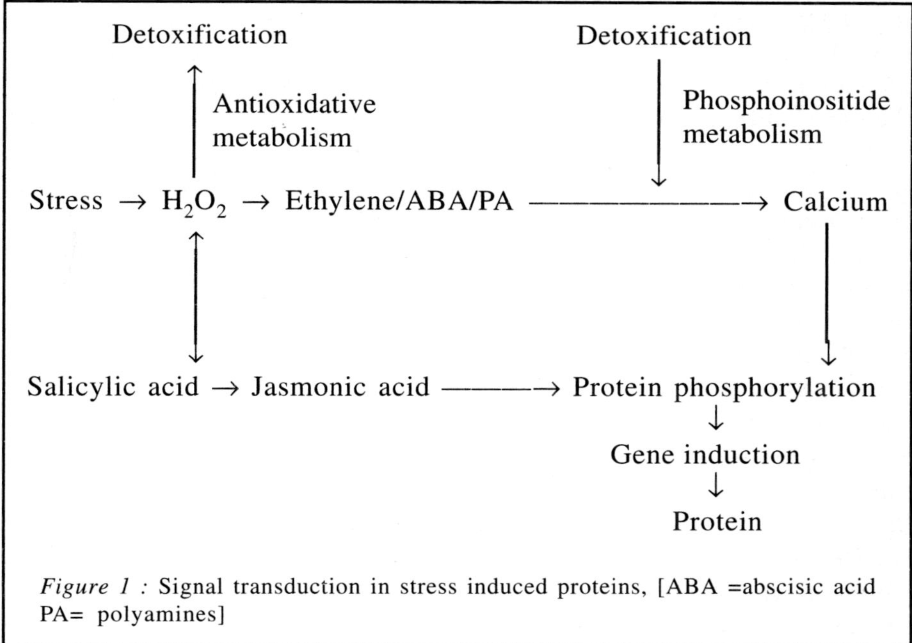

Figure 1 : Signal transduction in stress induced proteins, [ABA =abscisic acid PA= polyamines]

by the observation that supply of polyamines stimulated ethylene production (Pennazio and Roggero, 1990; Lutts *et al.*, 1999) and also the supply of ethylene precursor ACC induced polyamine biosynthesis (Lee and Chu, 1992: Tamai *et al.*, 1999). In the systems where NO might be acting as a signal molecule it perhaps acts between H_2O_2 and ethylene production (Leshem, 1996). However, the involvement of NO in signal transduction has to be authenticated.

Release of Ca^{2+} from its storage compartments causing an increase in cytoslic Ca^{2+} plays a central role in signal transduction. Anthony Trewavas (1999) in a thought provoking article remarks that "A calcium wave marked the onset of our existence, and will quite probably mark our demise: Irreversible failure of calcium wave generation in the heart is the most common cause of death". He aptly concludes the article by stating that "Life is complex, and 'le calcium c'est la vie' suggests that calcium signaling is as infinitely complex as life itself". Calcium wave is generated under the influence of cellular signals by opening the Ca^{2+} channels present in the membranes of its storage compartments. A single channel can transmit 10^6 atoms of Ca per second (Trewavas, 1999). Recently, Klusner and Weiler (1999) have characterized a Ca^{2+} channel from root tip endomembranes of garden cress (*Lepidium sativum*) (LCC1) which appears to have an important role in regulating cytosolic Ca^{2+} level. But how, the signal molecules or environmental stresses affect this channel is not known. One of the signals causing opening of these channels may be inositol 1,4,5 triphosphate, as demonstrated in beet root vacuoles (Alexandre *et al.*, 1990). Further, the presence of a phosphoinositide specific phopspholipase C activity, which is involved in phosphoinositide metabolism, has been demonstrated in oat roots (Huang *et al.*, 1995). That phosphoinositides participate in Ca^{2+} signaling is further strengthened by the fact that Ca^{2+} waves in pollen tubes of *Papaver rhoes* (Franklin-Tong *et al.*, 1996) and ABA induced oscillations in cytoplasmic levels of Ca^{2+} in the guard cells (Staxen *et al.*, 1999), are inhibited by phospholipase C inhibitors. Alternatively, the stress itself may act as the immediate signal for the elevation of cytoplasmic Ca^{2+} level, as in stresses such as in wind, and cold shock there is no lag in Ca^{2+} transients (Trewavas, 1999). In *Arabidopsis* (Polisensky and Braam, 1996) and in tomato (*Lycopersicon esculentum*) (Sebastiani *et al.*, 1999) also, an increase in Ca^{2+} in response to cold shock is noticed within seconds. However, in others, such as in heat shock, there may be lag of several minutes (see Trewavas, 1999). The elevated cytoplasmic level of Ca^{2+} may cause different types of responses. Perhaps the variability in the response depends upon the amplitude, duration, frequency and localization of the Ca^{2+} signal. The downstream responses of Ca^{2+} signals are not clearly understood in plants. But, protein phosphorylation involving Ca^{2+} dependent protein kinases (CDPKs) and gene induction are important events in response to many environmental stresses. The genes encoding CDPKs have been isolated from a variety of plant systems including *Arabidopsis*, maize, mungbean, rice and soybean (see Ellard-Ivey *et al.*, 1999 and references there in). Protein phosphorylation, which is observed in response to cold stress takes place within minutes (Monroy *et al.*, 1993). But altered gene expression including both changes in mRNA accumulation and enzyme activity is observed within a few hours (Guy, 1990; Thomashow, 1990). Thus, it is most likely that protein phosphorylation precedes gene expression (Monroy *et al.*, 1996). Some of these genes produce proteins which confer resistance to the stress. For example, a number of genes are known to be induced in response to the cold stress (Thomashow, 1990, 1998; Hughes and Dunn, 1996, Knight *et al.*, 1999). The product of some of these genes, such as cold responsive *COR 15a,* have been demonstrated to play a protective role in freezing stress (Artus *et al.*, 1996). The sfr6 mutant of *Arabidopsis*, which do not express cold inducible genes *LTI78, COR15a* and *Kin1* at normal wild type levels, are not cold acclimatized (Knight *et al.*, 1999)

However, the role of many other cold inducible proteins is not known. Increases in cytosolic Ca^{2+} as an upward signal for induction of gene has been demonstrated in response to cold, and osmotic and salt stress (Knight *et al.*, 1991, 1996, 1997).

As described earlier, salicylic acid has been demonstrated to act as a signal molecule in plant defense responses to biotic stresses and its role is abiotic stresses is also suspected. However, the mechanism of signal transduction is poorly understood. A mechanistic link between salicylic acid and H_2O_2, one of the major products of stress responses, has also been demonstrated in quite a few investigations. The treatment with this phenolic enhances the production of H_2O_2 in parsley suspension cultures (Kauss *et al.*, 1992). The increased level of H_2O_2 in some studies have been demonstrated to be due to reduced catalase activity (Chen *et al.*, 1993; Conrath *et al.*, 1995), although, in soybean cultured cells catalase activity along with the antioxidant enzymes glutathione reductase (GR, EC 1.6.4.2), glutathione s-transferase (GST, EC 2.5.1.18) and monodehydroascorbate reductase (MDAR, EC 1.6.5.4) increase in response to 0.6 to 1.0 mM salicylic acid (Knorzer *et al.*, 1999). Some studies have demonstrated that salicylic acid may serve as one-electron-donating substrate for catalases/peroxidases and in the process it is converted to salicylic acid free radical in the presence of H_2O_2 as electron acceptor (Durner and Klessig, 1996; Kvaratskhelia *et al.*, 1997). These and some other studies (Durner *et al.*, 1997) demonstrate that H_2O_2 acts upstream as compared to SA in the regulation of gene expression . The downstream action of salicylic acid is apparently through the modulation of some critical enzymes, which may be either through the phosphorylation of enzymic protein or through the production of ethylene, a well known plant regulator. The experimental evidences for the former possibility are more convincing than for the later. In fact, protein phosphorylation has already been proposed to be an essential component of the signal transduction pathway in elicitor induced responses (Chandra and Low, 1995; Suzuki *et al.*, 1995). In pathogen related responses, evidences for protein phosphorylation as a mechanism in salicylic acid induced signaling are also derived from the observations of suppression of salicylic acid action by protein kinases/phosphatase inhibitors and by the identification of an salicylic acid activated MAP kinase (Conrath *et al.*, 1997; Shirasu *et al.*, 1997; Zhang and Klessig, 1997). Protein phosphorylation might be involved in the expression of some of the salicylic acid responsive genes also (Stange *et al.*, 1997). Salicylic acid and also the jasmonates or wounding induce an increase in the transcription of a solanum pollinated pistil (*SPP2*) dioxygenase gene in *Solanum chacoense* (Lantin *et al.*, 1999). The dioxygenase enzyme is involved in the chemical guidance of the pollen tube in the ovary and also in the production of secondary metabolites having anti-stress role. Since both Ca^{2+} and salicylic acid appear to be acting through protein phosphorylation, it may be assumed that salicylic acid acts as a messenger between Ca^{2+} and protein phosphorylation in signal transductions, where it has a role (Fig. 1).

4. CONCLUSIONS AND FUTURE PROSPECTS

Induction of certain specific proteins, involved in stress management, by the stresses, seems to be a general phenomenon in plants. The induction of these proteins, which is at transcription levels involves several signal molecules and signal transducers. Although the involvement of several types of molecules and ions including hormones, secondary metabolites and Ca^{2+} ion has been demonstrated, their interrelationship is not clearly understood. The existence of cross-tolerance to different types of stress indicates a common pathway of siganl transduction in various types of stresses. However, some minor variations may be there because of the cellular and tissue specific locations and micro-environment of the signal molecules. It will require further experimentation, specially to determine the temporal sequence of various signal molecules identified so far and also to find out the possibility of the involvement of other molecules. It is possible that many of the stress inducible proteins are some already known proteins with some other functions and they perform the role of stress defensive proteins as an additional work.

REFERENCES

Alexandre J, Lassalles JP and Kado RT (1990). Opening of Ca^{2+} channels in isolated red beet root vacuole membrane by inositol 1,4,5-triphosphate. *Nature* **343** : 567-570.

Artus NN, Uemura N, Steponkus PL, Gilmour SJ, Lin C and Thomashow MF (1996). Constitutive expression of the cold regulated *Arabidopsis thaliana* COR 15a gene affects both chloroplast and protoplast freezing tolerance. *Proc. Natl. Acad. Sci. U.S.A.* **93** : 13404-13409.

Baker B, Zabryski P, Sraskawicz B and Dinesh-Kumar SP (1977). Signalling in plant microbe interactions. *Science* **276** : 726-733.

Baumlein H, Braun H, Kakhovskaya IA and Shutov AD (1995). Seed storage proteins of spermatophytes share a common ancestor with dessication proteins of fungi. *J. Mol. Evol.* **41** : 1070-1075.

Berna A and Bernier F (1999). Regulation by biotic and abiotic stress of a wheat germin gene encoding oxalate oxidase, a H_2O_2 producing enzyme. *Plant Mol. Biol.* **39** : 539-549.

Botella JR and Arteca RN (1994). Differential expression of two calmodulin genes in response to physical and chemical stimuli. *Plant Mol. Biol.* **24** : 757-766.

Braun H, Czihal A, Shutov AD and Baumlein H (1996). A viclin like seed protein of cycads: similarity to sucrose binding proteins. *Plant Mol. Biol.* **31** : 35-44.

Brill EM, Abrahams S, Kaynes CM, Jenkins CLD and Watson JM (1999). Molecular characterization and expression of a wound inducble cDNA encoding a novel cinnamyl-alcohol dehydrgenase enzyme in lucerne (*Medicago sativa* L.) *Plant Mol. Biol.* **41** : 279-291.

Chandra S and Low PS (1995). Role of phosphorylation in elicitation of the oxidative burst in cultured soybean cells. *Proc. Nat. Acad. Sci. USA,* **92** : 4120-123.

Chen C, Meyermans H, Burggraeve B, De Rycke RM, Inoue K, Vleesschauwer VD, Steenackers M, Van Montagu MC, Engler GJ and Boerjan WA (2000). Cell specific and conditional expression of caffeoyl-coenzyme A-3-0-methyl transferase in poplar. *Plant Physiol.* **123** : 853-867.

Chiron H, Drouet A, Lieutier F, Payer HD, Ernst D and Sandermann Jr H (2000). Gene induction of stilbene biosynthesis in Scots pine in response to ozone treatment, wounding and fungal infection. *Plant Physiol.,* **124** : 865-872.

Clemens S, Kim EJ, Neumann D and Schroeder JI (1999). Tolerance to toxic metals by a gene family of phytochelatin synthases from plants and yeast. *EMBO J.* **18** : 3325-3333.

Close TJ (1996). Dehydrins: energence of a biochemical role of a family of plant dehydration proteins. *Physiol. Plant* **97** : 795-803.

Close TJ (1997). Dehydrins: Emergence of a biochemical role of a family of plant dehydration proteins. *Physiol. Plant.,* **97** : 795-803.

Close TJ, Fenton RD and Monnan F (1993). A view of plant dehydrins using antibodies specific to the carboxy terminal peptide. *Plant Mol. Biol.,* **23** : 279-286.

Cobbett CS (2000). Phytochelatins and their role in heavy metal detoxification. *Plant Physiol.* **123** : 825-832.

Conrath U, Chen Z, Ricigliano JR and Klessig DF (1995). Two inducers of plant defense responses, 2,6 dichloroisonicotinic acid and salicylic acid inhibit catalase activity in tobacco. *Proc. Nat. Acad. Sci., U.S.A.* **92** : 7143-7147.

Conrath U, Silva H and Klessig DF (1997). Protein dephoshorylation mediates salicylic acid induced expression of PR1 genes in tobacco. *Plant J.* **11** : 747-757.

Danyluk J, Houde M, Rassart E and Sarhan F (1994). Differential expression of a gene encoding an acidic dehydrin in chilling sensitive and freezing tolerant graminae species. *FEBS Lett.* **344** : 20-24.

Delauney AJ and Verma DPS (1993). Proline biosynthesis and osmoregulation in plants. *Plant J.* **4** : 215-223.

Depege N, Verenne M and Boyer N (2000). Induction of oxidative stress and GPX like protein activation in tomato plants after mechanical stimulation. *Physiol. Plant.* **110** : 209-214.

Duman JG and Olsen TM (1993). Thermal hysteresis protein activity in bacteria, fungi and phylogentically diverse plants. *Cryobiology* **30** : 322-328.

Durner J and Klessig F (1996). Salicylic acid is a modulator of tobacco and mammalian catalases. *Science* **266** : 1247-1250.

Durner J, Shah J and Klessig F (1997). Salicylic acid and disease resistance in plants. *Trends Plant Sci.* **2** : 266-274.

Ellard-Ivey M, Hopkins RB, White TJ and Lomax TL (1999). Cloning expression and N-terminal myristoylation of CPCK1, a calcium dependent protein kinase from zucchini (*Cucurbita pepeo* L.). *Plant Mol. Biol.,* **39** : 1999-208.

Ellis RS and Van der Vries SM (1991). Molecular chaperones. *Annu. Rev. Biochem.,* **60** : 321-327.

Evart KV, Lin Q and Hew CL (1999). Structure, function and evolution of antifreeze proteins. *Cell. Mol. Life Sci.* **55** : 271-283.

Forreiter C and Nover L (1998). The heat stress response and the molecular chaperones. *J. Biosci.* **23** : 287-302.

Franklin-Tong VE, Drobak BK, Allan AC, Watkins PAC and Trewavas AJ (1996). Growth of pollen tubes of *Papaver rhoes* is regulated by a slow moving calcium wave propagated by inositol 1,4,5-triphosphate. *Plant Cell* **8** : 1305-1321.

Griffith MV, Ala P, Yang DS, Hon WC and Moffatt BA (1992) Antifreeze protein produced endogenouslly in winter rye leaves. *Plant Physiol.* **100** : 593-586.

Griffith M and Evart KV (1995) Antifreeze proteins and their potential use in frozen foods. *Biotechnol. Adv.* **13** : 271-283.

Guy CL (1990). Cold acclimation and freezing stress tolerance: role of protein metabolism. *Annu. Rev. Plant Physiol. Plant Mol. Biol.* **41** : 187-223.

Ha SB, Smith AP, Howden R, Dietrich WM, Bugg S, O'Connell MJ, Goldsbrough PB and Cobbett CS (1999). Phytochelatin synthesis genes from *Arabidopsis* and yeast *Scizosaccharomyces pombe. Plant Cell* **11** : 11531163.

Hon WC, Grifith M, Chong P and Yang DSC (1994). Extraction and isolation of antifreeze proteins from winter rye (*Secale cereale*) leaves. *Plant Physiol.* **104** : 971-980.

Hon WC, Grifith M, Mlyanaez A, Kwok YC and Yang DSC (1995). Antifreeze proteins in winter rye are similar to pathogenesis related proteins. *Plant Physiol.* **109** : 879-889.

Hu ACA, Delauney AJ and Verma DPS (1992). A bifunctional enzyme (Δ_1-pyrroline-5-carboxylate synthetase) catalyses the first two steps in ptoline biosynthesis in plants. *Proc. Natl. Acad. Sci. USA* **89** : 9354-9358.

Huang CH, Tate BF, Crain RC and Cote GG (1995). Multiple phosphoinositide specific phospholipase C in oat roots: Characterization and partial purification. *Plant J.* **8** : 257-267.

Hughes MA and Dunn MA (1996). The molecular biology of plant acclimation to low temperature. *J. Exp. Bot.* **47** : 291-305.

Hurkman WJ and Tanaka CK (1995). Effect of salt stress on germin expression in barley (*Hordeum vulgare* L.) roots. *Plant Physiol.* **110** : 971-977.

Jang HJ, Pih KT, Kang SG, Limm JH, Jin JB, Pio HL and Hwang I (1998). Molecular cloning of a novel Ca^{2+} binding protein that is induced by NaCl stress. *Plant Mol. Biol.,* **37** : 839-847.

Jarvis SB, Taylor MA, McLeod MR and Davies HV (1996). Cloning and characterization of the cDNA clones of three genes that are differentially expressed during dormancy-breakage in the seeds of Douglas fir (*Psedutsuga menzeisii*). *J. Plant Physiol.,* **147** : 559-566.

Kauss H, Theisinher H, Hinkel E, Minderman R and Conrath U (1992). Dichloroisonocotinic acid and salicylic acid inducers of systemmic acquired resistance enhance fungal elicitor responses in parsely. *Plant J.* **2** : 655-660.

Keeler SJ, Boettger CM, Haynes JG, Kuches KA, Johnson MM, Thureen DL, Keelr Jr, CL and Kitto SL (2000). Aquired thermotolerance and expression of the HSP100/ClpB genes of lima bean. *Plant Physiol.* **123** : 1121-1132

Klusener B and Weiler EW (1999). A calcium selective channel from root-tip endomembranes of garden cress. *Plant Physiol.* **119** : 1399-1405.

Knight MR, Campbell AK, Smith SM and Trewas AJ (1991). Transgenic plant aequorin reports the effect of touch and cold shock and elicitors on cytoplasmic calcium. *Nature* **352** : 524-526.

Knight H, Trewavas AJ and Knight MR (1996). Cold calcium signalling in *Arabidopsis* involves two cellular pools and a change in calcium signature after acclimation. *Plant Cell* **8** : 489-503.

Knight H, Trewavas AJ and Knight MR (1997). Calcium signalling in *Arabidopsis thaliana* responding to drought and salinity. *Plant J.* **12** : 1067-1078.

Knight H, Veale EL, Warren GJ and Knight MR (1999). The sfr6 mutation in *Arabidopsis* suppresses low temperature induction of genes dependent on the CRT/DRE sequence motif. *Plant Cell* **11** : 875-886.

Knorzer OC, Lederer B, Durner J and Boger P (1999). Antioxidative defense activation in soybean cells. *Physiol. Plant.* **10** : 294-300.

Koch JR, Scherzer AJ, Eshita SM and Davis KR (1998). Ozone sensitivity in hybrid popular is correlated with a lack of defense gene activation. *Plant Physiol.,* **118** : 1243-1252.

Kvaratskhelia M, George SJ and Thorneley RN (1997). Salicylic acid is a reducing substrate and not an effective inhibitor of ascorbate peroxidase. *J. Biol. Chem.* **272** : 20998-21001.

Lamb C and Dixon RA (1997). The oxidative burst in plant disease resisance. *Annu. Rev. Plant Physiol. Plant Mol. Biol.* **48** : 251-257.

Lane BG (1991). Cellular dessication and hydration: developmental regulated proteins, and the maturation and germination of embryos. *FASEB J.* **5** : 2893-2901.

Lantin S, O'Brien M and Matton DP (1999). Pollination, wounding and jasmonate treatments induce the expression of a developmentally regulated pistil deoxygenase at a distance, in the ovary, in the wild potato *Solanum chacoense* Bitt. *Plant Mol. Biol.,* **41** : 371-386.

Lee TM and Chu C (1992). Ethylene induced polyamine accumulation in rice (*Oryza sativa*) coleoptiles. *Plant Physiol.,* **100** : 238-245.

Leopold I, Gunther D, Schmidt J and Neumann D (1999). Phytochelatins and heavy metal tolerance. *Phytochemistry* **50** : 1323-1328.

Leshem YY (1996). Nitric oxide in biological systems. *Plant Growth Regul.* **18** : 155-159.

Lieutier F, Sauvard D, Brignolas F, Picron V, Yart A, Bastien Cand Jay-Allemand C (1996). Changes in phenolic metabolites of Scots pine phloem induced by *Ophiostoma brunneo-ciliatum,* a bark beetle associated fungus. *Eur. J. For. Pathol.* **26** : 145-158.

Ludwig-Muller J, Krishna P and Forreter C (2000). A glucosinolate mutant of *Arabidopsis* is thermosensitive and defective in cytosolic HSP90 expression after heat stress. *Plant Physiol.* **123** : 949-958.

Luo M, Liu JH, Mohapatra S, Hill RD and Mohapatra SS (1992). Characterization of a gene family encoding abscisic acid and environmental stress inducible proteins of Alfalfa. *J. Biol. Chem.* **267** : 15367-15374.

Lutts S, Kinet JM and Boucharmont J (1991). Ethylene production by leaves of rice (*Oryza sativa* L.) in relation to salinity tolerance and exogenous putrescine accumulation. *Plant Sci.,* **116** : 15-25.

McKersie BD, Bowley SR and Jones KS (1999). Winter survival of transgenic alfalfa overexpressing superoxide dismutase. *Plant Physiol.* **119** : 839-847.

Michalwski CB and Bohnert HJ (1992). Nucleotide sequence of a root specific transcript encoding a germin like protein from the halophyte *Mesembryanthemum crystallinum. Plant Physiol.* **100** : 537-538.

Monroy AF, Sangwan V and Dhindsa RS (1998). Low temperature signal transduction during cold acclimation: Protein phosphatase 2A as an early target for cold inactivation. *Plant J.* **13** : 653-660.

Monroy AF, Sarhan F and Dhindsa, RS (1993). Cold induced changes in freezing tolerance protein phopshorylation and gene expression. Evidence for a role of calcium. *Plant Physiol.* **102** : 1227-1235.

Montero E, Cabot C, Barcelo J and Poschenrieder C (1997). Endogenous abscisic acid levels are linked to decreased growth of bushbean plants treated with NaCl. *Physiol. Plant.* **101** : 1722.

Pareek A, Singla SL and Grover I (1995). Immunological evidence for accumulation of two high molecular weight (104 and 90 kDa) HSPs in response to different stresses in rice in response to high temperature stress in diverse plant genera. *Plant Mol. Biol.* **29** : 293-301.

Parsell DA, Kowel A, Singer MA and Lindquist S (1994). Protein disaggregation mediated by heat shock protein HSP104. *Nature* **372** : 475-478.

Peng Z, Lu Q and Verma DPS (1996). Reciprocal regulation of Δ_1-pyrroline-5 carboxylate synthetase and proline dehydrogenase genes control proline levels during and after osmotic stress in plants. *Mol. Gen Gent.* **253** : 334-341.

Pennazio S and Roggero P (1990). Exogenous polyamines stimulate ethylene biosynthesis by soybean leaf tissues. *Ann. Bot.* **65** : 45-50.

Polisensky DH and Braam J (1996). Cold shock regulation of the *Arabidopsis TCH* genes and the effects of modulating intracellular calcium levels. *Plant Physiol.* **111** : 1271-1279.

Prasad TK, Anderson MD, Martin BA and Stewart CR (1994). Evidence for chilling induced oxidative stress in maize seedlings and a regulatory role for hydrogen peroxide. *Plant Cell* **6** : 65-74.

Prasad TK, Anderson MD and Stewart CR (1995) Localization and characterization of peroxidases in the mitochondria of chilling-acclimated maize seedlings. *Plant Physiol.* **108** : 1597-1605.

Queitsch C, Hong SW, Vierling E and Lindquist S (2000) Heat shock protein plays a critical role in the thermotolerance in *Arabidopsis. Plant Cell* **12** : 479-492.

Polidoros AN and Scandalios JG (1999) Role of hydrogen peroxide and different classes of antioxidants in the regulation of catalase and glutathione S-transferase gene expression in maize (*Zea mays*). *Physiol. Plant.* **106** : 112-120.

Raymond P, Weber H, Damond M and Farmer FE (2000) Differential gene expression in response to mechanical wounding and insect feeding in *Arabidopsis. Plant Cell* **12** : 707-799.

Rauser WE (1995) Phytochelatins and related peptides. *Plant Physiol.* **109** : 1141-1149.

Richard S, Morency MJ, Drevet C, Jouanin L and Seguin A (2000) Isolation and characterization of a dehydrin gene from white spruce induced upon wounding, drought and cold stresses. *Plant Mol. Biol.,* **43** : 1-10.

Rosemann D, Heller W and Sandermann Jr H (1991) Biochemical plant responses to ozone II. Induction of stilbene biosynthesis in Scots pine (*Pinus sylvestris* L.) seedlings. *Plant Physiol.* **97** : 1280-1286.

Rouse DT, Marotta R and Parish RW (1996) Promoter and expression studies on an *Arabidopsis thaliana* dehydrin gene. *FEBS Lett.* **381** : 252-256.

Sabehat A, Weiss D and Lurie S (1998) Heat shock proteins and cross tolerance in plants. *Physiol. Plant.,* **103** : 437-441.

Savoure A, Jaqua S, Hua XJ, Ardiles W, Van Montagu M and Verbruggen N (1991) Isolation, characterization and chromosomal location of a gene encoding the Δ_1-pyrolline-5-carboxylase synthetase in *Arabidopsis thaliana. FEBS Lett.* **372** : 13-19.

Schoffl F, Prandl R and Reindl A (1998) Regulation of heat shock response. *Plant Physiol.* **117** : 1135-1145.

Schopnner A and Kindl H (1979) Stilbene synthase (pinosylvin synthase) and its induction by ultra-violet light. *FEBS Lett.* **108** : 349-352.

Sebastiani L, Kindberg S and Vitagliano C (1999) Cytoplasmic free Ca^{2+} dynamics in single tomato (*Lycopersicon esculentum*) protoplasts subjected to chilling temperatures. *Physiol. Plant.* **105** : 239-245.

Shirasu K, Nakajima H, Rajasekhar VK, Dixon RA and Lamb C (1997) Salicylic acid potentiates an agonist-dependent gain control that amplifies pathogen signals in the activation of defense mechanisms. *Plant Cell* **9** : 1315-1324.

Srivastava HS (1999) Biochemical defense mechanisms of plants to ozone and other atmospheric pollutants. *Curr. Sci.* **76** : 525- 533.

Stange C, Ramirez I, Gomez I, Jordana X and Holuigue L (1997) Phosphorylation of nuclear proteins directs binding to salicylic acid -responsive elements. *Plant J.* **11** : 1315-1324.

Staxen R, Pical C, Montgomery LT, Gray JE, Hetherington AM and McAinsh MR (1999) Abscisic acid induces oscillations in guard cell cytosolic free calcium that involve phosphoinositide -specific phospholipase C. *Proc. Natl. Acad. Sci. USA* **96** : 1779-1784.

Sumaryati S, Negrutiu I and Jacobs M (1992) Characterization and regeneration of salt and water stress mutants from protoplast culture of *Nicotiana plumbaginifolia* (viviana). *Theor. Appl. Genet.* **83** : 613-619.

Suzuki K, Fukuda Y and Shinshi H (1995) Studies on elicitor-signal transduction leading to differential expression of defense genes in cultured tobacco cells. *Plant Cell Physiol.* **36** : 281-289.

Takezawa D (2000) Arapid induction of elicitors of mRNA encoding CCD-1, a14 kDa Ca^{2+} -binding protein in wheat cultured cells. *Plant Mol. Biol.* **42** : 807-817.

Tamai T, Inoue M, Sugimoto T, Sueyosi K, Shiraishi N and Oji Y (1999) Ethylene induced putrescine accumulation modulates K^+ partitioning beteen roots and shoots in barley seedlings. *Physiol. Plant.* **106** : 296-301.

Thomashow MF (1990) Molecular genetics of cold acclimation in higher plants. *Adv. Genet.,* **28** : 99-131.

Thomashow MF (1998) Role of cold responsive genes in plant freezing tolerance. *Plant Physiol.* **118** : 1-7.

Trewavas AJ (1999) Le calcium, c'est la vie: Calcium makes waves. *Plant Physiol.* **120** : 1-6.

Trofimova MS, Andreev IM and Kuznetsov VV (1999) Calcium is involved in regulation of synthesis of HSPs in supension -cultured sugar beet cells under hyperthermia. *Physiol. Plant.* **105** : 67-73.

Vatamanuk OK, Mari S, Lu YP and Rea PJ (1999) At PCS1, a phytochelatin synthetase from *Arabidopsis* : isolation and *in vitro* reconstitution. *Proc. Natl. Avad. Sci. USA,* **96** : 7110-7115.

Verbruggen N, Villaroel R and Van Montagu M (1993) Osmoregulation of proline Δ_1-pyrolline-5-carboxylase reductase gene in *Arabidopsis thaliana. Plant Physiol.,* **103** : 771-788.

Vierling E (1991) The roles of heat shock proteins in plants. *Annu. Rev. Plant Physiol. Plant Mol. Biol.,* **42** : 579-620.

Willekens H, Inze D, Van Montagu M and Van Camp W (1995) Catalases in plants. *Mol. Breed.* **1** : 207-228.

Yeh S, Moffatt BA, Griffith M, Xiong F, Yang DSC, Wiseman SB, Sarhan F, Danyluk J, Xue YQ, Hew CL, Dohrtey- Kirkby A and Lajoie G (2000) Chitinase genes responsive to cold encode antifreeze proteins in winter cereals. *Plant Physiol.* **124** : 1251- 1263.

Yoshiba Y, Kiyosue T, Katagiri T, Ureda H, Mizoguchi T, Yamaguchi-Shinozaki K, Wada K, Harada Y and Shinozaki K (1995) Correlation between the induction of gene for Δ_1-pyrolline -5-carboxylase synthetase and the accumulation of proline in *Arabidopsis thaliana* under osmotic stress. *Plant J.* **7** : 751-763.

Zhang S and Klessig DF (1997) Salicylic acid activates a 48-kD MAP kinase in tobacco. *Plant Cell* **9** : 809-824.

Zhu YL, Pilon-Smits ELH, Tarun AS, Weber SU, Jouanin L and Terry N (1999) Cadmium tolerance and accumulation in Indian mustards enhanced by overexpressingg-glutamylcysteine synthetase. *Plant Physiol.* **121** : 1169-1177.

Zinser C, Ernst D and Sandermann Jr H (1998) Induction of stilbene synthase and cinnamoyl alcohol dehydrogenase mRNAs in Scots pine (*Pinus sylvestris* L.) seedlings. *Planta* **204** : 169-176.

Chapter 6

ENGINEERING ABIOTIC STRESS TOLERANCE IN CROP PLANTS

Linagraj Sahoo, Twinkle Sugla, Anila Baloda, Rana P Singh and Pawan K Jaiwal*

Department of Biosciences, MD University, Rohtak - 124001, India

Summary

Plants encounter a wide range of environmental stresses that detrimentally affect the agricultural productivity. Involvement of seemingly complex signal transduction network, elicited by abiotic stresses, has been the area of arduous exploration. Identification of a numerous genes and putative functional transcripts associated with threshold tolerance to abiotic stresses has given new dimension to the rational engineering of crop plants against the environmental challenges. A host range of signal transduction pathways with potential cross-talk flux and specific responses is discussed. Paradoxical functional genomics in model plants Arabidopsis and tobacco is innumerated and comprehensive approach to develop transgenics of crop plants with tolerance to abiotic stresses is presented. The future concern on field performance of transgenics under drastic environmental stresses has been discussed.

Keywords : Environmental stresses, osmolytes, scavenging systems, stress responsive genes, signal transduction, transcription factor

---

*Corresponding author : E-mail : pkjaiwal@yahoo.com

1. INTRODUCTION

Growth and development of crop plants are adversely affected by an array of abiotic stresses, which include drought, desiccation, hypoxia/ anoxia, temperature extremes, heavy metal toxicity etc. (Grover *et al.*, 2001), leading to substantial reduction in crop productivity. Improvement of crop plants for the tolerance to abiotic stresses has been the major thrust of research. Classical agronomic strategies based on genetic variations, inter-specific or inter-generic hybridization, induced mutations and somaclonal variations enjoyed notable progress, however, introgression of complex traits often limited due to their complex polygenic nature, absence of efficient selection techniques and low genetic variance of yield components (Ashraf, 1994; Bohnert *et al.*, 1995; Yeo, 1998; Frova *et al.*, 1999; Jain *et al.*, 2002).

Armoured with the cutting edge tools and techniques, genetic engineering approach has been perfected to a great extent in several crops (Sahoo *et al.*, 2001) and the first major breakthrough was transgenic tobacco with increased low temperature tolerance (Murata *et al.*, 1992). Since then, significant progress has been made in understanding the prime targets for manipulation of stress tolerance in plants with the impetus from functional genomics research unraveling the molecular mechanism(s) and candidate gene(s) involved in such complex traits.

2. SIGNAL SENSING AND TRANSDUCTION

Plants exhibit a variety of responses to abiotic stresses that enable them to tolerate and survive adverse conditions through both physical adaptations and interactive molecular and cellular changes that begin after the onset of stress (Knight and Knight, 2001; Xiong *et al.*, 2001). The molecular responses begin with perceival of stress as it occurs and subsequently the information is relayed through a signal transduction pathway. These pathways eventually lead to physiological changes or expression of genes and resultant modification of molecular and cellular processes.

Dissection of abiotic stress signaling networks in plants has been addressed under conditions of specific stresses as well as multiple stresses. The stresses being clearly different from each other in their physical nature, each elicits specific plant responses, they also activate some common reactions in plants (Shanker and Srivastava, 2002; Zhu, 2001). Increasing evidence implicates the existence of a general stress response system in plants, and overlapping responses to different environmental stresses may be mediated by common cellular signal transduction pathways (Hare *et al.*, 1999). Consequently there exists a complex signaling network underlying plant adaptation to these adverse conditions.

Apparently these signaling pathways clearly divided into two main categories, those that link a particular stimulus to a particular end response are linear and specific whereas others that constitute a network, interconnected at many levels. Thus, opportunity for both cross-talk and specificity can occur within a particular signaling pathway (Knight and Knight, 2001).

3. OPPORTUNITY FOR CROSS-TALK

3.1. ABA dependent and ABA-independent pathways

One important regulator of plant responses to abiotic stress environments is the phytohormone abscisic acid (ABA). ABA is involved in plant responses to abiotic stresses such as low temperature, drought, and salinity (Koornneef *et al.*, 1998; Leung and Giraudat, 1998; Mc Court, 1999; Rock, 2000). Evidences for a role of ABA in stress responsive gene regulation in plants have been in two ways. First, under cold, drought, or salt stress conditions, plants accumulate increased amounts of ABA, with drought stress having the most prominent effect on ABA accumulation. Second, the expression of many stress-responsive genes is induced by exogenous ABA, and their stress inducibility is decreased in mutant plants defective in ABA biosynthesis or responsiveness (Xiong *et al.*, 2001). Plant species, such as *Arabidopsis*, that normally experience freezing temperatures during vegetative growth, have the ability to increase their freezing tolerance by a process called cold acclimation (for review see Thomashow, 1999). The use of *Arabidopsis* ABA-deficient-mutants along with ABA-responsive mutants in stress gene regulation studies led to the notion that stress-responsive gene expression in plants is mediated by both ABA-dependent and ABA-independent pathways (Shinozaki and Yamaguchi-Shinozaki, 1997; Leung and Giraudat, 1998; Thomashow, 1999; Rock, 2000). Although the molecular mechanisms underlying the difference between ABA-dependent and ABA-independent gene regulation are unclear, analysis of promoters of stress-responsive genes and the isolation of transcription factors that activate these genes support the notion that there are distinct regulatory mechanisms for the different pathways. The *ABRE* (ABA-responsive element) complex in these promoters mediates gene induction by ABA

(Guiltinan *et al.*, 1990; Yamaguchi-Shinozaki and Shinozaki, 1994; Shen and Ho, 1995; Vasil *et al.*, 1995), whereas the *DRE/CRT* (dehydration-responsive element) mediates cold and osmotic stress responsiveness independently of ABA (Yamaguchi-Shinozaki and Shinozaki, 1994; Stockinger *et al.*, 1997). Despite these differences in transcriptional activation, genetic analysis has indicated that the ABA-dependent and ABA-independent pathways have extensive interactions in controlling gene expression under abiotic stress (Ishitani *et al.*, 1997; Xiong *et al.*, 1999a).

Extensive studies with *aba1* (ABA-deficient) or *abi1/2* (ABA-insensitive) mutants have yielded considerable, yet sometimes conflicting, information. For example, Savoure *et al.* (1997) reported that the expression of *P5CS* genes is independent of ABA, because they observed that the expression level is similar in the wild type and in *aba1* under cold or drought treatments. They suggested that ABA might affect proline biosynthesis post-transcriptionally (Savoure *et al.*, 1997). On the other hand, Yoshiba *et al.* (1999) reported that *AtP5CS1* induction by drought and salt stress is regulated by both ABA-dependent and ABA-independent pathways.

Although low temperature treatment can trigger a transient increase in ABA and application in ABA can induce the expression of cold-responsive genes at warm temperatures and increase plant freezing tolerance, a general consensus is that ABA does not have an important role in regulating the expression of the *DRE/CRT* class of genes (Thomashow, 1999).

In contrast to low temperature, drought stress can dramatically stimulate *de novo* ABA biosynthesis; thus ABA is more closely involved in drought/salt stress responses (Bray, 1993; Ingram and Bartel, 1996).

The genetic analysis using *RD29A-LUC* as a molecular marker (Xiong *et al.*, 2001) has shown that ABA-dependent and ABA-independent signaling pathways may not function independently of each other, rather, there exist extensive connections between them (Ishitani *et al.*, 1997; Xiong *et al.*, 1999a).

Liu *et al.* (1998), suggested that DREB2A activity in activating stress-responsive genes may require post-transcriptional modifications. Thus it is possible that phosphorylation/dephosphorylation of DREB2A or the functions of its cofactors may be dependent on ABA-regulated molecules such as ABI1, ABI2, Ca^{2+}-dependent protein kinases, or numerous other ABA-responsive regulatory factors (Leung *et al.*, 1997; Leung and Giraudat, 1998; Finkelstein and Lynch, 2000; Rock, 2000; Merlot *et al.*, 2001). This interdependence of ABA and stress signaling may underlie the mechanisms for the synergistic effect of ABA and drought/salt stress on the regulation of stress responsive genes (Xiong *et al.*, 2001; Bostock and Quatrano, 1992; Xiong *et al.*, 1999b).

3.2. Scavenging systems

The formation of oxygen radicals by partial reduction of molecular oxygen is an unfortunate consequence of aerobic life. Active oxygen species (AOS), such as the superoxide anion and hydrogen peroxide (H_2O_2) are natural by-products of metabolism but they can accumulate to toxic levels (oxidative stress) during a wide range of environmental stresses (such as chilling, ozone, drought, salt stress). Adequate defense responses to environmental changes that provoke oxidative stress in plants are obviously crucial for plant growth and survival. However, the molecular and biochemical mechanisms that orchestrate these response are still poorly understood and the signaling networks involved remain elusive. A central role for reactive oxygen species (ROS) during both biotic and abiotic stress responses is well recognized, although under these situations ROS can either exacerbate damage or act as signal molecules that activate multiple defense responses. Aerobic organisms have a battery of enzymatic and non-enzymatic antioxidants that scavenge AOS. Superoxide dismuatse (SOD, ascorbate peroxidaes (APX) and catalses are the main players within the enzymatic defense system. Evidences pooled up have demonstrated the importance of these enzymes in the protection of plants against environmental stress (Van Breusegam *et al.*, 1998). Elucidating the molecular details of plant antioxidant defenses and subsequently applying this fundamental knowledge to economically important crops has taken a momentum. The chloroplasts are the main site of oxidative damage and the working hypothesis is hence to generate lines with an improved scavenging system for oxygen radicals in the chloroplasts would be better protected against oxidative stress, more specifically during chilling stress. Transgenic maize lines overproducing Manganese (Mn) SOD (from *N. plumbaginifolia*) were more tolerant to the superoxide-generating herbicide methyl viologen (paraquat) showing an increased antioxidant capacity in the chloroplasts (Van Breusegem *et al.*, 1999a) Transgenic lines with higher Iron (Fe) SOD (from *A. thaliana*) activities were also more tolerant towards methyl viologen and in addition they have increased growth rates during chilling stress (Van Breusegem *et al.*, 1999b). In contrary to the knowledge on the antioxidant defense systems, information on oxidative stress signal transduction in plants is scarce. A central role for AOS themselves (with H_2O_2 playing an important role) is becoming recognized in the signal transduction cascade of defense responses against pathogen attack and against abiotic stress (Van Camp *et al.*, 1998). The active oxygen species (AOS) have entrusted as second messengers in plant signal transduction and cell death (Dat *et al.*, 2000).

3.3. Calcium signaling

Calcium signaling has been implicated in the transduction of drought- and salt-stress signals in plants and may play a role in a number of responses to drought and changes in water potential (Davies *et al.*, 1981; Johansson *et al.*, 1996; Takahashi *et al.*, 1997). The discovery of drought- and high salt-induced expression of calcium-dependent protein kinases (Wimmers *et al.*, 1992; Urao *et al.*, 1994), hyperosmotic shock-induction of putative calcium-binding proteins (Ko and Lee, 1995; Pestacz and Erdei, 1996), and salt stress-induced expression of a putative Ca^{2+}-ATPase (Perez-Prat *et al.*, 1992) in plants and algae provides indirect evidence of the importance of calcium in these responses. The observation that salt stress causes elevations in intracellular calcium (Lynch *et al.*, 1989) in isolated maize root protoplasts constitutes more direct evidence. Knight *et al.* (1997) suggested a role for intracellular calcium as second messenger in the signaling pathways leading to the induction of at least three genes (*p5cs*, *Iti78* and *rab18*) in response to salinity stress (*p5cs*), and that influx of external calcium plays an important part in these events. Attempt to discriminate between drought and salinity stress (Knight *et al.*, 1997), which is known to occur in plants (Tsiantis *et al.*, 1996) possibly mediated via calcium signaling in *Arabidopsis*, indicated that some factor other than calcium must be involved in the transformation of and discrimination between salinity and drought signals in *Arabidopsis*. Discrimination between these two signals could occur via long-term signal transduction events rather than the short-term $[Ca^{2+}]_{cyt}$ elevation. The discrimination between mannitol and salt in *Arabidopsis* is suggested to be mediated by a calcium-independent signaling factors(s). A number of osmotically induced genes have been shown to be expressed preferentially in response to either drought or salt stress in wheat, tomato and *Mesembryanthemum crystallinum* (Erdei *et al.*, 1990; Galiba *et al.*, 1993; Chen and Tabaeizadeh, 1992; Tsiantis *et al.*, 1996), indicating that different pathways may be involved in the transformation of these two stresses in these particular species. The vacuole has previously been implicated in the perception of osmotic stress (Alexandre and Lassalles,

1991) and it has already been shown that osmotic stress enhances the competence of *Beta vulgaris* IP_3-sensitive vacuolar calcium channels to respond to IP_3 (Allen and Sanders, 1994a). The considerations make vacuolar calcium a strong candidate for organelle-specific signaling in response to osmotic stress.

Similarly calcium and several genes encoding signaling proteins (MAP-kinases, CDPK, and calmodulin regulated kinases) have been found to be upregulated in response to low temperatures (Tahtiharju *et al.*, 1997; Monroy *et al.*, 1998; Jonak *et al.*, 1996; Knight *et al.*, 1996; Polisensky and Braam, 1996). In addition, a putative negative regulator loci (HOS1; Ishitani *et al.*, 1998; Ishitani *et al.*, 1997) has been identified by a genetic approach and is assumed to play a triggering role in signaling, low temperature, and ABA.

The histidine kinase that functions as a sensor molecule, in bacteria, transduces extracellular signals to the cytoplasm. The simple signaling unit is called a two-component system, which essentially involves two types of proteins, a sensory histidine kinase and a response regulator. The cDNA encoding a novel hybrid-type histidine kinase, *ATHK1* cloned from dehydrated *Arabidopsis* (Urao *et al.*, 1999) is transcriptionally upregulated in response to changes in external osmolarity. The *ATHK1* has been shown to possess the sequence similarity with the previously known osmosensor (*SLN1*) from yeast (Urao *et al.*, 1999), and functions as an osmosensor and transmits the signal to a downstream protein phosphorylation cascade, probably MAPK cascade. A MAPK cascade is also thought to be involved in osmotic stress response in higher plants (Hirt, 1997; Mizoguchi *et al.*, 1997). Expression of genes for MAPKs of various plants, such as *ATMPK3* in *Arabidopsis* and *MMK4* in alfalfa, is transcriptionally induced by drought, low temperature, and mechanical stress. More over, two genes for protein kinases, MAPKKK (*ATMEKK1*) and ribosomal S6 kinase (*ATPK19*) are induced by similar stresses (Mizoguchi *et al.*, 1996).

The mRNAs of several genes involved in signal transduction pathways, such as protein kinases, protein phosphatses, a phospholipase C, calmodulins, monomeric small GTP binding proteins, and transcription factors, also accumulate in response to environmental stimuli or stresses (Shinozaki and Yamaguchi-Shinozaki, 1996, 1997). Under stress conditions, it appears that the transcriptional regulation of genes for such signal transducers confers higher sensitivity and signaling efficiency to the cellular transducing processes.

Though soil salinity creates both ionic and osmotic stress for plants, the ionic aspect of salt stress is clearly distinct from other abiotic stresses such as drought and cold, and there are signaling pathways dedicated specifically to deal with ionic stress (i.e. excess sodium and associated potassium deficiency (Zhu, 2001). The molecular analysis of diverse stress gene regulation patterns has suggested a network of multiple signaling pathways that mediate salt, water and cold stress responses in plants. Genetic dissection of salt tolerance in *Arabidopsis* established the involvement of the SOS (salt-overly-sensitive) pathway in the response to the ionic aspect of the salt stress (Zhu, 2001). This novel protein kinase pathway is activated by calcium signaling and regulates ion transporters, which bring about homeostasis. Biochemical studies identified several protein kinases that are activated by the osmotic aspect of salt and water stresses. Cold induction of gene expression and acclimation is interveined with salt, water and ABA regulation.

Modulation of signaling regulators for improving stress tolerance in plants was successfully demonstrated in *Arabidopsis* for raising salt tolerant transgenic plants by altering the expression of a calcium stress signaling component, Ca^{2+}/calmodulin-dependent protein phosphatase calcineurin (PP2B) (Pardo *et al.*, 1998).

The transgenic tobacco plants, co-expressing the two catalytic and regulatory subunits of this protein (assuming the reconstitution of the two domains *in vivo*) were found to exhibit substantial NaCl tolerance linked with activated CaN (Pardo *et al.*, 1998). Remarkably, CaN functions mainly in roots to regulate the movement of ions from the apoplast to symplast, which controls the ion content of the xylem sap, which in turn, is transported to the shoots by the strength of the transpirational sink.

The calcineurin B-subunit (regulatory) like Ca^{2+} sensor, SOS3 (for salt overly sensitive 3) has been found to physically interact with and to activate a protein kinase SOS2 (Halfter *et al.*, 2000). The *SOS3* gene is predicted to encode a calcium binding protein with an N-myristoylation signature sequence. Ishitani *et al.* (2000) have shown that both N-myristoylation and calcium binding are crucial for SOS3 function in plant salt tolerance. The *SOS2* gene has been identified as a *serine/threonine* type protein kinase (Liu *et al.*, 2000), which forms another potential candidate for engineering abiotic stress tolerance. The functional domains in the protein kinase SOS2 that interacts with SOS3 and is required for plant salt tolerance have also been identified (Guo *et al.*, 2001). SOS3 interaction with and activation of SOS2 kinase is very consistent with genetic evidence that the two genes are both the positive regulators of salt tolerance and function in the same pathway (Hafter *et al.*, 2000). The SOS3 binds to accumulated cytosolic free Ca^{2+}, triggered by Na^+ stress, and activates the protein kinase SOS2. Activated SOS3-SOS2 kinase complex is necessary for increased expression of SOS1-a putative Na^+/H^+ antiporter in plasma membrane (Shi *et al.*, 2000), and perhaps the other transporter genes under salt stress. The SOS3/SOS2 pathway may also regulate the activities of SOS1 and other transporters at the post-translational level. This gene expression and transporter activity regulation brings about homeostasis of ions such as Na^+ and K^+ and consequently plant tolerance to Na^+ stress (Zhu, 2000).

The Ca^{2+} influx in response to multiple stress stimuli, including cold, salt and drought, emanates signals that are likely to be mediated by combinations of protein phosphorylation/dephosphorylation cascades. Presumably, majority of Ca^{2+} stimulated protein phosphorylation is performed predominantly by members of the Ca^{2+}-dependent protein kinases (CDPK) family in plants. The MAP kinase pathways are intracellular signal modules that mediate signal transduction from the cell surface to the nucleus (for review see Robinson and Cobb, 1997) and seem to be widely used as osmolarity signaling modules. The core MAPK cascades consist of 3 kinases that are activated sequentially by an upstream kinase. The MAP kinase kinase kinase (MAPKKK), upon activation, phosphorylates a MAP kinase kinase (MAPKK) on serine and threonine residues. This dual-specificity MAPKK in turn phosphorylates a MAP kinase (MAPK) on conserved tyrosene and threonine residues. The activated MAPK can then either migrate to the nucleus to activate transcription factor directly, or activate additional signal components to regulate gene expression, cytoskeleton-associated proteins or enzyme activities, or target certain signal proteins for degradation.

Transgenic rice plants with altered levels of OsCDPK7 have been shown to tolerate considerably to cold and salt/drought stresses (Saijo *et al.*, 2000). Further, overexpression of this protein in the transgenic rice enhanced induction of some-stress responsive genes in response to salinity/ drought, but not to cold. Thus the downstream pathways leading to cold and salt/drought tolerance are different from each other. Further, at least two disnict pathways commonly use a single CDPK, maintaining the signaling specificity through unknown post-translational regulation mechanisms. Thus, a simple manipulation of CDPK activity has great potential with regard to stress tolerance.

The H_2O_2, a central signaling molecule in stress and wounding responses, is a potent activator of cascades of mitogen-activated protein kinase (MAPKs). Hydrogen peroxide, in *Arabidopsis* leaf cells, more specifically activated the *Arabidopsis* NPK1 (*Nicotiana* protein kinase, which is mitogen activated protein kinase kinase kinase i.e. MAPKKK)-like protein kinase (ANP1), which then initiates a phosphorylation cascade involving MAPKs, causing the activation of stress responsive genes. Some of these genes code for the heat shock proteins and detoxification enzymes. Therefore, manipulation of key regulators of an oxidative stress signaling pathway, such as ANP1, provides a strategy for engineering multiple stress tolerance. Tobacco plants overexpressing an ANP1 homologue, NPK1 were found to tolerate heat, freezing, drought and high salt conditions (Kovtun *et al.*, 2000).

Different MAPK pathways may share common components; activation of one pathway may not necessarily affect another pathway. In alfalfa plants, a MAPK was activated within 10 mi of cold treatment. It was also activated by drought stress as well as mechanical stress, but not by heat, salt stress or exogenous ABA (Jonak *et al.*, 1996), suggesting that this MAPK mediates drought and cold signaling via an ABA-independent pathway. A salicyclic acid-induced protein kinase (SIPK) belonging to the MAP kinase family that is activated by salicyclic acid, pathogen attack and wounding was also found to be activated within 5-10 min after osmotic stress (Mikolajczyk *et al.*, 2000). Similarly, in tobacco cells, the S1PK (a MAPK) and another protein kinase, HOSAK, were activated by osmotic stress and this activation is independent of Ca^{2+} or ABA (Hoyos and Zhang, 2000). In *Arabidopsis*, the transcription of a MAPK gene, *ATMPK3*, is induced by drought, low temperature, salinity and touch (Mizoguchi *et al.*, 1996). To further explore the role of MAPK module in stress signaling in plants, Kovtun *et al.* (2000) over expressed a tobacco ANP ortholog, NPK1, which activates the H_2O_2-regulated gene expression in plants, and found that *Arabidopsis* plants overexpressing NPK1 showed an increased tolerance to freezing, heat shock and salt stress.

3.4. Transcription factors

The over-expression of genes for transcription factors is emerging as an proposition to confer stress tolerance. Since the cis-acting promoter sequences of different stress responsive genes induced in response to the same stress are similar to an extent, and thus can be possibly governed at the same time by modulating the transcription factor genes (Thomashow *et al.*, 2001; Grover *et al.*, 2001).

The upstream regulatory elements of some of the stress inducible genes have been reported to contain certain conserved elements, which play a major role in their expression (Singla-Pareek *et al.*, 2001). Dehydration responsive elements (DRE) has been implicated to play an important role in regulating gene expression in response to various stresses (Yamaguchi-Shinozaki and Shinozaki, 1994). Incidentally, the transcription factor DREB1A specifically interacts with the DRE and induces expression of the stress tolerance genes under normal growth conditions and resulted in improved tolerance to drought, salinity and freezing stress (Kasuga *et al.*, 1999). Ectopic expression of a seed specific transcriptional activator-*ABI3* gene confers on *Arabidopsis* vegetative tissues the enhanced ability to accumulate seed–specific transcripts (such as *RAB18* and *LT178*) in response to ABA, and also influences some ABA mediated vegetative responses (Tamminen *et al.*, 2001), further increased tolerance to freezing stress as well as enhanced responsiveness to ABA was observed when these genes were overexpressed.

The overexpression of a cold inducible zinc finger protein from soybean (SCOF-1) resulted in overexpression of *COR* (cold responsive proteins) genes and enhanced cold tolerance of non-acclimatized transgenic *Arabidopsis* and tobacco plants (Kim *et al.*, 2001). The SCOF-1 interacts with another protein SGBF-1, which has a DNA binding activity. Thus SCOF-1 has been assigned to function as a positive regulator of *COR* gene expression, which in turn enhances the tolerance of transgenic plants against cold stress.

The transcription factor CBF1 (CRT/DRE binding factor) has been implicated in the regulation of the *COR* (cold-regulated) genes. Transgenic plants overexpressing CBF1 in *Arabidopsis*, exhibited transcripts greater than the normal amounts of some of the low temperature inducible genes and increased tolerance to the freezing stress in non-acclimated plants (Jaglo-Ottosen *et al.*, 1998). Constitutive expression of transcriptional activator –CBF3 in transgenic *Arabidopsis* plants had elevated levels of proline and total soluble sugars, including glucose, sucrose, raffinose and fructose, which showed increased freezing tolerance of cold-acclimated plants (Gilmour *et al.*, 2000). The transcript levels of proline biosynthetic enzyme P5CS were also found to be enhanced in transgenic plants. It was proposed that CBF3 integrates the activation of multiple components of the cold acclimation response.

The heat shock transcription factor (*HSF*) genes, which regulate heat shock promoter, have been identified, cloned and characterized from diverse systems (Nover *et al.*, 1996; Grover *et al.*, 2001). The constitutive expression of *AtHSF1* in transgenic *Arabidopsis* led to constitutive expression of certain HSPs (Lee *et al.*, 1995) and enhanced thermotolerance. Parallel tolerance was indicated in *Arabidopsis* constitutively overexpressing *HSF3* (Prandl *et al.*, 1998). Another *HSF*, *HSF101* has been identified whose overexpression was instrumental in acquired thermotolerance in *Arabidopsis* (Queitsch *et al.*, 2000) and Lima bean (Keeler *et al.*, 2000).

Overexpression of *Tobacco Tsi1* gene encoding *EREBP/AP*$_2$ type transcription factor enhanced resistance against pathogen attack and osmotic stress in tobacco, which indicates a possible role of *Tsi1* as a transcription factor in two separate signal transduction pathways under biotic and abiotic stresses (Park *et al.*, 2001).

Winicov and Bastola (1999) reported improved salinity tolerance in transgenic alfalfa plants over-expressing the *Alfin1* gene, a putative transcription factor associated with salt-tolerance.

3.5. Compatible solutes

Accumulation of non toxic low molecular weight organic molecules in plants under stress is the most common form of adaption to abiotic stress at cellular level (Hanson *et al.*, 1994). These are termed as compatible solutes, since they do not interfere with regular cellular metabolism even when present at high concentrations (Bohnert and Jensen, 1999). The genes for synthesis/overproduction of compatible solutes viz. proline, glycinebetaine, mannitol, trehalose, D-ononitol and fructans have been successfully used to import tolerance to various stress (Tarczynski *et al.*, 1993; Holmstorm *et al.*, 1994, 1996; KaviKishore *et al.*, 1995; Pilon Smits *et al.*, 1995; Thomas *et al.*, 1995; Lilius *et al.*, 1996; Hayashi *et al.*, 1997; Shen *et al.*, 1997; Sheveleva *et al.*, 1997; Grover *et al.*, 1999; Prasad *et al.*, 2000a, b). Increased levels of these solutes often enhanced the tolerance of the plants to multiple stresses viz. salt, drought and low temperature stress. Opportunity for cross-talk in different signaling pathways involving accumulation of single osmolyte, has been discussed in details in recent reviews (Pardha Saradhi and Sharmila, 2002).

3.6. Specificity

Salt tolerance is a complex trait and the long list of salt-responsive genes seems to support this (Zhu, 2000). Arguably, thought in this direction has concluded that salt tolerance will be achieved only after pyramiding several characteristics in a single genotype, whereas each one alone could not confer a significant increase in salt tolerance (Yeo *et al.*, 1988; Cuartero and Fernandez, 1999). The detrimental effects of salt on plants are a consequence of both a water deficit resulting in osmotic stress and the effects of excess sodium ions on key biochemical processes.

Plants preferentially use three strategies for the maintenance of a low cytosolic sodium concentration: active Na^+ uptake and its compartmentation in the vacuole, restriction of Na^+ influx and active Na^+ efflux (Niu *et al.*, 1995; Serrano *et al.*, 1999). The presence of large, acidic-inside, membrane-bound vacuoles in plant cells allows the efficient compartmentation of sodium into the vacuole through the operation of vacuolar Na^+/H^+ antiports (Apse *et al.*, 1999; Blumwald and Poole, 1985). These antiports use the protonmotive force generated by the vacuolar H^+-translocating enzymes, H^+-adenosine triphosphatase (ATPase), and H^+-inorganic pyrophosphatase (PP_iase) to couple the downhill movement of H^+ (down its electrochemical potential) with the uphill movement of Na^+ (against its electrochemical potential) (Blumwald and Gelli, 1997).

Contrary to the notion that multiple traits needed to be introduced into crop plants to obtain salt-tolerant plants, the modification of a single trait significantly improved salt tolerance implying the specificity of salt tolerance events. The overexpression of *AtNHX1*, a vacuolar Na^+/H^+ antiport from *Arabidopsis thaliana*, in *Arabidopsis* plants allowed the transgenic plants to grow in 200 mM NaCl (Apse *et al.*, 1999). Transgenic salt-tolerant tomato- (Zhang and Blumwald, 2001) and *Brassica napus* (Zhang *et al.*, 2001) plants that over-expressed *AtNHX1* allowed the plants to grow and produce fruits even at 200 mM NaCl. The high sodium and chloride content in the leaves of transgenic plants grown in saline environment demonstrated that enhanced vacuolar accumulation of Na^+ ions, mediated by the Na^+/H^+ antiport, allowed the transgenic plants to ameriolate the toxic effects of Na^+. The seed yield, quality and quantity of oil of transgenic canola plants, grown at 200 mM NaCl, were similar to that of wild type of plants grown at low salinity.

Plants follow some short-term and long-term tolerance strategies to cope to the heavy metal toxicity: avoidance, reduced translocation to aerial parts, chillating by the metal binding peptides, phytochelatins, induction of antioxidant system, volatilization of toxic substances and exudation etc. (Ha *et al.*, 1999; Cobett, 2000).

Hyperaccumulators (plant species that accumulate extremely high concentration of heavy metals in shoots), fast growing species with large biomass production (as the members of *Brassicaceae*, e.g. *Brassica juncea*, *Alyssum* and *Thlaspi* species) with a relatively high trace element accumulation capacity, and trees in particular have emerged as an important tool for the remediation of the contaminated environments due to their extensive root systems that ensure an efficient uptake of the pollutants (cadmium, mercury, nickel and pesticides) from soil and provide a possibility of several cycles of decontamination with same plants.

Glutathione (GSH) and its derivatives play the major role in plant defense against these pollutants. Pesticides are detoxified by conjugation with GSH by glutathione-S-transferase and subsequent excretion of these conjugates in the vacuoles. Heavy metals induce synthesis of a wide range of cystine-rich peptides and proteins including metallothioneins and phytochelatins (PC). The later are synthesized enzymatically from glutathione, bind the metals with high affinity and the PC-metal complexes is sequestered to the vacuoles. Increase in glutathione content in poplars by overexpression of bacterial enzymes for GSH synthesis lead to an increased accumulation of cadmium. Specific glutathione-S-transferase and phytochelatin synthase overexpression in plants alone or in combination with bacterial enzymes for GSH synthesis are emerging as consensus strategy for engineering hyperaccumulator.

Transgenic Indian mustard overexpressing ATP sulfurylase (Pilon-Smits *et al.*, 1999), glutathione synthetase (Zhu *et al.*, 1999a), g-glutamylcysteine synthetase (Zhu *et al.*, 1999b) and glutathione reductase (Pilon-Smits *et al.*, 2000). Many other potential hyperaccumulator plants have also been engineered for enhanced efficiency of phytoremediation.

Transgenic *Arabidopsis* plants expressing bacterial (*merA* for mercuric reductase and *merB* for organomercurial lyase) genes, for an organic mercury detoxification pathway, grow on 50-fold higher methylmercury concentrations than wild-type plants and up to 10-fold higher concentrations than plants that express *merB* alone (Bizily *et al.*, 2000).

Overexpression of a bacterial citrate synthase gene in transgenic tobacco and papaya was reported to confer Al tolerance (de la Fuente *et al.*, 1997). Ezaki *et al.* (2001) characterized the mechanism of action of four transgenes (*AtBCB* [Arabidopsis blue copper-binding protein], *parB* [tobacco {*Nicotiana tabacum*} glutathione S-transferase], *NtPox* [tobacco peroxidase], and *NtGDH* [tobacco GDP dissociation inhibitor]) that independently conferred Al resistance on transgenic *Arabidopsis*. Overexpression of malate dehydrogenase in transgenic alfalfa enhanced organic acid synthesis, which conferred tolerance to Aluminium (Tesfaye *et al.*, 2001).

The ability of *Pelargonium* (scented geranium) to tolerate and accumulate more than one metal pollutant is of major significance, as most known metal accumulators are very metal specific (KrishnaRaj *et al.*, 2000). It is interesting to identify and characterize key steps and factors affecting heavy metal uptake, transport, tolerance, partitioning and accumulation among these rich biodiversity.

The use of plants to extract and detoxify TNT (2,4,6-Trinitrotoluene), one of the most recalcitrant and toxic of all the military explosives, is gaining impetus. The nitroreductase which catalyzes the reduction of TNT to hydroxyaminodinitrotoluene (HADNT), which is subsequently reduced to aminodinitro toluene derivatives (ADNTs). Transgenic plants expressing nitroreductase show a striking increase in ability to tolerate, take up, and detoxify TNT. French *et al.* (1999) demonstrated that the introduction and expression of a bacterial enzyme PETN reductase resulted in the enhancement of tobacco plants' ability to detoxify the explosive nitroglycerin. The expression of the nitroreductase (*nfsI*) gene (from *E. cloacae*) in tobacco lead to tolerance to TNT at concentration up to the aqueous solubility limit of TNT (Hannink *et al.*, 2001). The expression of microbial metal resistance genes and a mammalian cytochrome P450 has enabled plants to transform methylmercury (Bizily *et al.*, 1999) and TCE (Doty, 2000) respectively.

4. CONCLUSIONS AND FUTURE PROSPECTS

Dissection of diverse gene regulation patterns has concluded a network of multiple signaling pathways. Though the molecular components of these regulatory pathways have been identified, to a large extent, with information of large number of genes and proteins that mediate plant abiotic stress responses, understanding in several of these pathways is still scarce, and knowledge in some of them is still at a rudimentary state. Most important task in the coming years remains to identify pathway components and to establish their function by genetic approaches. Powerful molecular techniques such as high-throughput analysis of expressed sequence tags, cDNA microarrays, genome sequencing, T-DNA or transposon insertional mutagenesis, gain-of-function or mutant complementation etc. has accelerated the process of identification of stress-responsive genes by several-fold.

While there are innumerable reports showing production of abiotic stress tolerant transgenics, none of the abiotic stress tolerant transgenic plants has come close to field trials. These are unlikely scenario that with the present day understanding of abiotic stress defense mechanisms and laboratory results will immediately translate into stress tolerant crops standing in adverse field conditions. Transgenic technology combined with the genome sequencing, functional genomics and proteomics has the potential to advance crop breeding beyond imagination. Several gene transfer approaches have been employed to improve the stress tolerance of plants (Holmber and Bulow, 1998) which involves either product of single gene or a product of a regulatory gene (that activates the whole cascade of other gene products in the plant in response to stress). Though genetically engineered plants for single gene products cornered attention which include those encoding for enzymes required for biosynthesis of osmoprotectants (Tarczynski *et al.*, 1993; Kavikishore *et al.*, 1995; Hayashi *et al.*, 1997), or modifying membrane lipids (Kodama *et al.*, 1994; Ishizaki-Nishizawa *et al.*, 1996), LEA protein (Xu *et al.*, 1996), detoxification enzyme (McKersie *et al.*, 1996) and many more, complete tolerance is unlikely with the observed low magnitude of tolerance in most of the cases.

Thus, there should be a comprehensive approach to pyramid a number of genes involved in stress response or simultaneous regulation of many stress responsive genes by using a single gene encoding stress inducible transcription factor (Kasuga *et al.*, 1999), so as to enhance tolerance towards more complex and multiple stresses including drought, salinity, and freezing.

Further, dissection of signaling pathways as well as holistic approach for transgenic with stress tolerance have been restricted mostly to model crops with little economic significance. Crop plants facing dire consequences of hostile environment must be considered to priority to realm the fruitful the demand.

REFERENCES

Alexandre J and Lassalles JP (1991) Hydrostatic and osmotic pressure activated channel in plant vacuole. *Biophys. J.,* **60** : 1326-1336

Allen GJ and Sanders D (1994a) Osmotic stress enhances the competence of *Beta vulgaris* vacuoles to respond to inositol 1,4,5-triphosphate. *Plant J.,* **6** : 687-695

Apse MP, Aharon GS, Snedden WS and Blumwald E (1999) Salt tolerance conferred by overexpression of a vacuolar Na^+/H^+ antiport in *Arabidopsis. Science,* **285** : 1256-1258

Ashraf M (1994) Breeding for salinity tolerance in plants. *Crit. Rev. Plant Sci.,* **13** : 17-42

Bizily SP, Clayton LR and Richard BM (2000) Phytodetoxification of hazardous organomercurials by genetically engineered plants. *Nature Biotechnol.,* **18** : 213-217

Bizily SP, Rugh CL, Summers AO and Meagher RB (1999) Phytoremediation of methylmercury pollution: *merB* expression in *Arabidopsis thaliana* confers resistance to organomercurials. *Proc. Natl. Acad. Sci. USA,* **96** : 6808-6813

Blumwald E and Gelli A (1997) Secondary inorganic ion transport in plant vacuoles. *Adv. Bot. Res.,* **25** : 401-407

Blumwald E and Poole RJ (1985) Na^+/H^+ antiport in isolated tonoplast vesicles from storage tissue of *Beta vulgaris. Plant Physiol.,* **78** : 163-167

Bohnert HJ and Jensen RG (1996) Strategies for engineering water-stress tolerance in plants. *Trends Biotech.,* **14** : 89-97

Bohnert HJ, Nelson DE and Jensen RG (1995) Adaptations to environmental stresses. *Plant Cell,* **7** : 1099-1111

Bostock RM and Quatrano RS (1992) Regulation of *Em* gene expression in rice: Interaction between osmotic stress and abscisic acid. *Plant Physiol.,* **98** : 1356-1363

Bray EA (1993) Molecular responses to water deficit. *Plant Physiol.,* **103** : 1035-1040

Chen RD and Tabaeizadeh Z (1992) Alteration of gene expression in tomato plants (*Lycopersicon esculentum*) by drought and salt stress. *Genome,* **35** : 385-391

Cobbett CS (2000) Phytochelatins and their roles in heavy metal detoxification. *Plant Physiol.* **123** : 825-832

Cuartero J and Fernandez-Munoz R (1999) Tomato and salinity. *Sci Hortic.,* **78** : 83-125

Davies WJ, Wilson JA, Sharp RE and Osonubi O (1981) Control of stomatal behaviour in water stressed plants. In *Stomatal Physiology* (Jarvis P G and Mansfield T A eds) Cambridge, UK: Cambridge University Press, pp. 163-185

De la Fuente JM, Ramirez-Rodriguez V, Cabrera-Ponce JL and Herrera-Estrella L (1997) Aluminium tolerance in transgenic plants by alteration of citrate synthesis. *Science,* **276** : 1566-1568

Doty SL, Shang TQ, Wilson AM, Tangen J, Westergreen AD, Newman LA, Strand SE and Gordon MP (2000) Enhanced metabolism of halogenated hydrocarbons in transgenic plants containing mammalian cytochrome P450 2E1. *Proc. Natl. Acad. Sci. USA,* **97** : 6287-6291

Erdei L, Trivedi S, Takeda K and Matsumoto H (1990) Effects of osmotic and salt stresses on the accumulation of polyamines in leaf segments from wheat varieties differing in salt and drought tolerance. *J. Plant Physiol.,* **13** : 165-168

Ezaki B, Katsuhara M, Kawamura M and Matsumoto H (2001) Different mechanisms of four aluminium (Al)-resistant transgenes for Al toxicity in Arabidopsis. *Plant Physiol.,* **127** : 918-927

Finkelstein RR and Lynch TJ (2000) The *Arabidopsis* abscisic acid response gene *ABI5* encodes a basic leucine zipper transcriptional factor. *Plant Cell,* **12** : 599-609

French CE, Rosser SJ, Davies GJ, Nicklin S and Bruce NC (1999) Biodegradation of explosives by transgenic plants expressing pentaerythritol tetranitrate reductase. *Nature Biotechnol.,* **17** : 491-494

Frova C, Caffulli A and Pallavera E (1999) Mapping quantitative trait loci for tolerance to abiotic stresses in maize. *J. Exp. Bot.,* **49** : 915-929

Galiba G, Kocsy G, Kaur Sawhney R, Sutka J and Galston AW (1993) Chromosomal localization of osmotic and salt stress induced differential alterations in polyamine content in wheat. *Plant Sci.,* **92** : 203-211

Gilmour SJ, Sebolt AM, Salazar MP, Everard JD and Thomashaw MF (2000) Overexpression of the *Arabidopsis* CBF3 transcriptional activator mimics multiple biochemical changes associated with cold acclimation. *Plant Physiol.,* **124** : 1854-1856

Grover A, Kapoor A, Lakshmi OS, Agrawal S, Sahi C, Katiyar-Agrawal S, Agrawal M and Dubey H (2001) Understanding molecular alphabets of the plant abiotic stress responses. *Curr. Sci.,* **80** : 206-216

Guiltinan MJ, Marcotte WR and Quatrano RS (1990) A plant leucine zipper protein that recognizes an abscisic acid response element. *Science,* **250** : 267-271

Guo Y, Halfter U, Ishitani M and Zhu JK (2001) Molecular characterization of functional domains in the protein kinase SOS2 that is required for plant salt tolerance. *Plant Cell,* **13** : 1383-1400

Ha SB, Smith AP, Howden R, Dietrich WM, Bugg S, O'connell MJ, Goldsbrough PB and Cobbett CS (1999) Phytochelatin synthase genes from *Arabidopsis* and the yeast *Schizosaccharomyces pombe. Plant Cell,* **11** : 1153-1163

Halfter U, Ishitani AM and Zhu JK (2000) The *Arabidopsis* SOS2 protein kinase physically interacts with and is activated by the calcium-binding protein SOS3. *Proc Natl Acad Sci USA,* **97** : 3735-3740

Hannink N, Susan JR, French CE, Basran A, Murray JA, Nicklin S and Bruce N C (2001) Phytodetoxification of TNT by transgenic plants expressing a bacterial nitroreductase. *Nature Biotech.,* **19** : 1168-1172

Hanson AD, Rathinasabapathi B and Rivoal J (1994) Osmoprotective compounds in the *plumbaginaceae*: A natural experiment in metabolic engineering of stress tolerance. *Proc. Natl. Acad. Sci. USA,* **91** : 306-310

Hayashi H, Alia, Mustardy L, Deshnium P, Ida M and Murata N (1997) Transformation of *Arabidopsis thaliana* with the cod A gene for choline oxidase: accumulation of glycine betaine and enhanced tolerance to salt and cold stress. *Plant J.,* **12** : 133-142

Hirt H (1997) Multiple roles of MAP kinases in plant signal transduction. *Trends Plant Sci.,* **2** : 11-15

Holmber N and Bulow L (1998) Improving stress tolerance in plants by gene transfer. *Trends Plant Sci.,* **3** : 61-66

Hoyos ME and Zhang S (2000) Calcium-independent activation of salicyclic acid-induced protein kinase and a 40-kilodalton protei kinase by hyperosmotic stress. *Plant Physiol.,* **122** : 1355-1363

Ishitani M, Liu J, Halfter U, Kim CS, Shi W and Zhu JK (2000) SOS3 function in plant salt tolerance requires N-myristoylation and calcium binding. *Plant Cell,* **12** : 1667-1678

Ishitani M, Xiong L, Stevenson B and Zhu JK (1997) Genetic analysis of osmotic and cold stress signal transduction in *Arabidopsis*: Interactions and convergence of abscisic acid –dependent and abscisic acid –independent pathways. *Plant Cell,* **9** : 1935-1949

Ishizaki-Nishizawa O, Fujii T, Azume M, Sekiguchu K, Murata N, Ohtani T and Toguri T (1996) Low temperature resistance of higher plants is significantly enhanced by a nonspecific cyanobacterial destaurase. *Nature Biotechnol.,* **14** : 1003-1006

Jaglo-Ottosen KR, Gilmour SJ, Zarka DG, Schabenberegr O and Thomashow MF (1998) *Arabidopsis* CBF1 overexpression induces COR genes and enhances freezing tolerance. *Science,* **280** : 104-106

Jain RK, Saini N and Jain S (2002) Engineering salinity tolerance in crop plants: A reality. *Physiol. Mol. Biol. Plants* (In press)

Johansson I, Larsson C, Ek B and Kjellbom P (1996) The major integral proteins of spinach leaf plasma membranes are putative aquaporins and are phosphorylated in response to Ca^{2+} and apoplastic water potential. *Plant Cell,* **8** : 1181-1191

Jonak C, Kiegerl K, Ligterink W, Barker PJ, Huskisson NS and Hirt H (1996) Stress signaling in plants: A mitogen-activated protein kinase pathway is activated by cold and drought. *Proc. Natl. Acad. Sci. USA,* **93** : 11274-11279

Kasuga M, Liu Q, Miura S, Yamaguchi-Shinozaki K and Shinozaki K (1999) Improving plant drought, salt and freezing tolerance by gene transfer of a single stress-inducible transcription factor. *Nature Biotechnol.,* **17** : 287-291

Kavi Kishore PB, Hong Z, Miao GH, Hu CA and Verma DPS (1995) Overexpression of pyrroline-5-carboxylate synthatase increases proline production and confers osmotolerance in transgenic plants. *Plant Physiol.,* **108** : 1387-1394

Keeler SJ, Boettger CM, Haynes JG, Kuches KA, Johnson MM, Thureen DL, Keeler CL and Kitto SL (2000) Acquired thermotolerance and expression of the HSP100/C1pB genes of Limabean. *Plant Physiol.,* **123** : 1121-1132

Kim SA, Kwak JM, Jae SK, Wang MH and Nam HG (2001) Overexpression of the AtGluR2 gene encoding an *Arabidopsis* homolog of mammalian glutamate receptors impairs calcium utilization and sensitivity to ionic stress in transgenic plants. *Plant Cell Physiol.,* **42** : 72-84

Knight H and Knight MR (2001) Abiotic stress signaling pathways: specificity and cross-talk. *Trends Plant Sci.,* **6** : 262-267

Knight H, Trewavas AJ and Knight MR (1997) Calcium signaling in *Arabidopsis thaliana* responding to drought and salinity. *Plant J.,* **12** : 1067-1078

Ko JH and Lee SH (1995) Role of calcium in the osmoregulation under salt stress in *Duniella salina. J. Plant Biol.,* **38** : 243-250

Kodama H, Hamada T, Horiguchi G, Nishimura M and Iba K (1994) Genetic enhancement of cold tolerance by expression of a gene for chloroplast Ω-3 fatty acid desaturase in transgenic tobacco. *Plant Physiol.,* **105** : 601-605

Koornneef M, Leon-Kloosterziel K, Schwartz SH and Zeevaart JAD (1998) The genetic and molecular dissection of abscisic acid biosynthesis and signal transduction in *Arabidopsis. Plant Physiol. Biochem.,* **36** : 83-89

Kovtun Y, Chiu WL, Tena G and Sheen J (2000) Functional analysis of oxidative stress-activated mitogen-activated protein kinase cascade in plants. *Proc. Natl. Acad. Sci. USA,* **97** : 2940-2945

KrishnaRaj S, Dan TV and Saxena PK (2000) A fragrant solution to soil remediation. *Int. J. Phytorem.,* **2** : 117-132

Lee JH, Hubel A and Schoffl F (1995) Derepression of the activity of genetically engineered heat shock protein and increased thermotolerance in transgenic *Arabidopsis. Plant J.,* **8** : 603-612

Leung J and Giraudat J (1998) Abscisic acid signal transduction. *Annu. Rev. Plant Physiol. Plant Mol. Biol.,* **49** : 199-222

Leung J, Merlot S and Giraudat J (1997) The Arabidopsis *ABSCISIC ACID-INSENSITIVE 2* (*ABI2*) and *ABI1* genes encode homologous protein phosphatase 2C involved in abscisic acid signal transduction. *Plant Cell,* **9** : 759-771

Lilius G, Holmberg N and Bulow L (1996) Enhanced NaCl stress tolerance in transgenic tobacco expressing bacterial choline dehydrogenase. *BioTechnol.,* **14** : 177-180

Liu J, Ishitani M, Halfter U, Kim CS and Zhu ZK (2000) The *Arabidopsis thaliana* SOS2 gene encodes a protein kinase that is required for salt tolerance. *Proc. Natl. Acad. Sci. USA,* **97** : 3730-3734

Liu Q, Kasuga M, Sakuma Y, Abe H, Miura S, Yamaguchi-Shinozaki K and Shinozaki K (1998) Two transcription factors, DREB1 and DREB2, with an EREBP/AP2 DNA binding domain separate two cellular signal transduction pathways in drought- and low-temperature-responsive gene expression, respectively, in *Arabidopsis. Plant Cell,* **10** : 1391-1406

Lynch J, Polito VS and Lauchli A (1989) Salinity stress increases cytoplasmic Ca activity in maize root protoplasts. *Plant Physiol.,* **90** : 1271-1274

McCourt P (1999) Genetic analysis of hormone signaling. *Annu. Rev. Plant Physiol. Plant Mol. Biol.,* **50** : 219-243

Mckersie BD, Bowley SR, Harjanto E and Leprince O (1996) Water deficit tolerance and field performance of transgenic alfalfa overexpressing superoxide dismutase. *Plant Physiol.,* **111** : 1177-1181

Merlot S, Costi F, Guerrier D, Vavasseur A and Giraudat J (2001) The ABI1 and ABI2 protein phosphatases 2C act in a negative feedback regulatory loop of the abscisic acid signaling pathway. *Plant J.,* **25** : 295-303

Mikolajczyk M, Awotunde OS, Muszynska G and Klessig DF (2000) Osmotic stress induces rapid activation of a salicyclic acid-induced protein kinase and a homolog of protein kinase ASK1 in tobacco cells. *Plant Cell,* **12** : 165-178

Mizoguchi T, Ichimura K and Shinozaki K (1997) Environmental stress response in plants: The role of mitogen-activated protein kinases. *Trends in Biotech.,* **15** : 15-19

Mizoguchi T, Irie K, Hirayam T, Hayashida N, Yamaguchi-Shinozaki K, Matsumoto K and Shinozaki K (1996) A gene encoding a MAP kinase kinase kinase is induced simultaneously with genes for a MAP kinase and an S6 kinase by touch, cold and water stress in *Arabidopsis thaliana. Proc. Natl. Acad. Sci. USA,* **93** : 765-769

Mizoguchi T, Irie K, Hirayama T, Hayashide N, Yamaguchi-Shinoguchi K, Matsumoto K and Shinozaki K (1996) A gene encoding a mitogen-activated protein kinase kinase kinase is induced simultaneously with genes for a mitogen-activated protein kinase and an S6 ribosomal protein kinase by touch, cold, and water stress in *Arabidopsis thaliana. Proc. Natl. Acad. Sci. USA,* **93** : 765-769

Murata N, Ishizaki-Nishizawa O, Higashi S, Hayashi H, Tasaka Y and Nishida I (1992) *Nature,* **356** : 710-713

Nover L, Scharf KD, Gagliardi D, Vergne P, Czarneka-Verner E and Gurley WB (1996) The Hsf world: classification and properties of plant heat stress transcription factors. *Cell Stress Chaperones,* **1** : 215-223

Pardha Saradhi P and Sharmila P (2002) Improvement of crop plants for abiotic stress tolerance through introduction of glycinebetaine pathway. In : *Plant Genetic Engineering Vol.1 Application and Limitation* (Eds Singh RP and Jaiwal PK) Sci-Tech. Pub. Houstan USA (In press)

Pardo JM, Reddy MP, Yang S, Maggio A, Huh GH, Matsumoto T, Coca MA, Paino-D'Urzo M, Koiwa H, Yun DJ, Watad AA, Bressan RA and Hasegawa PM (1998) Stress signaling through Ca^{2+}/calmodulin-dependent protein phosphatase calcineurin mediates salt adaptation in plants. *Proc. Natl. Acad. Sci. USA,* **16** : 9681-9686

Park JM, Park CJ, Lee SB, Ham BK, Shin R and Paek KH (2001) Overexpression of the tobacco Tsi1 gene encoding an EREBP/AP2-type transcription factor enhances resistance against pathogen attack and osmotic stress in tobacco. *Plant Cell,* **13** : 1035-1046

Perez-Prat E, Narasimhan ML, Binzel ML, Botella MA, Chen Z, Valpuesta V, Bressan RA and Hasegawa PM (1992) Induction of a putative calcium ATPase mRNA in sodium chloride adapted cells. *Plant Physiol.,* **100** : 1471-1478

Pestacz A and Erdei L (1996) Calcium-dependent protein kinase in maize and sorghum induced by polyethylene glycol. *Physiol. Plant.,* **97** : 360-364

Pilon-Smits EAH, Ebskamp MJM, Paul MJ, Jeuken MJW, Weisbeek PJ and Smeekens SCM (1995) Improved performance of transgenic fructan-accumulating tobacco under drought stress. *Plant Physiol.,* **107** : 125-130

Pilon-Smits EAH, Hwang S, Lytle CM, Zhu YL, Tai JC, Rogelio CB, Chen Y, Leustek T and Terry N (1999) Overexpression of ATP sulfurylase in Indian Mustard leads to increased selenate uptake, reduction and tolerance. *Plant Physiol.,* **119** : 123-132

Pilon-Smits EAH, Zhu LY, Sears T and Terry N (2000) Overexpression of glutathione reductase in *B. juncea*: Effects on cadmium accumulation and tolerance. *Physiol. Plant.,* **110** : 455-460

Prandl R, Hinderhofer K, Eggers-Schumacher G and Schoffl F (1998) HSF3, a new heat shock factor from *Arabidopsis thaliana*, derepresses the heat shock response and confers thermotolerance when overexpressed in transgenic plants. *Mol. Gen. Genet.,* **258** : 269-278

Queitsch C, Hong SW, Vierling E and Lindquist S (2000) Heat shock protein 101 plays a crucial role in thermotolerance in *Arabidopsis*. *Plant Cell,* **12** : 479-492

Robinson MJ and Cobb MH (1997) Mitogen-activated protein kinase pathways. *Curr. Opin. Cell Biol.,* **9** : 180-186

Rock CD (2000) Pathways to abiotic acid-regulated gene expression. *New Phytol.,* **148** : 357-396

Sahoo L, Singh ND, Sonia, Sugla T, Singh RP and Jaiwal PK (2001) Genetically modified crops : a bane or boon to green revolution. *Physiol. Mol. Biol. Plants* **7** : 1-2.

Saijo Y, Hata S, Kyozuka J, Shimamoto K and Izui K (2000) Over-expression of a single Ca^{2+}-dependent protein kinase confers both cold and salt/drought tolerance on rice plants. *Plant J.,* **23** : 319-327

Sakamoto A and Murata N (2001) The use of bacterial choline oxidase, a glycinebetaine-synthesizing enzyme to create stress-resistant transgenic plants. *Plant Physiol.,* **125** : 180-188

Shankar N and Srivastava HS (2002) "Abiotic stress induced proteins and signal transduction". In Plant Genetic Engineering Vol. 1 Application and Limitation (Eds. Singh RP and Jaiwal PK) Sci-Tec Pub. Houstan USA (in press)

Shen Q and Ho THD (1995) Functional dissection of an abscisic acid (ABA)-inducible gene reveals two independent ABA-responsive complexes each containing a G-box and a novel cis-acting element. *Plant Cell,* **7** : 295-307

Shen V, Jensen RG and Bohnert HJ (1997) Increased resistance to oxidative stress in transgenic plants by targeting mannitol biosynthesis to chloroplasts. *Plant Physiol.,* **113** : 1177-1183

Sheveleva E, Chamara W, Bohnert HJ and Jensen RG (1997) Increased drought and salt tolerance by D-ononitol production in transgenic *Nicotiana tabacum L. Plant Physiol.,* **115** : 1211-1219

Shi H, Ishitani M, Kim C and Zhu JK (2000) The *Arabdopsis thaliana* salt tolerance gene SOS1 encodes a putative Na^+/H^+ antiporter. *Proc Natl Acad Sci USA,* **97** : 6896-6901

Shinozaki K and Yamaguchi-Shinozaki K (1997) Gene expression and signal transduction in water-stress response. *Plant Physiol.,* **115** : 327-334

Singla Pareek SN, Reddy MK and Sopory SK (2001) Transgenic approach towards developing abiotic stress tolerance in plants. *Proc. Ind. Natl. Sci. Acad.,* **5** : 265-284

Stockinger EJ, Gilmour SJ and Thomashow MF (1997) *Arabidopsis thaliana CBF1* encodes an AP2 domain-containing transcriptional activator that binds to the C-repeat/DRE, a cis-acting DNA regulatory element that stimulates transcription in response to low temperature and water deficit. *Proc. Natl. Acad. Sci. USA,* **94** : 1035-1040

Takahashi K, Isobe M, Knight MR, Trewavas AJ and Muto S (1997) Hypoosmotic shock induces increases in cytosolic Ca^{2+} in tobacco suspension-culture cells. *Plant Physiol.,* **113** : 587-594

Tamminen I, Makela P, Heino P and Palva ET (2001) Ectopic expression of ABI3 gene enhances freezing tolerance in response to abscisic acid and low temperature in *Arabidopsis thaliana. Plant J.,* **25** : 1-8

Tarczynski MC, Jensen RG and Bohnert HJ (1993) Stress protection of transgenic tobacco by production of osmolyte mannitol. *Science,* **259** : 508-510

Tesfaye M, Temple SJ, Deborah LA, Carroll PV and Deborah AS (2001) Overexpression of malate dehydrogenase in transgenic alfalfa enhances organic acid synthesis and confers tolerance to aluminium. *Plant Physiol.,* **127** : 1836-1844

Thomas JC, Sepahi M, Arendall B and Bohnert HJ (1995) Enhancement of seed germination in high salinity by engineering mannitol expression in *Arabidopsis thaliana. Plant Cell Environ.,* **18** : 801-806

Thomashow MF (1999) Plant cold acclimation: Freezing tolerance genes and regulatory mechanisms. *Annu. Rev. Plant Physiol. Plant Mol. Biol.,* **50** : 571-599

Thomashow MF (2001) So what's new in the field of plant cold acclimation? Lots! *Plant Physiol.,* **125** : 89-93

Tsiantis MS, Bartholomew DM and Smith JAC (1996) Salt regulation of transcript levels for the c subunit of a leaf vacuolar H^+-ATPase in the halophyte *Mesembryanthemum crystallinum. Plant J.,* **9** : 729-736

Urao T, Katagiri T, Mizoguchi T, Yamaguchi-Shinozaki K, Hayashida N and Shinozaki K (1994) Two genes that encode Ca^{2+}-dependent protein kinases are induced by drought and high-salt stresses in *Arabidopsis thaliana. Mol. Gen. Genet.,* **244** : 331-340

Urao T, Yakubov B, Satoh R and Yamaguchi-Shinozaki K, Seki M, Hirayama T and Shinozaki K (1999) A transmembrane hybrid-type histidine kinase in *Arabidopsis* functions as an osmosensor. *Plant Cell,* **11** : 1743-1754

Wimmers LE, Ewing NN and Bennett AB (1992) Higher plants Ca^{2+}-ATPase: primary structure and regulation by salt. *Proc. Natl. Acad. Sci. USA,* **89** : 9205-9209

Winicov II and Bastola DR (1999) Transgenic overexpression of the transcription factor alfin 1 enhances expression of the endogenous *MsPRP2* gene in alfalfa and improves salinity tolerance of the plants. *Plant Physiol.,* **120** : 473-480

Xiong L, Ishitani M, Lee H and Zhu JK (1999a) HOS5: A negative regulator of osmotic stress induced gene expression in *Arabidopsis thaliana. Plant J.,* **19** : 569-578

Xiong L, Ishitani M and Zhu JK (1999b) Interaction of osmotic stress, temperature, and abscisic acid in the regulation of gene expression in *Arabidopsis. Plant Physiol.,* **119** : 205-211

Xiong L, Ishitani M, Lee H and Zhu JK (2001) The *Arabidopsis LOS5/ABA3* Locus Encodes a Molybdenum Cofactor Sulfurase and Modulates Cold Stress- and Osmotic Stress-responsive Gene Expression. *Plant Cell,* **13** : 2063-2083

Xu D, Duan X, Wang B, Hong B, Ho THD and Wu R (1996) Expression of a late embryogenesis abundant protein gene, *HVA1*, from barley confers tolerance to water deficit and salt stress in transgenic rice. *Plant Physiol.,* **110** : 249-257

Yamaguchi Shinozaki K and Shinozaki K (1994) A novel cis-acting element in an *Arabidopsis* gene is involved in responsiveness to drought, low-temperature, or high-salt stress. *Plant Cell,* **6** : 251-264

Yeo A (1998) Molecular biology of salt tolerance in the context of whole-plant physiology. *J. Exp. Bot.,* **49** : 915-929

Yeo AR, Yeo ME, Flowers SA and Flowers TJ (1988) Screening of rice (*Oryza sativa* L.) genotypes for physiological characters contributing to salinity resistance, and their relationship to overall performance. *Theor. Appl Genet.,* **79** : 377-384

Zhang HX, Hodson J, Williams JP and Blumwald E (2001) Engineering salt-tolerant *Brassica* plants: Characterization of yield and seed oil quality in transgenic plants with increased vacuolar sodium accumulation. *Proc. Natl. Acad. Sci. USA,* **98** : 12832-12836

Zhu JK (2000) Genetic analysis of plant salt tolerance using *Arabidopsis. Plant Physiol.,* **124** : 941-948

Zhu JK (2001) Plant salt tolerance. *Trends Plant Sci.,* **6** : 66-71

Zhu LY, Pilon-Smits EAH, Jouanin L and Terry N (1999a) Overexpression of glutathione synthetase in Indian Mustard enhances cadmium accumulation and tolerance. *Plant Physiol.,* **119** : 73-79

Zhu LY, Pilon-Smits EAH, Tarun AS, Weber SU, Jouanin L and Terry N (1999b) Cadmium tolerance and accumulation in Indian Mustard is enhanced by over expressing γ-glutamylcystein synthetase. *Plant Physiol.,* **121** : 1169-1177.

Chapter 7

IMPROVEMENT OF CROP PLANTS FOR ABIOTIC STRESS TOLERANCE THROUGH INTRODUCTION OF GLYCINEBETAINE PATHWAY

P Pardha Saradhi* **and P Sharmila**

Department of Environmental Biology, University of Delhi, Delhi – 110 007, India

Summary

Inspite of success in enhancing the yield potential of crop plants through plant breeding, the losses experienced due to water and salt stress are increasingly being felt. One of the best means to tackle this problem is to introduce or over express the genes that enhance tolerance of crop plants to these stresses. In this chapter efforts have been made to review the efforts that have so far been laid to enhance tolerance of higher plants to various abiotic stresses by introduction of the genes associated with the synthesis of glycinebetaine (one of the compatible solutes that accumulate in wide range of organisms as an adaptive strategy to counteract salt/ water stress). The data presented here clearly show that the introduction of biosynthetic pathway for glycinebetaine into plants (that otherwise lack any means to synthesize glycinebetaine) can enhance their tolerance to abiotic stresses such as salinity, drought, low temperature, and freezing.

Keywords : Abiotic stress tolerance, genetic manipulation, glycinebetaine

---

*Corresponding author : E-mail : saradhi@ndf.vsnl.net.in

1. INTRODUCTION

In nature, plants are often exposed to variety of abiotic stresses such as drought, salinity, low temperature, high temperature, heavy metal and high CO_2 etc. (Aspinall and Paleg, 1981; Nikolopolus and Manetas, 1991; Alia and Pardha Saradhi, 1993; Pardha Saradhi *et al.*, 1995; Prasad *et al.*, 1999 and references therein). Due to ever increasing population the pressure on the land is continuously increasing. This is resulting in an extensive damage to the overall environment of the earth. Salinity and drought are the most wide spread soil constraints that limit plant growth and crop productivity. These environmental stresses are thus serious threats to sustainable food production. However, the world is finite but has to provide food and energy to meet the needs of the growing population. Hence, the significant reduction in potential crop yield due to abiotic stresses is being increasingly felt and has received serious attention of scientists throughout the globe (Mc Cue and Hanson, 1990; Bohnert and Jensen, 1996; Prasad *et al.*, 2000a,b).

Plants respond to various abiotic stresses by altering their metabolic events, in order to adapt maximally to the changed environmental conditions. Some of the common metabolic alterations observed in plants exposed to the abiotic stresses are i) a decrease in net photosynthesis (Dubey, 1997; Giardi *et al.*, 1997); ii) an increase (25-70%) in respiration (Livine and Levin, 1967; Rains, 1972); iii) decrease in the activities of some important enzymes of nitrogen metabolism such as nitrate reductase (Srivastava, 1980; Aryan *et al.*, 1983; Campbell and Smarrelli, 1986; Rao and Gnanam, 1990); iv) synthesis and accumulation of low molecular weight organic solutes such as proline, glycinebetaine, trehalose, mannitol, spermine etc (McCue and Hanson, 1990; Alia and Pardha Saradhi, 1991; Pardha Saradhi *et al.*, 1995; Hayashi *et al.*, 1997; Rajam *et al.*, 1998; v) excessive generation of toxic oxygen species (Foyer, *et al.*, 1994; Alscher *et al.*, 1997) and vi) increase in membrane permeability (Leopld and Willing, 1984).

In this chapter attempts have been made to summarize the work carried out so far in enhancing tolerance of various organisms to abiotic stresses by manipulating/introduction of glycinebetaine pathway. A brief information on the success in improving potential of plants to withstand abiotic stress through genetic transformation/manipulation is also provided in the beginning of this chapter.

2. IMROVEMENT OF PLANTS FOR ABIOTIC STRESS TOLERANCE

In order to attain/obtain maximum out put from the agricultural fields that are increasingly being exposed to the conditions that affect crop plant growth and productivity, efforts are being made to develop genotypes of crop plants that can tolerate/survive well under the adverse conditions prevalent in these areas. Therefore, identification of the genes for imparting abiotic stress tolerance from various sources and transferring them to the crop plant species in order to develop drought, salt and low temperature tolerant genotypes has been the major goal of the plant breeders all over the globe. Although conventional plant breeding methods have been encouraging, overall progress has been very slow because of i) lengthy and laborious procedures; ii) difficulty in modifying a single trait; and iii) its dependence on the availability of required gene(s) in a close relative/compatible genotype (Holmberg and Bülow, 1998). Identifying genes regulating abiotic stress tolerance and in transferring them from closely related species to cultivated crop plants has so far been a tough job for the plant breeders. The second approach, which has been successful to a larger extent is i) to identify the gene(s) that impart stress tolerance from plants/bacteria or some other organisms growing well under the conditions that are otherwise adverse for crop plant; and ii) transfer/introduction of the gene(s) into susceptible organisms using genetic engineering technology (Tarczynski *et al.*, 1993).

In recent years genetic engineering of biosynthetic pathways associated with stress responses, has evolved as one of the most promising methods for improving stress tolerance in plants (Mc Cue and Hanson, 1990; Bohnert and Shen, 1999). At cellular level, the most common type of adaptation to tackle abiotic stress, involves the accumulation of non toxic low molecular weight organic molecules widely known as compatible solutes (Alia and Pardha Saradhi, 1993; Alia *et al.*, 1993; Hanson *et al.*, 1994). These are termed compatible solutes, as they do not interfere with regular cellular metabolism even when present at higher concentrations (Alia and Pardha Saradhi, 1993; Bohnert and Shen, 1999).

Most common compatible solutes that get accumulated in plants under stress are proline, glycinebetaine, mannitol, trehalose, fructans and D-ononitol (Arora and Pardha Saradhi, 1995; Pardha Saradhi *et al.*, 1995; Sivakumar *et al.*, 1998, 2000, 2001a,b; Bohnert and Shen, 1999). Accumulation of these compatible solutes results in enhancing the tolerance of plants to various type of abiotic stresses (Alia and Pardha

Saradhi, 1993; Arora and Pardha Saradhi, 1995; Bohnert and Jensen, 1996; Bohnert and Shen, 1999). The compatible solutes have been shown to improve tolerance of plants by i) osmotic adjustment through mass action, which results in increased water retention and/or sodium exclusion; ii) acting like low molecular weight chaperones through the association with lipids or proteins prevent membrane disintegration or the dissociation of protein complexes/inactivation of enzymes (Arakawa and Timasheff, 1985; Bohnert and Jensen, 1996; Sivakumar *et al.*, 1998); iii) scavenging toxic oxygen species/free radicals (Smirnoff and Cumbes, 1989; Alia *et al.*, 1991, 1993; 1995; 1997; Pardha Saradhi *et al.*, 1995); iv) regulating the ratio of $NAD(P)^+$ to NAD(P)H (Alia and Pardha Saradhi, 1995; Pardha Saradhi *et al.*, 1993); and v) regulating cytosolic pH (Venekamp *et al.*, 1989; Alia and Pardha Saradhi, 1991, 1993). In fact, compatible solutes like proline and glycinebetaine have been demonstrated to protect photosynthetic machinery (PS II in particular) against photodamage/photoinhibition by scavenging toxic oxygen species and/or by preventing dissociation of 18 and 23 kDa polypeptides of water oxidation complex (Alia *et al.*, 1991, 1993, 1997; Papageorgiou and Murata, 1995).

One of the most successful approach, for enhancing the tolerance of plant to abiotic stresses such as salt, drought and low temperature stresses, so far has been through identification and successful transfer of the gene(s) for various compatible solutes through genetic engineering technology (Table 1). The genes for synthesis/over-production of compatible solutes viz. proline, glycinebetaine, mannitol, trehalose, D-ononitol and fructans (which are known to impart stress tolerance in plants) have so far been used successfully (Tarczynski *et al.*, 1993; Holmstrom *et al.*, 1994, 1996; Kavi Kishore *et al.*, 1995; Pilon Smits *et al.*, 1995; Thomas *et al.*, 1995; Lilius *et al.*, 1996; Hayashi *et al.*, 1997; Shen and Jensen, 1997; Sheveleva *et al.*, 1997; Grover *et al.*, 1999). Further, increased levels of these solutes often enhanced the tolerance of plants to many abiotic stresses viz. salt, drought and low temperature stress, implying that genetic engineering for biosynthesis/over-production of compatible solutes is a fruitful approach for obtaining genotypes of crop plants with enhanced tolerance to various kinds of abiotic stresses.

As described earlier, the plants exposed to abiotic stresses generate excess of toxic oxygen species. To detoxify these highly reactive oxygen species which can disrupt entire cellular metabolism, the cells of plants exposed to stress quickly switch on or over express the genes responsible for the synthesis of enzymes and other components that are associated

Table 1. Successful introduction of genes for abiotic stress tolerance

Plant species transformed	Gene /source/ gene product	Product synthesized	Enhanced tolerance to
Compatible solutes			
Nicotiana tabacum	*mtlD/E. coli/* Mannitol-1-phosphate dehydrogenase	mannitol	salt stress (Tarczynski *et al.*, 1993)
Arabidopsis thaliana	*mtlD/E. coli/* Mannitol-1-phosphate dehydrogenase	mannitol	salt stress (Thomas *et al.*, 1995)
N. tabacum	*mtlD/E. coli/* Mannitol-1-phosphate dehydrogenase	mannitol	oxidative stress (Shen *et al.*, 1997)
N. tabacum	*bet B/E. coli/* Betaine aldehyde dehydrogenase	glycinebetaine	osmotic stress (Holmstorm *et al.*, 1994)
N. tabacum	*bet A/E. coli/* Choline dehydrogenase	glycinebetaine	drought (Lilius *et al.*, 1996)
N. tabacum	*p5cs/Vigna Aconitifolia/*Pyrroline 5-carboxylate synthase	proline	drought (Kishor *et al.*, 1995)
N. tabacum	*sac B/B. subtilis/* Levan sucrase	fructan	drought (Pilon-Smits *et al.*, 1995)
N. tabacum	*tps 1/A. thaliana/* Trehalose 6-phosphate synthase	trehalose	drought and salinity (Holmstrom *et al.*, 1996)
A. thaliana	*cod A /Arthrobacter globiformis/*Choline oxidase	glycinebetaine	salt and cold (Hayashi *et al.*, 1997)
Oryza sativa	*cod A/Arthrobacter globiformis/*Choline oxidase	glycinebetaine	salt and cold (Sakamoto *et al.*, 1998)
Brassica juncea	*cod A/Arthrobacter globiformis/*Choline oxidase	glycinebetaine	salt (Prasad *et al.*, 2000a,b)

Table 1. Continued

Plant species transformed	Gene /source/ gene product	Product synthesized	Enhanced tolerance to
Diospyros kaki	*cod A/Arthrobacter globiformis/*Choline oxidase	glycinebetaine	salt (Gao *et al.*, 2000)
N. tabacum	*imt 1/M. crystallinum/* Myo-inositol-*o*-methyl transferase	D-ononitol	drought and salt stress (Sheveleva *et al.*, 1997)
N. tabacum	*ODC* / Yeast/ Ornithine decarboxylase	putresine	nd (Hamill *et al.*, 1990)
N. tabacum	*ODC* / Mouse/ Ornithine decarboxylase	putresine	nd (De Scenzo and Minocha, 1993)
Daucus carota	*ODC* / Mouse/ Ornithine decarboxylase	putresine	nd (Anderson *et al.*, 1998)
Solanum tuberosum	*SAMDC* (sense)/ Potato/SAM decarboxylase	spermidine	nd (Kumar *et al.*, 1996)
S. tuberosum	*SAMDC* (antisense)/ Potato/SAM decarboxylase	——	nd (Kumar *et al.*, 1996)
N. tabacum	*ADC/Oat/* Arginine decarboxylase	putresine	nd (Masgrau *et al.*, 1997)
N. tabacum	*SAMDC/*Human/ SAM decarboxylase	spermidine	nd (Noh and Minocha, 1994)
Scavenging enzymes			
Medicago sativa	*sod/N.plumbaginifolia/* Superoxide dismutase	superoxide dismutase	freezing stress (McKersie *et al.*, 1993)
N. tabacum	*GR/E. coli/*Glutathione reductase	glutathione reductase	Ozone (Aono *et al.*, 1993)

Table 1. Continued

Plant species transformed	Gene /source/ gene product	Product synthesized	Enhanced tolerance to
N. tabacum	*SOD (Cu/Zn)/* Pea/ Superoxide dismutase	superoxide dismutase	chilling and drought (Gupta *et al.*, 1993)
N. tabacum *Lycopersicon esculentum*	*SOD (Cu/Zn)/* Petunia/ Superoxide dismutase	superoxide dismutase	remained sensitive to low temperature and high light (Tepperman and Dunsmuir, 1990)
S. tuberosum	*SOD (Cu/Zn)/* Tomato/Superoxide dismutase	superoxide dismutase	oxidative stress (Perl *et al.*, 1993)
N. tabacum	*SOD (Cu/Zn) /* Petunia/Superoxide dismutase	superoxide dismutase	No tolerance to ozone (Pitcher *et al.*, 1991)
M. sativa	*SOD(Mn)/* *N. plumbaginiofolia/* Superoxide dismutase	superoxide dismutase	drought (Mckersie *et al.*, 1996)

nd = not determined

with instantaneous removal/detoxification of toxic oxygen species and their toxic products. Activities of the enzymes viz. catalase, peroxidase, superoxide dismutase increase in plants exposed to abiotic stresses (Bowles *et al.*, 1992; Scandalios, 1993, Foyer *et al.* 1994, 1997). Activities of enzymes associated with ascorbate glutathione cycle viz. ascorbate peroxidase, monodehydroascrobate reductase, dehydroascorbate reductase and glutathione reductase, which play a vital role in the detoxification of toxic oxygen species, also increase in plants under stress.

Owing to importance of these antioxidant enzymes in providing protection (to cellular metabolism) against the toxic oxygen species several attempts have been successfully made to introduce the gene(s) for these enzymes in plants (Table 1) to enhance their tolerance to abiotic stresses which results in an oxidative stress. Earlier, over-expression of the gene for plastidic Cu/ZnSOD has been shown to increase tolerance of transgenic tobacco plants to highlight and chilling stress (Gupta *et al.* 1993a,b). In contrast, studies with transgenic tobacco plants expressing 30-50 fold higher SOD than controls failed to display increased protection against paraquat or photoinhibition (Tepperman and Dunsmuir, 1990).

Although genetic engineering approach has now become a very useful technique for transferring the gene for abiotic stress tolerance, the success was mostly achieved in either *Nicotiana tabacum* or *Arabidopsis thaliana* (Table 1). As shown in Table 1, amongst important crop plants only Japonica rice (Sakamoto *et al.*, 1998) and Indian mustard (Prasad *et al.*, 2000a,b) have so far been successfully transformed with a gene for compatible solute imparting abiotic stress tolerance. Introduction of a gene encoding HVAI (group 3 LEA protein) from barley into Japonica rice also increased tolerance of rice to water deficit and salt stress (Xu *et al.*, 1996).

3. METABOLIC ENGINEERING FOR INTRODUCTION OF GLYCINEBETAINE BIOSYNTHETIC PATHWAY

3.1. General properties of glycinebetaine

Glycinebetaine is a zwitterionic, fully N-methyl-substituted derivative of glycine that is found in a large variety of microorganisms, higher plants, and animals (Rhodes and Hanson, 1993). It is structurally the simplest of the betaines. It is always present as a free dipole, but this is not the case with the other betaines. Unlike amino acids it has only one titratable group, the carboxyl, which has a pKa of 1.83 at 25^0C. Thus over the physiological pH range, glycinebetaine will have no net charge and no buffering capacity. Glycinebetaine is extremely water soluble (157g 100 ml^{-1} at 19^0C), but insoluble in ether (Dawson *et al.*, 1969). Glycinebetaine does not inhibit a number of plant and animal enzymes. It is also found to be non-toxic. In contrast to proline, it is slowly metabolized in plants and is a semi- permanent end product of metabolism. Thus, glycinebetaine (N-trimethyl glycine) has been presented as a solute notable for its lack of toxicity and intracellular metabolic stability. Soil microorganisms however, rapidly degrade it extracellularly (Kortstee, 1970;Wyn Jones *et al.*, 1973).

It is the smallest amino acid zwitterion. Hence, cells can achieve osmoregulation with comparatively low metabolic energy. Being a zwitterion, glycinebetaine requires no counter ions for electrical neutrality in the cytoplasm. It is also an amphiphilic compound with both hydrophobic and hydrophilic ends. This property helps to neutralize anionic groups situated within the hydrophobic protein domains. As glycinebetaine lacks the lone electron pair, it cannot interact with the electrophilic centres (eg. The Mn cluster), and therefore, it does not alter electric properties (Papageorgiou and Murata, 1995). Nakaya *et al.* (1991) have reported that glycinebetaine does not react with water although it is strongly hydrated (the ratio of water to glycinebetaine at the rate of 8.6:1, mol/mol) in aqueous media.

3.2. Distribution

Glycinebetaine, a quaternary ammonium compound, is amongst the most effective osmoprotectants found in a wide variety of organisms viz. bacteria, cyanobacteria, algae, animals and higher plants (Hayashi *et al.*, 1997; Prasad *et al.*, 2000a,b; Sakamoto and Murata, 2001). In general, higher plant species adapted to dry and saline environments (eg. members of Chenopodiaceae and Poaceae) accumulate glycinebetaine as an adaptive means (Rhodes and Hanson, 1993). The most extensively studied plants are spinach and barley (Papageorgiou *et al.*, 1991).

Among plants, glycinebetaine exhibit sporadic distribution. Taxonomically distinct plants are found to be its accumulators (Weretilnyk *et al.*, 1989). Chenopodiaceae, Amaranthaceae, Portulacaceae, Malvaceae, Convolvulaceae, Cuscutaceae, Solanaceae, Avicenniaceae and Asteraceae are the dicotyledon families in which its accumulation have been reported. All the members of Chenopodeaceae show its accumulation and more critical evaluation is required for confirming about accumulation reports in other families. Highest levels of glycinebetaine content in plant kingdom were recorded in *Pachycornia tenuis* (533 μmol. g dry wt^{-1}) and *Atriplex canescens* (418 μmol. g dry wt^{-1}) (Wyn Jones and Storey, 1981).

There are indications that glycinebetaine accumulation is associated mainly in halophytes, Xerophytes and the other plant species that grow under salty and arid habitats. However, not all the xerophytic or halophytic species accumulate glycinebetaine or other betaines under field conditions. So far no salt-sensitive or drought sensitive plant with low osmotic pressure has been found to accumulate glycinebetaine. As glycinebetaine was found in halophytes, much of the later works were confined to saline and dry habitats giving relatively less importance to glycophytes and mesophytes. In fact, keeping in mind probable adaptive role played by glycinebetaine, several scientific groups have successfully attempted to enhance the tolerance of non-glycinebetaine accumulating plants to salinity and/or drought (Hayashi *et al.*, 1997; Alia *et al.*, 1998, 1999; Sakamoto *et al.*, 1998; Nucio *et al.*, 1998; Prasad *et al.*, 2000a,b).

3.3. Adaptive significance of glycinebetaine

Glycinebetaine, a quaternary ammonium compound, is among the most effective osmoprotectant found in a wide variety of organisms viz. bacteria, cynobacteria, algae, animals and higher plants. In general, higher plant species adapted to dry and saline environment (eg. Member of Chenopodiaceae and Poaceae) accumulate glycinebetaine as an adaptive mean (Rhodes and Hanson, 1993). Glycinebetaine is known to protect cell from osmotic stress by a) maintaining an osmotic balance with the environment (Robinson and Jones, 1986, Deshnium et.al., 1995; 1997); b) stabilizing several; functional units such as oxygen evolving photosystem II complex; c) protecting the structure of proteins by facilitating hydration and thereby retaining its native conformational state (Anjum *et al.*, 1999) including that of Rubisco; d) maintaining structural integrity of membranes (lncharoenskid *et al.*, 1986; Bernard *et al.*, 1988; Santoro *et al.*, 1992; winzor *et al.*, 1992; ; pepageorgious and Murata, 1995; Mansour, 1998; Sakamoto and Murata, 2000); e) protect complex II of mitochondria against Na^+ toxicity (Hamilton *et al.*, 2001); f) protects the transcriptional and translational machinery under stress (Allard *et al.*, 1998; Houry *et al.*, 1999; Bourot *et al.*, 2000).

3.4. Biosynthetic pathways

Four distinct biosynthetic pathways for glycinebetaine have so far been reported in living organisms.

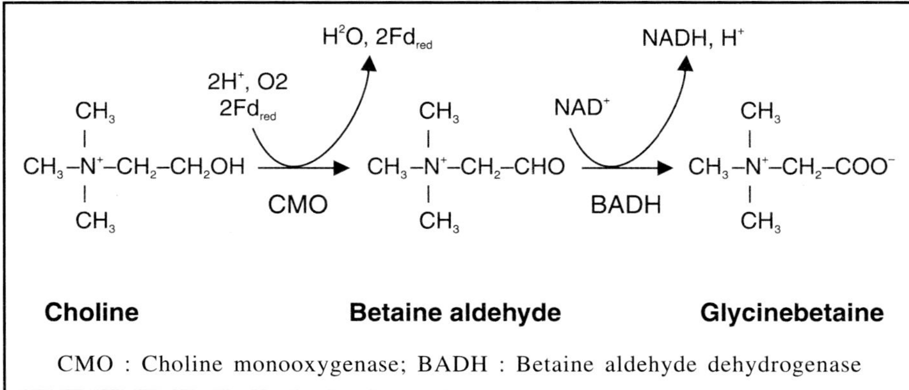

CMO : Choline monooxygenase; BADH : Betaine aldehyde dehydrogenase

Pathway I : In the most extensively studied group of glycinebetaine-accumulating higher plants, the Chenopodiaceae, glycinebetaine is synthesized from choline in a two-step oxidation catalyzed by a ferredoxin (Fd)-dependent choline monooxygenase (CMO) (Brouquisse *et al.*, 1989) and a betaine aldehyde dehydrogenase (BADH) with a strong preference for NAD^+ (Weretilnyk and Hanson, 1989; Weigel *et al.*, 1986; Rhodes and Hanson, 1993).

$^{18}O_2$ tracer studies confirm that CMO utilizes molecular oxygen as a substrate (Lerma *et al.*, 1988). The immediate product of the CMO reaction may be betaine aldehyde hydrate $[(CH_3)_3N^+\text{-}CH_2\text{-}CH(OH)_2]$ which is in spontaneous equilibrium with betaine aldehyde $[(CH_3)_3N^+\text{-}CH_2\text{-}C(OH)=O]$ (Lerma *et al.*, 1988)

Both CMO and BADH enzymes are predominantly localized in the chloroplast stroma of chenopods (Weigel *et al.*, 1986; 1988; Brouquisse *et al.*, 1989), although in spinach leaves approximately 10% of the BADH may exist as a cytosolic isoform (Weretilnyk and Hanson, 1988). Glycinebetaine is also localized predominantly in the chloroplasts of salinized spinach leaves, where it provides osmotic adjustment (Robinson and Jones, 1986). The spinach CMO is not membrane bound, requires ferredoxin and Mg^{2+}, and unlike the bacterial choline oxidizing enzymes (CDH and COX) does not appear to catalyze the oxidation of betaine aldehyde to glycinebetaine (Brouquisse *et al.*, 1989).

Both CMO (Burnet *et al*, 1995) and BADH (Weretilnyk and Hanson, 1989) have been purified to homogeneity from spinach. CMO appears to be a homodimer (or possibly homotrimer) of subunits of MW 42.9 kDa (Burnet *et al*, 1995; Rathinasabapathi *et al*, 1997), whereas BADH is a homodimer with a subunit MW of ~ 60 kDa (Weretilnyk and Hanson, 1989).

cDNAs encoding BADH have been isolated from spinach (Weretilnyk and Hanson, 1990) and sugarbeet (McCue and Hanson, 1992). The spinach and sugarbeet BADHs show significant sequence identity to the human and bacterial BADHs described above (Boyd *et al.*, 1991; Chern and Pietruszko, 1995). The spinach BADH has an unusual chloroplast transit target sequence (Rathinasabapathi *et al*, 1994). The chenopod BADH cDNAs have facilitated cloning of BADHs from a number of other higher plant sources (reviewed in Rhodes and Hanson, 1993), including several monocotyledons (barley, rice and sorghum) (Ishitai *et al.*, 1995; Nakamura *et al.*, 1997; Wood *et al.*, 1996). In leaves of both chenopods and grasses BADH transcripts and enzyme level increase in response to salinity stress (Arakawa *et al.*, 1992; McCue and Hanson, 1992a). A biochemical signal from the roots is implicated in this response in sugar beet (McCue and Hanson, 1992a). Transgenic tobacco plants over-expressing chloroplast localized BADH have been obtained (Rathinasabapathi *et al*, 1994). However, these are unable to accumulate glycinebetaine in the absence of exogenously supplied betaine aldehyde because tobacco lacks CMO (Rathinasabapathi *et al.*, 1994).

It should be cautioned, however, that BADH may be a general aldehyde dehydrogenase, acting on other aldehyde substrates in addition to betaine aldehyde (Trossat *et al.*, 1997). BADH will utilize 3-dimethylsulfoniopropionaldehyde, an intermediate in DMSP synthesis, and certain aldehydes (3-aminopropionaldehyde and 4-aminobutyraldehyde) involved in polyamine metabolism (Trossat *et al.*, 1997). Transgenic plants engineered for BADH expression (Holmstrom *et al.*, 1994; Rathinasabapathi *et al.*, 1994) may possibly exhibit phenotypes related to perturbed polyamine metabolism in addition to phenotypes attributable to alterations in the glycinebetaine synthesis pathway (Trossat *et al.*, 1997). The multiple substrate specificities of BADH may explain the occurrence of BADH in plants that do not accumulate glycinebetaine (Ishitani *et al.*, 1993; Weretilnyk *et al.*, 1989), and in organs of glycinebetaine-accumulating plants which do not contain glycinebetaine (e.g. roots of cereals) (Ishitani *et al.*, 1995).

In monocotyledonous plants, recent studies suggest that BADH is localized in peroxisomes (Nakamura *et al.*, 1997). All monocotyledonous BADHs have a C-terminal tripeptide (SKL) that is known to be a signal for targeting preproteins to microbodies (Nakamura *et al.*, 1997). This raises the question of whether the subcellular localization of the choline oxidation pathway is the same in grasses as in chenopods. If glycinebetaine synthesis occurs in the peroxisome in grasses, then CMO is unlikely to be the choline-oxidizing enzyme of grasses, because its Fd requirement would not be met in the peroxisome (Russell *et al.*, 1998).

A cDNA encoding CMO has recently been cloned from spinach. The cDNA sequence confirms that CMO is an Fe-S protein containing one [2Fe-2S] cluster per subunit, and that CMO has a typical chloroplast transit peptide sequence (Rathinasabapathi *et al.*, 1997). Spinach CMO does not show any homology with the other choline oxidizing enzymes (viz. CDH and COX) (Rathinasabapathi *et al.*, 1997). The activity of CMO rises incrementally in response to salinity levels probably as a result of increased transcription (Brouquisse *et al.*, 1989; Rathinasabapathi *et al.*, 1997). The magnitude of salt induction of CMO mRNA is comparable to that reported for BADH (Rathinasabapathi *et al.*, 1997). The spinach CMO has been used to clone the sugar beet CMO, and to detect salt-inducible CMO mRNA in leaves of *Amaranthus caudatus* (Amaranthaceae) (Russell *et al.*, 1998).

Pathway II : In *E.coli* and certain animals, synthesis of glycinebetaine from choline is mediated by two dehydrogenases viz. membrane bound choline dehydrogenase and NAD^+-dependent betainealdehyde dehydrogenase.

In mammals, where glycinebetaine plays an important role as a compatible solute, particularly in the kidney (Bagnasco *et al.*, 1986; Garcia-Perez and Burg, 1991), the pathway of glycinebetaine synthesis resembles that in *E. coli*, and is catalyzed by a particulate, mitochondrial CDH (Haubrich and Gerber, 1981) and a soluble BADH (Chern and Pietruszko, 1995).

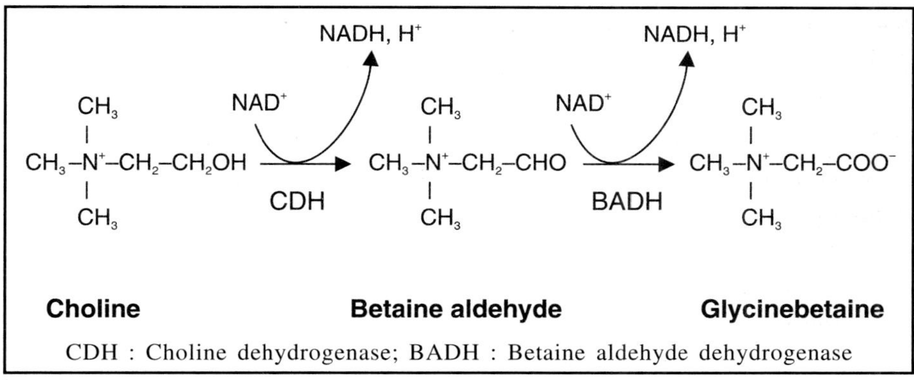

CDH : Choline dehydrogenase; BADH : Betaine aldehyde dehydrogenase

Pathway III : The third pathway unlike the above two pathways, converts choline to glycinebetaine in a single step and is operational only in certain soil bacteria such as *Arthrobacter globiformis*, *A. pascens* and *Alcaligenes* sp (Ikuta *et al.*, 1977; Rozwadowski *et al.*, 1991).

A pathway of choline oxidation similar to *Arthobacter* sp. has also been described for the fungus *Cylindrocarpon didymum* M-1. A choline oxidase (a dimer of two subunits of MW. 64 kDa, containing two mol of FAD per mol of enzyme), utilizes O_2 as electron acceptor and generates H_2O_2 as product. Like the *Arthobacter* enzyme, the fungal

$$\underset{\text{Choline}}{CH_3{-}\overset{\displaystyle CH_3}{\underset{\displaystyle CH_3}{N^+}}{-}CH_2{-}CH_2OH} \xrightarrow[\text{COD}]{2O_2,\ 2H^+ \ \to\ 2H_2O_2} \underset{\text{Glycinebetaine}}{CH_3{-}\overset{\displaystyle CH_3}{\underset{\displaystyle CH_3}{N^+}}{-}CH_2{-}COO^-}$$

COD : Choline oxidase

enzyme is also capable of oxidizing betaine aldehyde (K_m for betaine aldehyde = 5.8 mM, cf. K_m for choline = 1.3 mM), but occurs together with a NAD^+-dependent BADH (Yamada *et al.*, 1979; Tani *et al.*, 1977; 1979).

Pathway IV : In the extreme halophytic phototrophic bacterium *Ectothiorhodospira halochloris*, glycinebetaine is synthesized from glycine through direct *N*-methylation (Galinksi and Truper, 1994; Nyyssola *et al.*, 2000). Three *N*-methylations as shown in the following reaction are catalyzed by two methyltransferases, glycine sarcosine methyltransferase and sarcosine dimethylglycine methyltransferase, with partially overlapping substrate specificity.

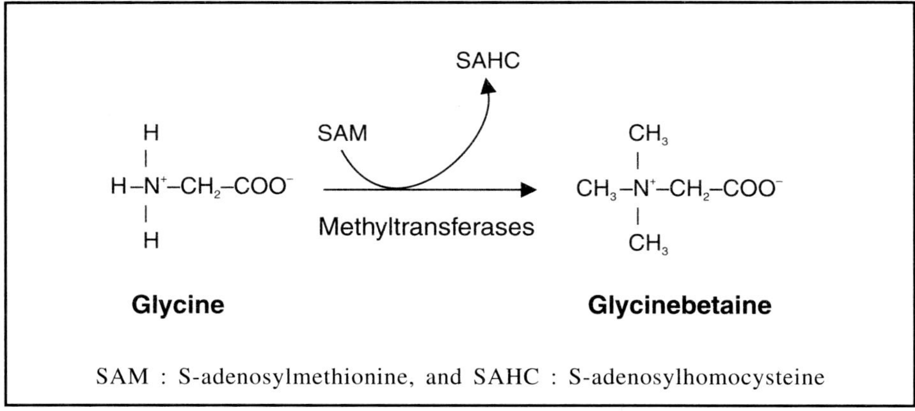

SAM : S-adenosylmethionine, and SAHC : S-adenosylhomocysteine

The methyltransferases have been successfully expressed in *E. coli*, where they confer betaine accumulation and improved salt tolerance.

4. SUCCESSFUL ENGINEERING OF PLANTS FOR PRODUCTION OF GLYCINEBETAINE

Owing to the importance of glycinebetaine, there have been increased efforts to introduce its biosynthetic pathway into plant species, which lack the potential to synthesize glycinebetaine. Gene(s) associated with synthesis of glycinebetaine through the above pathways using choline as the precursor could be successfully introduced into higher plants. Summary of the work carried out in this direction so far is presented in Table 2.

Introduction of the gene coding for betainealdehyde dehydrogenase into tobacco either from spinach or *E. coli* failed to synthesize glycinebetaine unless betaine aldehyde was exogenously supplied (Holmstrom *et al.*, 1994; Rathinasabapathi *et al.*, 1994). Moreover, neither of these groups could demonstrate any enhancement in stress tolerance in transgenic plants. However, introduction of the gene encoding choline dehydrogenase from *E. coli* into tobacco resulted in enhanced tolerance of transgenic plants to salt stress probably due to the inherent potential of tobacco plants to convert betaine aldehyde to glycinebetaine. However, they did not demonstrate the actual synthesis/accumulation of glycinebetaine in the transgenic plants (Lilius *et al.*, 1996).

As introduction of one of the genes from two-step pathway of glycinebetaine biosynthesis has so far not been (with the exception of Nuccio *et al.*, 1998a,b) clearly demonstrated to help in the synthesis of glycinebetaine, it was worthwhile to introduce the pathway, which converts choline to glycinebetaine in a single step. Earlier, Hayashi *et al.* (1997) and Alia *et al.* (1998, 1999) were successful in enhancing tolerance of Arabidopsis to salt, low, high temperature and highlight intensity stresses by transforming Arabidopsis with the gene for choline oxidase. With the exception of the introduction of *codA* gene into a Japonica rice cultivar (Sakamoto *et al.*, 1998) and mustard (Prasad *et al.*, 2000a,b) there has been no published information on the introduction of genes associated with glycinebetaine synthesis into crop plants.

Seed germination and early seedling growth have been reported to be the most vulnerable stages in the life-cycle of plants that are sensitive to osmotic stress (Bewly and Black, 1982; Bliss *et al.*, 1986). Unlike wild-type, the seed of transgenic lines with *codA* gene germinated well both under salt (NaCl) (Table 3) as well as drought (mannitol) stresses (Fig. 1). Further, even growth (measure in terms of shoot length, fresh weight and dry weight) of seedlings of the transgenic lines were

Table 2. An overview on the transgenic expression of the gene(s) responsible for glycinebetaine synthesis

Enzyme	Gene	Source	Plant/Cyanobacterial species transformed	Product synthesized	Result/Enhanced tolerance to	Reference
BADH	*betB*	*Spinacea oleracea* *Beta vulgaris*	*Nicotiana tabaccum*	glycinebetaine	not determined	Rathinasabapathi *et al.*, 1994
BADH	*betB*	*Hordium vulgare*	*Nicotiana tabaccum*	not determined	not determined	Ishitani *et al.*, 1995 Nakamura *et al.*, 1997
BADH	*betB*	*Escherichia coli*	*Nicotiana tabaccum*	glycinebetaine	osmotic stress	Holmstorm *et al.*, 1994
CDH, BADH	*betA*, *betaB*	*Escherichia coli*	*Synechococcus* spp.	glycinebetaine	salt	Nomura *et al.*, 1995
CDH	*betA*	*Escherichia coli*	*Nicotiana tabaccum*	betaine aldehyde	salt and drought	Lilius *et al.*, 1996
CDH	Modified *betA*	*Escherichia coli*	*Oryza sativa*	betaine aldehyde	salt and drought	Takabe *et al.*, 1998
CDH	Modified *betA*	*Escherichia coli*	*Nicotiana tabaccum*	betaine aldehyde	salt and cold	Holmstorm *et al.*, 2000

Table 2. Continued

Enzyme	Gene	Source	Plant/Cyanobacterial species transformed	Product synthesized	Result/Enhanced tolerance to	Reference
CMO	Not assigned	*Spinacea oleracea*	*Nicotiana tabaccum*	betaine aldehyde	not determined	Nuccio *et al.*, 1998
COD	*codA*	*Arthrobacter globiformis*	*Synechococcus* spp.	glycinebetaine	salt and cold	Deshnium *et al.*, 1995
COD	*codA*	*Arthrobacter globiformis*	*Arabidopsis thaliana*	glycinebetaine	salt, cold, heat and freezing	Hayashi *et al.*,1997; Alia *et al.*, 1998 a,b; Sakamoto *et al.*, 2000
COD	*codA*	*Arthrobacter globiformis*	*Oryza sativa*	glycinebetaine	salt and cold	Sakomoto *et al.*, 1998
COD	*codA*	*Arthrobacter globiformis*	*Diospyros kaki*	glycinebetaine	salt	Gao *et al.*, 2000
COD	*codA*	*Arthrobacter globiformis*	*Brassica juncea*	glycinebetaine	salt	Prasad *et al.*, 2000a,b
COD	*cox*	*Arthrobacter pascens*	*Arabidopsis thaliana*	glycinebetaine	salt, drought and freezing	Huang *et al.*, 2000

Table 2. Continued

Enzyme	Gene	Source	Plant/Cyanobacterial species transformed	Product synthesized	Result/Enhanced tolerance to	Reference
COD	*cox*	*Arthrobacter pascens*	*Brassica napus*	glycinebetaine	salt and drought	Huang *et al.*, 2000
COD	cox	*Arthrobacter pascens*	*Nicotiana tabaccum*	glycinebetaine	salt	Huang *et al.*, 2000

BADH : Betaine aldehyde dehydrogenase; CDH : Choline dehydrogenase; CMO : Choline monooxygenase; COD : Choline oxidase.

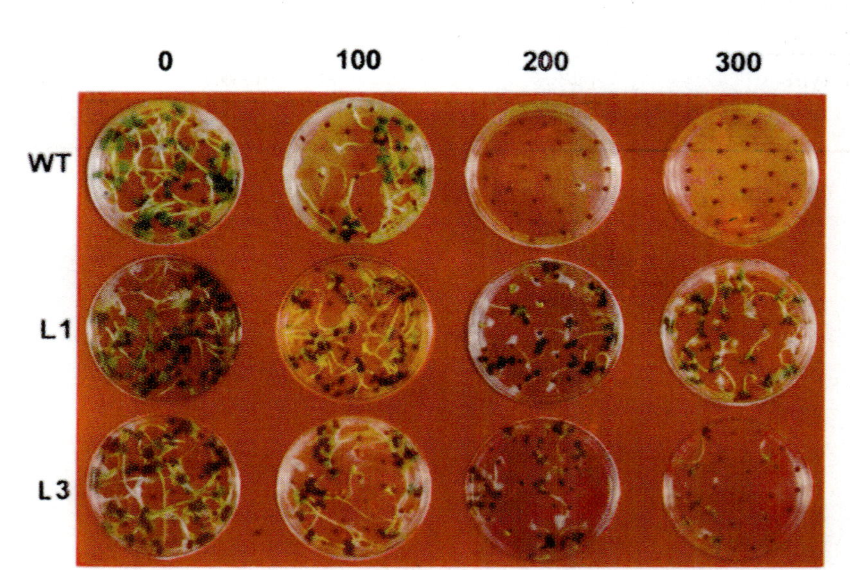

Figure 1 : Effect of mannitol on the seed germination of wild type and transformed line 1 and 3. The seeds were inoculated on medium supplemented with 0, 100, 200 and 300 mM mannitol. Seeds were allowed to germinate under an illumination of 120 mmol m^{-2} s^{-1} for 16 h and a dark period of 8 h for 10 d.

significantly superior to that of wild-type under salt (Fig. 2) as well as drought stress (unpublished).

The photosynthetic machinery (in particular photosystem II activity) of the transgenic genotypes of *Brassica juncea* (Fig. 3), Arabidopsis as well as Japonica rice with the *codA* gene insert, exhibited enhanced tolerance to salt, drought (unpublished), low as well as high temperature stresses in comparison to wild type during all the developmental stages (Hayashi *et al.*, 1997; Sakamoto *et al.*, 1998; Prasad *et al.*, 2000a; Sakamoto and Murata, 2000).

In summary, the investigations carried out by our research team and that of Murata clearly observed enhancement in the tolerance of *Brassica juncea* (Indian mustard), *Arabidopsis* and Japonica rice (*Oryza sativa*) to salt, drought and low temperature stresses (Prasad *et al.*, 2000a,b; Prasad *et al.*, unpublished; Sakamoto and Murata, 2002) at all the developmental stages (viz. seed germination, seedling growth, one and three month old plants both at vegetative as well as reproductive stages).

Table 3. Effect of NaCl on germination of seeds from wild-type and transgenic genotypes {L1 (Line-1) and L3 (Line-3)} of *Brassica juncea* cv. Pusa Jaikisan.

Plant type	Days	% germination on NaCl (mM)				
		0	50	100	150	200
WT	2	44	40	0	0	0
L1	2	48	40	0	0	0
L3	2	49	42	0	0	0
WT	4	100	80	0	0	0
L1	4	100	100	48	28	12
L3	4	100	88	40	24	8
WT	6	100	92	12	0	0
L1	6	100	100	52	36	32
L3	6	100	100	76	48	20
WT	8	100	94	12	0	0
L1	8	100	100	68	40	36
L3	8	100	100	76	56	20
WT	10	100	94	12	0	0
L1	10	100	100	76	56	36
L3	10	100	100	76	56	20

Half- strength MS medium was used for germination of seeds. Values presented here are the mean of three independent experiments. 60 seeds of each plant type were taken for each set experiment.

5. IS CHOLINE REALLY A LIMITING FACTOR?

Most importantly, the level of choline in transformed plants of mustard (Prasad *et al.*, 2000a,b), *Arabidopsis* (Hayashi *et al.*, 1997) and Japonica rice (Sakamoto *et al.*, 1998) did not differ significantly from that of wild-type plants suggesting that the transformed lines seems to have potential to maintain the required level of choline inspite of the introduction of a new pathway which uses choline for the synthesis of glycinebetaine.

Earlier, Nuccio *et al.* (1998) have shown that the low levels (than that observed in natural glycinebetaine accumulators) of the glycinebetaine

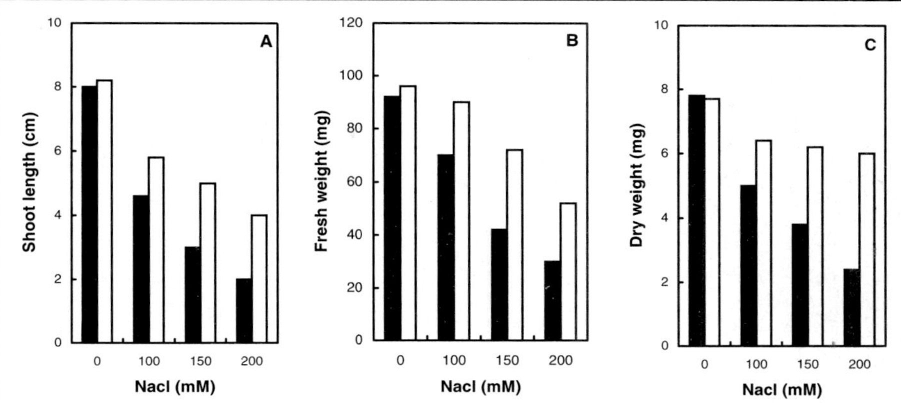

Figure 2 : Increased tolerance of seedlings of transgenic line to salt stress. Effect of NaCl on growth of 3- day old seedling of wild type (■) and transformed line-3 (□) of *Brassica juncea* cv. pusa Jaikisan. The young seedlings were initially raised on MS medium and subsequently transformed to MS medium consisting of 0, 50, 100, 150 and 200 mM NaCl. The growth was measured in terms of shoot length (a), fresh weight (b), and dry weight (c). The results obtained are the average from three independent experiments each with a minimum of four replicates.

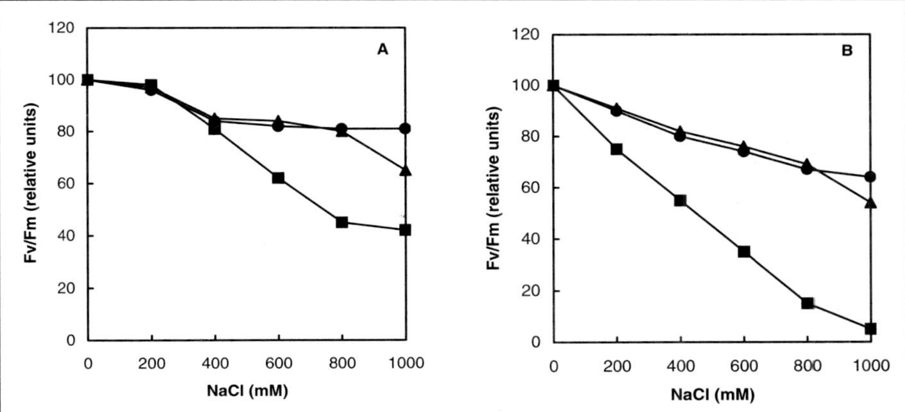

Figure 3 : Changes in PSII activity in the leaf discs of wild type, transgenic lines 1 and 3 of *Brassica juncea* cv. Pusa Jaikisan exposed to various concentration of NaCl for (A) 12 and (B) 24 h. PSII activity was measured in terms of Fv/Fm ratio. The initial value of Fv/Fm was taken as 100%. The initial value of Fv/Fm was 0.820 ± 0.02 for both wild type as well as transgenic lines. The experiments each with a minimum of four replicate. (-■-) Wild type, (-●-) transgenic line 1 and (-▲-) transgenic line 3.

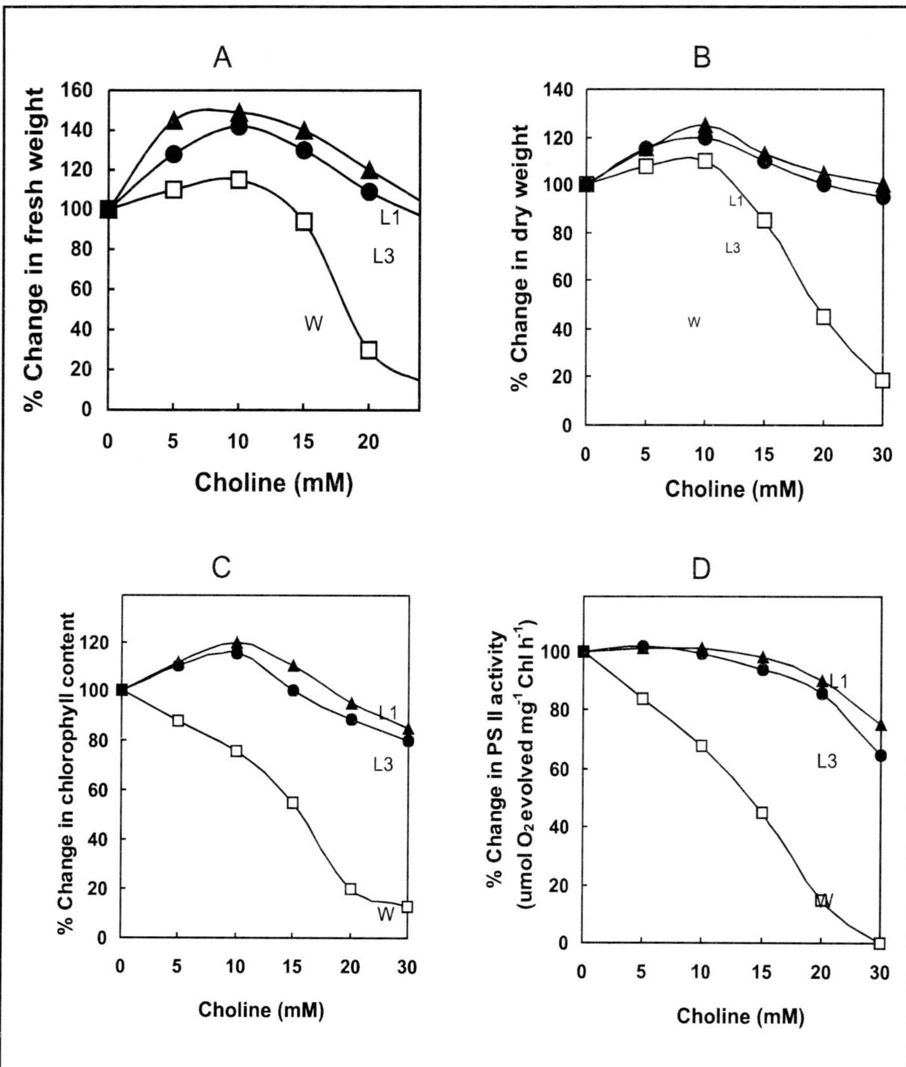

Figure 4 : Change in fresh weight (A), dry weight (B), chlorophyll content (C), PSII activity of shoot of wild type and two transgenic lines (viz. L1 and L3) of *B. juncea* grown on medium with various concentration of choline. The initial values of fresh weight, dry weight, chlorophyll content, PSII activity for wild type and transformed lines were taken as 100%. The initial value of fresh weight of the shoot was 380 ± 10, dry weight was 38 ± 3 mg, chlorophyll content was 2.19 ± 0.21 mg g-1 and oxygen evolution by thylakoid was 210.9 ± 18.6 μmol O_2 evolved mg-1 chl h-1. The results are the average of data obtained from three independent experiments. (□) Wild Type; (▲) transgenic line-1 and (●) transgenic line-3.

in the transgenic lines of tobacco might be due to limitation of endogenous choline. These investigators managed to enhance the level of glycinebetaine from 50 nmol g-1 fresh weight to a maximum of 800 nmol g-1 fresh weight by exogenous application of choline (5 mM). In contrast, even in the absence of any exogenously applied choline, the level of glycinebetaine in plants of *B. juncea* transformed with the *codA* gene ranged between 815 and 950 nmol g-1 fresh weight. Moreover, the level of choline in *B. juncea* was several times higher than that reported in tobacco (Nuccio *et al.*, 1998; Prasad *et al.*, 2000a,b). Hayashi *et al.* (1997) and Sakamoto *et al.* (1998) reported that the genotypes of *Arabidopsis* and *Oryza* bearing the *codA* gene also accumulate glycinebetaine at the levels similar to that observed by us in *B. juncea.* Accordingly Sakamoto and Murata (2000) also seem to believe that choline may not be a limiting factor.

Low glycinebetaine levels noted by Nuccio *et al.* (1998) in tobacco plants could be due to (a) low level of choline; (b) lower potential to convert toxic betaine aldehyde (Holmstrom *et al.* 1994; Rathinasabapathi *et al.* 1994) to non-toxic glycinebetaine due to low activity of betaine aldehyde dehydrogenase (if present) or by some other dehydrogenase with broad specificity; and/or (c) the site of production of betaine aldehyde in transgenic tobacco is the chloroplast, whereas betaine aldehyde dehydrogenase might be located in some other site such as peroxisomes as observed in rice (Nakamura *et al.*, 1997).

Earlier, we have demonstrated that exogenously supplied choline suppresses growth and development of plants even at concentrations as low as 5 mM (Prasad *et al.*, 2000b). As is clear from Figure 4, besides growth and chlorophyll content choline severely effects photosynthetic machinery. Accordingly, there seem to be perfect homeostasis with regard to choline biosynthesis in plants. Therefore, any efforts that are made to promote over production of choline can disrupt the essential cellular metabolic events.

6. HOW ARE GLYCINEBETAINE SYNTHESIZING TRANSGENICS SUPERIOR?

Earlier, glycinebetaine had been proposed to play a key role in osmoregulation (Robinson and Jones, 1986; McCue and Hanson, 1990). However, as the level of glycinebetaine in the leaves of transgenic genotypes of *B. juncea* was only 815 -950 nmoles g-1 fresh weight (~0.04% of dry matter) and it did not alter in plants exposed to drought, it does not seem to contribute significantly towards osmoregulation. In contrast, the level of total sugars in the leaves of *B. juncea* exposed to drought stress go to the levels as high as ~25% of the dry matter (unpublished data). Therefore, we believe that glycinebetaine might be playing an important role in enhancing drought tolerance in *B. juncea* by protecting biological components such as cell membranes and proteins/ enzymes (Hayashi *et al.*, 1997 and references therein). This facilitates protection and maintenance of major metabolic events in plant cells such as photosynthesis. Earlier, Papagerogiou and Murata (1995) had reported that glycinebetaine help in protection of PS II activity by stabilizing its machinery against stress.

7. CONCLUSIONS AND FUTURE PROSPECTS

In summary, our findings along with the findings of Murata's research team clearly show that irrespective of the developmental stage at which the plants were exposed to salt, drought or low temperature stress, the transformed lines show better tolerance than those of wild type (in terms of maintenance of turgidity of leaves, level of chlorophyll and photosynthetic functions). Moreover, we believe that the introduction of this pathway into the crop plants would be beneficial than introduction of other pathways for glycinebetaine synthesis.

REFERENCES

Alia and Pardha Saradhi P (1991) Proline accumulation under heavy metal stress. *J. Plant Physiol.* **138** : 554-558.

Alia, Pardha Saradhi P and Mohanty P (1991) Proline enhances primary photochemical activities in isolated thylakoid membranes of *Brassica juncea* by arresting photoinhibitory damage. *Biochem. Biophy. Res. Commun.* **181** : 1238-1244.

Alia and Pardha Saradhi P (1993) Suppression in mitochondrial electron transport is the prime cause behind stress induced proline accumulation. *Biochem. Biophy Res. Commun.* **193** : 54-58.

Alia, Pardha Saradhi P and Mohanty P (1993) Involvement of proline in protecting thylakoid membranes against free radical-induced photodamage. *J. Photochem. Photobiol.* **38** : 253-257.

Alia, Prasad KVSK and Pardha Saradhi P (1995) Effect of zinc on free radicals and proline in *Brassica juncea* and *Cajanus cajan*. *Phytochem.* **39** : 45-47.

Alia, Pardha Sardhi P and Mohanty P (1997) Involvement of proline in protecting thylakoid membranes against free radical-induced photodamage. *J. Photochem. Photobiol.* **38** : 253-257.

Alia, Hayashi H, Sakomoto A and Murata N (1998a) Enhancement of tolerance of *Arabidopsis* to high temperatures by genetic engineering of the synthesis of glycinebetaine. *Plant J.* **16** : 155-161.

Alia, Hayashi H, Chen THH and Murata N (1998b) Transformation with gene for choline oxidase enhances the cold tolerance of *Arabidopsis* during germination and early growth. *Plant Cell Environ.* **21** : 232-239.

Alia, Kondo Y, Sakamoto A, Nonaka H, Hayashi H, Pardha Saradhi P, Chen THH and Murata N (1999) Enhanced tolerance to light stress of transgenic *Arabidopsis* plants that express the *codA* gene for a bacrterial choline oxidase. *Plant Mol. Biol.* **40** : 279-288.

Allard F, Houde M, Krol M, Ivanov A, Hunter NPA and Sarhan F (1998) Betaine improves freezing tolerance in wheat. *Plant Cell Physiol.* **39** : 1194-1202.

Alscher RG, Donahue JL and Cramer CL (1997) Reactive oxygen species and antioxidants: Relationships in green cells. *Physiol. Plant* **100** : 224-233.

Anderson SE, Bastola DR and Minocha SC (1998) Metabolism of polyamines in transgenic cells of carrot expressing a mouse ornithine decarboxylase cDNA. *Plant Physiol.* **116** : 299-307.

Anjum F, Rishi V and Ahmad F (1999) Compatibility of osmolytes with Gibbs energy of stabilization of proteins, *Biochim. Biophys. Acta.* **1476** : 75-84.

Aono M, Kubo A, Saji H, Tanaka K and Kondo N (1993) Enhanced tolerance to photooxidative stress of transgenic *Nicotiana tabacum* with high chloroplastic glutathione reductase activity. *Plant Cell Physiol.* **34** : 129-135.

Arakawa T and Timasheff SN (1985) The stabilization of proteins by osmolytes. *Biophys. J.* **47** : 411- 414.

Arakawa K, Mizuno K, Kishitaini S and Takabe T (1992) Immunological studies of betaine aldehyde dehydrogenase in barley. *Plant Cell Physiol.* **108** : 581-588.

Arora S and Pardha Saradhi P (1995) Light-induced enhancement in proline levels in *Vigna radiata* exposed to environmental stresses. *Aust. J. Plant Physiol.* **22** : 383-386.

Aryan AP, Batt RG and Wallace W (1983) Reversible inactivation of nitrate rerductase NADH and the occurrence of partially inactive enzyme in wheat leaf. *Plant Physiol.* **71** : 582-587.

Aspinall D and Paleg LG (1981) Proline accumulation: Physiological aspects. In: *The Physiology and Biochemistry of Drought Resistance in Plants.* (Eds. Paleg LG and Aspinall D), Academic Press, Sydney, pp 215-228.

Bagnasco S, Balaban R, Fales HM, Yang YM and Burg M (1986) Predominant osmotically active organic solutes in rat and rabbitrenal medullas. *J. Biol. Chem.* **261** : 5872-5877.

Bewely JD and Black N (1982) *Physiology and Biochemistry of Seed in Relation to Germination.* 2, Berlin: Spring-Verlag.

Bliss RD, Platt-Aloka KA and Thomson WW (1986) The inhibitory effect of NaCl on barley germination. *Plant Cell Environ.* **9** : 727- 723.

Bohnert HJ and Jensen RG (1996) Strategies for engineering water stress tolerance in plants. *TIBTECH* **14** : 89-97.

Bohnert HJ and Shen B (1999) Transformation and compatible solutes. *Scientia Hort.* **78** : 237-260.

Bourot S, Sire O, Trautwetter A, Touze T, Wu LF, Blanco C and Bernard T 2000. Glycinebetaine assisted protein folding in a lysA mutant of *Escherichia coli. J. Biol. Chem.* **275** : 1050-1056

Bowler C, Montagu MV and Inze D (1992) Superoxide dismutase and stress tolerance. *Annu. Rev. Plant. Physiol. Plant Mol. Biol.* **43** : 83-116.

Brouquisse R, Weigel P, Rhodes D, Yocum CF and Hanson AD (1989) Evidence for a ferredoxin-dependent choline monoxygenase from spinach chloroplast stroma. *Plant Physiol.* **108** : 581-588.

Burnet M, Latontaine PJ and Hanson AD (1995). Assay, purification, and partial characterization of choline manooxygenase from spinach. *Plant Physiol.* **108** : 581-588.

Chern MK and Pietruszko R (1995) Human aldehyde dehydrogenase E3 isoenzyme is a betine aldehyde dehydrogenase (BADH): structural similarities to the mammalian ALDHs and a plant BADH. *Gene* **103** : 45-52.

Campbell WH and Smarelli J Jr. (1986) Nitrate reductase biochemistry and regulation In: *Biochemical Basis of Plant Breeding,* Vol. II. (Ed. Neyra CA) CRC Press Inc. Boca Raton, FL. pp 1-39.

Dawson RWC, Elliot DC, Elliot WH and Jones KM (1969). *Data for Biochemical Research.* 2nd edition, Clarendon Press, Oxford, p. 12.

DeScenzo RA and Minocha SC (1993) Modulation of cellular polyamines in tobacco by transfer and expression of mouse ornithine decarboxylase cDNA. *Plant Mol. Biol.* **22** : 113-127.

Deshnium P, Los DA, Hayashi H, Mustardy L and Murata N (1995) transformation of Synechococcus with a gene for choline oxidase enhances tolerance to salt stress. *Plant Mol. Biol.* **29** : 897-907.

Deshnium P, Gombos Z, Nishiyama Y and Murata N (1997) The action of glycine betaine in enhancement of tolerence of Synechococcus sp. strain PCC 7942 to low temperature. *J. Bacteriol.* **212** : 339-334.

Dubey RS (1997) Photosynthesis in plants under stressful conditions. In: *Handbook of Photosynthesis.* (Ed. Pessarakli M) Marcel Dekker, Inc. New York, pp 859-876.

Foyer CH, Lelandais M and Kunert KJ (1994) Photooxidative stress in plants. *Physiol. Plant.* **92** : 696-717.

Foyer CH, Lopez-Delgado H, Dat JF and Scott IM (1997) Hydrogen peroxide-and glutathione-associated mechanism of acclimatory stress tolerance and signalling. *Physiol. Plant.* **100** : 241-254.

Gao M, Sakamoto A, Miura K, Murata N, Sugiura A and Tao R (2000) Transformation of Japanese persimmon (*Diospyros kaki* Thunb.) with a bacterial gene for choline oxidase. *Mol. Breed.* **6** : 501-510.

Giardi MT, Masojidek J and Godde D (1997) Effects of abiotic stresses on the turnover of the D1 reaction centre II protein. *Physiol. Plant.* **101** : 635-642.

Garcia-Perez A and Burg MB (1991) Renal medullary organic osmolytes. *Physiol. Rev.* **71** : 1081-1115.

Grover A, Sahi C, Sanan N and Grover A (1999) Taming abiotic stress in plants through genetic engineering: current strategies and perspective. *Plant Sci.* **143** : 101-111.

Gupta AS, Heinen JL, Holaday AS, Bruke JJ and Allen RD (1993a) Increased resistance to oxidative stress in transgenic plants that over expresses chloroplastic Cu/Zn superoxide dismutase. *Proc. Natl. Acad. Sci. USA* **90** : 1629-1633.

Gupta AS, Webb RP, Holaday AS and Allen RD (1993b) Overexpression of superoxide dismutase protects plants from oxidative stress. Induction of ascorbate peroxidase in superoxide dismutase-overexpressing plants. *Plant Physiol.* **103** : 1067-1073.

Hamill JD, Robins RJ, Parr AJ, Evans DM, Furze JM and Rhodes MJC (1990) Overexpression of a yeast ornithine decarboxylase gene in transgenic roots of *Nicotiana rustica* can lead to enhanced nicotine accumulation. *Plant Mol. Biol.* **15** : 27-38.

Hamilton III EW and Heckathron SA (2001) Mitochondrial adaptations to NaCl. Complex I is protected by anti-oxidants and small heat shock proteins, whereas complex II is protected by proline and betaine. *Plant Physiol.* **126** : 1266-1274.

Hanson AD, Rathinasabapathi B, Revoal J, Burnet M, Dillon MO and Gage DA (1994) Osmoprotective compounds in plumbaginaceae: A natural experiment in metabolic engineering of stress tolerance. *Proc. Natl. Acad. Sci. USA* **91** : 306-310.

Hayashi H, Alia, Mustardy L, Deshnium P, Ida M and Murata N (1997) Transformation of Arabidopsis thaliana with codA gene for choline oxidase; accumulation of glycinebetaine and enhanced tolerance to salt and cold stress. *Plant J.* **12** : 133-142.

Holmstrom KO, Mantyla E, Welin B, Mandal A and Palva ET (1996) Drought tolerance in tobacco. *Nature* **379** : 683-684.

Holmberg N and Bulow L (1998) Improving stress tolerance in plants by gene transfer. *Trends Plant Sci.* **3** : 61-66.

Holmstrom KO, Weilin B, Mandal A, Kristiansdottir I, Teeri TH, Lamark T, Strom AR and Palva ET (1994) Production of the Escherichia coli betaine-aldehyde dehydrogenase, an enzyme required for the synthesis of osmoprotectant glycinebetaine in transgenic plants. *Plant J.* **6** : 749-758.

Houry WA, Frishman D, Eckerskorn C, Lottspeicch F and Hartl F (1999) Identification of in vivo substrates of the chaperonin GroEL. *Nature* **402** : 147-154.

Haubrich DR and Gerber NH (1981) Choline dehydrogenase: assay, properties and inhibitors. *Biochem. Pharmacol* **30** : 2993-3000.

Ikuta S, Imamura S, Misaki H and Horiuchi Y (1977) Purification and characterization of choline oxidase from Arthrobacter globiformis. *J. Biochem.* **82** : 1741-1749.

Incharoenskdi A, Takabe T and Akazawa T (1986) Effect of betaine on enzyme activity and subunit interaction of ribulose 1-5-bisphosphate carboxylase/oxygenase from Aphanothece halophytica. *Plant Physiol.* **81** : 1044-1049.

Ishitani M, Arakawa K, Mizuno K, Kishitani S and Takabe T (1993) Betaine aldehyde dehydrogenase in gramineae: levels in leaves of both betaine-accumulating and non accumulating cereal plants. *Plant Cell Physiol.* **34** : 493-495.

Ishitani M, Nakamura T, Han SY and Takabe T (1995) Expression of the betaine aldehyde dehydrogenase gene in barley in response to osmotic stress and abscise acid. *Plant Mol. Biol.* **27** : 307-315

Jones H, Flowers TJ and Jones MB (ed.) (1989) Plants under stress, Cambridge University Press, Cambridge.

Kishor PBK, Hong Z, Miao GH, Hu CAA and Verma DPS (1995) Overexpression of Δ^1-pyrroline-5-carboxylate synthetase increases proline production and confers osmotolerance in transgenic plants. *Plant Physiol.* **108** : 1387-1394.

Kortstee GJJ (1970). The aerobic decomposition of choline by microorganisms. *Arch. Microbiol.* **71** : 235-244.

Kumar A, Taylor MA, Arif SAM and Davies HV (1996) Potato plants expressing antisense and sense S-adenosyl methionine decarboxylase (SAMDC) transgenes show altered levels of polyamines and ethylene: antisense plants display abnormal phenotypes. *Plant J.* **9** : 147-158.

Leopold AC and Willing RP (1984) Evidence for toxicity effects of salt on membranes. In: *Salinity Tolerance in Plants,* (Eds. Staples RC and Toenniessen GH) Wiley, New York, pp 67-76.

Lerma C, Hanson AD and Rhodes D (1988) Oxygen-18 and deuterium labeling studies of choline oxidation by spinach and sugar beet. *Plant Physiol.* **88** : 695-702.

Lilius G, Holmberg N and Bulow L (1996) Enhanced NaCl stress tolerance in transgenic tobacco expressing bacterial choline dehydrogenase. *Biotechnology.* **14** : 177-180.

Livena A and Levin N (1967) Tissue respiration and mitochondrial oxidative phosphorylation of NaCl treated pea seedlings. *Plant Physiol.* **42** : 407-414.

Masgrau C, Altabella T, Fareas R, Flore D, Thompson AJ, Bestford RT and Tibucio AF (1997) Inducible overexpression of oat arginine decarboxylase in transgenic tobacco plant. *Plant J.* **11** : 465-473.

McCue KF and Hanson AD (1990) Drought and salt tolerance: towards understanding and application. *TIBTECH* **8** : 358-362.

McCue KF and Henson AD (1992) Salt - inducible betaine aldehyde dehydrogenase from sugar beet: cDNA cloning and expression. *Plant Mol. Biol.* **18** : 1-15.

McKersie BD, Bowley SR, Harjanto E and Leprince O (1996) Water deficit tolerance and field performance of transgenic alfa alfa overexpressing superoxide dismutase. *Plant Physiol.* **111** : 1177-1181.

Nakamura T, Yokota S, Muramoto Y, Tsutsui K, Oguri Y, Fukui K and Takabe T (1997) Expression of betaine aldehyde dehydrogenase gene in rice, a glycinebetaine nonaccumulator, and possible localization of its protein in peroxisomes. *Plant J.* **11** : 1115-1120.

Nakaya S, Tanaka Y, Yoshioka Y and Yasuda N (1991) Interaction of water with nitrogenous compounds in human urine as studied by nuclear magnetic resonance spectroscopy, *J. Iwat. Med. Asc.* **43** : 311-320.

Nikolopoulos D and Manetas Y (1991) Compatible solutes and in vitro stability of Salsola soda enzymes: proline incompatibility. *Phytochemistry* **30** : 411-413.

Noh E and Minocha SC (1994) Expression of a human S-adenosyl-methionine decarboxylase cDNA in transgenic tobacco and its effects on polyamine biosynthesis. *Transgenic Res.* **3** : 26-35.

Nuccio ML, Russel BL, Nolte KD, Rathinasabapathi B, Gag DA and Hanson AD (1998) The endogenous choline supply limits glycinebetaine synthesis in transgenic tobacco expressing choline monooxygenase. *Plant J.* **16** : 487-496.

Nyysola A, Kerovuo, Kaukinen P, Weymarn NV and Reinikainen T (2000) Extremes halophiles synthesize betaine from glycine by methylaton. *J. Biol. Chem.* **275** : 22196-22201.

Papageorgiou GC, Fujimura Y and Murata N (1991) Protection of oxygen-evolving Photosystem II complex by glycinebetaine. *Biochem. Biophys. Acta* **1057** : 361-366.

Papageorgiou GC and Murata N (1995) The unusually strong stabilizing effects of glycinebetaine on the structure and function in the oxygen-evolving photosystem-II complex. *Photosyn. Res.* **44** : 243-252.

Pardha Saradhi P, Alia and Vani B (1993) Inhibition of mitochondrial electron transport is the prime cause behind proline accumulation during mineral deficiency in *Oryza sativa. Plant Soil* **155/156** : 465-468.

Pardha Saradhi P, Alia, Arora S and Prasad KVSK (1995) Proline accumulates in plants exposed to UV radiation and protects them against UV induced peroxidation. *Biochem. Biophy Res. Commun.* **209** : 1-5.

Perl A, Trenes RP, Galili S, Aviv D, Shalgi E, Malkin S and Calun E (1993) Enhanced oxidative stress defense in transgenic plants expressing tomato Cu, Zn superoxide dismutases. *Theo. Appl. Genet.* **85** : 568-576.

Pilon-Smits EAH, Ebskamp MJM, Paul MJ, Jeuken MJW, Weisbeek PJ and Smeekens CM (1995) improved performance of transgenic fructan accumulating tobacco under drought stress. *Plant Physiol.* **107** : 125-130.

Pitcher LH, Brennan E, Hurlay A, Dunsmuir P, Tepperman JM and Zilinskas BA (1991) Overproduction of petunia Copper/Zinc superoxide dismutase does not confer ozone tolerance in transgenic tobacco. *Plant Physiol.* **97** : 452-455.

Prasad KVSK, Pardha Saradhi P and Sharmila P (1999) Concerted action of antioxidant enzymes and curtailed growth under zinc toxicity in Brassica juncea. *Environ. Exp. Bot.* **42** : 1-10.

Prasad KVSK, Sharmila P and Pardha Saradhi P (2000a) Enhanced tolerance of transgenic *Brassica juncea* to choline confirms successful expression of the bacterial *codA* gene. *Plant Sci.* **159** : 233-242.

Prasad KVSK, Sharmila P, Kumar PA and Saradhi PP (2000b) Transformation of *Brassica juncea* (L.) Czern with bacterial *codA* gene enhances its tolerance to salt stress. *Mol. Breed* **6** : 489-499.

Rains DW (1972) Salt transport by plants in relation to salinity. *Annu. Rev. Plant Physiol.* **23** : 516-525.

Rajam MV, Dagar S, Waie B, Yadav JS, Kumar PA, Shoeb F and Kumaria R (1998) Genetic engineering of polyamine and carbohydrate metabolism for osmotic stress tolerance in higher plants. *J. Biosci.* **23** : 473-482.

Rao KR and Gnanam A (1990) Inhibition of nitrate and nitrate reductase activities by salinity stress in *Sorghum vulgare. Phytochemistry* **29** : 1047-1050.

Rathinasabapathi B, McCue KF, Gage DA and Hanson AD (1994) Metabolic engineering of glycinebetaine synthesis: plant betaine aldehyde dehydrogenase lacking transit peptide are targeted to tobacco chloroplast where they confer betaine aldehyde resistance. *Planta* **193** : 155-162.

Rathinasabapati B, Burnet M, Russel BL, Gage DA and Hanson AD (1997) Choline monooxygenase, an unusual iron-sulphur enzyme catalyzing the first step of glycinebetaine synthesis in plant : prosthetic group characterization and cDNA cloning. *Proc. Natl. Acad. Sci. USA* **94** : 3454-3458.

Rhodes D and Hanson AD (1993) Quaternary ammonium and tertiary sulfonium compounds in higher plants. *Annu. Rev. Plant. Physiol. Plant Mol. Biol.* **44** : 357-384.

Robinson SP and Jones GP (1986) Accumulation of glycinebetaine in chloroplast provides osmotic adjustments during salt stress. *Aust. J. Plant Physiol.* **13** : 659-668.

Rozwadowski KL, Khachatourians GG and Selvaraj G (1991) Choline oxidase, a catabolic enzyme in *Arthrobacter pascens*, facilitates adaptation to osmotic stress in *Escherichia coli. J. Bacteriol.* **173** : 472-478.

Russell BL, Rathinasabapati B and Hanson AD (1998) Osmotic stress induced expression of choline monooxygenase in sugar beet and Amaranth. *Plant Physiol.* **116** : 859-865.

Sakamoto A, Alia and Murata N (1998) Metabolic engineering of rice leading to biosynthesis of glycinebetaine and tolerance to salt and cold. *Plant Mol. Biol.* **38** : 1011-1019.

Sakamoto A and Murata N (2000) Genetic engineering of glycinebetaine synthesis in plants: current status and implications for enhancement of stress tolerance. *J. Exp. Bot.* **51** : 81-88.

Santoro MM, Liu Y, Khan SM, Hou LX and Bolen DW (1992) Increased thermal stability of proteins in the presence of naturally occurring osmolytes. *Biochemistry* **31** : 5278-5283.

Scandalios JG (1993) Oxygen stress and superoxide dismutase. *Plant Physiol.* **101** : 7-12.

Shen B, Jensen RG and Bohnert HJ (1997a) Increased resisitance to oxidative stress in transgenic plants by targeting mannitol biosynthesis to chloroplasts. *Plant Physiol.* **113** : 1177-1183.

Shen B, Jensen RG and Bohnert HJ (1997b) Mannitol protects against oxidation by hydroxyl radicals. *Plant Physiol.* **115** : 527-532.

Sheveleva E, Chmara W, Bohnert HJ and Jensen RG (1997) Increased salt and drought tolerance by D-ononitol production in transgenic *Nicotiana tabacum* L. *Plant Physiol.* **115** : 1211-1219.

Sivakumar P, Sharmila P and Pardha Saradhi P (1998) Proline suppresses Rubisco activity in higher plants. *Biochem. Biophys. Res. Commun.* **252** : 428-432.

Sivakumar P, Sharmila P and Pardha Saradhi P (2000). Proline alleviates salt stress induced enhancement in ribulose 1,5-bisphosphate oxygenase activity. *Biochem. Biophys. Res. Commun.* **279** : 512-515.

Smirnoff N and Cumbes QJ (1989) Hydroxyl radical scavenging activity of compatible solutes. *Phytochemistry* **28** : 1057-1060.

Srivastava HS (1980) Regulation of nitrate reductase activity in higher plants. *Phytochemistry* **19** : 728-733.

Storey R and Wyn Jones RG (1975) Betaine and choline levels in plants and their relationship to NaCl. Stress. *Plant Sci. Lett.* **4** : 161-168

Storey R (1976) Salt resistance and quaternary ammonium compounds in plants. *Ph. D. Thesis,* University of Whales, UK.

Tani Y, Mori N, Ogata K and Yamada H (1979) Production and purification of choline oxidase from *Cylindrocarpon didymium* M-1. *Agric. Biol. Chem.* **43** : 815-820.

Tarczynski M, Jensen R and Bohnert HJ (1993) Stress protection of transgenic tobacco by production of osmolytes mannitol. *Science* **259** : 508-510.

Tepperman JM and Dunsmuir P (1990) Transformed plants with elevated levels of chloroplastic SOD are not more resistant to superoxide toxicity. *Plant Mol. Biol.* **14** : 501-511.

Thomas JC, Sephai M, Arendall B and Bohnert HJ (1995) Enhancement of seed germination in high salinity by engineering mannitol expression in *Arabidopsis thaliana*. *Plant Cell Environ.* **18** : 801-806.

Trossat C, Rathinasabapathi B and Hanson AD (1997) Transgenically expressed betaine aldehyde dehydrogenase efficiently catalyzes oxidation of dimethylsulphoniopropionaldehyde and ω-aminoaldehydes. *Plant Physiol.* **113** : 1457-1461.

Venekamp JH, Lampe JE and Koot JTM (1989) Organic acids as a source of drought-induced proline synthesis in field bean plants, *Vicia faba* L. *J. Plant Physiol.* **133** : 654-659.

Weigel P, Weretilnyk EA and Hanson AD (1986) Beatine aldehyde oxidation by spinach chloroplast. *Plant Physiol.* **82** : 753-759.

Weigel P, Lerma C and Hanson AD (1988) Choline oxidation by intact spinach chloroplast. *Plant Physiol.* **86** : 54-60.

Weretilnyk EA and Hanson AD (1988) Betine aldehyde dehydrogenase polymorphism in spinach: genetic and biochemical characterization. *Biochem. Genet.* **26** : 143-151.

Weretilnyk EA, Bednarek S, Kent F, McCue KF, Rhodes D and Hanson AD (1989) Comparative biochemical and immunological studies of the glycinebetaine synthesis pathway in diverse families of dicotyledons, *Planta* **178** : 342-352.

Weretilynk EA and Hanson AD (1990) Molecular cloning of a plant betaine aldehyde dehydrogenasc in an enzyme implicated in adaptation to salinity and drought. *Proc. Natl. Acad. Sci. USA* **87** : 2475-2749.

Winzor CL, Winzor DJ, Paleg LG, Jones GP and Naidu BP (1992) Rationalization of the effects of compatible solutes on protein stability in terms of thermodynamic non ideality. *Arch. Biochem. Biophys.* **296** : 102-107.

Wood AJ, Saneoka H, Rhodes D, Joly RJ and Goldsbrough PB (1996) Betaine Aldehyde dehydrogenase in sorghum. Molecular cloning and expression of two related genes. *Plant Physiol.* **110** : 1301-1308.

Wyn Jones RG, Rippin AJ and Storey R (1973). Metabolism of choline in the rhizosphere and its possible influence on plant growth. *Pestic Sci.* **4** : 375-383

Wyn Jones RG and Storey R (1981) Betaines. In: *The Physiology and Biochemistry of Drought Resistance in Plants* (Eds. Paleg LG, Aspinall D) Academic Press, Sydney New York London Toronto San Francisco, pp. 171-204.

Xu D, Duan B, Wang B, Hong T, Ho HD and Wu R (1996) Expression of a late embryogenesis abudant protein gene, *HVA1,* from barley confers tolerance to water deficit and salt stress in transgenic rice. *Plant Physiol.* **110** : 249-257.

Yamada H, Mori N and Tani Y (1979) Properties of choline oxidase of *Cylindrocarpon didymium* M-1. *Agric. Biol. Chem.* **43** : 2173-2177.

Chapter 8

GENETIC MANIPULATION OF POLYAMINE METABOLISM

MV Rajam★, R Kumria, B Waie and R Sharma

Plant Polyamine and Transgenic Research Laboratory, Department of Genetics, University of Delhi – South Campus, Benito Juarez Road, New Delhi - 110 021, India

Summary

Polyamines (PAs) are small ubiquitous molecules involved in various cellular and molecular processes, including plant morphogenesis and stress responses. The extent of involvement and mechanism of action of PAs are, however, not fully understood. Therefore, a number of strategies have been employed in order to understand their mechanism of action. These strategies include the use of inhibitors, mutants of PA genes and transgenics. These studies have given some clues about the role of PAs in growth and development and stress responses. The study of PA mutants mostly in bacterial and fungal systems, have given some insight into the role of different PA biosynthetic genes in various cellular and molecular processes. The PA mutants in higher plants although lesser in number, also suggest the role of PA genes in development of higher plants. Recently, transgenic plants over-expressing PA genes have been reported and such transgenics would be very important to study the functions of PAs.

Keywords : Polyamines, puterscine, spermidine, spermine, mutants, transgenics, plant development, plant stress.

Abbreviations : PA - Polyamine; PUT - Putrescine; CAD - Cadaverine; SPD - Spermidine; SPM - Spermine;

★Corresponding author : E-mail : rajam@bol.net.in

ADC - Arginine decarboxylase; ODC - Ornithine decarboxylase; SAMDC - S-Adenosylmethionine decarboxylase; DAO - Diamine oxidase; PAO - Polyamine oxidase; DFMA - Difluoromethylarginine; DFMO - Difluoromethylornithine; CHA - *bis* (Cyclohexylammonium) sulphate; MGBG - Methylglyoxal-*bis* (guanylhydrazone).

1. INTRODUCTION

Polyamines (PAs), the diamine putrescine (PUT), triamine spermidine (SPD) and tetraamine spermine (SPM) are naturally occurring polycationic compounds, which due to their ubiquity and versatility are involved in the regulation of various cellular and molecular processes, including the regulation of growth and differentiation, stress responses, membrane integrity and the synthesis and function of macromolecules (DNA, RNA and proteins) (Rajam, 1997; Rajam *et al.*, 1998). The role of PAs is much better studied in microbes and animals systems than plants, though they have been suggested to have a role as new plant growth regulators either by mediating the plant hormone effects or independently signalling other responses (Galston and Kaur-Sawhney, 1990; Smith, 19985; Evans and Malmberg, 1989; Rajam *et al.*, 1998).

PAs exist in three forms in the cell; as free cations, covalently bound to low molecular weight phenolic compounds like hydroxycinnamic acids (conjugated form of PAs) and bound to marcomolecules or membranes (bound form of PAs). Though the major form is the free cationic form of PAs, there are instances when the amounts of conjugated form exceed the free from and these are known to be critical in certain physiological processes including seed germination, flower development, defence responses and abiotic stress reactions (Martin-Tanguy, 1985; Rajam, 1997). Besides common PAs (PUT, SPD and SPM), certain unusual and novel PAs such as nor-SPD (caldine), homo-SPD, nor-SPM (thermine), homo-SPM and thermo-SPM have been found in some organisms like thermophilic bacteria and in plants under temperature stress (Rajam, 1997).

The present review deals with the role of PAs in cellular and molecular processes, with special emphasis on the genetic manipulation of PA metabolism in plants.

2. POLYAMINE METABOLISM

PUT can be derived from two independent pathways, the decarboxylation of ornithine and arginine by ornithine decarboxylase (ODC) and the arginine decarboxylase (ADC), respectively. In plants ADC forms the major pathway for PA biosynthesis. The PUT, hence derived from either of the two pathways is converted into higher PAs, SPD and tetra-amine SPM by the addition of an aminopropyl group obtained from decarboxylated S-adenosylmethionine (dcSAM). The dcSAM is in turn synthesized from S-adenosyl-methionine (SAM) by the action of S-adenosyl-methionine decarboxylase (SAMDC) (Rajam, 1997). SAM is in turn synthesized from methionine by SAM synthase. SAM is a major methyl donor in the cellular metabolism and also forms a part of ethylene biosynthesis. SAM is converted to 1-aminocyclopropane-1-carboxylic acid (ACC) by ACC synthase which is converted to ethylene by ACC oxidase (Fig. 1).

The catabolism of PAs involves the enzymes diamine oxidase (DAO) and polyamine oxidase (PAO). DAO preferentially acts on diamines (PUT, cadaverine) to form pyrroline, ammonia (NH_3) and hydrogen peroxide (H_2O_2), though it can break down SPD to aminopropyl pyrroline, NH_3 and H_2O_2 (Smith, 1985). PAO on the other hand breaks down SPD and SPM to pyrroline and aminopyrroline respectively, and diamine propane (DAP) and H_2O_2 (Smith, 1985). Pyrroline formed due to the activity of DAO and PAO is further converted to γ-aminobutyric acid (GABA) by a NAD (nicotinic acid diamine) dependent dehydrogenase (Flores and Filner, 1985).

The enzymes involved in PA biosynthesis and catabolism have been purified from different organisms (Smith, 1985; Rajam, 1997). The study of PAs owns much to the availability of specific, irreversible, substrate- or product- based inhibitors of its biosynthetic enzymes (Bey *et al.*, 1987). The substrate analogue of ODC, α-difluoromethylornithine (DFMO) was the first inhibitor to be synthesized (Metacalf *et al.*, 1978). Similarly, α-difluoromethylarginine (DFMA) has been used as a potent inhibitor of ADC (Kallio *et al.*, 1981). Another set of similar substrate-based analogoues for ODC and ADC were monofluoromethylornithine (MFMO) and monofluoromethylarginine (MFMA) respectively, these have been reported to be much more potent than DFMO and DFMA (Bey *et al.*, 1987). The substrate based inhibitor for lysine decarboxylase (LDC), DL-α-difluoromethyllysine (DFML) is also available (Bet *et al.*, 1987; Bitonti *et al.*, 1987). A very potent inhibitor of the enzyme SAMDC is

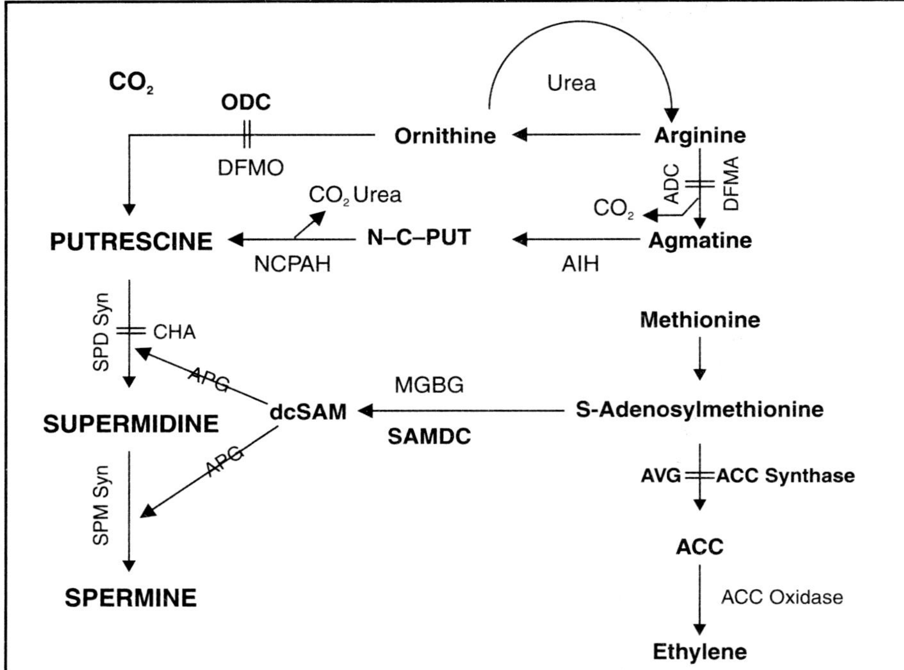

Figure 1 : Biosynthetic pathway of polyamines and its inter-relationship with ethylene biogenesis in plants. Biosynthetic inhibitors are indicated with double lines at the target site. ADC, Arginine decarboxylase; DFMA, difluoromethylarginine; AIH, Agmatine iminohydrolase, ODC; Ornithine decarboxylase; DFMO, difluoromethylornithine; CHA, *bis* (cyclohexyl-ammonium) sulphate; APG, Aminopropyl group; SAMDC, S-Adenosyl-methionine decarboxylase; MGBG, methylglyoxal-*bis* (guanylhydrazone); dc SAM, Decarboxylated S-adenosylmethionine; SPD Syn, Spermidine Synthase; SPM Syn, Spermine Synthase AVG, Amino ethoxyvinlyglycine; ACC, 1-Aminocyclopropane-1-Carboxylic acid; NCP, N-Carbomyl putrescine; NCOAH, N-Carbomyl- putrescine amidohydrolase

methylgloxyl *bis* (guanylhydrazone) (MGBG), though it has been put to limited use due to its non-specific effects on the respiratory enzymes (Pegg and Willams-Ashman, 1987). The inhibitors have provided insight into the dynamics of interconversion and regulation of the levels of PAs in the cell.

3. THE ROLE OF POLYAMINES IN CELLULAR AND MOLECULAR PROCESSES

The study of plants PAs has come a long way since the first report by Bagni (1966) regarding the stimulatory effect of PAs on growth of

Helianthus tuberous explants. PAs have been demonstrated to be associated with regulation of somatic embryogenesis (Minocha and Minocha, 1995; Sharma and Rajam, 1995; Yadav and Rajam, 1997, 1998), root and shoot formation (Galston and Flores, 1991; Sharma *et al.*, 1997; Watson *et al.*, 1998), flower and fruit development (Kakkar and Rai, 1993), stress responses (Evans and Malmberg, 1989; Galston, 1989; Rajam, 1997; Rajam *et al.*, 1998) and senescence (Galston and Kaur-Sawhney, 1987; Tiburcio *et al.*, 1994). Yet no clear picture of their mechanism of action or extent, period or stage of involvement, emerges.

The polycationic nature of PAs allows them to interact with various molecules whose functioning is modulated and regulated by them. PAs also have a role in free radical scavanging due to their polycationic nature (Drolet *et al.*, 1986). Although there are many other suggested possible mechanisms regarding the functions of PAs, the exact role, the extent of involvement and the mechanism of action of PAs is as yet not very clearly understood.

The chemical nature of PAs allows them to bind easily with different macromolecules. A very crucial binding of PAs is with membrane phospholipids thus stabilizing them. The binding was found to be specific and not due to their polycationic nature as Ca^{2+} and Mg^{2+} ions were not able to substitute the effect (Roberts *et al.*, 1986). In senescing *Hordeum* leaves, PAs were bound to thylakoid membranes and they were able to reduce chlorophyll loss (Besford *et al.*, 1993; Tiburcia *et al.*, 1994). The binding of SPM to mitochondrial membrane prevented the lowering of membrane potential and the Ca^{2+}, PO_4^{-3} fluxes under saline stress (Bueno *et al.*, 1993). Similarly PAs have been found associated with cell wall components like lignin and pectins and have been implicated in maintaining cell wall characteristics by strengthening the links between cell wall components (Berta *et al.*, 1997). PAs play a role in cell wall expansion and are part of modulators involved in host-pathogen interactions (Charney *et al.*, 1992; Moustacas *et al*; 1991). The binding of SPD to the plasma membrane proteins in zucchini hypocotyls has been characterized. SPD was found to have a specific binding to a 44 and a 66 kD protein (Tassoni *et al.*, 1998).

PAs share a common precursor with nucleic acids, carbamoyl phosphate, thus being linked in the biosynthetic chain at least at one point. Besides, the binding of PAs with DNA and RNA is also known to be responsible for playing some part in the regulation of gene expression. It had been proposed that PAs affect growth by interacting with DNA, a model for the interaction of SPM with DNA was proposed by performing conformational energy changes when SPM might be docked in the groove of B-DNA (Feuerstein *et al.*, 1986). SPD has been reported to be associated with Z-DNA (Ohishi *et al.*, 1996) and SPM plays a part in B to Z DNA transitions. The Z DNA formation in discrete regions in the DNA requires Z DNA forming sequences, negative supercoiling as well as multi-valent cations like SPM (Balasundarum and Tyagi, 1991; Howell *et al.*, 1996). SPM has been reported to stabilize the triplex DNA formation and aggregation (Musso *et al.*, 1997). PAs have been reported to stimulate DNA, RNA and protein synthesis (Auvinen *et al.*, 1992). PAs are involved in joining of okazaki fragments and the depletion of PAs leads to accumulation of short DNA pieces (Pohjanpelto and Holta, 1996). As a matter of fact *odc* has been suggested to be a proto-oncogene and its over expression leads to cell transformation (Auvinen *et al.*, 1992).

The binding of PAs to nucleic acids leads to compactation of macromolecules and also neutralization of negative charges of phosphate groups, the RNA molecules are thus protected from RNases, therefore RNA molecules are also reported to bind PAs (Serafini-Fracassini *et al.*, 1984). PUT and to some extent SPD and SPM are an obligate requirement for the replication of yeast double stranded RNA plasmids (Tyagi *et al.*, 1984). The antisenescence effects of PAs are in part due to the inhibition of ribonuclease synthesis and activity by PAs (Isola and Franzoni, 1989).

SPD and SPM are known to stimulate the reading of amber mutations and play an active role in the expression of specific genes (Morch and Benicourt, 1980; Tabor and Tabor, 1982). SPM also has a role in stimulating protein synthesis in ageing seeds, therefore prolonging the seed viability by affecting the active conformation of t-RNA (Mukhopathyay and Ghosh, 1986). PAs also inhibit the protease activity, thereby delaying the degradative processes initiated during stress or senescence (Kaur-Sawhney and Galston, 1991).

PAs regulate their own biosynthesis by inducing a ribosomal frame shifiting in the translation of ODC antizyme (Rom and Kahana, 1994). Besides, being involved at the DNA and RNA levels, PAs are also known to regulate protein synthesis and activity. SPM has been reported to have a specific role in activity of cyclic AMP independent caesin kinase (Datta *et al.*, 1997; Roux, 1993). A branched quarternary PA, tetrakis (3-aminopropyl) ammonium along with SPM has been reported to support protein synthesis at high temperature in a thermophilic bacteria (Uzawa *et al.*, 1993). SPD stimulates protein synthesis in chloroplast especially in light (Blatter *et al.*, 1992). PUT, SPD, SPM and CAD enhanced the phosphorylation of several plasma membrane proteins in tobacco, cucumber and *Arabidopsis* (Yet *et al.*, 1994). PUT has been reported to increase the phosphorylation of many soluble proteins too unlike SPD and SPM which decreased the phosphorylation of soluble proteins (Ye *et al.*, 1994). PAs are known to regulate cell division and also prolong the cell division phase by inhibiting the synthesis of phenylpropanoids. The conjugation of PAs with phenolics regulated the free PA levels and therefore, led to the cessation of cell division (Mader and Hanke, 1997).

The critical role played by PAs in growth and development along with their role in protein phosphorylation/dephosphorylation point towards the possibility of these being considered as signal transduction molecules. Though the high PA titres in the cell are quite unlike secondary messengers which increase rapidly and transiently in response to a stimuli. Therefore, this aspect of PAs warrants more attention to be able to pinpoint their role and mechanism of involvement, if any, in signal transduction (Ye *et al.*, 1994; Tiburcio *et al.*, 1993).

4. POLYAMINE BIOSYNTHESIS GENES

The PA biosynthetic genes were first isolated from animal systems, yeast and bacteria. *odc* gene has been cloned from human, rat, mouse, holstein, *Trypanosoma, Leishmania*, yeast, *Neurospora* and *E. coli* (van Kranen *et al.*, 1997; Katz and Kahana, 1988; Hickok *et al.*, 1990; Willams *et al.*, 1992; Yao *et al.*, 1995). In recent years PA biosynthetic genes have been cloned from plants too. The *odc* gene has been cloned from *Datura*, tobacco and tomato (Malik *et al.*, 1996; Michael *et al.*, 1996; Alabadi and Carbonell, 1998).

The *adc* gene has been cloned from oat, tomato, pea, *Arabidopsis* and soybean (Bell and Malmberg, 1990; Rastogi *et al.*, 1993; Perez-Amador *et al.*, 1995; Watson and Malmberg, 1996; Nam *et al.*, 1997). The *samdc* gene has been cloned from *Arabidopsis*, *Datura*, potato (Taylor *et al.*, 1992), spinach (Bolle *et al.*, 1995), *Catharanthus roseus* (Schroder and Schroder, 1995), *Tritordeum* (Dresselhaus *et al.*, 1996), *Pharbitis nil* (Yoshida *et al.*, 1998), tomato, tobacco (Kumar *et al.*, 1997), rice (Le and Chen, 2000) and also from human genome. The *spd syn* gene has been cloned from *N. sylvestris, Hyoscymus niger* and *Arabidopsis* (Hashimoto *et al.*, 1998).

The enzymes involved in formation of conjugated PAs have also been cloned. The gene for the enzyme homo-SPD synthase has been cloned from bacteria (*Acetobacter*) (Yamamoto *et al.*, 1993), *Senecia vernalis* (Ober, 1997) and *Eupatorium cannavulgaris* (Kaiser, 1999). Besides, the biosynthetic enzyme, the gene for the catabolic enzyme PAO has also been cloned from maize (Tavladoraki *et al.*, 1998) and DAO from lentil (Rossi *et al.*, 1992) and pea (Tavladoraki *et al.*, 1998).

5. GENETIC MANIPULATION OF POLYAMINE METABOLISM

The study of mutants is a powerful tool in understanding the physiological role and genetics of any gene in a living system. Therefore, this technique has been extensively used by workers in the PA area to understand the physiological and biochemical role of PAs in various systems such as *E. coli,* yeast and higher plants and also to study the genetics of PA biosynthesis.

Morris and Jorstad (1970, 1973) isolated *E. coli* mutant defective in PUT synthesis (ADC). The study of these mutants revealed that on unsupplemented medium *E. coli* preferentially uses ODC pathway for synthesis of PUT, whereas in arginine supplemented medium, exclusively the ADC pathway is used. The Partial mutants could grow on unsupplemented medium but complete mutants for ADC either required PUT or SPD for optimal growth. Also, it was concluded that the major requirement is for SPD and not for PUT as SPD analogues could restore growth but PUT analogues that cannot be converted into SPD, could not (Hirshfield *et al.*, 1970).

E. coli mutants of *SpeD* gene encoding SAMDC were isolated which lacked adenosyl-methionine and SPD. These mutants showed negligible changes in the growth rate, suggesting that other amines or cations may substitute SPD in *E. coli* (Tabor *et al.*, 1978). In further experiments, K12, a strain of *E. coli* deficient in Spe A, B, C and D coding for *Adc*, agmatine ureohydrolase, *ODC* and *SAMDC*, respectively were used. This strain could grow indefinitely at 1/3rd the growth rate but was normal with respect to its phage producing ability except for phage l which is known to require SPD. The structure and function of the pili was affected by the deficiency of PAs (Hafner *et al.*, 1979). This strain K12 was found to have some amount of CAD and therefore, another mutation in cad A was transduced into this strain. This modified strain also exhibits the same phenotype as the parent strain and thus the ability of K12 to grow could not be attributed to the presence of CAD (Tabor *et al.*, 1980). The role of PAs in protein synthesis machinery, especially the ribosomal complex was emphasised by Tabor and Tabor (1983). A rps L mutation was introduced in the PA deficient strain resulting in the failure of this strain to grow in the absence of PAs. It was suggested that rps L mutation causes an additional defect in the ribosomal complex, which is already impaired due to lack of PAs. With the use of these mutants, all the PA biosynthetic genes were mapped on the *E. coli* chromosome.

Saccharomyces cerevisiae was targeted next to understand the role of PAs in eukaryotic systems. Yeast mutants deficient in ODC were isolated in which the entire PA biosynthesis was blocked suggesting that the alternate pathway for PUT synthesis involving ADC and agmatine ureohydrolase was absent in yeast (Whitney and Morris, 1978). SPD mutants of yeast deficient in SAMDC lacked both SPD and SPM and showed very retarded growth. Addition of SPD or SPM could not be achieved with PUT or CAD. Sporulation in these mutants was entirely absent, suggesting the absolute requirement of SPD and / or SPM for sporulation (Cohn *et al.*, 1978). Tabor (1981) and Tabor *et al.* (1983) isolated individual mutants of yeast lacking ODC, SAMDC, SPD and SPM synthase activities and from the study of these mutants it could be concluded that PA requirement is absolutely necessary for yeast growth, separate enzymes are present for SPD and SPM synthesis and that ODC activity in yeast is controlled by cellular SPM.

Arabidopsis and toabcco, two model plants were also used for the study of PA action and genetics in higher plants. *Arabidopsis* mutants defective in PA biosynthesis showed low levels of ADC, ODC and PUT and displayed alteration in root, stem and floral morphology. Genetic analysis of ADC mutants revealed that Spe 1 and Spe 2 fall under 2 complementation groups; Spe 1 encodes the ADC structural gene and Spe 2 encodes a regulatory ADC protein (Kumar *et al.*, 1997).

Malmberg and McIndoo (1983) isolated PA mutants in tobacco cell lines. The growth and floral development of mutant mgr 3 was abnormal. The stigma became longer than the corolla and turned black, the anthers had non-viable pollen and died early. Most of the ovules inside the ovary were transformed into anthers. However, these transformed anthers did not show any pollen development. PA analysis showed that this mutant had significantly high titres of SPD and SPM. Earlier, two temperature sensitive mutants ts4 and rt1 were isolated (Malmberg, 1980). Both ts4 and rt1 had short internodes but ts was light green and could be maintained only in culture whereas rt1 was dark green and had flowers with a 2nd row of petals in place of anthers. These and a number of other mutants having floral abnormalities suggested the role of PAs in flower development. ODC might be involved in initiation of flowering as ODC mutants failed to flower, whereas SAMDC plays a role in anther development and in male-female balancing system (Malmberg and McIndoo, 1984; Malmberg, 1985).

Most recently, a novel technique called activation T-DNA tagging has been used to generate mutants. This method involves the use of a T-DNA driven by the transcriptional enhancer sequence of the CaMV35S RNA promoter which is cloned as a tetramer near the right border. After insertion of such a sequence in the plant genome, the flanking regions of the genome are expected to come under the influence of the transcriptional enhancers and are over expressed resulting in the production of dominant mutation. This method allows the isolation of the genes involved in the mutations. Genetically mutated transformed tobacco cell lines were selected on toxic concentrations of MGBG and two mutants were studied for their phenotype, SAMDC activity and PA content. mgbg, 2 had two T-DNA insertions and showed leaf wrinkling or elongation, abnormal flowers, elongated styles, male sterility and parthenocarpy mgbgr 3 plants were shorter than control, displayed smaller flowers having elongated styles and were male sterile. SAMDC activity of both mgbgr 2 and 3 were 25-38% higher than in control tissue. PA analysis of mgbgr 2 and 3 showed that although mgbgr 2 had double the amount of PUT and significantly higher SPD when compared to controls, mgbgr 3 has similar free PA levels with that of control but has a 50% reduced conjugated SPD levels. Flanking regions of T-DNA from mgbgr 3 were rescued and cloned. This DNA when introduced into untransformed tobacco protoplasts by PEG-mediated uptake, conferred ability to grow under selective levels of MGBG. However, the gene tagged was not identified (Fritze *et al.*, 1995).

More recently, transgenic plants expressing PA biosynthetic genes were generated to gain better understanding of the PA metabolism, reconfirm the effects of the modulation of PA titres caused by the inhibitors at the molecular level and also to overcome the limitations of the inhibitor-based experiments. The transgenic approach was highly specific to the target gene and moreover, it provided a tool for manipulating the metabolic flux with the persistent shift in the PA metabolism. Even though some of the plant PA biosynthesis genes have been isolated and characterised, most transgenics have been raised using genes from heterologous source as these were the first to be isolated. In most of the transgenics generated, CaMV35S promoter has been used to drive the transgene, though tetracycline inducible promoter has also been used in cases where extreme deleterious effects of the transgene were expected (Rajam, 1997; Rajam *et al.*, 1998).

The first report of the introduction of yeast *odc* gene was in root cultures of tobacco using *Agrobacterium rhizogenes* (Hamill *et al.*, 1990). The study was aimed to increase the nicotine content of the culture as PUT is a precursor for nicotine. Hence, over-production of PUT was attempted by using a double enhancer sequence containing promoter but only a 3-fold increase in ODC activity and a 2-fold increase in nicotine was observed. No significant increase in SPD and SPM levels was observed. No plants were regenerated from the transformed root cultures (Hamill *et al.*, 1990). The tobacco transgenic plants expressing the *odc* gene were the fist PA transgenic plants raised by DeScenzo and Minocha (1993). Two constructs were used for the transformation of tobacco, one having complete coding sequence of *odc* and the other in which 350 bp of 3 end were removed. The truncated gene produced a functional peptide of 37 amino acids less at the C terminus and with an increased half-life. The enzyme activity when checked at the pH optimum for mouse ODC and was much higher in transgenics compared to endogenous plant ODC activity in the controls, whereas at the pH optimum for the plant ODC, no significant change was observed. PUT was found to be 2-3 fold higher in leaves and 4-12 fold higher in callus, though no significant increase was observed in SPD and SPM content as the amounts of SAM were suggested to be limiting. The transgenics having high PUT titres were stunted with wrinkled leaves and reduced stamens.

Carrot cell lines are known to have no detectable ODC activity, only the ADC pathway is functional in them. Carrot cell lines were transformed with mouse *odc* gene driven by CaMV35S promoter (Bastola and Minocha, 1995) and the effect of the high PUT titres on somatic embryogensis was studied. The transformed cell lines showed improved somatic embryogenesis which could be correlated to higher PUT amounts. The somatic embryos were formed even in the presence of DFMA which inhibited the carrot ADC, therefore all the PA requirements of the embryos were fulfilled by the introduced mouse *odc* gene. Exogenous addition of PUT was not found to be helpful thus suggesting that a fast turn over of PUT is also essential besides the high concentration for somatic embryogenesis. These transformed carrot cell lines were used to study the shift in metabolic flux as compared to the control cell lines (Anderson *et al.*, 1998). ^{14}C labelled arginine, ornithine, methionine or PUT was fed to the cell cultures and amount of label incorporated in different PAs and their fractions was analyzed. ^{14}C labelled PUT was much higher in transgenic cell lines when ^{14}C ornithine was given as substrate and there was no difference in labelled PAs when ^{14}C arginine was fed to the cultures. In correlation the conversion of ^{14}C-methionine to ethylene was much lower in transgenics due to a shift in the dynamics towards PA metabolism as more of PUT was available to be converted to SPD and SPM (Anderson *et al.*, 1998).

Transgenic tobacco plants overexpressing human *samdc* gene driven by CaMV35S promoter were generated by Noh and Minocha (1994). These plants were found to have 2-6 fold higher SPD than the untransformed controls, there was an increase in SPM too though PUT understandably decreased. Since high amounts of SPD is cytotoxic, the regenerants obtained might have been moderate accumulators of SPD as the increase in SPD and SPM was not comparable to the dramatic decrease in PUT. The cytotoxicity of the high amounts of SPD did not allow the regeneration of any plants overexpressing potato *samdc* in potato, hence a tetracycline (tet)-inducible promoter was used to drive *samdc*. The sense *samdc* plants showed 7-fold increase in SPD, 3-fold in SPM and a decrease in PUT on tet-induction. Similarly potato plants expressing antisense *samdc* gene driven by both 35S promoter and tet-inducible promoter were raised (Kumar *et al.*, 1996). The antisense *samdc* plants were stunted, branched, necrotic with few small tubers, this was attributed to increase in ethylene levels caused by the channelling of SAM for the formation of ethylene due to down-regulation of SAMDC. Though a decrease in PUT levels was observed due to down-regulation of ODC, ADC by the elevated levels of ethylene. Similar plant morphology was observed in case of plants transformed with *Agrobacterium rhizogenes* (Martin Tanguy *et al.*, 1990).

The fact that elevated levels of PA were deterimental for plant regeneration and were cytotoxic, led to the use of tet-inducible promoter for driving oat *adc* gene introduced into tobacco (Masgrau *et al.*, 1997). The PUT levels were increased by 16-46% on tet induction and more significant increase was seen in the conjugated and bound fractions of PAs. A prolonged induction of the transgene at an early stage of development led to plant growth inhibition, necrotic, wrinkled leaves, but no such effects were seen in case of older plants on tet-induction of the transgene. These results clearly brought forward the differential role of PAs at various developmental stage (Masgrau *et al.*, 1997).

Rice transgenics overexpressing the oat *adc* gene have been raised (Capell *et al.*, 1998), these transgenics showed a 4-7 fold increase in the activity of the ADC enzyme, along with a 4-fold increase in the PUT titres. The high PUT titres were found to be inhibitory to the plant regeneration from the transformed calli. The effect of the strength of the promoter driving the *adc* gene on the PA metabolism as well as the morphogenic capacity of the transformed calli has also been analyzed (Bassie *et al.*, 2000). In a later study, an increase in PA levels in rice shoots and seeds has been observed (Noury *et al.*, 2000).

Since PAs are known to play a role in stress responses, particularly the activity of the enzyme ADC is known to increase under stress along with increase in PUT levels, rice transgenics with *adc* gene were tested for their tolerance to abiotic stress and it was observed that no chlorophyll loss was observed after 8-days of drought (Capell *et al.*, 1998).

Though the *spd syn* gene and PA catabolic genes (DAO and PAO) have been isolated, no transgenic plants have been raised with it, neither are there any transgenic reported expressing the *adc* and *odc* genes in antisense orientation for studying the effects of long-term down-regulation of these genes on plant development.

6. CONCLUSIONS AND FUTURE PROSPECTS

PA biochemistry is well worked out, though the biosynthetic pathways for synthesis of novel and unusual PAs is not studied in details. Since the 1970's a lot of progress has been made to understand the genetics of PAs as well as their physiological role. A lot of information has been gathered in *E. coli* and yeast on the maping of various PA genes in their genome and also the role of PAs in their life-cycle. Mutant analysis has given a lot of insight into the role and interaction of PAs with each other and with other biomolecules. Most of the genes involved in PA biosynthesis have been isolated, characterized, cloned and most of these have been used to create transgenics, further proving the vital role of PAs in various biological processes. The studies done so far reveal that besides the essential role of PAs in regular cellular functions and normal growth and development in plants, they play an important role under various biotic and abiotic stresses. This information is being used by various workers including us, to create transgenic overexpressing PA genes in order to confer stress tolerance into the plants. However, transgenics are not available for the genes involved in PA catabolism and with antisense *odc* and *adc* genes, and the availability of such transgenic plants is utmost important to understand the role of PA metabolism in various biological processes.

A lot of work still needs to be done to understand the mechanism of action of PAs in different processes so that PAs may be exploited further to improve the quality of crops.

ACKNOWLEDGEMENTS

Polyamine research in my laboratory has been generously supported by grants from the Department of Biotechnology (No. BT/R&D/08/40/96), the Department of Science and Technology (No. SP/SO/AO6/96) and the Indian Council of Agricultural Research (No. F-1 (21)/96-FFC), New Delhi. Senior Research Fellowships from the University Grants Commission, New Delhi to RK and BW are gratefully acknowledged.

REFERENCES

Alabadi D and Carbonell J 1998. Expression of ornithine decarboxylase is transiently increased by pollination, 2,4-Dicholorphenoxyacetic acid, gibberllic acid in tomato ovaries, *Plant Physiol.* **118** : 323-328.

Anderson SE, Bastola DR and Minocha SC 1998. Metabolism of polyamines in transgenic cells of carrot expressing a mouse ornithine decarboxylase cDNA, *Plant Physiol.* **116** : 299-307.

Auvinen M, Passinene A, Andersson LC and Holtta E 1992. Ornithine decarboxylase activity is critical for cell transformation, *Nature* **360** : 355-358.

Bagni N 1996. Aliphatic amines as a growth factor of coconut milk as stimulating cellular proliferation of *Helianthus tuerosus* (Jerusalem artichoke) *in vitro, Experimentia* **22** : 732-736.

Balasundarum D and Tyagi AK 1991. Polyamine-DNA nexus: structural ramifications and biological implication, *Mol. Cell. Biochem.* **100** : 129-140.

Bassie L, Noury M, Lepri O, Lahaye T, Christou P and Capell T 2000. Promoter strength influences polyamine metabolism and morphogenic capacity in transgenic rice tissues expressing oat *adc* cDNA constitutively, *Transgenic Res.* **9** : 33-42.

Bastola DR and Minocha SC 1995. Increaed putrescine biosynthesis through transfer of mouse ornithine decarboxylase cDNA in corrot promotes somatic embryogenesis, *Plant Physiol.* **109** : 63-71.

Bell E and Malmberg RL 1990. Analysis of a cDNA encoding arginine decarboxylase from oat reveals similarity to the *Escherichia coli* arginine decarboxylase and evidence of protein processing, *Mol. Gen. Genet.* **224** : 431-436.

Berta G, Altamura MM, Fusconi A, Cerruti F, Capitiani F and Bagni N 1997. The plant cell wall is altered by inhibition of polyamine biosynthesis, *New Phytol.* **137** : 569-577.

Besford RT, Richardson CM, Campos JL and tiburcio AF 1993. Effect of polyamines on stabilization of molecular complexes in thylakoid membranes of osmotically stressed leaves, *Planta* **189** : 201-206.

Bey Y, Dnazin C and Jung M 1987. Inhibition of basic amino acid decarboxylases involved in polyamine biosynthesis. In: McCann P P, Pegg A E and Sjoerdsma A (Eds) Academic Press, San Diego, pp 1032.

Bitonti AJ and McCann PP 1987. Inhibition of polyamine biosynthesis in microorganisms. In: McCann P P. Pegg A E and Sjoerdsma A (Eds) Academic Press, San Diego, pp 259-275.

Blattler R, Ochsenbein A and Boschetti A 1992. Polyamines and cytoplasmic preparations enhances light-driven protein synthesis in isolated chloroplasts of *Chlamydomonas reinhardtii, Plant Physiol. Biochem.* **30** : 743-752.

Bolle C, Herrmann RG, Oelmuller R 1995. A spinach cDNA with homology to S-adenosylmethionine decarboxylase, *Plant Physiol.* **107** : 1461-1462.

Bueno M, Garrido D and Matilla A 1993. Gene expression induced by spermine in isolated embryonic axes of chickpea seeds, *Plant Physiol.* **87** : 381-388.

Capell T, Escobar C, Liu H, Burtin D, Lepri O and Christou P 1998. Overexpressin of oat arginine decarboxylase cDNA in transgenic rice (*Oryza sativa* L.) affects normal development pattern *in vitro* and results in putrescine accumulation in transgenic plants, *Theor. Appl. Genet.* **97** : 246-254.

Charney D, Nari J and Noat G 1992. Regulation of plant cell wall pectin methyl esterase by polyamines-interaction with effects of metal ions, *Eur. J. Biochem.* **205** : 711-714.

Cohn MS, Tabor CW and Tabor H 1978. Isolation and characterization of *Saccharomyces cerevisiae* mutants deficient in S-Adenosylmethionine decarboxylase, Spermidine and Spermine, *J. Bacteriol.* **134** : 208-213.

Datta N, Schell MB and Roux SJ 1987. Spermine stimulation of a nuclear NII kinase from pea plumule and its role in the phosphorylation of a nuclear polypeptide, *Plant Physiol.* **84** : 1397-1401.

DeScenzo RA and Minocha SC 1993. Modulation of cellular polyamines in tobacco by transfer and expression of mouse ornithine decarboxylase cDNA, *Plant. Mol. Biol.* **22** : 113-127.

Dresselhaus T, Barcela P, Hagel C, Lorez H and Humbeek K 1996. Isolation and characterization of a Tritordeum cDNA encoding S-adenosylmethionine decarboxylase that is cireadian clock-regulated, *Plant Mol. Biol.* **30** : 1021-1033.

Drolet G, Dumbroff EB, Legge RL and Thompson JE 1986. Radical scavenging properties of polyamines, *Phytochemistry* **25** : 367-371.

Evans PT and Malmberg RL 1989. Do polyamines have roles in plant development? *Annu. Rev. Plant Physiol. Plant Mol. Biol.* **40** : 235-269.

Feuerstein BG, Pattabiraman N and Marton LJ 1986. Spermine-DNA interaction: A theoretical study, *Proc. Natl. Acad, Sci. USA* **83** : 5948-5952.

Flores HE and Filner P 1985. Polyamine catabolism in higher plants: characterization of pyrroline dehydrogenase, *Plant Growth Regul.* **3** : 277-291.

Fritze K, Czaja I and Walden R 1995. T-DNA tagging of genes influencing polyamine metabolism : isolation of mutant plant lines and rescue of DNA promoting growth in the presence of polyamines biosynthetic inhibitor, *Plant J.* **7(2)** : 261-271.

Galston AW and Flores HW 1991. Polyamines and plant morphogenesis. In: *Biochemistry and Physiology of Polyamines in Plants* (Eds. Slocum RD and Flores HE) CRC Press, Boca Raton, London, pp 175-186.

Galston AW and Kaur-Sawhney R 1987. Polyamines and senescence in plants. In: *Plant Senescence: Its Biochemistry and Physiology,* (Eds. Thomson W, Nothnagel EA and Huffaker RC). American Society of Plant Physiologists, Rockville, MD, pp 167-181.

Galston AW 1989. Polyamines. In : *The Physiology of Polyamines* (Eds. Bachrach U and Heimer YM) vol. 2 CRC Press, Boca Raton, Florids, pp 99-105.

Hafner EW, Tabor CW and Tabor H 1979. Mutants of *Escherichia coli* that do not contain 1,4-Diaminobutane (Putrescine) or Spermidine, *J. Bacteriol. Chem.* **254** : 12419-12426.

Hamasaki-Katagiri N, Tabor CW and Tabor H 1997. Spermidine biosynthesis in *Saccharomyces cerevisiae*: Polyamine requirement of a null mutant of the Spe 3 gene (Spermidine Synthase), *Gene.* **187** : 35-43.

Hamill JD, Robins RJ, Parr AJ, Evans DM, Furze JM and Rhodes MJC 1990. Over-expression of a yeast ornithine decarboxylase gene in transgenic roots of *Nicotiana rustica* can lead to enhanced nicotine accumulation, *Plant Mol. Biol.* **15** : 27-38.

Hashimoto T, Tamaki K, Suzuki K and Yamada Y 1998. Molecular cloning of plant spermidine synthase, *Plant Cell Physiol.* **39** : 73-79.

Hickok NJ, Wahlfors J, Crozat A, Halmekyto M, Alhonen L, Jahne J and Jahne OA 1990. Human ornithine decarboxylase- encoding loci : nucleotide sequence of the expressed gene and characterization of a psuedogene, *Gene* **93** : 257-263.

Hirshfield IN, Rosenfeld HJ, Leifer Z and Maas WK 1970. Isolation and characterization of a mutant of *Escherichia coli* blocked in the synthesis of Putrescine, *J. Bacteriol.* **101** : 725-730.

Howell ML, Schroth GP and Ho PS 1996. Sequence dependent effects of spermine on thermodynamics of B-DNA to Z-DNA transition. *Biochemistry* **35** : 15373-15382.

Isola CM and Franzoni L 1989. Inhibition of net synthesis of ribonuclease by polyamines in potato tuber slices, *Plant Sci.* **63** : 39-45.

Kaiser A 1999. Cloning and expression of a cDNA encoding homospermidine synthase from *Senecio vulgaris* (Asteraceae) in *Escherichia coli, Plant J.* **19** : 195-201.

Kakker RK and Rai KV 1993. Plant polyamines in flowering and fruit ripening, *Phytochemistry* **33** : 1281-1288.

Kallio A, McCann PP and Bey P 1981. DL-alpha-(Difluoromethyl) arginine: A potent enzyme activated irreversible inhibitor of bacterial arginine decarboxylase, *Biochemistry* **20** : 3163-3166.

Katz A and Kahana C 1988. Isolation and characterization of the mouse ornithine decarboxylase gene, *J. Biol. Chem.* **263** : 7604-7609.

Kaur-Sawhney R and Galston AW 1991. Physiological and biochemical studies on the anti-senescence properties of polyamines in plants. In : *Biochemistry and Physiology of Polyamines in Plants* (Eds. Slocum RD and Flores HE) CRC Press, Boca Raton, London, pp 201-211.

Kumar A, Altabella T, Taylor NA and Tiburcio AF 1997. Recent advances in poyamines research, *Trends Plant Sci.* **2** : 124-130.

Kumar A, Taylor MA, Arif SAM and Davis HV 1996. Potato plants expressing antisense and sense S-adenosylmethionine decarboxylase (SAMDC) transgenics show altered levels of polyamines and ethylene atntisense plants display abnormal phenotypes, *Plant J.* **9** : 147-158.

Li YZ sand Chen SY 2000. Differential accumulation of the S-adenosylmethionine decarboxylase transcript in rice seedlings in response to salt and drought stresses. *Mol. Breed.* **100** : 782-788.

Mader JC and Hanke DE 1997. Polyamine sparing may be involved in the prolongation of cell division due to inhibition of phenylpropanoid synthesis in cytokinin–starved soybean oils, *Plant Growth Regul.* **16** : 89-93.

Malik V, Watson MB and Malmberg RL 1996. A tobacco ornithine decarboxylase partial cDNA clone, *J. Plant Biochem. Biotech.* **5** : 109-112.

Malmberg RL 1980. Biochemical, cellular and development characterization of a temperature sensitive mutant of *Nicotiana tabaccum* and its second site revertant, *Cell* **22** : 603-609.

Malmberg RL and McIndoo J 1983. Abnormal floral development of a tobacco mutant with elevated polyamine levels, *Nature* **305** : 623-625.

Malmberg RL and McIndoo J 1984. Ultraviolet mutagenesis and genetic analysis of resistance to methylglyoxal-bis (guanylhyrazone) in tobacco, *Mol. Gen. Genet.* **196** : 28-34.

Malmberg RL, McIndoo J, Hiatt AC and Lowe BA 1985. Genetics of polyamine synthesis in tobacco, Development switches in flower, *Cold Spring Harbour symp.* **50** : 475-482.

Martin-Tanguy J 1985. The occurrence and possible function of hydroxycinnamoyl acid amides in lants, *Plant Growth Regul.* **3** : 381-399.

Martin-Tanguy J, Cabanne F, Perdrizet E and Martin C 1978. The distribution of hydroxycinnamic amides in flowering plants, *Phytochemistry* **17** : 1927-1928.

Martin-Tanguy J, Tepfer D, Paynot M, Burtin D, Heisler L and Martin C 1990. Inverse relationships between polyamine levels and the degree of phenotypic alternation induced by the root inducing, left hand transferred DNA from *Agrobacterium rhizogenes, Plant Physiol.* **92** : 912-918.

Masgrau C, Altabella T, Fareas R, Flores D, Thompson AJ, Besford RT and Tiburcio AF 1997. Inducible overexpression of oat arginine decarboxylase in transgenic tobacco plants, *Plant J.* **11** : 465-473.

Messiaen J, Cambier P and van Cutsem P 1997. Polyamines and pectins I. Ion exchange and selectivity, *Plant Physiol.* **113** : 387-395.

Metcalf BW, Bey P, Danzin C, Jung M, Casara P and Vevert JP 1978. Catalytic irreversible inhibition of mammalian ornithine decarboxylase by substrate and product analogues, *J. Amer. Chem. Soc.* **100** : 2251-2253.

Michael AJ, Furze JM, Rhodes MJC and Burtin D 1996. Molecular cloning and functional identification of a plant ornithine decarboxylase cDNA, *Biochem. J.* **314** : 241-248.

Minocha SC and Minocha R 1995. Role of polyamines in somatic embryogenesis. In: *Somatic Embryogenesis and Synthetic Seed I,* vol 30 (Ed. Bajaj YPS) Springer-Verlag, Berlin, pp 53-70.

Morch MD and Bencourt C 1980. Polyamines stimulate suppression of amber termination codons *in vitro* by normal tRNAs, *Eur. J. Biochem.* **105** : 445-451.

Morris DR and Jorstad CM 1970. Isolation of conditionally Putrescine deficient mutants of *Escherichia coli, J. Bacteriol.* **101** : 731-737.

Morris DR and Jorstad CM 1973. Growth and macro molecular composition of a mutant of *Escherichia coli* during polyamine limitation, *J. Bacteriol.* **113** : 271-277.

Moustacas AM, Nari J, Borel M, Noat G and Ricard J 1991. Pectin methyl esterase, metal ions and plant cell wall expansion, *Biochem. J.* **279** : 351-354.

Mukhopadhyay A and Ghosh B 1986. Protein synthesis and loss of viability of rice seeds: Effect of polyamines on *in vitro* translation, *Physiol. Plant.* **68** : 441-445.

Musso M, Thomas T, Shirahata A, Sigal LH, van Dyke MW and Thomas TJ 1997. Effects of chain length modification and bis(ethyl) substitution of spermine analogs on purine-pyrimidine triplex DNA stabilization, aggregation and conformational transitions, *Biochemistry* **36** : 1441-1449.

Nam KH, Lee SH and Lee JN 1997. Differential expression of the ADC mRNA during development and upon acid stress in soybean (*Glycine max*) hypocotyls, *Plant Cell Physiol.* **38** : 1156-1166.

Noh E and Minocha SC 1994. Expression of a human S-adenoxylmethionine decarboxylase cDNA in transgenic tobacco and its effects on polyamine biosynthesis, *Transgenic Res.* **3** : 26-35.

Noury M, Bassie L, Lepri O, Kurek I, Christou P and Capell T 2000. A transgenic rice cell lineage expressing that oat anginine decarboxylase (adc) cDNA constitutively accumulates putrescine in callus and seeds but not in vegetative tissues, *Plant Mol. Biol.,* **43** : 537-544.

Ober D 1997. Strategien zur immunologischen and molekularbiologischen untersuchung der homospermidin synthase, dem eingangsenzym der pyrrolizidinalkaloid biosnthase. *Dissertation,* TU Braunschweig.

Ohishi H, Nakanishi I, Inubishi K, van der Marel G, van Boom JH, Rich A, Wang AHJ, Hakoshima T and Tomita K 1996. Interactions between the left-handed Z-DNA and polymine- The crystal structure of the $d(CG)_3$ and spermidine complex, *FEBS Lett.* **391** : 153-156.

Pegg AE and William-Ashman HG 1987. Pharmacologic interference with enzymes of polyamine biosyntheis and of 5-methylthioadenosine metabolism. In: *Inhibition of Polyamine Metabolism: Biological Significance and the Basis for New Therapeis,* McCann (Eds. PP, Pegg AE and Sjoerdsma A), Academic Press, San Diego, pp 33-48.

Perez-Amador MA, Carbonell J and Granell A 1995. Expression of arginine decarboxylase is induced during early fruit development and in young tissues of *Pisum sativum* (L.), *Plant Mol. Biol.* **28** : 997-1009.

Pohjanpelto P, and Boltta E 1996. Phosphorylation of okazaki-like DNA fragments in mammalian cells and role of polyamines in the processing of this DNA, *EMBO J.* **15** : 1193-1200.

Rajam MV 1997. Polyamines. In : *Plant Ecophysiology* (Ed. MNV Prasad), John Wiley, New York, pp 343-374.

Rajam MV, Dagar S, Waie B, Yadav JS, Kumar PA, Shoeb F and Kumria R 1998. Genetic engineering of polyamine and carbohydrate metabolism for osmotic stress tolerance in higher plants, *J. Biosci.* **23** : 473-482.

Rastogi R, Dulson J and Rothstein SJ 1993. Cloning of tomato (*Lycopersicum esculentum* Mill) arginine decarboxylase gene and its expression during fruit ripening. *Plant Physiol.* **103** : 829-834.

Roberts DR, Dumbroff EB and Thompson JE 1986. Exogenous polyamines alter membrane fluidity in bean leaves – A basis for potential misinterpretation of their true physiological role. *Planta* **167** : 395-401.

Rom E and Kahana C 1994. Polyamines regulate the expression of ornithine decarboxylase antizyme *in vitro* by inducing ribosomal frame-shifting, *Proc. Natl. Acad. Sci. USA* **91** : 3959-3963.

Rossi A, Petruzzelli R and Agro AF 1992. cDNA- derived amino- acid sequence of lentil seedlings amine oxidase, *FEBS* **301** : 253-257.

Roux SJ 1993. Casein kinase-2- type protein kinases in plants: possible targets of polyamines action during growth regulation, *Plant Grow. Regul.* **12** : 189-193.

Schroder G and Schroder J 1995. cDNA for S-adenosyl-L-methionine decarboxylase from *Catharanthus roseus,* heterologous expression, identification of the proenzyme processing site, evidence for the presence of both subunits in the active enzyme and a conserved region in the 5 messenger RNA leader, *Eur. J. Biochem.* **228** : 74-78.

Serafini-Fracassini D, Torrigiani P and Branca C 1984. Polyamines bound to nucleic acids during dormancy and activation of tuber cells of *Helianthus tuberosus, Physiol. Plant* **60** : 351-357.

Sharma P and Rajam MV 1995a. Genotype, explant and position effects on organogenesis and somatic embryogenesis in eggplant (*Solanum melongena* L.), *J. Exp. Bot.* **46** : 135-141.

Sharma P and Rajam MV 1995b. Spatial and temporal changes in endogenous polyamines levels associated with somatic embryogenesis from different phypocotyl segments of eggplant (*Solanum melongena* L.), *J. Plant Physiol.* **146** : 658-664.

Sharma P, Yadav JS and Rajam MV 1997. Induction of laterals in root cultures of eggplant (*Solanum melongena* L.) in hormone-free liquid medium: A novel system to study the role of polyamines, *Plant Sci.* **125** : 103-111.

Smith TA 1985. Polyamines. *Annu. Rev. Plant Physiol.* **36** : 117-143.

Tabor CW 1981. Mutants of *Saccharomyces cerevisiae* deficient in polyamine biosynthetis: studies on the regulation of ornithine decarboxylase, *Mol. Biol.* **59** : 272-278.

Tabor CW, Tabor H and Hafner EW 1978. *Escherichia coli* mutants completely deficient in Adenosylmethionine decarboxylase and in spermidine biosnthesis, *J. Bacteriol Chem.* **253** : 3671-3676.

Tabor CW, Tabor H and Tyagi AK 1983. Biochemical and genetic studies on polyamines in *Saccharomyces cerevisiae,* Adv. Polyamine Res. **4** : 467-478.

Tabor H and Tabor CW 1983. Polyamine biosynthesis and fucntion in Escherichia coli, *Adv. Polyamine Res.* **4** : 455-565.

Tabor H, Hafner EW and Tabor CW 1980. Construction of an *Escherichia coli* strain unable to synthesize Putrescine, spermidine or cadaverine: Characterization of two genes controlling lysine decarboxylase, *J. Bacteriol.* **144** : 952-956.

Tassoni A, Antognoni F, Battistini ML, Sanvido O and Bagni N 1998. Characterization of spermidine binding to solubilized plasma membrane proteins from Zucchini hypocotyls, *Plant Physiol.* **117** : 971-977.

Tavladoraki P, Schinia ME, Cecconi F, Di Agostino S, Manera F, Rea G, Mariottini P, Federico R and Angelini R 1998. Maize polyamine oxidase: primary structure from protein and cDNA sequencing, *FEBS Lett.* 62-66.

Taylor MA, Madarif SA. Kumar A, Davies HV, Scobie LA, Pearce SR and Flavell AJ 1992. Expression and sequence analysis of cDNAs induced during the early stages of tuberization in different organs of potato plant (*Solamum tuberosum* L.), *Plant Mol. Biol.* **20** : 641-651.

Tiburcio AF, Besford RT, Capell, Borell A, Testillano PA and Risuene MC 1994. Mechanism of polyamine action during senescence responses induced by osmotic stress, *J. Exp. Bot.* **45** : 1789-1800.

Tiburcio AF, Campos JL, Gifueras X and Besford RT 1993. Recent advances in the understanding of polyamine functions during plant development, *Plant Growth Regul.* **12** : 331-340.

Tipping AJ and McPherson MJ 1995. Cloning and molecular analysis of the pea seedling copper amine, J. Biol. Chem.

Tyagi AK, Wickner RB, Tabor CW and Tabor H 1984. Specificity of polyamine requirements for the replication and maintenance of different double- stranded RNA plasmids in *Saccharomyces cerevisiae, Proc. Natl. Acad, Sci. USA* **81** : 1149-1153.

Uzawa T, Hamasaki N and Oshima T 1993. Effects of novel polyamines on cell-free polypeptide synthesis catalyzed by *Thermus thermophilus* HB8 extract, *J. Biochem.* **114** : 478-486.

Van Kranene HJ, Van de Zen L, Van Kreijl CF, Bisschop A and Wieringa B 1987. Cloning and nucleotide sequence of rat ornithine decarboxylase cDNA, *Gene* **60** : 145-155.

Watson MB and Malmberg RL 1996. Regulation of *Arabidopsis thaliana* (L.) heynh arginine decarboxylase by potassium deficiency stress, *Plant Physiol.* **111** : 1077-1083.

Watson MB, Emory KK, Piatak RM and Malmberg RL 1998. Arginine decarboxylase (Polyamine synthesis) mutants of *Arbaidopsis thaliana* exhibit altered root growth, *Plant J.* **13** : 231-239.

Whitney PA and Morris D 1978 Polyamine auxotrophs of *Saccharomyces cerevisiae, J. Bacteriol.* **134** : 214-220.

Willaims LJ, Barnett GR, Ristow JL, Pitkin J, Perriere M and Davis RH 1992. Ornithine decarboxylase gene of *Neurospora crassa*: Isolation, sequence and polyamine mediated relation of its mRNA, *Mol. Cell. Biol.* **12** : 347-359.

Yadav JS and Rajam MV 1997. Spatial distribution of free and conjugated polyamines in leaves of *Solamum melongena* L. associated with differential morphogenetic capacity; efficient somatic embryogenesis with putrescine. *J. Exp. Bot.* **48** : 135-141.

Yadav JS and Rajam MV 1998. Temporal regulation of somatic embryogenesis by adjusting cellular polyamine content in eggplant. *Plant physiol.* **116** : 617-625.

Yamamtoto S, Nagata S and Kusaba K 1993. Purification and characterization of homospermidine synthase in *Acetobacter tartarogenes* ATCC 31105, *J. Biochem.* **114** : 45-49.

Yao J, Zadworny D, Kuhnlein U and Yayes JF 1995. Molecular cloning of a bovine ornithine decarboxylase cDNA and its use in the detection of restriction fragment length polymorphsim in holsteins, *Genome* **38** : 325-331.

Ye XS, Avdishko SA and Kuc J 1994. Effects of polyamines on *in vitro* phosphorylation of soluble and plasma membrane proteins in tobacco, cucumber and *Arabidopsis thaliana, Plant Sci.* **97** : 109-118.

Yoshida I, Yamagata H and Hirasawa E 1998. Light regulated gene expression of S-adenosylmethionine decarboxylase in *Pharbitis nil, J. Exp. Bot.* **49** : 617-620.

Chapter 9

ENHANCING THE POTENTIAL OF BIOLOGICAL NITROGEN FIXATION BY GENETIC MANIPULATIONS OF DIAZOTROPHIC BACTERIA FOR SUSTAINABLE AGRICULTURE

SS Sindhu, DK Malik and KR Dadarwal*

Department of Microbiology, CCS Haryana Agricultural University, Hisar - 125 004, India

Summary

Biological nitrogen fixation is becoming increasingly important in sustainable agriculture production system. Quantitative inputs of biologically fixed nitrogen derived from free-living, associative or symbiotic microorganisms vary from 0.1-600 Kg N ha^{-1} annually. The nitrogenase enzyme complex, which mediates the conversion of dinitrogen to ammonia, could be manipulated genetically to enhance biological nitrogen fixation. After inoculation of diazotrophic bacteria in the soil or rhizosphere, its functioning depends on the survival, establishment and competitive ability of the introduced bacteria. The selection of efficient N_2-fixing bacteria from local niches, their genetic manipulation and further introduction in the same soil environment could help in achieving the potential benefits. The possibilities of extending the host range of rhizobia to nodulate the root systems of rice, wheat, maize and oilseed rape could provide the opportunities for endophytic nitrogen fixation. An improved understanding of the genetics and biochemistry of the biological N_2 fixation process is required to enhance the soil fertility leading to improved crop productivity.

*Corresponding author : E-mail : dadarwal@hau.nic.in

Keywords : Biological nitrogen fixation, diazotrophic bacteria, legume-rhizobium symbiosis.

1. INTRODUCTION

The availability of a useful nitrogen source for the plant is the major limiting factor in agricultural productivity. The striking rise in cereal grain yields in developed countries between 1950-1990 is directly attributable to the intensive agricultural cultivation practices including introduction of high yielding varieties and a 10-fold increase in nitrogen fertilizer use. However, the increased use of chemical fertilizers and pesticides in agriculture that helped in achieving self-sufficiency in food grains production has polluted the environment and is causing slow deterioration in soil health. High cost of fertilizers, rapid depletion of non-renewable energy sources and release of pollutants during fertilizer production, volatilization of N oxides (green house gases) into the atmosphere, leaching of excess NO_3^- into ground water and an imbalance in the global N cycle has emphasized the need to consider novel ways of increasing food grain production that are compatible with sustainability and the retention of environmental quality. To achieve these objectives, biological nitrogen fixation has emerged as an alternative or supplemental technology in sustainable agriculture. The beneficial N_2-fixing microorganisms, therefore, should be widely applied into the soil as microbial inoculants for cereals and legumes to replace chemical nitrogenous fertilizers.

Although nitrogen makes up approximately 80% of the elements present in the earth's atmosphere, the dinitrogen (N_2) is chemically inert and, therefore, is not utilized as a source of nitrogen by most of the living organisms. Some prokaryotes - a few bacteria and cyanobacteria have acquired the ability to reduce dinitrogen from atmosphere to ammonia using the bacterial enzyme nitrogenase and furnish this essential nutrient into agricultural soils. Biological N_2 fixation occurs in the free-living state as well as in association or in symbiosis with plants. These diazotrophic bacteria reduce the dependency of agricultural crops on fossil fuel-derived nitrogenous fertilizers.

A variety of N_2-fixing bacteria have been used in soil inoculations intended to improve the supply of fixed nitrogen as nutrients to crop plants. Rhizobia have been successfully used worldwide to permit an effective establishment of the N_2-fixing symbiosis with leguminous crop plants (Eaglesham, 1989). Other N_2-fixing symbionts, such as *Frankia* spp. have also been successfully introduced into soil (Sougoufara *et al.*, 1989). On the other hand, large areas of arable land in Australia, India, Russia, and United Kingdom have been inoculated with non-symbiotic N_2-fixing bacteria such as *Azotobacter, Azospirillum, Bacillus* and *Klebsiella* spp. for enhancing plant productivity (Lynch, 1983). This review briefly introduces the different strategies or biotechnological approaches undertaken for enhancing biological nitrogen fixation. The various constraints faced to improve crop productivity by inoculation with genetically engineered bacterial strains and the possibilities of deriving desired benefits by ensuring the establishment and survival of introduced microbial inoculants in soil has also been explored.

2. NITROGEN-FIXING ORGANISMS

The diazotrophy is widely distributed in nature. The only confirmed nitrogen-fixing organisms belong to two major domains of living organisms - Bacteria and Archaea, which are free-living as well as symbiotic. Nitrogen fixation generally occurs only under anaerobic or micro-aerophilic conditions. Exceptions are the members of the Azotobacteraceae, which have evolved several mechanisms of nitrogenase protection from oxygen while growing in an aerobic environment. Facultative anaerobes like *Klebsiella* can grow in both aerobic and anaerobic conditions but they need anaerobic environment for nitrogen fixation. Cyanobacteria can fix nitrogen in both aerobic and anaerobic conditions. The aerobic nitrogen fixation by cyanobacteria is either by special cells known as heterocysts that have anaerobic environment or by temporal separation of nitrogenase from oxygen during dark as observed in some unicellular cyanobacteria such as *Gloeothece* (Haselkorn and Buikema, 1997).

Many nitrogen-fixing bacteria are heterotrophic and need a supply of reduced carbon e.g. *Azotobacter* and *Azospirillum.* A few, such as *Bradyrhizobium japonicum* and *Arthrobacter fluorescens* can grow autotrophically on hydrogen and carbon dioxide. Most photosynthetic bacteria are anaerobic autotrophs. Diazotrophs like *Herbaspirillum* spp. grow endophytically in the stems and leaves of sugar cane. A saccharophilic bacterium *Acetobacter diazotrophicus* has been reported recently as the main endophytic N_2-fixer in sugarcane and sweet sorghum (Dobereiner *et al.*, 1993). Another N_2-fixing endophyte *Azoarcus* inhabits the roots of Kallar grass (*Leptochloa fusa*). Recently Hurek *et al.* (1994) showed that endophyte *Azoarcus* strain BH72 has the ability to invade and colonize rice roots. Diazotrophs of the genera *Azospirillum* have the ability to associate with growing root system of a variety of crop plants (Okon, 1994).

Four important N_2-fixing associations i.e. *Rhizobium*-leguminous plants, *Frankia*-actinorhizal plants, *Anabaena-Azolla* symbiosis and lichen symbiosis involving cyanobacteria have been studied in great detail. In these symbiotic systems, the host plant supplies the reduced carbon compounds and other nutrients to support bacterial growth and the nitrogenase system of the symbiont reduces dinitrogen. Due to these symbiotic associations, plants are generally able to grow in nitrogen-poor soils.

3. CONTRIBUTION OF FIXED NITROGEN BY VARIOUS DIAZOTROPHIC BACTERIA TO CEREAL AND LEGUME CROPS

Legume-*Rhizobium* symbiosis is the most extensively studied nitrogen-fixing system. On the global basis, this association reduces about 70-80% of the total of 17.2×10^7 tones biologically fixed nitrogen per year (Ishizuka, 1992). The symbiotic rhizobia have been found to fix nitrogen ranging from 24-600 Kg N ha^{-1} annually (Elkan, 1992). The effectiveness of different legume species and their microsymbionts has been found variable (Table 1). In general, faba bean (*Vicia faba*), and pigeon pea (*Cajanus cajan*) have been found to be very efficient; soybean (*Glycine max*), ground nut (*Arachis hypogaea*) and cowpea (*Vigna unguiculata*) to be average; and common bean (*Phaseolus vulgaris*) and pea (*Pisum sativum*) poor in fixing atmospheric nitrogen (Hardarson, 1993). The *Azolla-Anabaena* symbiotic system has been reported to contribute 12-313 Kg N ha^{-1} and *Frankia*-actinorhizal symbiosis provide 2-300 Kg N ha^{-1} (Stevenson, 1982).

The free-living bacteria, including *Azotobacter, Beijerinckia, Klebsiella, Pseudomonas* and enterobacters are of great significance in the gradual accumulation of nitrogen in undisturbed environments and contribute a significant input of nitrogen to plants. Free-living bacteria contribute 15 Kg N ha^{-1} $year^{-1}$ and the free-living cyanobacteria, on an average add 7-80 Kg N ha^{-1} $year^{-1}$ in rice cropping system (Elkan, 1992). Potential of the root-associative *Azospirillum* or *Azotobacter paspali* is mainly judged from the gain in crop productivity when these bacteria are colonized in the crop rhizosphere. These microbes have the potential to fix atmospheric nitrogen upto the level of 15-36 Kg N ha^{-1}. N-balance studies indicated that endophytic *Acetobacter*

Table 1. Estimated average rate of biological nitrogen fixation for specific organisms and associations.

Organism or system	Dinitrogen fixed (kg ha^{-1} yr^{-1})
Free-living microorganisms :	
Cyanobacteria ('blue-green algae')	25
Azotobacter	0.3
Clostridium pasteurianum	0.1 - 0.5
Grass-bacteria associative symbioses	5 - 25
Plant-cyanobacterial associations :	
Gunnera	12 - 21
Azolla	313
Lichens	39 - 84
***Rhizobium* –Legume symbioses :**	
Soybeans (*Glycine max* L. Merr.)	57 - 94
Cowpea (*Vigna*, *Phaseolus,* and others)	84
Clover (*Trifolium hybridum* L.)	104 - 160
Alfalfa (*Medicago sativa* L.)	128 - 600
Lupines (*Lupinus* sp.)	150 - 169
Nodulated non-legumes :	
Alnus (alders, e.g. red and black alders)	40 - 300
Hippophae (sea buckthorn)	2 - 179
Coriaria ('tutu' in New Zealand)	60 - 150
Casuarina (Australian pine)	58

Adapted from Stevenson (1982).

diazotrophicus contribute as high as 150-200 Kg N ha^{-1} $year^{-1}$ in sugarcane (Dobereiner *et al.*, 1993).

Symbiotically fixed nitrogen also becomes available to an intercrop or subsequent crop and therefore, more than 50% of the crops grown in

Africa, India and Latin America are either intercropped or rotated with nitrogen-fixing species (Fujiata *et al.*, 1992). Thus nitrogen supplied to crops by biological nitrogen fixation reduces our dependence on nitrogenous fertilizers and also builds up soil fertility for succeeding crops. However, various soil environmental factors have been found to affect nitrogen fixation which include temperature, moisture, acidity, available nitrogen, phosphorus, calcium and molybdenum content (Somasegaran and Bohlool, 1990; Zhang *et al.*, 1996).

N_2-fixing bacteria *Azotobacter* and *Azospirillum* have been widely tested as inoculants in cereal crops under field conditions in several countries. Plant responses to inoculation with azotobacters and azospirilla are often reported in terms of increased grain yield, plant biomass yield, nutrient uptake, grain and tissue N contents, nitrogenase activity, early flowering, tiller numbers, greater leaf size, increase in number of grains per spike, increased root length, increased enzyme levels in plant parts and reduced incidences of insect and disease infestation (reviewed in Pandey and Kumar, 1989; Okon and Labendra, 1994; Wani, 1994). The results indicated that in many cases, inoculations increased plant yields but such increases were variable from statistically significant increases to negative effects. A variety of endophytic nitrogen-fixing bacteria have already been found that colonize the interior roots of rice, maize and grass (Barraquio *et al.*, 1997) and are capable of contributing directly to the nitrogen requirement in rice and wheat (Webster *et al.*, 1997; Sturz *et al.*, 2000). Nitrogen balance studies showed that BNF in wetland rice fields could yield upto 50 Kg N ha^{-1} $year^{-1}$ (Roger and Ladha, 1992). Boddey *et al.* (1995) have reported high rates of BNF and high yields of sugarcane in fields by inoculation of *Acetobacter diazotrophicus* on various sugarcane varieties. Inoculation of legumes with effective strains of rhizobia has also often resulted in significant increases in yields of various legume crops (Eaglesham, 1989; Thies *et al.*, 1991). However, many failures or inconsistencies in achieving the yield increases following inoculation with rhizobial strains have been reported as well (Miller and May, 1991). The responses to inoculation with N_2-fixing bacteria have been found to vary with crops, cultivars, locations, seasons, agronomic practices, bacterial strains, levels of soil fertility and interactions with native soil microflora.

The algal biofertilizer is an easily manageable input forming a perpetually self-generating system, which adds to the nutrient status as well as health of the soil. In addition to contributing 20-25 Kg N ha^{-1} $season^{-1}$, they add organic matter to soil, excrete growth-promoting substances and increase the fertilizer utilizing efficiency of the crop plants. Using ^{15}N technique, Mac Rae and Castro (1969) demonstrated an addition of 10-15 Kg N ha^{-1} in rice fields due to cyanobacteria. Based on crop responses to cyanobacterial inoculation, Venkataraman (1981) found a gain of 25-30 Kg N ha^{-1} $season^{-1}$ in long term algalization trials.

Coinoculation of *Azotobacter vinelandii* and *Rhizobium* spp. has been found to increase the number of nodules on the roots of soybean, pea and clover (Burns *et al.*, 1981). Similarly, coinoculation of *Azospirillum brasilense* with *Rhizobium* strains showed synergistic effect in soybean and groundnut (Iruthayathas *et al.*, 1983; Ravekar and Konde, 1988). Recently, coinoculation of *Pseudomonas* and *Bacillus* spp. with *Rhizobium* has been found to increase nodule number, nitrogen fixation and plant biomass of green gram, chickpea and other legume crops (Parmar and Dadarwal, 1999; Sindhu *et al.*, 1999; 2001; 2002; Goel *et al.*, 2000). Therefore, coinoculation of diazotrphic bacteria could be exploited to enhance nitrogen fixation in the rhizosphere of cereals as well as legumes.

4. STRATEGIES FOR IMPROVING N_2 FIXATION

Considerable research efforts are being made recently for improving the efficiency of biological nitrogen fixation because symbiotic and free-living organisms have the potential to reduce our overall dependence on nitrogenous fertilizers. The strategy selected depends primarily on whether the diazotrophic bacteria under study are capable of fixing N_2 under free-living or symbiotic conditions and whether nitrogen fixation (*nif, fix*) or nodulation (*nod*) genes are targetted. To date, most efforts of improving N_2 fixation have been made on symbiotic N_2-fixing organisms of the genera *Rhizobium* and *Bradyrhizobium* because they form symbiotic associations with agronomically important legume crops (Shantharam and Mattoo, 1997). Cyanobacteria of the genus *Anabaena* (Golden, 1988; Haselkorn and Buikema, 1997), *Frankia* (Lalonde *et al.*, 1988; Lawson *et al.*, 1998) and *Azospirillum* (Okon, 1994) have also been studied.

Several strategies have been proposed to optimize endophytic nitrogen fixation in nonlegume crops either by (i) altering the receptivity of the host plant to colonization by nitrogen-fixing bacteria (de Bruijn *et al.*, 1995; Cristiansen-Weniger, 1998); (ii) exploiting naturally occurring stable plant-diazotrophic endophytic bacterial associations in cereals such as rice and wheat to fix nitrogen endophytically (Kennedy *et al.*, 1997; Swensen and Mullin, 1997) or (iii) through the genetic alteration of selected endophytic bacteria. Alternative approach used involves the genetic manipulation of nonlegumes to incorporate *nif* genes from bacteria (Dixon *et al.*, 1997; Gough *et al.*, 1997) or extension of the host range of symbiosis between rhizobia and nonlegume crops (Trinick and Hadobas, 1995; Sindhu and Dadarwal, 2001).

5. ENHANCEMENT OF NODULATION AND BROADENING OF HOST RANGE

Rhizobium-legume associations are usually host specific and a given rhizobial strain can infect a limited number of hosts (Brewin, 1991). Some specialized rhizobia associated with the tribes Cicereae, Trifolieae and Vicieae have restricted host ranges (Broughten and Perret, 1999). In other symbiotic associations, specificity varies greatly amongst the symbionts. *Azorhizobium caulinodans* nodulates only *Sesbania rostrata; Rhizobium meliloti* can initiate nodule formation on few host plants (*Medicago, Melilotus and Trigonella*), whereas *Rhizobium* sp. NGR234, nodulates more than 110 genera of legumes, as well as the nonlegume *Parasponia andersonii* (Young and Johnston, 1989).

Under field conditions, legumes are confronted with a diversity of rhizobial strains and opportunities for legume-mediated genetic exchange among the rhizobia or genetic exchange between rhizobia and other types of rhizosphere bacteria do exist. Kinkle *et al.* (1993) have reported plasmid transfer between populations of *R. leguminosarum* bv. *viciae* and *B. japonicum* respectively, in non-sterile soil. Souza *et al.* (1994) presented evidence that gene exchange is frequent among local soil populations of *R. etli*. Due to gene transfer and genomic rearrangements between bacteria in soil, some times rhizobial populations appear that are different from the original inoculants (Sullivan *et al.*, 1995; Vlassak *et al.*, 1996). These genetic exchanges could construct highly adaptable rhizobial population that will dominate nodule formation in subsequent years.

5.1. Transfer of symbiotic plasmid

Many *Rhizobium* strains harbor plasmids and the genes affecting nodulation (*nod, nol, noe* genes), nitrogen fixation (*nif* and *fix* genes), polysaccharide production (*exo* and *lps* genes) and other cellular functions are located on the plasmids (Denarie *et al.*, 1992; Fischer, 1994). The symbiotic (*sym*) plasmids of *R. leguminosarum* and *R. meliloti* vary in size from 140 kb to 1500 kb (Beynon *et al.*, 1980; Long, 1989). The number and size of these plasmids varies among different isolates.

The transfer of *sym* plasmid of *R. leguminosarum* to other closely related rhizobia belonging to either bv. *trifolii*, bv. *viciae* or bv. *phaseoli* normally results in the formation of normal nitrogen-fixing nodules on the host plants of donor strains (Beynon *et al.*, 1980; Brewin *et al.*, 1980) but when the *sym* plasmid of *R. leguminosarum* is transferred to distantly related species of *R. meliloti,* the transconjugants induce non-nitrogen-fixing root nodules on pea and vetch (Kondorosi *et al.*, 1980; Young and Johnston, 1989). Similarly, Kondorosi *et al.* (1982) also observed that transconjugants of *Lotus* rhizobia or tropical cowpea miscellany rhizobia, carrying the symbiotic magaplasmid pRme41b of *R. meliloti* strain 41, formed white non-nitrogen-fixing nodules on *Medicago sativa.* When the *sym* megaplasmid (pRme41b) of *R. meliloti* was mobilized into *Agrobacterium tumefaciens* by cloning a *mob* region into the *sym* megaplasmid (Kondorosi *et al.*, 1982), the transconjugants were able to induce ineffective nodule-like deformations on alfalfa roots.

Introduction of *R. leguminosarum* or *R. trifolii sym* plasmid into *Agrobacterium tumefaciens* conferred the ability to nodulate pea and clover, respectively but the nodules formed were ineffective without formation of bacteroids (Hooykass *et al.*, 1981; 1982). Djordjevic *et al.* (1983) showed that transfer of plasmid pBRIAN (encoding clover specific nodulation and nitrogen fixation functions) to *A. tumefaciens* strain ANU109 enabled the strain to nodulate white clovers, whereas the same strain carrying the plasmid pJB5JI (encoding pea-specific nodulation and nitrogen fixation functions) failed to nodulate peas.

Truchet *et al.* (1984) mobilized the *sym* megaplasmid of *R. meliloti* strain 2011 into *A. tumefaciens* with the help of plasmid RP4 or PGM142. The resulting transconjugants induced root deformations on the homologous hosts *Medicago sativa* and *Melilotus alba* but not on the heterologous hosts *Trifolium repens* and *T. pratense*. Cytological observations indicated that bacteria penetrated only the superficial layers of the host tissue by an atypical infection process. Sindhu and Dadarwal (1993) constructed recombinant strains by protoplast fusion between *R.* sp. *Vigna* and *R.* sp. *Cicer* that formed effective nodules on green gram but ineffective pseudonodules on chickpea. These results indicated that infection and nodule initiation genes could be expressed in heterologous rhizobia leading to broadening of host range but bacteroid development leading to formation of effective nitrogen-fixing nodules is difficult to achieve.

A cryptic plasmid, pRmeGR4b, has been reported to affect nodulation efficiency and competitiveness in *R. meliloti* GR4 (Sanjuan and Olivares, 1989). Mutations in the relevant locus, spanning 5kb region, delayed nodule formation and also reduced nodulation competitiveness. Nucleotide sequence analysis revealed the occurrence of two contiguous genes, *nfe*1 and *nfe*2 (nodule formation efficiency), preceded by a functional σ54 and a NifA-dependent promoter (Soto *et al.*, 1993). The *nfe* genes were not present in 4 other strains of *R. meliloti* and transfer of *nfe* genes by conjugation in these strains was found to improve the nodulation competitiveness in two of the strains (Sanjuan and Olivares, 1991a). The expression of both *nfe*1 and *nfe*2 is probably activated during infection and nodule development by the switch to microaerobic conditions that induce NifA synthesis. Addition of multiple copies of *nif*A from *Klebsiella pneumoniae* also conferred increased nodulation competitiveness of constructed *R. meliloti* strains (Sanjuan and Olivares, 1991b). However, Dillewijn *et al.* (1998) reported that this observed increase in nodulation was not dependent on plasmid borne *nif*A activity but was dependent on the sensitivity of nonresistant strains to streptomycin carried over from growth cultures. It was reported that the *nif*A of *K. pneumoniae* on an *Inc*P vector does not have an effect on competitiveness.

5.2. Transfer of cloned *nod* genes

In rhizobia, the nodulation genes are located either on large symbiotic plasmids (p*sym*) or on chromosome, which are organized into several coordinately regulated operons. So far, over 50 different nodulation genes have been characterized in different rhizobia (Sindhu and Dadarwal, 2001). These nodulation genes are basically categorized into three main classes, namely: (a) the regulatory *nod*D and *nod*VW genes which activate the transcription of other common and host specific *nod* genes, (b) the common *nod*ABC, *nod*M and *nod*IJ genes which are functionally and structurally conserved among different rhizobia and (c) the host specific *nod* genes which are variable with bacterial species and strains. Mutations in these host specific *nod* genes usually do not complement with cloned genes from other rhizobia. The structural organization and regulation of the nodulation genes of *Rhizobium, Bradyrhizobium* and *Azorhizobium* has been reviewed recently (Schultze *et al.*, 1994; Long, 1996; Spaink, 1996; Hanin *et al.*, 1999).

The construction of bacterial strains with an increased copy number of specific genes has been widely used for biotechnological applications. In some cases natural gene amplification, a common feature of the genome of prokaryotic organisms is associated with adaptive responses in bacteria (Romero and Palacios, 1997). Amplification is usually achieved by cointegrating a plasmid carrying the region of interest into the homologous region of the genome. Homologous recombination between the DNA-repeat regions leads to duplication and further amplification of the whole amplicon structure. Castillo *et al.* (1999) used specific DNA amplification (SDA) strategy to construct *Sinorhizobium meliloti* strains CFNM101 and CFNM103, which contained an average of 2.5 to 3 copies of the symbiotic region (containing *nod*D1, *nod*ABC and *nif*N of p*sym* plasmid). The inoculation of alfalfa with these strains resulted in an increase in nodulation, nitrogen fixation and growth of alfalfa plants under environmentally controlled conditions. Similarly, Mavingui *et al.* (1997) used random DNA amplification (RDA) in the symbiotic plasmid of *R. tropici* to obtain strains with enhanced competitiveness for nodulation.

Spaink *et al.* (1989) constructed chimeric *nod*D gene, consisting of 75% from *nod*D1 gene of *R. meliloti* at the 5' end and 27% of the *nod*D gene from *R. leguminosarum* bv. *trifolii*. Its expression in *R. leguminosarum* bv. *trifolii* and *R. meliloti* resulted in an extension of the host range for nodulation to the tropical legumes *Macroptilium atropurpureum*, *Lablab purpureus* and *Leucaena leucocephala*. The expression of chimeric *nod*D gene in *R. leguminosarum* bv. *trifolii* and *R. leguminosarum* bv. *viciae* also resulted in a significant increase in nitrogen fixation during symbiosis with *Vicia sativa* and *Trifolium repens*. Bender *et al.* (1988) transferred *nod*D1 gene from *Rhizobium* strain NGR234 to a restricted host range *R. leguminosarum* bv. *trifolii* strain and this transfer extended the nodulation capacity of the recipients to new hosts including the nonlegume *Parasponia andersonii*. The point mutations in the *nod*D of *R. leguminosarum* bv. *trifolii* strain was also found to extend the host range even to the nonlegume *Parasponia* (McIver *et al.*, 1989).

The transfer of a 14 kb HindIII fragment on recombinant plasmid pRt032 (carrying *nod*ABC and *nod*D genes from *sym* plasmid of *R. leguminosarum* bv. *trifolii* strain ANU843) to other *Rhizobium* species, or to *A. tumefaciens* conferred the capacity to nodulate clover by these recipients (Schofield *et al.*, 1984). The conjugal transfer of this 14 kb HindIII fragment on plasmids pRt032 and pRKR9032, to *R. fredii*

USDA192 strain, extended the host range of *R. fredii* even to clover (Yamato *et al.*, 1997). However, nodules induced by the transconjugant strains NA102 and YA101 on clover roots were small and whitish, and nitrogen fixation was not detected. The Nod factors produced by the transconjugants in the presence of apigenin and genistein flavonoids also differed from those of their recipient strains. Concomitant treatment of *Glycine max* and *Vigna unguiculata* roots with NodNGR factors and *nod*ABC mutants of strain NGR234 or *B. japonicum* USDA110 permitted these strains to nodulate and fix nitrogen in their respective hosts (Relic *et al.*, 1994). NodNGR factors also allowed the entry of *R. fredii* USDA257 into the roots of nonhost *Calopogonium caeruleum* (Relic *et al.*, 1994) and of the *nod*ABC mutant of NGR234 into *Macroptilium atropurpureum* (Relic *et al.*, 1993).

The transfer of the *nod*FEGHPQ genes of *R. meliloti* to strains of *R. leguminosarum* bv. *trifolii* or bv. *viciae* conferred to these strains the ability to nodulate alfalfa (Putnoky and Kondorosi, 1986) but strongly inhibited nodulation on the normal host plants, white clover and vetch, respectively (Debelle *et al.*, 1988; Faucher *et al.*, 1989). Mutations in the *nod*H gene of *R. meliloti* (involved in the transfer of sulfate on lipo-oligosaccharide Nod factor) strongly inhibited nodulation on the normal host *Medicago sativa*, and led to delayed nodulation on *Melilotus alba* but conferred the ability to nodulate the non- host plant, vetch (Faucher *et al.*, 1988; Roche *et al.*, 1991). Mutation in *nod*Q gene also extended the host range of *R. meliloti* to vetch (Schwedock and Long, 1992). Introduction of the *R. meliloti nod*HPQ genes into *R. leguminosarum* bv. *trifolii* or *R. leguminosarum* bv. *viciae*, neither of which possesses these genes, leads to the production of sulphated Nod signals and extended the host range of these strains to alfalfa (Denarie *et al.*, 1996; Long, 1996). Mutation of strain NGR234 *noe*E gene (involved in fucose-specific sulfotransferase) blocked the nodulation of *Pachyrhizus tuberosus*, while its introduction into closely related strain USDA257 extended the host range of *R. fredii* to include *Calopogonium caeruleum* (Hanin *et al.*, 1997).

The *nod*L gene is required for the addition of an O-acetyl residue at the terminal non-reducing glucosamine residue in *R. meliloti* Nod factors (Ardourel *et al.*, 1994). In strain NGR234, disruption of the flavonoid-inducible *nol*L gene results in the synthesis of NodNGR factors that lack the 3-O- or 4-O- acetate group (Berck *et al.*, 1999). The transconjugants of *R. fredii* strain USDA257 containing *nol*L of NGR234 produced acetylated Nod factors and nodulated the nonhosts

Calopogonium caeruleum, *L. leucocephala* and *Lotus halophilus*. Acetylation of the fucose of Nod factors of *R. etli* also conferred efficient nodulation on some *P. vulgaris* cultivars and on the alternate host *Vigna umbellata* (Corvera *et al.*, 1999).

The *nod*Z gene, encodes a fucosyltransferase, which is required for nodulation of the legume siratro by *B. japonicum,* but the mutation in *nod*Z of *B. japonicum* does not affect significantly nodulation in soybeans (Nieuwkoop *et al.*, 1987; Stacey *et al.*, 1994). NodZ$^-$ mutants of NGR234 lost the capacity to nodulate *Pachyrhizus tuberosus* (Quesada-Vincens *et al.*, 1997). Transfer of *nod*Z gene to *R. leguminosarum* bv. *viciae* resulted in the production of fucosylated Nod signals and an extension of host range to include *Macroptillium* (Lopez-Lara *et al.*, 1996). Inactivation of gene *nod*S, involved in methylation of Nod factors of *A. caulinodans*, NGR234 and *R. tropici* abolished nodulation of *Leucaena leucocephala* and *Phaseolus vulgaris* (Lewin *et al.*, 1990; Waelkens *et al.*, 1995). Introduction of either *nod*S or *nod*U into *R. fredii* USDA257 extended the host range to include *Leucaena* spp. (Krishnan *et al.*, 1992; Jabbouri *et al.*, 1995). These results indicate that various substitutions or modifications at the reducing or non-reducing terminus of Nod factors could widen the host range.

6. ENHANCEMENT OF NITROGEN FIXATION

The structural or regulatory *nif* genes of the nitrogenase enzyme complex can be manipulated to increase the efficiency of N_2 fixation. It has been hypothesized that increasing NifA production, which is the transcriptional activator of other *nif* genes, would enhance the expression of the whole N_2-fixing system (Szeto *et al.*, 1990). Preliminary green house studies indicated that some *R. meliloti* strains having enhanced *nif*A gene products showed a 7-15% increase in biomass compared with the wild-type parent (Williams *et al.*, 1990). Since the rate-limiting step in nitrogen fixation appears to be the cycle of binding reduced dinitrogenase reductase (Fe-protein, the *nif*H gene product) to dinitrogenase (MoFe-protein) followed by one electron transfer, it has been proposed that increased copies of the *nif*H gene and its product may result in increasing the turnover rate of nitrogenase. This may be the basis for the presence of more than one copy of the *nif*H gene in some diazotrophs such as *Azotobacter vinelandii* (Jacobson *et al.*, 1986), *Rhizobium phaseoli* (Quinto *et al.*, 1985) and *Azorhizobium sesbaniae* (Norel and Elmerich, 1987).

Improved substrate transport through modified expression of the C4-dicarboxylate transport (*dct*) genes has also been used to enhance nitrogen fixation (Ronson *et al.*, 1990). Root nodules are the largest sink of photosynthetic energy and consume approximately 10% of the plant's net photosynthetic output for N_2 fixation. Thus, N_2 fixation in the *Rhizobium*-legume symbiosis is presumed to be limited by the amount of plant-derived photosynthate available to bacteroids (Hardy and Havelka, 1975). Birkenhead *et al.* (1988) suggested that increasing the ability of the endosymbiont to utilize photosynthate in the nodule may lead to increased N_2 fixation rates. Recombinant strains of *R. meliloti* and *B. japonicum* designed to have increased expression of *dct*A (structural gene for dicarboxylate transport) and *nif*A genes were found to show 15% increase in the rate of N_2 fixed (Ronson *et al.*, 1990).

Some diazotrophic bacteria including *Rhizobium*, *Azotobacter, Azospirillum* etc. are able to improve the efficiency of nitrogen fixation by oxidizing hydrogen using uptake hydrogenase enzyme, which is simultaneously produced and evolved during nitrogen fixation. This oxidation of H_2 increases the rate of ATP biosynthesis. Enhanced nitrogen fixation rates have been reported in nodules and bacteroids of soybean, pea and *Vigna* group of hosts formed by inoculation of Hup^+ strains (Emerich *et al.*, 1979; Dadarwal *et al.*, 1985). Uptake hydrogenase activity in root nodule bacteroids has been shown to improve yield of soybean using near-isogenic strains of *B. japonicum* (Hanus *et al.*, 1991; Hungaria *et al.*, 1989). Another approach to improve yields could be to increase the activity of the uptake hydrogenase in bacterial strains that already possess it. Mutants were obtained for increased hydrogenase activity in Hup^+ *B. japonicum* strains (Merberg and Maier, 1983) or *Rhizobium* sp. strains (Sindhu and Dadarwal, 1992). The mutants of *Rhizobium* sp. strains showed increase in dry matter yield when inoculated onto green gram and black gram. The *hup* genes, encoding the biosynthesis of uptake hydrogenase, have been cloned and used to transform Hup^- strains. These Hup^+ recombinants have been shown to exhibit increased nitrogen fixation (Pau, 1991).

7. BREEDING OF LEGUME CULTIVARS FOR ENHANCED NODULATION WITH EFFECTIVE STRAINS

Altering the genetic make-up of plants to manipulate both endophytic and external populations offers the possibility of creating preferred rhizosphere communities (O'Connell *et al.*, 1996). Plant breeding strategy could be used to combine preference traits from several sources to generate plant genotypes capable of excluding nodulation by most suboptimally effective indigenous rhizobia. Hardarson *et al.* (1982) showed that the selection of alfalfa for physiological and morphological traits associated with nitrogen fixation capability altered the preference of the host plant for effective strains of *R. meliloti*. Nutman (1984) reported that red clover bred for improved nitrogen fixation maintained its superiority against a range of *R. leguminosarum* bv. *trifolii* strains. These studies illustrated the potential for developing broad-spectrum effectiveness for genetically diverse indigenous rhizobia in some legume species. Mytton *et al.* (1984) assessed genetic variation in nitrogen fixation using a range of *M. sativa* cultivars inoculated with different strains of *R. meliloti*; one of these cultivars was found relatively insensitive to changes in *Rhizobium* genotype and maintained high average yield.

The specific compatibility between *nod*X of *R. leguminosarum* bv. *viciae* strain TOM and *sym*2 of *Pisum sativum* cv. Afghanistan could be used to prevent indigenous rhizobia from nodulating and to allow inoculant strains to nodulate. The *sym*2 gene has already been crossed into a desirable pea cultivar (Trapper) and *nod*X was transferred in effective N_2-fixing *Rhizobium* strain. Performance of these manipulated host cultivar and rhizobial strains appeared promising enough to be exploited further in field studies (Fobert *et al.*, 1991). A similar combined approach involving alteration of both soybean host and *Bradyrhizobium* strains has also been proposed as a means to improve symbiotic N_2 fixation in soybean-*B. japonicum* symbiosis (Cregan *et al.*, 1989; Sadowsky *et al.*, 1991). This strategy involves the use of soybean genotypes that restrict the nodulation of indigenous competitive strains and allow nodulation only with desired added strains (Sadowsky *et al.*, 1995). In this way, improved strains produced by genetic engineering or other techniques can be targeted to specifically improve soybean varieties. Thus development of legume cultivars with broad-spectrum effectiveness for genetically diverse indigenous rhizobia could be an alternative beneficial plant breeding strategy to obviate the requirement for legume inoculation (Brockwell and Bottomley, 1995). This requires an understanding of the genetics of host and microbe, and offers real promise for genetically well-defined systems, such as alfalfa and soybean.

Another strategy of increasing nodule number by manipulations of host genome is also employed for enhancing nitrogen fixation capacity in symbiotic microorganisms. The basis for these approaches lies in several assumptions, including that nodulation in legume is suboptimal and increasing the nodule number will lead to an increase in total fixed nitrogen. Research efforts to increase the number of root nodules were tested with hypernodulating mutants of soybean. Such mutants produced 100-times more nodules than the parent plant (Carroll *et al.*, 1985; Betts and Herridge, 1987). Several soybean mutants with the ability to form large number of nodules even in the presence of nitrate have been isolated (Carroll *et al.*, 1985). These nitrate-tolerant symbiotic (*nts*) mutants usually formed 3 to 40 times as many nodules as the parent and showed increased nitrogen fixation ability in the presence of nitrate (Hansen *et al.*, 1989). Unfortunately, these mutants turned out to be poor agronomic performers (Pracht *et al.*, 1994). The apparent reason for the failure of this approach was that plant spent considerable amount of energy in bearing root nodules, thus limiting energy needed for the nitrogen fixation process. Nitrogen fixation is a highly energy intensive process consuming 6.5 g carbon g^{-1} of the nitrogen fixed (Kennedy *et al.*, 1997). Sato *et al.* (1999) manipulated source-sink relationship in hypernodulating soybeans by decreasing infection dose in such a way that nodulation is reduced to normal level. They concluded that autoregulatory control may play an important role in optimizing the nodule numbers in soybeans to maximize the nodule growth and total nitrogen fixation activity.

8. NODULATION AND NITROGEN FIXATION IN NONLEGUME HOSTS

Some nonlegume plants are capable of establishing a nitrogen-fixing symbiosis. The *Frankia*-induced nodules on predominately woody angiosperms such as *Alnus* or *Casuarina* are of largest significance. These nodules have primitive, branched structures, reminiscent of thickened lateral roots, yet their ability to fix nitrogen is equivalent to that in legumes (Clawson *et al.*, 1998). Similarly, nonlegume nodulation and nitrogen fixation was observed with *Bradyrhizobium* inoculants in the genus *Parasponia* (Trinick and Hadobas, 1995; Webster *et al.*, 1995). *Parasponia* nodules induced by *Bradyrhizobium* fixes nitrogen at highly efficient rates and are structurally similar to actinorhizal nodules.

The *nod*D gene of rhizobia has been demonstrated to control the first level of host specificity (Denarie *et al.*, 1992). The mobilization of *nod*D1 of NGR234 into *R. leguminosarum* bv. *trifolii* caused the extension of host range to nodulate nonlegume *Parasponia andersonii* (Bender *et al.*, 1988). Plasmids bearing *nod*DABC genes of *R. leguminosarum* bv. *trifolii* were transferred to *A. tumefaciens, Pseudomonas aeruginosa, Lignobacter* sp., *Azospirillum brasilense, E. coli* and different non-nodulating mutant rhizobia (possessing *sym* plasmid deletions). It conferred on these strains the ability to cause root hair curling and distortions on clover and a range of other non-host legumes (Plazinski *et al.*, 1994), suggesting that the expression of *nod*DABC genes in diverse range of soil bacteria may extend or affect the normal growth pattern of plant root hairs of a wide range of host and non-host legumes.

Although, a large amount of research for enhancing nitrogen fixation through alteration of macrosymbiont host plant has been done with leguminous crops such as soybean and alfalfa, improvement of nitrogen fixation is also being examined in nonleguminous systems. Particularly of interest, is improvement of biological nitrogen fixation in actinorhizal (*Frankia*)-nonleguminous tree associations (Sprent and Sprent, 1990) and in cereal plant-*Azospirillum* interactions (Dobereiner *et al.*, 1988). The genetics of host and microbes in both of these systems, however, are largely uncharacterized, and more research efforts are needed to enhance biological nitrogen fixation. Recently, two model legumes *Medicago truncatula* and *Lotus japonicus* have been identified for genetic dissection of formation and functioning of root nodules, by which transgenics can routinely be obtained. These legumes will act as tools for the identification and genetic characterization of plant genes involved in nodule formation and understanding the mechanisms that control root nodule development.

Recent studies to transfer the nitrogen fixation ability or the capacity to form symbiotic associations to nonleguminous plants have shown that nodule-like structures could be induced on rice and wheat roots with *Rhizobium* strains under certain artificially created conditions using hormones or cell wall degrading enzymes (Al-Mallah *et al.*, 1989; Cocking *et al.*, 1994). A critical examination of these nodule-like structures revealed that the bacteria accumulated at the site of injury due to emergence of lateral roots. Recently, *Rhizobium* strains isolated from *Aeschynomene indica* (strain ORS310) and from *Sesbania rostrata* (strain ORS571) were reported to form nodule-like structures on emerging lateral roots of rice, wheat and maize (Cocking *et al.*, 1994). The nodules showed significant levels of nitrogen fixation activity using acetylene reduction assay. A high level of acetylene reduction activity was reported in lateral roots of wheat plants inoculated with *Azorhizobium caulinodans* when grown in pots under aseptic conditions. Nitrogenase activity, however, was not observed in uninfected plants or plants infected with Nif- strain of *A. caulinodans* (Sabry *et al.*, 1997). *Azorhizobium caulinodans* strain ORS571 carrying a *lac*Z reporter gene was shown to be present within the cracks associated with emerging lateral roots of rice and wheat (Webster *et al.*, 1997). A high proportion of inoculated rice and wheat plants showed colonized lateral root cracks. The *lac*Z-tagged *Azorhizobium caulinodans* strain ORS571 was also able to enter *Arabdiopsis thaliana* roots after colonizing lateral root cracks at the point of emergence of lateral roots. The flavonoids, naringenin and diadzein at low concentration significantly stimulated the frequency of lateral root cracks and intercellular colonization of *Arabdopsis thaliana* roots by *Azorhizobium caulinodans.*

Tchan and Kennedy (1989) showed the induction of 'para nodules' on wheat following treatment with the herbicide 2, 4-dichlorophenoxyacetic acid (2, 4-D) or with the auxins IAA (indole-3-acetic acid) and NAA (Naphthyl-acetic acid), and by inoculation with rhizobia or *Azospirillum.* Low levels of nitrogenase activity were found only after inoculation with *Azospirillum* and there was little evidence for bacterial infection. Rolfe and Bender (1991) have reported the formation of similar structures on rice roots following inoculation with a strain of *Rhizobium* containing a *nod*D allele whose gene product interacts with rice root exudate. True infection was apparent in few nodules that were formed, but nitrogenase activity could not be detected. It is argued that the ability to induce new genes into rice by transformation has created an excellent opportunity to investigate the possibility for nitrogen fixation in rice. Besides, some genes homologous to the early nodulin genes in legumes, have been detected in the rice genome.

The understanding of the various physiological and genetic processes within the legume plants and the bacteria, and the identification of the essential characters that are present in legumes might result in the design of strategies by which nonlegumes like rice, can be given the ability to enter into symbiosis with nitrogen-fixing bacteria(Kennedy and Tchan, 1992). Therefore, extensive basic studies are needed to understand interactions between *Rhizobium* and cereal plants with special emphasis on signal-exchange mechanisms. Furthermore, these modified lateral roots of cereals having nodule-like structures must contain some sort of a micro-aerobic environment for plant protection of the O_2-sensitive nitrogenase. For this requirement, the plant could be engineered so that the intercellular space becomes filled with polysaccharides or other O_2-excluding material upon infection. Much efforts and coordination are required in genetics, molecular biology and developmental biology to achieve a complete understanding of the *Rhizobium*-legume symbiosis and to explore potential avenues for achieving the ultimate goal of expressing active nitrogenase in cereal crops (Dixon *et al.*, 1997; Shantharam and Mattoo, 1997).

9. PERFORMANCE OF GENETICALLY MANIPULATED INOCULANT STRAINS : LIMITATIONS

Several releases of beneficial microorganisms into soils have been successful; i.e. they resulted in the colonization of soil and plant roots to a level sufficiently high for the intended purpose. However, the desired impact of biofertilizer application is usually not achieved under certain field conditions and inoculation of legume or cereal plants with commercial inoculant strains often fails to improve crop productivity (van Elsas and Heijnen, 1990; Akkermans, 1994). The problem is of the survival of inoculant diazotrophic bacteria under field conditions that has been identified as super nitrogen-fixing strain under axenic and laboratory conditions. For each introduction, abiotic soil factors such as texture, pH, temperature, moisture content and substrate availability need critical assessment, since these factors largely determine the survival and activity of the introduced microorganisms (Hegazi *et al.*, 1979; van Veen *et al.*, 1997). The response of the inoculant diazotrophic bacteria to the prevailing soil conditions depends on its genetic and physiological constitution (Brockwell *et al.*, 1995). The use of genetic markers like resistance to drugs/antibiotics or introduction of metabolic markers from other bacterial species could help in tracing the introduced strains, whether it is rhizobia, cyanobacteria, azotobacters or azospirilla (Wilson *et al.*, 1995). The microbiologists are, therefore, required to go for in-depth analysis of the competitive and survival aspects (Sindhu and Dadarwal, 2000).

A key factor involved in the lack of success of rhizobial inoculants is its failure to compete with the indigenous, ineffective and built in populations of homologous strains for nodulation. Production of bacteriocins by rhizobia have been shown to suppress growth as well as nodulation by the indigenous non-producer strains, thus improving nodulation competitiveness of bacteriocins producing inoculant strains (Goel *et al.*, 1999; Sindhu and Dadarwal, 2000). Transfer and expression of *tfx* genes (involved in trifolitoxin production) in various rhizobia showed stable trifolitoxin production and restricted nodulation by indigenous trifolitoxin-sensitive strains on many leguminous species (Triplett, 1988; 1990). However, attempts to manipulate certain rhizobial genes in specific legume rhizosphere niches for improving competition have not produced impressive results (Nambiar *et al.*, 1990; Sitrit *et al.*, 1993; Krishnan *et al.*, 1999).

Biotechnological approaches used to enhance nitrogen fixation and crop productivity had been of limited usefulness under field conditions. For example, recombinant constructs of *R. meliloti* and *B. japonicum* having increased expression of *nif*A and *dct*A genes although showed increase in the rate of N_2 fixed but under the field conditions, the same constructs were unsuccessful in showing any increase in N_2 fixation or agronomically significant yield improvement (Ronson *et al.*, 1990). Manipulations of common nodulation genes to improve the bacterial competition have usually resulted in either no nodulation, delayed nodulation or inefficient nodulation (Devine and Kuykendall, 1996). Mendoza *et al.* (1995) enhanced NH_4^+ assimilating enzymes in *R. etli* through genetic engineering, by adding an additional copy of glutamate dehydrogenase (GDH), which resulted in total inhibition of nodulation on bean plants. However, nodule inhibition effect was overcome when *gdh*A expression was controlled by NifA and thereby delaying the onset of GDH activity after nodule establishment (Mendoza *et al.*, 1998). Similarly, attempts to engineer hydrogen uptake (Hup^+) ability by cloning hydrogenase genes into Hup^- strains of *Rhizobium* resulted in experimental successes only in areas where soybeans are cultivated and the photosynthetic energy is limited (Evans *et al.*, 1987). Attempts to develop self-fertilizing crops for nitrogen had also been a failure, mainly because of the complexity of the nitrogenase enzyme complex to be expressed in absence of an oxygen protection system in eukaryotes (Dixon *et al.*, 1997). Induction of nodule-like structures or pseudonodules with the help of lytic enzymes or hormones treatment in wheat and rice crops showed nitrogenase activity and $^{15}N_2$ incorporation. However, the activity expressed is > 1% of the value observed in legumes (Cocking *et al.*, 1994).

10. CONCLUSIONS AND FUTURE PROSPECTS

Although striking advances have been made in understanding the molecular and biochemical components regulating N_2 fixation, this has yet to be translated into applied environments. To overcome the problem of establishment of inoculated microbes, the diazotrophic bacteria chosen for the purpose of inoculation should be selected from local ecological niches and re-inoculated into the same environment to ensure the desired benefits. The effects of soil on the physiology and ecology of introduced microorganisms are still poorly understood at the micro scale level. Future research should aim for a better understanding of the *in situ* physiology of inoculant cells, as well as for possible ways to manipulate it. For example, the use of reporter genes inserted either randomly or directly into the bacterial genome allows the specific detection and possible enhancement of *in situ* gene expression in inoculant cells. On the applied side, the concept of application of mixtures of ecologically diverse strains with similar functions should be tested instead of single strains. The coinoculation of diazotrophic bacteria with rhizosphere bacteria or the inoculation of microbial consortia is preferable because these microorganisms might express beneficial functions more continually in a soil or rhizosphere system, even under ecologically different and/or variable conditions. Therefore, both traditional and biotechnical approaches can be employed to increase rates of N_2 fixation and crop yield in sustainable agricultural crop production.

REFERENCES

Akkermans ADL (1994). Application of bacteria in soils: problems and pitfalls. *FEMS Microbiol. Rev.,* **15 :** 185-194.

Al-Mallah MK, Davey MR and Cocking EC (1989). Formation of nodular structures on rice seedlings by rhizobia. *J. Expt. Bot.,* **40 :** 473-478.

Ardourel M, Demont N, Debelle F, Maillet F, de Billy F, Prome JC, Denarie J and Truchet G (1994). *Rhizobium meliloti* lipo-oligosaccharide nodulation factors: different structural requirements for bacterial entry into target root hair cells and induction of plant symbiotic development responses. *Plant Cell,* **6 :** 1357-1374.

Barraquio WL, Revilla L and Ladha JK (1997). Isolation of endophytic diazotrophic bacteria from wetland rice. *Plant Soil,* **194 :** 15-24.

Bender GL, Nayudu M, Le Strange KK and Rolfe BG (1988). The *nod*DI gene of *Rhizobium* strain NGR234 is a key determinant in the extension of host range to non-legume *Parasponia. Mol. Plant-Microbe Interact.,* **1 :** 254-256.

Berck S, Perret X, Quesada-Vincens D, Prome JC, Broughten WJ and Jabbouri S (1999). NolL of *Rhizobium* sp. NGR234 is required for O-acetyltransferase activity. *J. Bacteriol.,* **181 :** 957-964.

Betts JH and Herridge DF (1987). Isolation of soybean lines capable of nodulation and nitrogen fixation under high levels of nitrate supply. *Crop Sci.,* **27 :** 1156-1161.

Beynon JL, Beringer JE and Johnston AWB (1980). Plasmids and host range in *Rhizobium leguminosarum* and *Rhizobium phaseoli. J. Gen. Microbiol.,* **120 :** 421-429.

Birkenhead K, Manian SS and O'Gara F (1988). Dicarboxylic acid transport in *Bradyrhizobium japonicum*: Use of *Rhizobium meliloti* dct gene(s) to enhance nitrogen fixation. *J. Bacteriol.,* **170 :** 184-189.

Boddey RM, de Oliveira OC, Urquiaga S, Reis VM, de Oliveres FL, Baldani VLD and Dobereiner J (1995). Biological nitrogen fixation associated with sugarcane and rice: Contributions and prospects for improvement. *Plant Soil,* **174 :** 195-209.

Brewin NJ (1991). Development of the legume root nodule. *Annu. Rev. Cell Biol.,* **7 :** 191-226.

Brewin NJ, Beringer JE and Johnston AWB (1980). Plasmid mediated transfer of host range specificity between two strains of *Rhizobium leguminosarum. J. Gen. Microbiol.,* **120 :** 413-420.

Brockwell J and Bottomley PJ (1995). Recent advances in inoculant technology and prospects for the future. *Soil Biol. Biochem.,* **27** : 683-697.

Brockwell J, Bottomley PJ and Thies JE (1995). Manipulation of rhizobia microflora for improving legume productivity and soil fertility: A critical assessment. *Plant Soil.,* **174** : 143-180.

Broughten WJ and Perret X (1999). Geneology of legume-Rhizobium symbioses. *Curr. Opinion in Plant Biol.,* **2** : 305-311.

Burns TA Jr, Bishop PE and Israel DW (1981). Enhanced nodulation of leguminous plant roots by mixed cultures of *Azotobacter vinelandii* and *Rhizobium*. *Plant Soil,* **62** : 399-412.

Carroll BJ, McNeil DL and Gresshoff PM (1985). A supernodulation and nitrate tolerant symbiotic (*nts*) soybean mutant. *Plant Physiol.,* **78** : 34-40.

Castillo M, Flores M, Mavingui P, Martinez-Romero E, Palacios R and Hernandez G (1999). Increase in alfalfa nodulation, nitrogen fixation and plant growth by specific DNA amplification in *Sinorhizobium meliloti. Appl. Environ. Microbiol.,* **65** : 2716-2722.

Christiansen-Weniger C (1998). Endophytic association of diazotrophic bacteria in auxin induced tumors of cereal crops. *CRC Crit. Rev. Plant Sci.,* **17** : 55-76.

Clawson ML, Caru M and Benson DR (1998). Diversity of *Frankia* strains in root nodules of plants from the families Elaegnaceae and Rhamnaceae. *Appl. Environ. Microbiol.,* **64** : 3539-3543.

Cocking EC, Webster G, Batchelor CA and Davey MR (1994). Nodulation of non-legume crops: A new look. *Agro-Industry Hi-Tech.,* pp. 21-24.

Corvera A, Prome D, Prome JC, Martinez-Romero E and Romero D (1999). The *nol*L gene from *Rhizobium etli* determines nodulation efficiency by mediating the acetylation of the fucosyl residue in the nodulation factor. *Mol. Plant-Microbe Interact.,* **12** : 236-246.

Cregan PB, Keyser HH and Sadowsky MJ (1989). Host plant effect on nodulation and competitiveness of the *Bradyrhizobium japonicum* serotype strains constituting serocluster 123. *Appl. Environ. Microbiol.,* **55** : 2532- 2536.

Dadarwal KR, Sindhu SS and Batra R (1985). Ecology of Hup^+ *Rhizobium* strains of cowpea miscellany: native frequency and competence. *Arch. Microbiol.,* **141** : 255-259.

De Bruijn FJ, Jing Y and Dazzo FB (1995). Potential pitfalls of trying to extend symbiotic associations of nitrogen-fixing organisms to presently non-nodulated plants, such as rice. *Plant Soil,* **174** : 225-240.

Debelle F, Maillet F, Vasse J, Rosenberg C, de Billy F, Truchet G, Denarie J and Ausubel FM (1988). Interference between *Rhizobium meliloti* and *Rhizobium trifolii* nodulation genes: Genetic basis of *R. meliloti* dominance. *J. Bacteriol.,* **170** : 5718-5727.

Denarie J, Debelle F and Prome JC (1996). *Rhizobium* lipo-chitooligosaccharide nodulation factors: signalling molecules mediating recognition and morphogenesis. *Annu. Rev. Biochem.,* **65** : 503-535.

Denarie J, Debelle F and Rosenberg C (1992). Signaling and host range variation in nodulation. *Annu. Rev. Microbiol.,* **46** : 497-531.

Devine TE and Kuykendall LD (1996). Host genetic control of symbiosis in soybean (*Glycine max* L.). *Plant Soil,* **186** : 173-187.

Dillewijn P van, Martinez-Abarca F and Toro N (1998). Multicopy vectors carrying the *Klebsiella pneumoniae nif*A gene do noi enhance the nodulation competitiveness of *Sinorhizobium meliloti* on alfalfa. *Mol. Plant-Microbe Interact.,* **11** : 839-842.

Dixon R, Cheng Q, Shen GF, Day A and Day MD (1997). Nif gene and expression in chloroplasts: Prospects and problems. *Plant Soil,* **194** : 193-203.

Djordjevic MA, Zurkowski W, Shine J and Rolfe BG (1983). Sym plasmid transfer to symbiotic mutants of *Rhizobium trifolii, Rhizobium leguminosarum* and *Rhizobium meliloti. J. Bacteriol.,* **156** : 1035-1045.

Dobereiner J, Reis MV and Lazarani AC (1988). New N_2-fixing bacteria in association with certain cereals and sugarcane. In: *Nitrogen Fixation: Hundred Years After.* (Eds. Bothe H, de Bruijn, FJ and Newton WE). Gustav Fischer, Stuttgart. pp. 671-680.

Dobereiner J, Reis MV, Paula MA and Olivares F (1993). In: *New Horizons in Nitrogen Fixation.* (Eds. Palacois R, Mora J and Newton WE). Kluwer Academic Publishers, The Netherlands. pp. 61.

Eaglesham ARJ (1989). Global importance of *Rhizobium* as an inoculant. In: *Microbial Inoculation of Crop Plants.* Chap. 3 (Eds. Campbell R and Macdonald R). Oxford University Press, Oxford.

Elkan GH (1992). Biological nitrogen fixation systems in tropical ecosystems: An overview. In: *Biological Nitrogen Fixation and Sustainability of Tropical Agriculture.* (Eds. Mulongoy K, Gueye M and Spencer DSC). John Wiley and Sons, Chichester, U.K. pp. 27-40.

Emerich DW, Ruiz-Argueso T, Ching TM and Evans HJ (1979). Hydrogen dependent nitrogenase activity and ATP formation in *Rhizobium japonicum* bacteroids. *J. Bacteriol.,* **137** : 153-160.

Evans HJ, Harker AR, Papen H, Russell SA, Hanus FJ and Zuber M (1987). Physiology, biochemistry anf genetics of the uptake hydrogenase in rhizobia. *Annu. Rev. Microbiol.,* **41** : 355-361.

Faucher C, Camut S, Denarie J and Truchet G (1989). The *nod*H and *nod*Q host range genes of *Rhizobium meliloti* behave as virulence genes in *R. leguminosarum* bv. *viciae* and determine changes in the production of plant-specific extracellular signals. *Mol. Plant-Microbe Interact.,* **2** : 291-300.

Faucher C, Maillet F, Vasse J, Rosenberg C, van Brussel AN, Truchet G and Denarie J (1988). *Rhizobium meliloti* host range *nod*H determines production of an alfalfa-specific extracellular signal. *J. Bacteriol.,* **170** : 5489-5499.

Fischer HM (1994). Genetic regulation of nitrogen fixation in rhizobia. *Microbiol. Rev.,* **58** : 352-386.

Fobert PR, Roy N, Nash JH and Iyer VN (1991). Procedure for obtaining efficient root nodulation of a pea cultivar by a desired *Rhizobium* strain and preempting nodulation by other strains. *Appl. Environ. Microbiol.,* **57** : 1590-1594.

Fujiata K, Ofosu-Budu KG and Ogata S (1992). Biological Nitrogen Fixation in mixed legume-cereal cropping systems. *Plant Soil,* **141** : 155-175.

Goel AK, Sindhu SS and Dadarwal KR (1999). Bacteriocin producing native rhizobia of green gram (*Vigna radiata)* having competitive advantage in nodule occupancy. *Microbiol. Res.,* **154** : 43-48.

Goel AK, Sindhu SS and Dadarwal KR (2000). Pigment diverse mutants of *Pseudomonas* sp.: inhibition of fungal growth and stimulation of growth of *Cicer arietinum. Biol. Plant.,* **43** : 563-569.

Golden JW (1988). Genome rearrangements during Anabena heterocyst differentiation. *Can. J. Bot.,* **66** : 2098-2102.

Gough C, Vasse J, Galera C, Webster G, Cocking E and Denarie J (1997). Interactions between bacterial diazotrophs and non-legume dicots: *Arabdiopsis thaliana* as a model plant. *Plant Soil,* **194** : 123-130.

Hanin M, Jabbouri S, Broughten WJ, Fallay R and Quesada-Vincens D (1999). Molecular aspects of host specific nodulation. In: *Plant-Microbe Interactions.* Vol.4. (Eds, Stacey G and Keen NT). APS Press, St. Paul, Minnesota, pp 1-37.

Hanin M, Jabbouri S, Quesada-Vincens D, Freiberg C, Perret X, Prome JC, Broughten WJ and Fallay R (1997). Sulphation of *Rhizobium* sp. NGR234 Nod factors is dependent on noeE, a new host specificity gene. *Mol. Biol.,* **24** :1119-1129.

Hansen AP, Peoples MB, Gresshoff PM, Atkins CA, Pate JS and Carroll BJ (1989). Symbiotic performance of supernodulating soybean [*Glycine max* (L.) Merr.] mutants during development on different nitrogen regimes. *J. Expt. Bot.,* **40** : 715-724.

Hanus FJ, Albrecht SL, Zablotowicz RM, Emerich DW, Russell SA and Evans HJ (1981). Yield and Ncontent of soybean seed as influenced by *Rhizobium japonicum* inoculants possessing the uptake hydrogenase characteristics. *Agron. J.*, **73** : 368-372.

Hardarson G (1993). Methods for enhancing symbiotic nitrogen fixation. *Plant Soil,* **152** : 1-17.

Hardarson G, Heichel GH, Barnes DK and Vance CP (1982). Rhizobial strain preference of alfalfa populations selected for characteristics associated with N_2 fixation. *Crop Sci.,* **22** : 55-58.

Hardy RWF and Havelka UD (1975). Nitrogen fixation research: A key to world food? *Science,* **188** : 633-643.

Haselkorn R and Buikema WJ (1997). Heterocyst differentiation and nitrogen fixation in cyanobacteria. In: *Biological Fixation* of *Nitrogen for Ecology and Sustainable Agriculture*. (Eds. Legocki A, Bothe H and Puhler A). Springer-Verlag, Berlin, Heidelberg. pp 163-166.

Hegazi NA, Vlassak K and Monib M (1979). Effect of amendments, moisture and temperature on acetylene reduction in Nile Delta soil. *Plant Soil,* **51** : 27-37.

Hooykaas PJJ, Snijdewint FGM, and Schilperoort RA (1982). Identification of the *sym* plasmid of *Rhizobium leguminosarum* strain 1001 and its transfer to and expression in other rhizobia and *Agrobacterium tumefaciens. Plasmid,* **8** : 73-82.

Hooykaas PJJ, van Brussel AAN, Den Dulk Ras H, van Slogteren GMS and Schilperoort RA (1981). Sym-plasmid of *Rhizobium trifolii* expressed in different rhizobial species and *Agrobacterium tumefaciens. Nature,* **291** : 351-353.

Hungria M, Neves MCP and Dobreiner J (1989). Relative efficiency, ureide transport and harvest index in soybeans inoculated with isogenic Hup^- mutants of *Bradyrhizobium japonicum. Biol. Fertil. Soils,* **7** : 325-329.

Hurek T, Reinhold-Hurek B, van Montagu M and Kellenberger E (1994). Root colonization and systemic spreading of *Azoarcus* sp. strain BH72 in grasses. *J. Bacteriol.*, **178** : 1913-1923.

Iruthayathas EE, Gunasekaran S and Vlassak K (1983). Effect of combined inoculation of *Azospirillum* and *Rhizobium* on nodulation and N_2-fixation of winged bean and soybean. *Sci. Hort.* (Amsterdam), **20** : 231-240.

Ishizuka J (1992). Trends in biological nitrogen fixation research and application. *Plant Soil,* **141** : 197-209.

Jabbouri S, Fallay R, Telmont F, Kamalapriya P, Burger U, Relic B, Prome JC and Broughten WJ (1995). Involvement of *nod*S in N-methylation and *nod*U in 6-O-carbomylation of *Rhizobium* sp. NGR234 Nod factors. *J. Biol. Chem.,* **270** : 22968-22973.

Jacobson MR, Premakumar R and Bishop PE (1986). Transcriptional regulation of nitrogen fixation by molybdenum in *Azotobacter vinelandii. J. Bacteriol.,* **167** : 480-486.

Kennedy IR and Tchan YT (1992). Biological nitrogen fixation in non-leguminous field crops: Recent advances. *Plant and Soil,* **141** : 93-118.

Kennedy IR, Pereg-Gerk LL, Wood C, Deaker R, Gilcrest K and Katupitia S (1997). Biological nitrogen fixation in non-leguminous field crops: facilitating the evolution of an effective association between *Azospirillum* and wheat. *Plant Soil,* **194** : 65-79.

Kinkle BK, Sadowsky MJ, Schmidt EL and Koskinen WC (1993). Plasmids pJP4 and R68.45 can be transferred between populations of bradyrhizobia in nonsterile soil. *Appl. Environ. Microbiol.,* **59** : 1762-1766.

Kondorosi A, Kondorosi E, Pankhurst CE, Broughton WJ and Banfalvi Z (1982). Mobilization of *Rhizobium meliloti* megaplasmid carrying nodulation and nitrogen fixation genes in other rhizobia and *Agrobacterium. Mol. Gen. Genet.,* **188** : 433-439.

Kondorosi A, Vincze E, Johnston AWB and Beringer JE (1980). A comparison of three *Rhizobium* linkage maps. *Mol. Gen. Genet.,* **178** : 403-408.

Krishnan HB, Kim KY and Krishnan AH (1999). Expression of a Serratia marcescens chitinase gene in *Sinorhizobium fredii* USDA191 and *Sinorhizobium meliloti* RCR2011 impedes soybean and alfalfa nodulation. *Mol. Plant-microbe Interact.,* **12** : 748-751.

Krishnan HB, Lewin A, Fallay R, Broughten WJ and Pueppke SG (1992). Differential expression of nodS accounts for the varied abilities of *Rhizobium fredii* USDA257 and *Rhizobium* sp. NGR234 to nodulate Leucaena spp. *Mol. Biol.,* **6** : 3321-3330.

Lalonde M, Simon L, Bousquet J and Sequin A (1988). Advances in taxonomy of *Frankia*. Recognition of species *alni* and *elaeagni* and novel subspecies *pommerie* and *ijkii*. In: *Nitrogen Fixation: Hundred Years After.* (Eds. Bothe H, de Bruijn FJ and Newton WE). Gustav Fischer, Stuttgart. pp. 671-680.

Lewin A, Cervantes E, Wong CH and Broughton WJ (1990). *nod*SU, two new *nod* genes of the broad host-range *Rhizobium* strain NGR234 encode host specific nodulation of the tropical tree *Leucaena leucocephala. Mol. Plant-Microbe Interact.,* **3** : 317-326.

Long SR (1989). *Rhizobium* genetics. *Annu. Rev. Genet.,* **23** : 483-506.

Long SR (1996). *Rhizobium* symbiosis: Nod factors in perspective. *Plant Cell,* **8** : 1885-1898.

Lopez-Lara IM, Blok-Tip L, Quinto C, Garcia ML, Bloemberg GV, Lamers GEM, Kafetzopoulos D, Stacey G, Lugtenberg BJJ, Thomas-Oates JE and Spaink HP (1996). NodZ of *Bradyrhizobium* extends the nodulation host range of *Rhizobium* by adding a fucosyl residue to nodulation factors. *Mol. Microbiol.,* **21** : 397-408.

Lynch JM (1983). *Soil Biotechnology.* Blackwell Scientific Publications, Oxford, U.K.

Mac Rae IC and Castro TF (1969). Nitrogen fixation in some tropical rice soils. *Soil sci.*, **103** : 277- 282.

Mavingui P, Flores M, Romero D, Martinez-Romero E and Palacios R (1997). Generation of *Rhizobium* strains with improved symbiotic properties by random DNA amplification (RDA). *Nature Biotechnol.,* **15** : 564-569.

McIver J, Djordjevic MA, Weinman JJ, Bender GL and Rolfe BG (1989). Extension of host range of *Rhizobium leguminosarum* bv. *trifolii* caused by point mutations in *nod*D that result in alterations in regulatory function and recognition of inducer molecules. *Mol. Plant-Microbe Interact.,* **2** : 97-106.

Mendoza A, Leija A, Martinez-Romero E, Hernandez G and Mora J (1995). The enhancement of ammonium assimilation in *Rhizobium elti* prevents nodulation of *Phaseolus vulgaris. Mol. Plant-Microbe Interact.,* **8** : 584-592.

Mendoza A, Valderrama B, Leija A and Mora J (1998). NifA-dependent expression of glutamate dehydrogenase in *Rhizobium etli* modifies nitrogen partitioning during symbiosis. *Mol. Plant-Microbe Interact.,* **11** : 83-90.

Merberg D and Maier RJ (1983). Mutants of *Rhizobium japonicum* with increased hydrogenase activity. *Science,* **220** : 1064-1065.

Miller RH and May S (1991). legume inoculation: sucesses and failures In: *The Rhizosphere and Plant Growth.* (Eds. Keister DL and Cregan PB). Kluwer, Dordrecht. pp 123-134.

Mytton LR, Brockwell J and Gibson AH (1984). The potential for breeding an improved legume-*Rhizobium* symbiosis.1. Assessment of genetic variation. *Euphytica,* **33** : 401-410.

Nambiar PTC, Ma SW and Iyer YN (1990). Limiting insect infestation of nitrogen fixing root nodules of the pigeon pea (*Cajanus cajan* L.) by engineering an entomocidal gene in its root nodules. *Appl. Environ. Microbiol.,* **56** : 2866-2869.

Nieuwkoop AJ, Banfalvi Z, Deshmane N, Gerhold D, Schell MG, Sirotkin KM and Stacey G (1987). A locus encoding host range is linked to the common nodulation genes of *Bradyrhizobium japonicum. J. Bacteriol.,* **169** : 2631-2638.

Norel F and Elmerich C (1987). Nucleotide sequence and functional analysis of the two *nif*H copies of *Rhizobium* ORS571. *J. Gen. Microbiol.,* **133** : 1563-1576.

Nutman PS (1984). Improving nitrogen fixation in legumes by plant breeding: The relevance of host selection experiments in red clover (*Trifolium pretense* L.) and subterraneum clover (*T. subterraneum* L.). *Plant and Soil,* **82** : 285-301.

O'Connell KP, Goodman RM and Handelsman J (1996). Engineering the rhizosphere: a expressing a bias. *Trends Biotechnol.,* **14** : 83-86.

Okon Y (1994). *Azospirillum/Plant Associations.* CRC Press, Boca Raton, Fla.

Okon Y and Labandera C (1994). Agronomic application of *Azospirillum*: An evaluation of 20 years worldwide field application. *Soil Biol. Biochem.,* **26** : 1591-1601.

Pandey A and Kumar S (1989). Potential of *Azotobacter* and *Azospirillum* as biofertilizers for upland agriculture. *J. Sci. Indus. Res.,* **48** : 134-144.

Parmar N and Dadarwal KR (1999). Stimulation of nitrogen fixation and induction of flavonoid-like compounds by rhizobacteria. *J. Appl. Microbiol.,* **86** : 36-44.

Pau AS (1991). Improvement of Rhizobium inoculants by mutation, genetic engineering and formulation. *Biotechnol. Adv.,* **9** : 173-184.

Plazinski J, Ridge RW, McKay IA and Djordjevic MA (1994). The nod ABC genes of Rhizobium leguminosarum biovar trifolii confer root-hair curling ability to a diverse range of soil bacteria and the ability to induce novel root swellings on beans. *Australian J. Plant Physiol.,* **21** : 311-325.

Pracht JE, Nickell CD, Harper JE and Bullock DG (1994). Agronomic evaluation of non-nodulating and hypernodulating mutants of soybean. *Crop Sci.,* **34** : 738-740.

Putnoky P and Kondorosi A (1986). Two gene clusters of *Rhizobium meliloti* code for early essential nodulation functions and a third influences nodulation efficiency. *J. Bacteriol.,* **167** : 881-887.

Quesada-Vincens D, Fallay R, Nassim T, Viprey V, Burger U, Prome JC, Broughten WJ and Jabbouri S (1997). *Rhizobium* sp. NGR234 NodZ protein is a fucosyltransferase. *J. Bacteriol.,* **179** : 5087-5093.

Quinto C, de la Vega H, Flores M, Leemans J, Cevallos MA, Pardo MA, Azpiroz R, de Lourdes GM, Calva E and Palacois R (1985). Nitrogenase reductase: a functional multigene family in *Rhizobium phaseoli. Proc. Natl. Acad. Sci. USA.,* **82** : 1170-1174.

Raverkar KP and Konde BK (1988). Effect of *Rhizobium* and *Azosprillum lipoferum* inoculation on the nodulation, yield and nitrogen uptake of peanut cultivars. *Plant and Soil* **106** : 249-252.

Relic B, Perret X, Estrada-Garcia J, Kopcinska J, Golinowsky W, Krishnan HB, Pueppke SG and Broughten WJ (1994). Nod-factors of *Rhizobium* are a key to legume door. *Mol. Microbiol.,* **13** : 171-178.

Relic B, Talmont F, Kopcinska J, Golinowsky W, Prome JC and Broughten WJ (1993). Biological activity of *Rhizobium* sp. NGR234 Nod-factors on *Macroptillium atropurpureum. Mol. Plant-microbe Interact.,* **6** : 764-774.

Roche P, Debelle F, Maillet F, Lerouge P, Faucher C, Truchet G, Denarie J and Prome JC (1991). Molecular basis of symbiotic host specificity in *Rhizobium meliloti*: *nod*H and *nod*PQ genes encode the sulfation of lipo-oligosaccharide signals. *Cell,* **67** : 1131-1143.

Roger PA and Ladha JK (1992). Biological nitrogen fixation in wetland rice fields: estimation and contribution to nitrogen balance. *Plant Soil,* **141** : 41-55.

Rolfe BG and Bender GL (1991). Evolving a *Rhizobium* for non-legume nodulation. In: *Nitrogen fixation: Achievements and Objectives.* (eds. Gresshoff PM, Roth LE, Stacey G and Newton WE). Chapman and Hall, New York. pp. 779-780.

Romero D and Palacios R (1997). Gene amplification and genomic plasticity in prokaryotes. *Annu. Rev. Genet.,* **31** : 91-111.

Ronson CW, Bosworth A, Genova M, Gudbrandsen S, Hankinson T, Kwaitowski R, Ratcliffe H, Robie C, Sweeney P, Szeto W, Williams M and Zablotowicz R (1990). Field release of genetically engineered *Rhizobium meliloti* and *Bradyrhizobium japonicum* strains. In: *Nitrogen Fixation: Achievements and Objectives.* (Eds. Gresshoff PM, Roth LE, Stacey G and Newton WE). Chapman and Hall, New York, U.S.A. pp. 397-403.

Sabry SRS, Saleh SA, Batchelor CA and Davey MR (1997). In: *Biological nitrogen fixation, novel association with nonleguminous crops.* (Eds. Xanfu V, Kennedy IR and Tinagwei C). Qungdao Ocean University Press, China. pp. 59.

Sadowsky MJ, Cregan PB, Gottfert M, Sharma A, Gerhold D, Rodriquez-Quinones F, Keyser HH, Hennecke H and Stacey G (1991). The *Bradyrhizobium japonicum nol*A gene and its involvement in the genotype-specific nodulation of soybeans. *Proc. Natl. Acad. Sci. USA.,* **88** : 637-641.

Sadowsky MJ, Kosslak RM, Madrzak CJ, Golinska B and Cregan PB (1995). Restriction of nodulation by *Bradyrhizobium japonicum* is mediated by factors present in the root of *Glycine max*. *Appl. Environ. Microbiol.,* **61** : 832-836.

Sanjuan J and Olivares J (1989). Implication of *nif*A in regulation of genes located on a *Rhizobium meliloti* cryptic plasmid that affect nodulation efficiency. *J. Bacteriol.,* **171** : 4154-4161.

Sanjuan J and Olivares J (1991a). Multicopy plasmids carrying the *Klebsiella pneumoniae nif*A gene enhances *Rhizobium meliloti* nodulation competitiveness on alfalfa. *Mol. Plant-Microbe Interact.,* **4** : 365-369.

Sanjuan J and Olivares J (1991b). NifA-NtrA regulatory system activates transcription of *nfe*, a gene locus involved in nodulation competitiveness of *Rhizobium meliloti*. *Arch. Microbiol.,* **155** : 543-548.

Sato T, Yashima H, Ohtake N, Sueyoshi K, Akao S and Ohyama T (1999). Possible involvement of photosynthetic supply in changes of nodule characteristics of hypernodulating soybeans. *Soil Sci. and Plant Nutr.,* **45** : 187-196.

Schofield PR, Ridge RW, Rolfe BG, Shine J and Watson JM (1984). Host-specific nodulation is encloded on a 14 kb fragment in *Rhizobium trifolii*. *Plant Mol. Biol.,* **3** : 3-11.

Schultze M, Kondorosi E, Ratet P, Buire M and Kondorosi A (1994). Cell and molecular bology of *Rhizobium*-plant interactions. *Int. Rev. Cytol.,* **156** : 1-75.

Schwedock J and Long SR (1992). *Rhizobium meliloti* genes involved in sulfate activation - The two copies of *nod* PQ and a new locus, *saa*. *Genetics,* **132** : 899-909.

Shantharam S and Mattoo AK (1997). Enhancing biological nitrogen fixation: An appraisal of current and alternative technologies for N input into plants. *Plant and Soil,* **194** : 205-216.

Sindhu SS and Dadarwal KR (1992). Symbiotic effectivity of cowpea miscellany *Rhizobium* mutants having increased hydrogenase activity. *Indian J. Microbiol.,* **32** : 411-416.

Sindhu SS and Dadarwal KR (1993). Broadening of host range infectivity in cowpea miscellany *Rhizobium* by protoplast fusion. *Indian J. Expt. Biol.,* **31** : 521-528.

Sindhu SS and Dadarwal KR (2000). Competition for nodulation among rhizobia in legume-*Rhizobium* symbiosis. *Indian J. Microbiol.,* **40** : 211-246.

Sindhu SS and Dadarwal KR (2001). Genetic manipulation of rhizobia to improve nodulation and nitrogen fixation in legumes. In: *Recent Advances in Biofertilizer Technology* (Eds. Yadav, A.K., Motsara, M.R. and Ray Choudhary S.). Society for Promotion and Utilization of Resources and Technology, New Delhi. pp 1-97.

Sindhu SS, Gupta SK and Dadarwal KR (1999). Antagonistic effect of *Pseudomonas* spp. on pathogenic fungi and enhancement of growth of green gram (*Vigna radiata*). *Biol. Fertil. Soils,* **29** : 62-68.

Sindhu SS, Gupta SK, Sunita Suneja and Dadarwal KR (2001). Enhancement of green gram nodulation and plant growth by *Bacillus* species. *Biol. Plant.,* **44** : (in press).

Sindhu SS, Sunita Suneja, Goel AK, Parmar N and Dadarwal KR (2002). Plant growth promoting effects of *Pseudomonas* sp. on coinoculation with *Mesorhizobium* sp. *Cicer* strain under sterile and "wilt sick" soil conditions. *Appl. Soil Ecol.,* **19** : 57-64.

Sitrit Y, Barak Z, Kapulnik Y, Oppenheim AB and Chet I (1993). Expression of *Serratia marcescens* chitinase gene in *Rhizobium meliloti* during symbiosis on alfalfa roots. *Mol. Plant Microbe Interact.,* **6** : 293-298.

Somasegaran P and Bohlool BB (1990). Single strain versus multistrain inoculation: Effect of soil mineral N availability on rhizobial strain effectiveness and competition for nodulation on chickpea, soybean and dry bean. *Plant Soil,* **170** : 351-358.

Soto MJ, Zorzano A, Mercado-Blanco J, Lepek V, Olivares J and Toro N (1993). Nucleotide sequence and characterization of *Rhizobium meliloti* nodulation competitiveness genes *nfe. J. Mol. Biol.,* **229** : 570-579.

Sougoufara B, Diem HG and Dommergues YR (1989). Response of field grown *Casuarina equisetifolia* to inoculation with *Frankia* strain ORS021001 entraped in alginate beads. *Plant Soil,* **118** : 133-137.

Souza V, Eguiarte L, Avila G, Cappello R, Gallardo C, Montoya J and Pinero D (1994). Genetic structure of *Rhizobium etli* biovar *phaseoli* associated with wild and cultivated bean plants (*Phaseolus vulgaris* and *Phaseolus coccineus*) in Morelos. *Appl. Environ. Microbiol.* **60** : 1260-1268.

Spaink HP (1996). Regulation of plant morphogenesis by lipo-chitinoligosaccharides. *Crit. Rev. Plant Sci.* **15** : 559-582.

Spaink HP, Okker RJH, Wijffelman CA, Tak T, Roo LG, Pees E, van Brussel AAN and Lugtenberg BJJ (1989). Symbiotic properties of rhizobia containing a flavonoid-independent hybrid *nod*D product. *J. Bacteriol.* **171** : 4045-4053.

Sprent JI and Sprent P (1990). *Nitrogen Fixing Organisms: Pure and Applied Aspects.* Chapman Press, Cambridge.

Stacey G, Luka S, Sanjuan J, Banfalvi Z, Nieuwkoop AJ, Chun JY, Forsberg S and Carlson RW (1994). *nod*Z, a unique host-specific nodulation gene, is involved in the fucosylation of the lipo-oligosaccharide nodulation signal of *Bradyrhizobium japonicum. J. Bacteriol.* **176** : 620-633.

Stevenson FJ (1982). Origin and distribution of nitrogen in soil. In*: Nitrogen in Agricultural Soils, Agronomy No. 22.* (Ed. Stevenson, F.J.). American Society of Agronomy, Madison, Wisconsin. pp. 1-42 .

Sturz AV, Christie BR and Nowak J (2000). Bacterial endophytes: potential role in developing sustainable systems of crop production. *Crit. Rev. Plant Sci.,* **19** : 1-30.

Sullivan JT, Patrick HN, Lowther WL, Scott DB and Ronson CW (1995). Nodulating strains of *Rhizobium loti* arise through chromosomal symbiotic gene transfer in the environment. *Proc. Natl. Acad. Sci. USA.* **92** : 8985-8989.

Swensen SM and Mullin BC (1997). The impact of molecular systematics on the hypotheses for the evolution of root nodule symbiosis and implications for expanding symbiosis to new host plant genera. *Plant Soil,* **194** : 185-192.

Szeto W, Kwiatkowski R, Cannon FC and Ronson CW (1990). The enhancement of symbiotic nitrogen fixation in *Bradyrhizobium japonicum*. In: *Abstacts of fifth International Symposium on the Molecular Genetics of Plant-Microbe Interactions*. Interlaken, Switzerland. pp 152.

Tchan YT and Kennedy IR (1989). Possible nitrogen-fixing root nodules induced in non-legumes. *Agric. Sci.* **2** : 57-59.

Thies JE, Singleton PW and Bohlool BB (1991). Influence of the size of indigenous rhizobial populations on establishment and symbiotic performance of introduced rhizobia on field grown legumes. *Appl. Environ. Microbiol.* **57** : 19-28.

Trinick MJ and Hadobas PA (1995). Formation of nodular structures on the non-legumes *Brassica napus*, *B. campestris*, *B. juncea* and *Arabdiopsis thaliana* with *Bradyrhizobium* and *Rhizobium* isolated from *Parasponia* spp. or legumes grown in tropical soils. *Plant Soil*, **172** : 207-219.

Triplett EW (1988). Isolation of genes involved in nodulation competitiveness from *Rhizobium leguminosarum* bv. *trifolii* T24. *Proc. Natl. Acad. Sci. USA.*, **85** : 3810-3814.

Triplett EW (1990). Construction of a symbiotically effective strain of *Rhizobium leguminosarum* bv. *trifolii* with increased nodulation competitiveness. *Appl. Environ. Microbiol.*, **56** : 98-103.

Truchet G, Rosenberg C, Vasse J, Julliot JS, Camut S and Denarie J (1984). Transfer of *Rhizobium meliloti sym* genes into *Agrobacterium tumefaciens:* Host specific nodulation by atypical infection. *J. Bacteriol.*, **157** : 134-142.

van Elsas JD and Heijnen CE (1990). Methods for the introduction of bacteria in soil: A review. *Biol. Fertil. Soils.*, **10** : 127-133.

van Veen JA, van Overbeek LS and van Elsas JD (1997). Fate and activity of microorganisms introduced into soil. *Microbiol. Mol. Biol. Rev.*, **61** : 121-135.

Venkataraman GS (1981). Blue green algae for rice production. In: *FAO Soil Bull.*, **46** : 102- 109.

Vlassak KM, Vanderleyden J and Franco A (1996). Competition and persistence of *Rhizobium tropici* and *R. etli* in tropical soil during successive bean (*Phaseolus vulgaris* L.) cultures. *Biol. Fertil. Soils*, **21** : 61- 66.

Waelkens F, Voets T, Vlassak K, Vanderleyden J and van Rhizn P (1995). The *nod*S gene of *Rhizobium tropici* CIAT899 is necessary for nodulation of *Phaseolus vulgaris* and *Leucaena leucocephala*. *Mol. Plant-Microbe Interact.*, **8** : 147-154.

Wang SP and Stacey G (1990). Ammonia regulation of *nod* genes in *Bradyrhizobium japonicum*. *Mol. Gen. Genet.*, **223** : 329-331.

Wani SP (1994). Role of free-living and associative diazotrophs in non-legumes. In: *Biotechnological Approaches in Soil Microrganisms for Sustainable Crop Production.* (Ed. Dadarwal KR). Scientific Publishers, Jodhpur. pp 109-127.

Webster G, Davey MR and Cocking EC (1995). Parasponia with rhizobia: a neglected non-legume nitrogen-fixing symbiosis. *AgBiotech. News Info.* **7** : 119N-124N.

Webster G, Gough C, Vasse J, Batchelor CA, O'Callaghan KJ, Kothari SL and Davey MR (1997). Interactions of rhizobia with rice and wheat. *Plant Soil*, **194** : 115-122.

Williams MK, Beynon JL, Ronson CW and Cannon FC (1990). In: *Abstracts of fifth International Symposium on the Molecular Genetics of Plant-Microbe Interactions.* Interlaken, Switzerland. pp 152.

Wilson KJ, Peoples MB and Jefferson RA (1995). New techniques for studying competition by rhizobia and for assessing nitrogen fixation in the field. *Plant Soil*, **174** : 241-253.

Yamato M, Nakayama Y, Yokoyama T, Ueno O and Akao S (1997). Nodulation of *Rhizobium fredii* USDA192 containing *Rhizobium leguminosarum* bv. *trifolii* ANU843 *nod* genes on homologous host soybean and heterologous host clover. *Soil Sci. Plant Nutr.*, **43** : 51-61.

Young JPW and Johnston AWB (1989). The evolution of specificity in the legume-*Rhizobium* symbiosis. *Tree*, **4** : 341-349.

Zhang F, Dashti N, Hynes RK and Smith DL (1996). Plant growth promoting rhizobacteria and soybean {Glycine max (L.) Merr.} nodulation and nitrogen fixation at suboptimal root zone temperatures. *Annals Bot.* **77** : 453-459.

Chapter 10

GENETIC ENGINEERING FOR ENHANCING LEVELS OF CAROTENOIDS

Peter M Bramley

School of Biological Sciences, Royal Holloway, University of London, Egham, Surrey, TW20 0EX, UK.

Summary

Carotenoids are a group of plant pigments that have important roles in the plant itself and also in the diet. β-Carotene is a precursor of vitamin A, whilst carotenoids are also used for camouflage in crustacea, for visual attraction in birds and as antioxidants in humans. There is increasing evidence that a diet rich in carotenoid-containing fruits and vegetables is linked to a reduction in the incidence of age-related diseases such as coronary heart disease, macular degeneration and certain cancers. The genetic modification of carotenoid levels in crop plants offers one route to increasing dietary intake of carotenoids. The ready availability of carotenoid biosynthetic genes, appropriate promoters and our ability to transform most crops, presents the opportunity to enhance the carotenoids of fruits and vegetables. This article describes our current understanding of carotenoid biosynthesis and its regulation, summarises the carotenoid genes available for transformation and discusses the range of plants that have been transformed with carotenoid genes and the strategies used to produce them.

Keywords : carotenoid, biosynthesis, genetic engineering, functional foods

Abbreviations : AMD, age-related macular degeneration; CHD,

*Corresponding author : E-mail : p.bramley@hu.ac.uk

coronary heart disease; DMAPP, dimethylallyl diphosphate; DOXP, deoxy-D-xylulose-5-phosphate; GGPP, geranyleranyl diphosphate; HMGCoA, hydroxymethlglutaryl coenzyme A; IPP, isopentenyl diphosphate; MVA, mevalonic acid.

1. INTRODUCTION

Carotenoids are a class of structurally related isoprenoids, of which some 600 have been characterised (Pfander, 1987; Britton, 1991). They are one of the largest groups of pigments in nature, and can be seen most prominently in yellow to red flowers, fruits and vegetables. The characteristic colours of crustacea and some birds are also due to the presence of carotenoids. Animals are unable to synthesis carotenoids *de novo,* and so rely upon the diet as the source of these compounds. Over recent years there has been considerable interest in dietary carotenoids with respect to their potential as antioxidants that alleviate chronic diseases. The large amount of research in this topic has been paralleled by the significant advances in cloning most of the genes involved in carotenogenesis and our increased understanding of factors that regulate carotenoid formation and deposition in plants. Taken together, it is these advances in knowledge that open the way to enhancing carotenoid levels in crop plants.

2. STRUCTURE, NOMENCLATURE AND DISTRIBUTION

Carotenoids are isoprenoids, consisting of a symmetrical isoprenoids chain, typically C_{40} in length and typified by lycopene and β-carotene (Fig. 1). All carotenoids are derived from this basic structure, by hydrogenation, dehydrogenation, cyclization and oxidation reactions. The extent of the chromophore of conjugated double bonds influences their physical, biochemical and chemical properties, including their ability to act as antioxidants (Britton, 1991, 1995). The carotenoids are subdivided into two groups; the hydrocarbon carotenes (*e.g.* lycopene and β-carotene) and the oxygenated xanthophylls (*e.g.* lutein, zeaxanthin; Fig. 1). The majority of carotenoids are oxygenated, whilst some have less than 40 carbon atoms and are known as apocarotenoids. The nomenclature of carotenoids has been established by a semi-systematic convention (IUAPC Commission on Nomenclature 1975). In this article the trivial names of carotenoids will be used (Weedon and Moss, 1995).

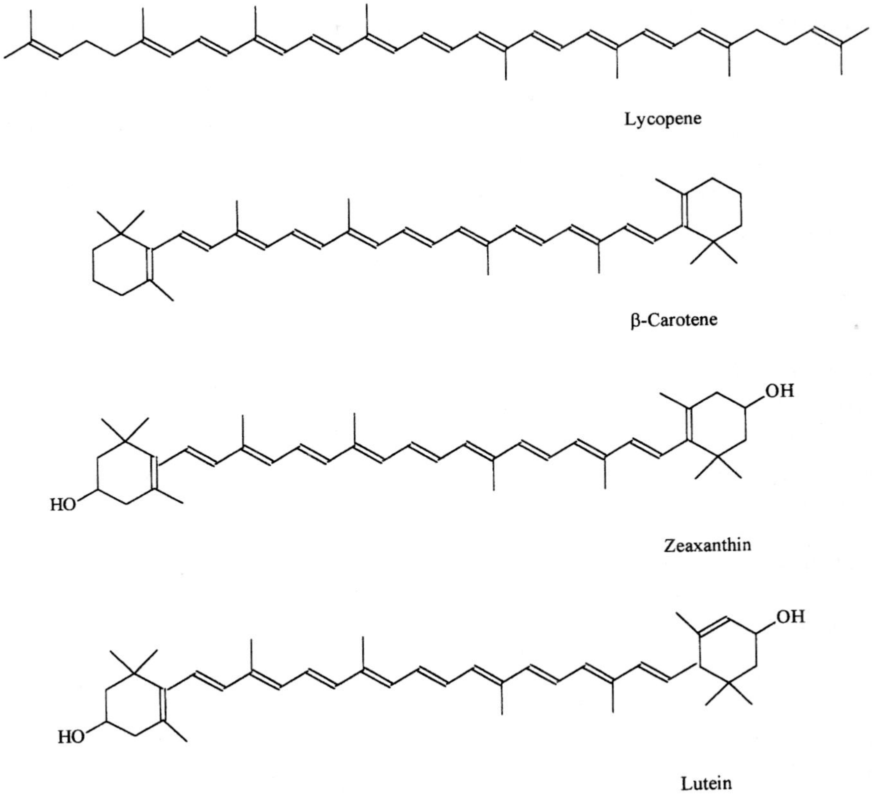

Figure 1 : Structure of typical carotenoids.

The large numbers of double bonds in carotenoid molecules allow a range of geometrical *cis/trans* isomers to exist. Generally, carotenoids are found predominantly in the all-*trans* configuration, although exceptions do occur, *e.g.* the Tangella variety of tomato contains poly-*cis* carotenoids. Many carotendoids have chiral centres due to the presence of asymmetric carbon atoms (Weedon and Moss, 1995). Normally, however, natural carotenoids exist in only one of the possible enantiomeric forms due to the specificity of the biosynthetic enzymes.

Carotenoids are universally found in photosynthetic tissues and organisms. The precise carotenoids that are present vary between organisms (Young, 1993) but in higher plants there is a limited number of carotenoids within the chloroplasts, with lutein, violaxanthin, zeaxanthin, neoxanthin α-carotene and β-carotene predominating. Carotenoids serve two principal functions during photosynthesis: photoprotection and light collection (light harvesting). These two photofunctions involve interactions with chlorophyll; in photoprotection energy is channelled away from chlorophyll, whilst in the light harvesting energy is passed from carotenoids to chlorophyll. These functions are essential to plant life in an aerobic atmosphere, as shown by the effectiveness of bleaching herbicides that inhibit carotenoid biosynthesis (Bramley, 1993). Both functions are possible because of the participation of carotenoids in an abundance of photochemical reactions, *e.g.* singlet-singlet energy transfer, triplet-triplet energy transfer, oxidation, reduction and isomerization. Frank and Cogdell (1993), Britton (1995) and Frank *et al.* (1997) have reviewed details of the energy states of carotenoid and the experimental approaches taken to elucidate these transitions.

Literature reports of carotenoid levels in plants and foods are often misleading and inaccurate, as the conditions of growth, genotype, analytical procedures, etc. are often not the same. Early studies, before the advent of more rigorous identification and chromatographic protocols, should also be viewed with some caution. The development of HPLC systems, especially those able to separate *cis* and *trans* isomers of carotenoids has enabled the accurate quantitation of carotenoids and such isomers in fruits and vegetables (*e.g.* Bramley, 1992; Vecchi *et al.*, 1981; Khachik *et al.*, 1991).

2.1. Main plant food sources

All leaves of green vegetables contain qualitatively similar carotenoids, as part of the photosynthetic apparatus (Young, 1993), although there are signification quantitative differences between species (Table 1). It is noticeable that those with the highest concentrations of carotenoids, such as parsley, spinach and watercress, tend to be rather minor components of the human diet.

Visually, carotenoids are most obvious in fruits (*e.g.* tomato, orange and pepper), seeds (*e.g.* maize) and roots (*e.g.* carrot), where they are concentrated in chromoplasts or amyloplasts (Gross, 1991). It is noticeable that the qualitative differences between species are far more marked

Table 1. Carotenoid content of raw leafy green vegetables

Species	Carotenoid (μg/g fresh weight)			
	Total	Lutein	α-Carotene	β-Carotene
Brussel sprout	1163	610	-	553
Green bean	940	494	70	376
Broad bean	767	506	-	261
Broccoli	2533	1614	-	919
Cabbage	139	80	-	59
Lettuce	201	110	-	91
Parsley	10335	5812	4523	
Pea	2091	1633	-	458
Spinach	9890	5869	-	4021
Watercress	16632	10713	-	5919

Taken from Scott and Hart (1994). Values include *cis* and *trans* isomers

than for the equivalent photosynthetic tissues from the same species. Some fruits, such as the tomato, contain a predominance of one carotenoid that is absent from leaf tissue, although citrus fruits contains over 50 carotenoids, many of which are apocaroenoids.

The distribution of cartenoids in fruits is complex and subject to variation. Patterns characteristic of each species and variety occur often at each stage of ripening. A higher rate of accumulation can occur in the peel than the pulp, leading to differences and so will continue after harvest from the parent plant. Unripe (green) fruit contain the same carotenoids as leaf tissue, but upon ripening there is often, but not always, *de novo* synthesis of carotenoids. According to Goodwin (1980), eight main groups of fruits can be distinguished according to carotenoid content:

1. Insignificant amounts, *e.g.* strawberry.
2. Large amounts of chloroplast carotenoids, *e.g.* blueberry.
3. Large amounts of lycopene and its hydroxy derivatives, *e.g.* tomato.
4. Large amounts of β-carotene and its hydroxy derivatives, *e.g.* peach.

5. Large amounts of epoxides, particularly furanoid epoxides, *e.g.* carambola.

6. Unusual carotenoids, *e.g.* capsanthin in red pepper.

7. Poly-*cis* carotenoids, *e.g.* prolycopene in tangerine tomatoes.

8. Apocarotenoids, *e.g.* Citrus species.

Comprehensive tables detaining carotenoid levels in fruits have been published (Goodwin, 1980; Gross, 1991; Scott and Hart, 1994). Representative examples are shown in Table 2. The carotenoid levels in fruit at various stages of ripening have been studied in considerable detail. A typical set of data for the Ailsa Craig tomato variety is shown in Table 3.

Table 2. Carotenoid content of fresh fruits, roots and seeds

Species	Carotenoid (μg/g fresh weight)						
	Total	Lutein	Zea	β-Cryp	Lyco	α-Car	β-Car
Apricot	2196	101	31	231	-	37	1766
Banana	126	33	4	Nd	-	50	39
Carrot (May)	11427	170	-	-	2660	8597	-
Carrot (Sept)	14693	283	-	-	3610	10800	-
Orange	211	64	50	83	-	Nd	14
Pepper	2784	503	1608	90	-	167	416
Peach	309	78	42	86	-	Trace	103
Sweet corn	1978	522	437	-	-	60	59
Tomtao	3454	78	-	-	2937	-	439

Taken from Scott and Hart (1994).
Nd = not detected; β-cryp = β-cryptoxanthin; lyco = lycopene; α-carotene; β-carotene; Zea = zeaxanthin

3. CAROTENOID BIOSYNTHESIS AND GENES

Carotenoid biosynthetic pathways in bacteria, fungi, algae and plants have been mainly elucidated from biochemical analyses using labelled precursors, specific inhibitors and characterisation of mutants, especially the tomato. Spurgeon and Porter (1983) have reviewed these classical

Table 3. Carotene changes during Ailsa Craig tomato fruit ripening

Fruit	Carotenes					Xanthophylls				Total[a]
	P	PF	ζ	L	γ-C	β-C	Lut	Viol	Veo	
			C					A		
Mature green	0	0	0	0.1	0	2.1	4.7	1.7	1.5	10.1
Breaker	1.9	0	0	3.7	0.4	5.6	1.5	0.4	1.1	15.9
24 d.p.b	40.6	22.2	7.5	70.5	11.3	36.8	6.4	2.3	6.4	207

p, phytoene; PF, phytofluene; ζ-C, z-carotene; L, lycopene; γ-C, carotene; β-C, β-carotene; Lut, lutein; Viola, violaxanthin; Neo, neoxanthin; D.p.b., days post breaker. All values as μg/g fresh weight, includes all minor carotenoids.[a]
Taken from Fraser et al., 1994.

studies. However, the lack of *in-vitro* assays for some of the carotenogenic enzymes has limited investigations of the pathway at the molecular and physiological levels. The advent of recombinant DNA technology has facilitated studies of the pathways through characterisation of genes that encode the biosynthetic enzymes. This approach has provided a better insight into the structure and function of carotenogenic enzymes and elucidated some of the mechanisms that regulate carotenoid biosynthesis and deposition. In this article, the salient features of the carotenoid pathways in plants will be discussed, with emphasis on those steps relevant to the manipulation of carotenoids by genetic engineering. There are several extensive reviews of carotenoid biosynthesis, its regulation and the genes that encode the enzymes. The most recent include those by Bartley and Scolnik, 1995; Hirschberg *et al.*, 1997; Cunningham and Gantt, 1998 and Harker and Hirschberg, 1998.

Most genes encoding enzymes that catalyse specific steps in the biosynthesis of carotenoids from photosynthetic bacteria, fungi, cyanobacteria, green algae and higher plants have been cloned (reviewed by Armstrong, 1994; Bartley *et al.*, 1994; Armstrong and Hearst, 1996; Hirschberg *et al.*, 1997; Harker and Hirschberg, 1998 and Cunningham and Gantt, 1998). Hirshhberg (1999) has described the approaches to cloning these genes. Nomenclature of the genes or cDNA that have been cloned has yet to be standardised. The first genes, from the purple photosynthetic bacterium *Rhodobacter capsulatus,* were designated by the suffix *crt*, with the corresponding protein CRT (Marrs, 1981). However, subsequent workers have also used other suffixes that are acronyms of the enzymic step (*e.g. Psy* for phytoene synthase, *Pds* for phytoene desaturase). Both types of nomenclature are currently in use and hence can lead to some confusion. Wherever possible, both types of nomenclature have been included in this article. Sequences of the genes/cDNAs can be retrieved from databases on the Internet at various web sites. For example, Gene Bank (at the National Center for Biotechnology Information, NCBI) is at http://www.ncbi.nlm.nih.gov/Web/Search/index.html. The EMBL Nucleotide Sequence database (through the European Bioinformatic Institute, EBI) can be found at http://www.ebi.ac.uk/queries/queries.html, whilst that of the DNA Data Bank of Japan is at http://www.ddbj.nig.ac.jp/.

Though carotenogenesis in plants takes place in plastids, all of the carotenoid biosynthesis genes are nuclear encoded and their polypeptide products are imported into the plastids. For example, the size of the transit peptide of PSY from ripe tomato fruit is approximately 9 kDa, corresponding to about 80 amino acid residues (Misawa *et al.*, 1994a).

3.1. Early steps in carotenoid biosynthesis

Since carotenoids are isoprenoids, they share a common early pathway with other biologically important isoprenoids such as sterols, gibberellins, phytol and the terpenoid quinones (Fig. 2). In all cases, these compounds are derived from the C_5 isoprenoid, isopentenyl diphosphate (IPP). Until a few years ago it was believed that a single pathway from the C_6 precursor mevalonic acid (MVA) formed IPP, which itself was synthesised from HMG CoA by the action of HMG CoA reductase. However, it has now been shown, using ^{13}C and ^{2}H-labelling techniques, NMR spectroscopy and mass spectroscopy that an alternative pathway exists from glyceraldehyde 3-phosphate and pyruvate. This is known as the non-MVA or deoxyxylulose phosphate (DOXP) pathway and is found in higher plants and algae. The DXP reductoisomerase has been cloned from peppermint (Lange and Croteau, 1999) and *Arabidopsis* (Schwender *et al.*, 1999). The complete set of reactions still remains to be elucidated (Fig. 3). It has now been established that the DOXP pathway is responsible for producing IPP that is then converted into carotenoids within the plastid, whereas the MVA pathway to IPP results in the synthesis of other isoprenoids (reviewed by Lichtenthaler, 1999). The discovery is significant for the manipulation of the early reactions of carotenogenesis. It has been shown that expression of *dxps* genes in lycopene-producing *E. coli* significantly increases carotenoid and ubiquinone formation (Harker and Bramley, 1999).

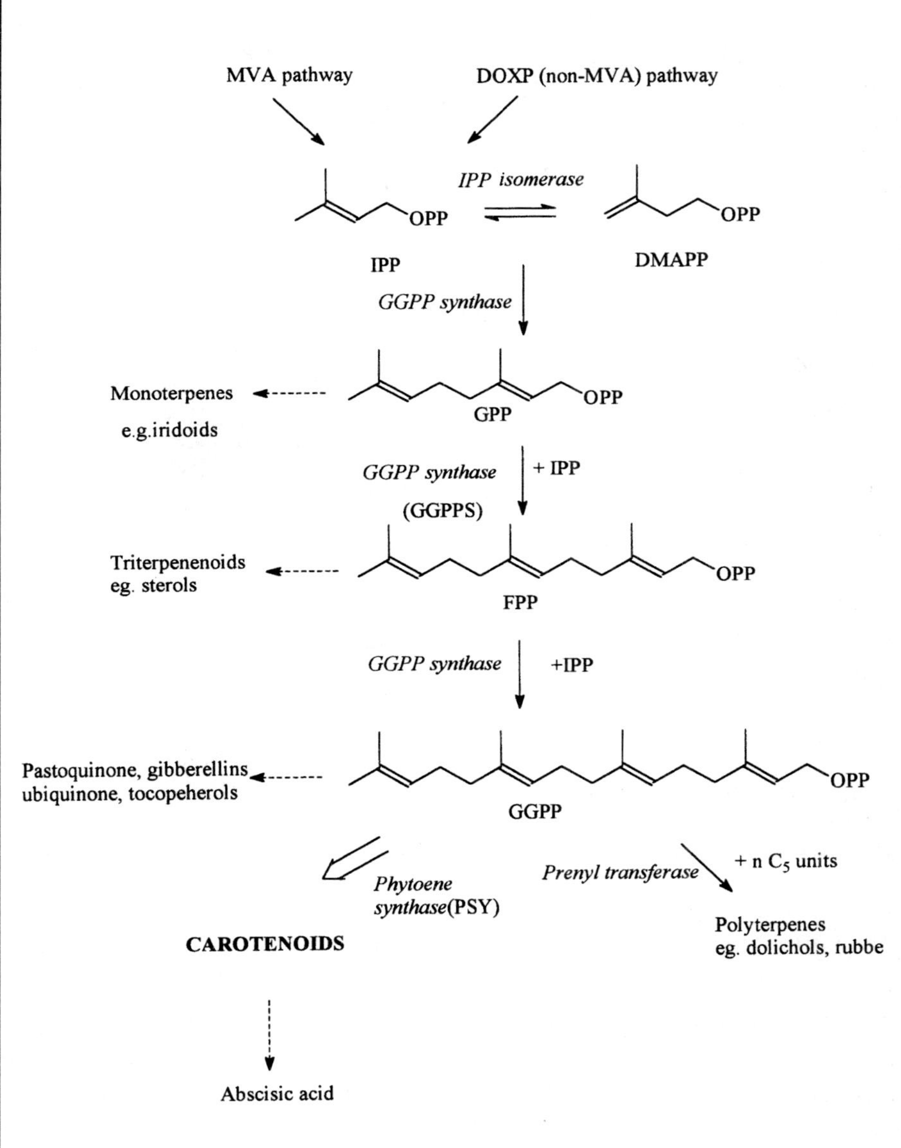

Figure 2 : Biosynthetic relationships of isoprenoids.
Abbreviations : MVA = mevalonic acid; DOXY = 1-deoxy-D-xylulose-5-phosphate; IPP = isopentenyl diphosphate; DMAPP = dimethylallyl diphosphate; GPP = geranyl diphosphate; FPP = farnesyl diphosphate; GGPP, geranylgeranyl diphosphate.

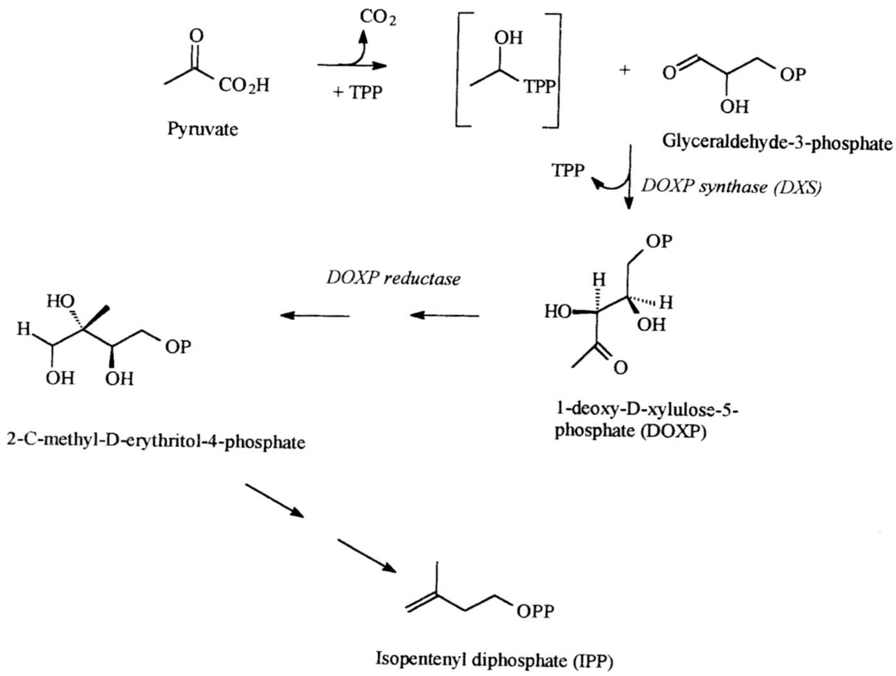

Figure 3 : The deoxyxylulose phosphate pathway for IPP formation. Abbreviation : TPP = thiamine pyrophosphate.

Once IPP is formed, it is isomerised to dimethylallyl diphosphate (DMAPP). This soluble enzyme is rate-limiting in yeast, and may be so in higher plants (Kajiwara *et al.*, 1997). Sequential addition of three IPP molecules yields the C_{20} isoprenoid isoprenoid geranylgeranyl diphosphate (GGPP), the reactions being catalysed by GGPP synthase (Fig. 2). At least five isoforms of GGPP synthase are present in *Arabidopsis*, but their individual roles are unknown. It is possible that individual isoenzymes catalyse the metabolism of GGPP to different products, *e.g.* carotenoids, phytol, terpenoid quinones. GGPP synthases from higher plants exhibit a significant sequence similarity to one another with over 70% identical nucleotides (Kuntz *et al.*, 1992; Aitken *et al.*, 1995), but individual GGPP synthases exhibit different affinities with respect to the chain length of the allylic substrate (Math *et al.*, 1992).

3.2. Formation of phytoene and desaturation reactions

The first committed step to the carotenoids is the head-to-head condensation of 2 GGPP molecules to produce phytoene, a colourless molecule with 3 conjugated double bonds (Fig. 4). Phytoene synthase (PSY, CRTB) is loosely associated with plastid membranes. At least two isoforms exist in tomato (PSY-1 and 2), of which *Psy-1* is upregulated in ripening fruit, whilst *Psy-2* appears to predominate in green tissues (Fraser *et al.*, 1999). The enzyme forms a functional, soluble complex with GGPP synthase in tomato (Fraser *et al.*, 2000). Multiple copies of *Psy* are also present in melon (Karvouni *et al.*, 1995). Whether there is a *Psy* gene family in other plants has not been established. Comparison of the amino acid sequence of phytoene synthases from various organisms indicates that these enzymes are both structurally and functionally conserved in eukaryotes (plants) and prokaryotes (reviewed by Armstrong, 1994).

The desaturation of 15-*cis* phytoene into lycopene occurs in 4 stepwise dehydrogenations, yielding phytofluence, ζ-carotene, neurosporene and lycopene (Fig. 4). The introduction of double bonds extends the chromophore to 11 conjugated double bonds in lycopene, giving the molecule its characteristic red colour and also high antioxidant activity (Miller *et al.*, 1996). The four dehydrogenations are catalysed by a single enzyme in bacteria (CRTI, AL-1), but by two in higher plants; phytoene desaturase (PDS, CRTP), forming ζ-carotene from phytoene (Pecker *et al.*, 1992), and ζ-carotene desaturase (ZDS, CRTQ), which catalyses the remaining two reactions to lycopene (Linden *et al.*, 1994; Albrecht *et al.*, 1995). A dincleotide-binding motif has been found in all phytoene and ζ-carotene desaturases. The conservation in amino acid sequences is 81% among various PDS enzymes and 34-60% among CRTI enzymes. However, comparison of PDS/CRTP-type to CRTI-type enzymes reveals less than 22% sequence conservation, most of which is in the FAD/NAD(P) binding motif found in the amino termini of the two types of enzymes. Both desaturases use quinones as redox cofactors (Schulz *et al.*, 1993). They are membrane bound, possibly in both thylakoids and envelope of plastids (reviewed by Bramley *et al.*, 1993).

Isomerization of 15-*cis* phytoene to the all-*trans* configuration must occur during the desaturation steps, since most desaturated carotenes are in the all-*trans* form. This occurs in *Erwinia* at the phytoene stage, suggesting the involvement of CRTI in the process (Fraser *et al.*, 1992) but it is still unclear whether a specific isomerase exists in plants. It has

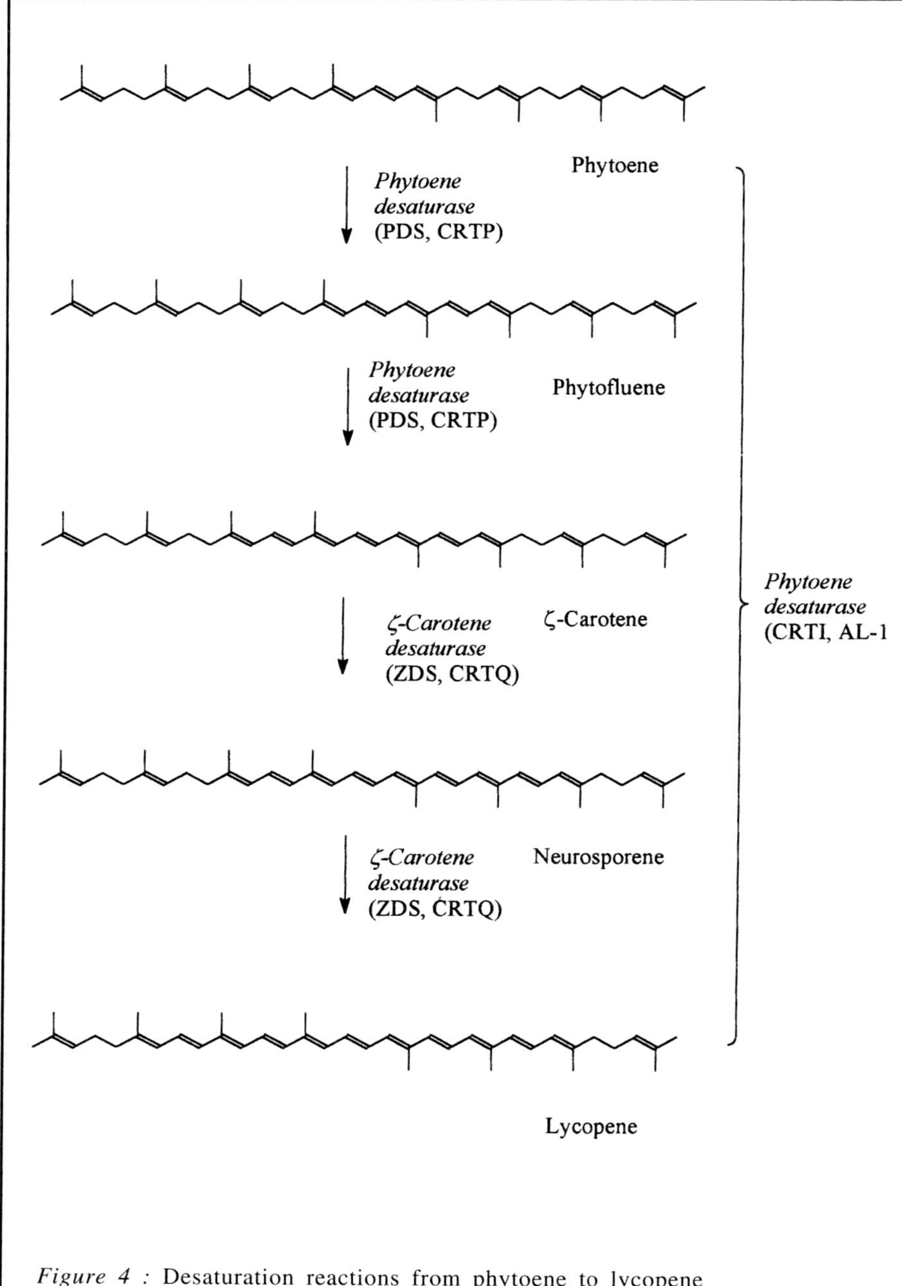

Figure 4 : Desaturation reactions from phytoene to lycopene
Abbreviations : PDS, CRTP = phytoene desaturase (plant); CRTI, AL-1 = phytoene desaturase (microbial); ZDS, CRTQ, CRTQ = ζ-carotene desaturase (plant).

been suggested that 7,9,7',9'-tetra-*cis* lycopene (pro-lycopene) is produced by the two carotene desaturases in *Arabidopsis* (Bartley *et al.*, 1999).

3.3. Cyclisation reactions

The cyclization reactions in plants represent an important branching point in the biosynthetic pathway of carotenoids; one route leads to β-carotene, zeaxanthin and violaxanthin, while the other route leads to α-carotene (ε, β-carotene) and lutein (Fig. 5). Cyclization of the acyclic precursor lycopene is initiated by proton attack at C-2 and C-2'. The gene *crtY,* which encodes lycopene β-cyclase, has been cloned from *Erwinia* (Misawa *et al.*, 1990). The cDNA of the higher plant equivalent, *crtL-b,* has been cloned from several plants including tomato (Pecker *et al.*, 1996), daffodil (Al-Babili *et al.*, 1996a) and *Arabidopsis* (Scolnik and Bartley, 1995). In both prokaryotes and eukaryotes this enzyme catalyzes two cyclization reactions that convert lycopene to β-carotene. While the lycopene β-cyclase enzymes from cyanobacteria and higher plants (CRTL-B or LCY) exhibit a high degree of amino acid sequence similarity along their entire length (35% identities and 55% similarities), they are distinct from the eubacterial lycopene β-cyclases (CRTY). There are three short regions of sequence similarity between CRTL-B and CRTY, one of them is a dinucleotide-binding motif and the other two may define regions that are important for substrate recognition or catalysis. The gene for lycopene ε-cyclase (*crtL-e*) has been cloned from *Arabidopsis* (Scolnik and Bartley, 1995) and tomato (Accession, Y14387).

3.4. Xanthophyll formation

The oxidation of carotenes results in the formation of a diverse array of xanthophylls (Figs. 6,7). Zeaxanthin is synthesised from β-carotene by the hydroxylation of C-3 and C-3' of the β-rings via the mono hydroxylated intermediate β-cryptoxanthin, a process requiring molecular oxygen in a mixed-function oxidase reaction. The gene encoding β-carotene hydroxylase (*crtZ*) has been cloned from a number of non-photosynthetic prokaryotes (reviewed by Armstrong, 1994) and from *Arabidopsis* (Sun *et al.*, 1996). Zeaxanthin is converted to violaxanthin by zeaxanthin epoxidase which epoxidises both β-rings of zeaxanthin at the 5,6 positions (Fig. 6). The zeaxanthin epoxidase gene of *Nicotiana plumbaginifolia* has been cloned from an abscisic acid-deficient mutant (Marin *et al.*, 1996). The homologous gene of pepper has also been cloned (Bouvier *et al.*, 1996). The violaxanthin de-epoxidase (*Vde*) cDNA has been cloned from lettuce (Bugos and Yamamoto, 1996) and the

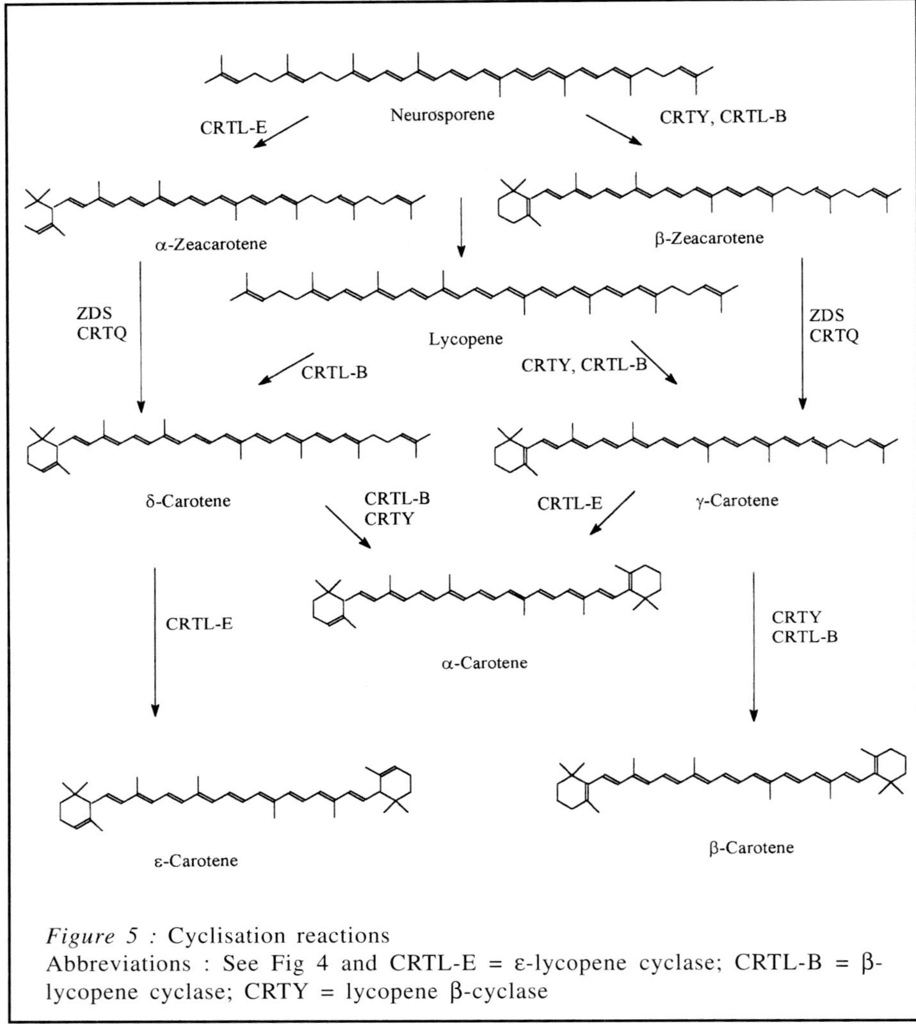

Figure 5 : Cyclisation reactions
Abbreviations : See Fig 4 and CRTL-E = ε-lycopene cyclase; CRTL-B = β-lycopene cyclase; CRTY = lycopene β-cyclase

properties of this enzyme reviewed by Rockholm and Yamamoto (1996). The products of violaxanthin/neoxanthin cleavage are the substrates for ABA synthesis (reviewed by Bramley, 1997a). The gene for the cleaving enzyme has been cloned (Schwartz *et al.*, 1997).

In the ripening chromoplasts of *Capsicum* fruit, two unusual carotenoids, capsanthin and capsorubin accumulate as the major pigments. These oxo-carotenoids are formed from antheraxanthin in the case of capsanthin and violaxanthin via capsanthin-5,6-epoxide in the case of capsorubin (Fig. 8). The gene product of capsanthin-capsorubin synthase

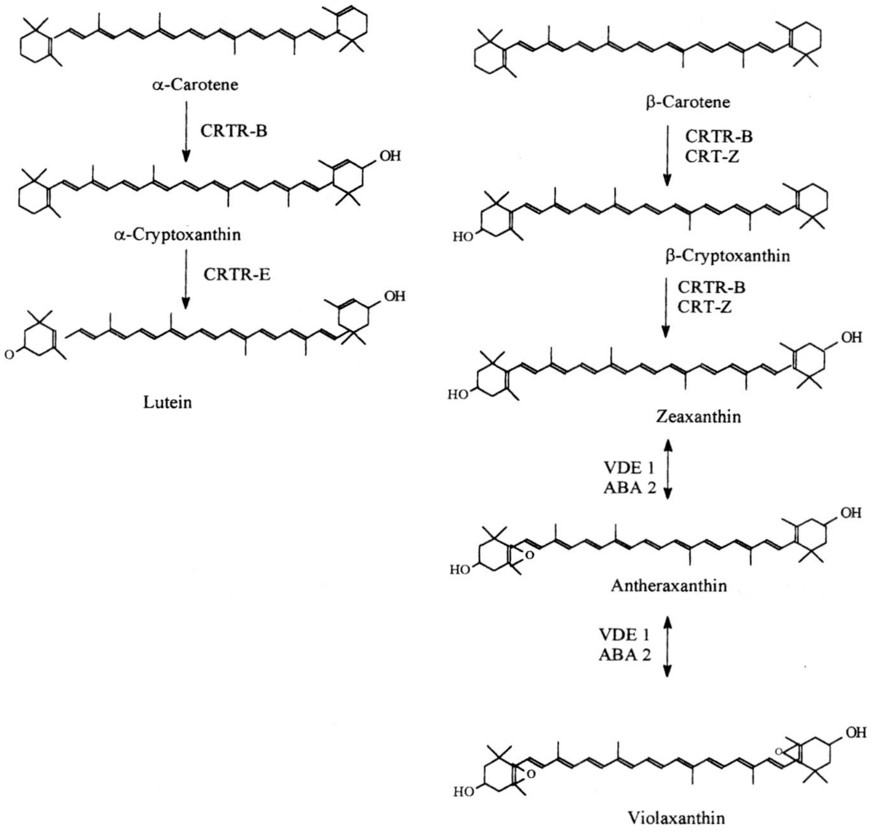

Figure 6 : Xanthophyll formation
CRTR-B = β-ring hydroxylase (plant); CRTC-E, ε-ring hydroxylase; CRTZ, β-ring hydroxylase (bacterial); VDE1, ABA2 = violaxanthin de-epoxidase

(*Ccs*) catalyzes this reactions (Bouvier *et al.*, 1994). The amino acid sequence of CCS is highly conserved (58% similarities, 75% identities) with lycopene β-cyclases from higher plants (Hugueney *et al.*, 1995).

Ketocarotenoids, *e.g.* echinenone, canthaxanthin, adonirubin and astaxanthin are oxygenated derivatives of β-carotene (Fig. 7). The cDNA encoding β-carotene-4-oxygenase have been cloned from *Agrobacterium auranticum* and *Alcaligenes PC-1* (Misawa *et al.*, 1995) and the green alga *Haematococcus pluvialis* (*CrtO, Bkt*) (Lotan and Hirschberg, 1995; Kajiwara *et al.*, 1995), which accumulates astaxanthin under stress

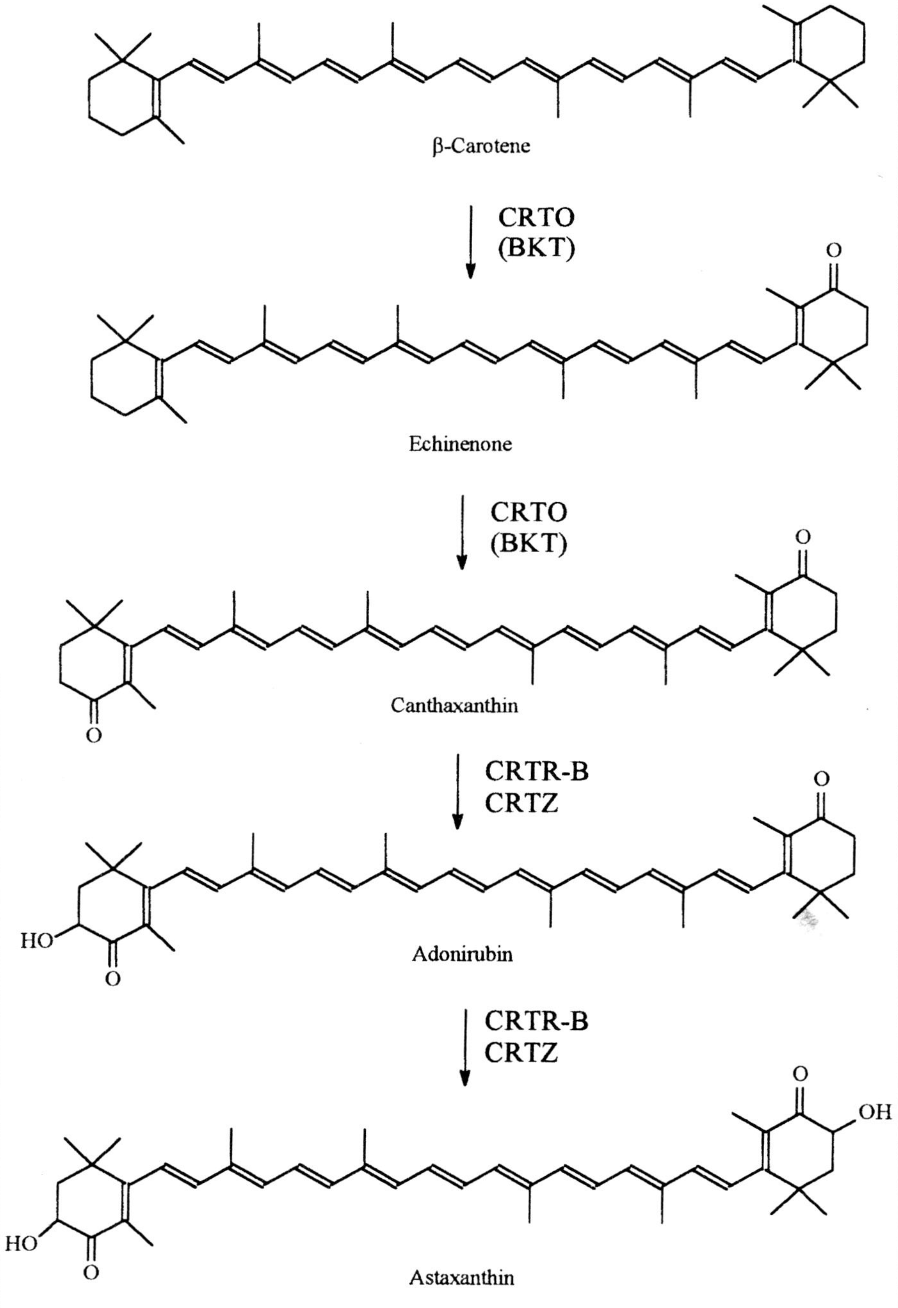

β-Carotene

CRTO
(BKT)

Echinenone

CRTO
(BKT)

Canthaxanthin

CRTR-B
CRTZ

Adonirubin

CRTR-B
CRTZ

Astaxanthin

Figure 7 : Biosynthesis of ketocarotenids
See Fig 6 and CRTO, BKT, carotene oxygenase.

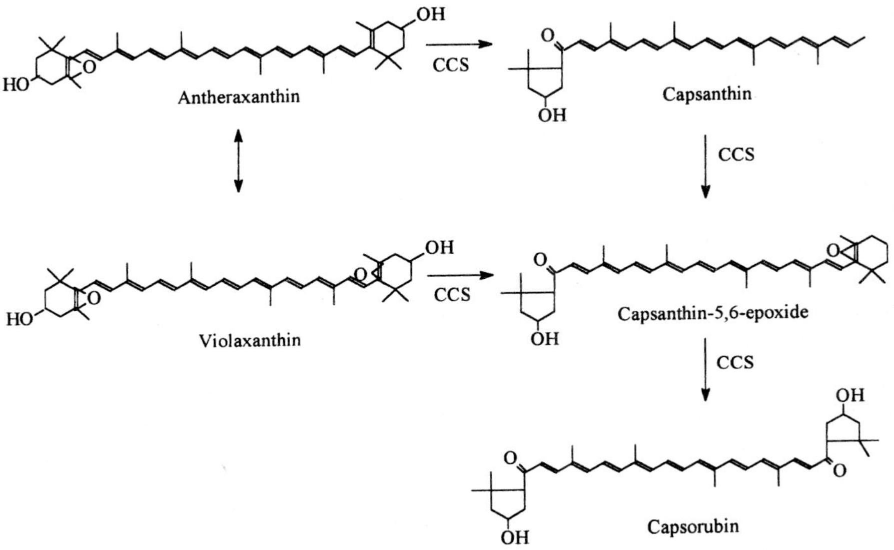

Figure 8 : Formation of capsanthin and capsorubin in *Capsicum annuum*. Abbreviation : CSS = capsanthin-capsorubin synthase

conditions. The enzyme catalyzes the introduction of keto groups to C4 and C4' of β-carotene to form canthaxanthin via the mono keto intermediate echinenone (Fig. 7).

4. REGULATION OF CAROTENOID BIOSYNTHESIS

Since carotenoids are derived for the central isoprenoid pathway (Fig. 2), the regulation of their formation must involve a co-ordinated flux of isoprenoid units into the C_{40} carotenoids as well as into other branches such as the biosynthesis of sterols, gibberellins, phytol and terpenoid quinones. An understanding of the complexities of regulation of the central pathway as well as within the carotenogenic pathways themselves is necessary to avoid misjudgements in attempts to genetically manipulate changes in plant carotenoids.

The discovery of gene families for several isoprenoid and carotenoid steps (see section 3) implies that there may be unique roles for each member of a family. This has been well documented for the multiple forms of HMG CoA reducatse (reviewed by Chappell, 1995), but our understanding of the roles of isoenzymes for later steps, *e.g.* GGPP synthase, is poor. These findings, together with the newly discovered DOXP pathway to IPP formation (Fig. 3) as a specific precursor of carotenoids, suggest that the traditional view on subcellular compartmentation of isoprenoid formation (Gray, 1987) may be an oversimplification. Although carotenoids are produced exclusively in plastids, it is likely that exchanges of cytoplasmic and plastidic metabolites, especially prenyl diphosphates occur and that these exchanges vary depending upon the type and developmental stage of the tissue (reviewed by McCaskill and Croteau 1998). The concept of metabolic channelling by isoenzymes dedicated to the formation of specific classes of isoprenoids, which may be regulated independently form one another has been suggested by Chappell (1995). It has been shown, from work with transgenic tomatoes, that perturbations of the carotenoid pathway using either up-or down-regulation of *Psy-1* lead to changes in the levels of gibberellins. Increases in phytoene formation cause a reduction of gibberellin levels and dwarfing of the progeny (Fray *et al.*, 1995). Such cross talk between classes of isoprenoids suggests that any metabolic channels must have regulatory mechanisms that affect each of the channels.

Carotenoid formation itself in plants is a highly regulated process. For example, the concentration and composition of leaf xanthophylls is affected by light intensity (Demmig-Adams *et al.*, 1996) and the accumulation of specific carotenoids in chromoplasts of fruits and flowers is developmentally regulated (reviewed by Bramley *et al.*, 1993). The signalling pathways mediating these transformations are unknown, but reactive oxygen species act as second messengers mediating carotenoid formation during chromoplast differentiation (Bouvier *et al.*, 1998). Although carotenoid biosynthesis is necessary for the development of chloroplasts, expression of carotenoid genes is neither light dependent nor controlled by cryptochrome1 genes (Ninu *et al.*, 1999). During de-etiolation in tomato seedlings carotenoid levels increase several-fold, whilst the transcript levels of *Pds* and *Psy-1* remain relatively unchanged (Giuliano *et al.*, 1993). This indicates that transcriptional control of *Pds* and *Psy-1* does not play a major role in the regulation of carotenoid biosynthesis during greening and chloroplast development. Although it has been thought that the structural requirements for carotenoids present in light-harvesting complexes was extremely rigid, it has now been demonstrated that other carotenoids can replace those normally present with no apparent harmful effects on the plant (Pogson *et al.*, 1996; Kumaga *et al.*, 1998).

Carotenoid accumulation during fruit ripening in tomato has been studied extensively and is good model system to elucidate the regulation of the process. During ripening the concentration of carotenoids increases between 10 to 15-fold due mainly to a 500-fold increase in the concentration of lycopene (Fraser *et al.*, 1994, Table 3). Accumulation of lycopene begins at the breaker stage of fruit ripening after the fruit has reached the mature green stage. The mRNA levels of *Psy* and *Pds* increase significantly during the breaker stage (Pecker *et al.*, 1992; Fraser *et al.*, 1994; Giuliano *et al.*, 1993) due to transcriptional control that is developmentally regulated (Corona *et al.*, 1996). In contrast, the mRNA of lycopene β-and ε-cyclase genes decreases at this stage, probably accounting for the reduction in cyclic carotenoids in ripe fruit (Pecker *et al.*, 1996; Ronen *et al.*, 1999). A similar increase in mRNA was found in the genes *Ggps, Psy* and *Pds* during fruit ripening of *Capsicum annuum* (Kuntz *et al.*, 1992; Romer *et al.*, 1993). The involvement of ethylene in carotenogenesis has been implied from studies with a pleiotropic mutant of tomato, affected in ethylene production, which does not express *Psy-1* and has an albino pericarp (Thompson *et al.*, 1999). Evidence for transcriptional regulation of carotenoid genes has also been found in flowers where the steady-state levels of mRNA for *Psy, Pds* and *crtL-b* increase dramatically in the petals of tomato flowers (Giuliano *et al.*, 1993).

These reports indicate that control of gene expression at the transcriptional level is a key regulatory mechanism controlling carotenogenesis *in vivo*. However, post-transcriptional regulation of carotenoid biosynthesis enzymes has been found in chromoplasts of the daffodil. The enzymes PSY and PDS were detected in an inactive form in the soluble fraction and in an active from when membrane-bound (Al-Babili *et al.*, 1996b; Schledz *et al.*, 1996). The presence of inactive proteins indicates that a post-translation regulation mechanism is present and is linked to the redox state of the membrane-bound electron acceptors. In addition, substrate specificity of the β-and ε-lycopene cyclases may control the proportions of the β, β and β, ε carotenoids in plants (Cunningham *et al.*, 1996).

Regulation of carotenoid accumulation in non-photosynthetic tissues (as compared with formation) is governed by the sequestration of carotenoids in chromoplasts. For example, the expression of chromoplast-specific carotenoid-associated genes from the corolla of cucumber (CHRC and CHRD) shows that two regulatory factors control carotenoid sequestration: (i), floral tissue-specific transcriptional regulators and (ii), post-transcriptional regulators related to the amount or type of cartoenoids. Phytohormones act as transcriptional regulators of CHRC expression (Vishnevetsky *et al.*, 1999). In *Capsicum annuum,* the fibrillin protein is essential for carotenoid accumulation in the fruit (Deruere *et al.*, 1994).

The carotenoud pathway may also be regulated by feedback inhibition by the end products. Inhibition of lycopene cyclization in leaves of tomato caused increase in the expression of *Pds* and *Psy-1* (Guiliano *et al.*, 1993; Corona *et al.*, 1996). This hypothesis is supported by other studies using carotenoid biosynthesis inhibitors where treated photosynthetic tissues accumulated higher concentrations of carotenoids than untreated tissues (reviewed by Bramley, 1993). The mechanism of this regulation is unknown. Although abscisic acid (ABA), has also been suggested as a possible activator, *Arabidopsis* mutants containing cyclic carotenoids, but inhibited in ABA synthesis, did not exhibit a significant increase in carotenoids (Rock and Zeevart, 1991). A contrary view, however, comes from studies on the *immutans* mutant of *Arabidopsis*, which contains phytoene-accumulating tissue, that show there is no feedback inhibition of phytoene desaturase gene expression (Wetzel and Rodermel, 1998).

5. WHY ENHANCE CAROTENOID LEVELS IN PLANTS?

The prime reason for elevating or altering carotenoids in crop plants is linked to their functions in human health.

The best documented function of carotenoids in humans is their provitamin A activity, which is restricted to β-carotene and other carotenoids with β-end groups, *e.g.* β-cryptoxanthin and zeaxanthin. Evidence has been documented to suggest that humans with low carotenoid levels are more susceptible to degenerative diseases (Mayne, 1996). Much of the evidence for the alleviation of chronic diseases such as CHD and certain cancers has come epidemiological studies (*e.g.* Giovannucci, 1995; Gann *et al.*, 1999; Ascherio *et al.*, 1999). Lycopene has been particularly implicated in the reduction of such diseases, although the evidence indicates that a high tomato diet, rather than lycopene alone, is beneficial (reviewed by Giovannucci, 1999; Bramley, 2000). The benefits of carotenoids in the diet are thought to be linked to their powerful antioxidant capacity, although whether this is their role *in vivo* remains uncertain (Rice-Evans *et al.*, 1997). Epidemiological studies have also shown an inverse association between antioxidant/carotenoid intake or blood level and the risk of cataract and age-related macular degeneration (AMD). Cataract risk was reported to decrease with higher intakes of lutein and zeaxanthin, but not other carotenoids (Brown *et al.*, 1999). Lutein and zeaxanthin (Fig. 1) are the principal components of the macula pigments and their contents can be modified by dietary changes, thus reducing the amount of blue light reaching those eye tissues vulnerable to AMD by 30-40% (Landrum *et al.*, 1997).

Based upon such reports, there is an increasing view that functional foods (nutraceuticals), in which carotenoid levels are high or modified qualitatively, would be beneficial to human health, especially with respect to degenerative disease in an ageing population or as dietary sources of provitamin A. One strategy for increasing the levels in plants is by genetic engineering of the parthway in crops (Hirschberg, 1999; Grusak and DellaPenna, 1999).

6. STRATEGIES ENABLING THE GENETIC ENGINEERING OF CAROTENOID LEVLES

The ready availability of most genes encoding carotenoid biosynthetic enzymes (section 3), the elucidation of the pathway in plants (section 3) and the ability to transfer genes to plants by *Agrobacterium*-based protocols provides the basis for the genetic modification of crops with respect to carotenoid content. Such modifications are now well established in micro-organisms (reviewed by Cunningham and Gantt, 1998; Misawa and Shimada, 1998). However, the complexity and plasticity of higher plant metabolism, combined with the branched nature of the carotenoid (isoprenoid) pathways, has meant that such modifications have proved to be more difficult than in microorganisms and the phenotypes somewhat unpredictable. The current range of crops transformed with carotenoid genes is shown in Table 4.

Certain prerequisites need to be addressed before such transgenic crops can be successfully produced. These include:

6.1. Cloning and choice of genes

Most of the genes encoding the biosynthetic enzymes for carotenoid formation have now been cloned from a range of organisms, including higher plants. Therefore, these are available for plant transformation. The remaining structural genes (*e.g.* those for apocarotenoid formation) are likely to be isolated within the next 2-3 years. Similarly, genes encoding carotenoid-associated proteins are also being cloned (section 3). At the present time, however, no regulatory genes have been isolated.

Table 4. Higher plants transformed with carotenoid genes

Plant	Inserted gene/cDNA	Promoter	Phenotype	Reference
Rice	*Psy* cDNA from daffodil	CaMV 35S	Phytoene accumulation in endosperm.	Burkhardt *et al.*, 1997
	Psy, crtI, crtL-B	CaMV 35S, glutelin	β-carotene accumulation in endosperm	Ye *et al.*, 2000
Tomato	Antisense *Psy-1* from tomato	CaMV 35S	Reduced carotenoids, increased gibberellins	Bird *et al.*, 1991
Tomato	*crtI* from *E. uredovora*	CaMV 35S	Increased β-carotene, decreased lycopene	Bramley, 1997
Tomato	*Psy-1* cDNA from tomato	CaMV 35S	Sense suppression, premature lycopene accumulation, dwarf plants	Fray *et al.*, 1995
Oil seed rape	*crtB* genes from *E. uredovora*	napin	50-fold increase in seed carotenoids	Shewmaker *et al.*, 1990
Carrot	*crt* genes from *E. herbicola*	CaMV 35S	2-5 fold increase in root carotenoids	Ausich *et al.*, 1991; Haupt-mann *et al.*, 1997
Tobacco	*crtI* gene from *Erwinia*	CaMV 35S	Tolerance to bleaching herbicides, reduction in lutein	Misawa *et al.*, 1994b
Tobacco	*CCS* cDNA (RNA viral vector)	Tobacco-virus	Capsanthin increase	Kumagai *et al.*, 1995, 1998
Tobacco	*CrtO*	?	Ketocarotenoids in nectary tissue	Hirschberg, 1999

The selection of which gene(s) (or cDNA) to use depends upon the end product required, the carotenoid content of the host and the possible need to avoid sense suppression. For tissues that do not normally contain carotenoids it is necessary to transfer genes that encode enzymes that convert prenyl diphosphates into phytoene as well as any subsequent steps needed (Figs. 2, 4-8). For example, the formation of β-carotene in rice endosperm, where carotenoids are normally absent, required the simultaneous transfer of three genes: *Psy* from daffodil, *crtI* from *Erwinia herbicola* and *crtL-B* from daffodil (Ye *et al.*, 2000). In contrast, a single gene transformation (*crtB* from *Erwinia*) was sufficient to increase carotenoid levels some 50-fold in seeds of canola (*Brassica napus*), which already contained carotenoids (Shewmaker *et al.*, 1999). In some cases, a single gene transfer does not elevate carotenoid levels, but does alter their composition. Transformation of tobacco with the *crtI* gene from *Erwinia* caused an alteration in xanthophylls, but no increase in total carotenoids (Misawa *et al.*, 1994b).

Transformation of a plant with the endogenous gene usually leads to the phenomenon of co-suppression (sense suppression, gene silencing). Under these circumstances, expression of the transgene causes down regulation of the endogenous gene and hence a phenotype similar to that shown in antisense transgenic plants. For example, transformation of tomato with its own *Psy-1* gene caused a reduction in ripe fruit carotenoids (Truesdale, 1993). In order to overcome this problem genes from different species, and specially bacteria such as *Erwinia spp.,* are used as they have very low homologies to the plant carotenoid genes and hence do not cause co-suppression. However, since the bacterial genes have no leader sequence for targeting the protein to the plastid, a chimeric gene construct is used, typically with the small sub-unit of Rubisco as the leader. The transgenic canola and rice described above are good examples of this approach. Alternatively, a synthetic cDNA can be made, with every third base of the triplet code changed to reduce homology by 33%. This approach has been successful with a synthetic *Psy* transformed into tomato (Bramley, 1997b). Clearly the bacterial enzymes are active in the plant plastid and must integrate into the carotenogenic enzyme complex.

A further consideration in the choice of gene relates to those that have gene families. Only one member of such a family may be involved in carotenogenesis in a particular tissue. A good example of this is *Psy 1* and *–2* of tomato. PSY-1 is responsible for phytoene synthesis in ripening fruit (Fraser *et al.*, 1999) and PSY-2 is not functional in chromoplasts, even if the protein is produced. Gene families for enzymes in the common isoprenoid pathway (Fig. 2) may also be pathway-specific.

6.2. Choice of promoter

A range of promoters that are not from carotenoid genes have been used successfully (Table 4). These include constitutive promoters such as CaMV 35S, or tissue specific promoters such as napin and glutelin. The *Pds* promoter has also been used (Corona *et al.*, 1996) and all the promoters of carotenoid genes from *Arabidopsis* have now been cloned (Giuliano, personal communication). Choice of promoter is a very important part of the strategy, since it controls the dynamics, level and tissue specificity of expression of the gene of interest. The advantage of tissue-specific promoters is that transgene expression can be focused on a particular tissue or at the time of development, thus minimising unwanted effects on isoprenoid metabolism elsewhere in the plant or at the wrong time of development. The use of PG rather than CaMV 35S with phytoene synthase in tomato has changed a dwarf phenotype with 35S (Truesdale, 1993) to one with no dwarfism and elevated carotenoids (Bramley, Romer and Fraser, unpublished). The strength of the promoter in triggering transcription may also influence the mRNA levels and phenotype.

6.3. Determination of rate limiting steps and fluxes through the pathway

Elucidation of the rate-limiting (regulatory) steps of the isoprenoid pathway has proved to be significantly more difficult than cloning the genes (see section 4). Although there is some evidence that phytoene synthase is rate-limiting in ripening tomato fruit (Fraser *et al.*, 1994), the complexity of the carotenoid pathway, with its multiple branch points makes elucidation of the fluxes through the pathway intractable at the present time. Although end products can be quantified accurately, measurements of intermediates and the dynamics of flow still require detailed study. Therefore, most studies have used a variety of genes in an empirical fashion.

7. CONCLUSIONS AND FUTURE PROSPECTS

Plant biotechnology is at an exciting stage with respect to the alteration of carotenoid levels by genetic engineering. The ready availability of genes from plants and microorganisms, together with an array of suitable promoters, has facilitated the successful transformation of several key crops in recent years (Table 4). These transgenic crops provide the basis for producing functional foods across the world with elevated or altered carotenoid levels that have the potential to alleviate vitamin A deficiency (rice, oil seed rape), increase lycopene intake (tomato) or provide ketocarotenoids in animals or fish feeds (tobacco). In the near future, it should be anticipated that other transformants will be produced, perhaps with increased levels of zeaxanthin and lutein (for prevention of AMD) or greater increases than those presently reported. The latter will require us to understand the regulation of carotenogenesis more thoroughly than we do at present and also perhaps require the isolation of transcription factors and also genes that control the deposition of carotenoids in the plastids. It is also likely that multiple gene transformations will be needed in order to overcome bottlenecks in the pathway and the transfer of genes from the DOXP pathway into plants is very likely in the near future.

At present, we do not understand why changes in carotenoid content in transgenic plants vary so dramatically between species, some of which may have been transformed with the same gene. These differences may be linked to the type of tissue in which expression is targeted (*e.g.* embryo, endosperm, vegetative) or the photosynthetic capacity of the tissue (chloroplast or chromoplast). In addition, the availability of precursors of the carotenoids may differ in these tissues.

In conclusion, the advances in transformation technology, the cloning of relevant genes and the increasing awareness of the health benefits of carotenoids in the diet has provided a large impetus to this field of plant biotechnology. Over the next few years one should expect significant improvements in the transgenic crops that produce elevated levels of carotenoids.

ACKNOWLEDGEMENTS

The author would like to thank Dr Paul Fraser for helpful advice in preparing the manuscript. Studies in the author's laboratory are partly funded by the BBSRC, EU, MAFF and Syngenta.

REFERENCES

Aitken SM, Attucci S, Ibrahim RK and Gulick PJK (1995) A cDNA encoding GGPP synthase from white lupin. *Plant Physiol.,* **108** : 837-838.

Al-Babili S, Hobeika E and Beyer P (1996a) A cDNA encoding lycopene cyclase from *Narcissus pseudonarcissus*. *Plant Physiol.,* **112** : 1398.

Al-Babili S, von Lintig J, Haubruck H and Beyer P (1996b) A novel, soluble form of phytoene desaturase in *Narcissus pseudonarcissus* chromoplasts is Hsp 70-complexed and competent for flavinylation, membrane association and enzymatic activation. *Plant J.,* **9** : 601-612.

Albrecht M, Klein A, Hugueney P, Sandmann G and Kuntz M (1995) Molecular cloning and functional expression in *E. coli* of a novel plant enzyme mediating ζ-carotene desaturation. *FEBS Lett.,* **372** : 199-202.

Armstrong GA (1994) Eubacteria show their true colours: genetics of carotenoid pigment biosynthesis from microbes and plants. *J. Bacteriol.,* **45** : 4795-4802.

Armstrong GA and Hearst JE (1996) Genetics and molecular biology of carotenoid pigment biosynthesis. *FASEB J.,* **10** : 228-237.

Ascherio SA, Rimm EB, Hernan MA, Giovannucci E, Kawachi I, Stampfer MJ and Willett WC (1999) Relation of consumption of vitamin E, vitamin C and carotenoids to the risk of stroke in men in the US. *Annals Internal Med.,* **130** : 963-970.

Ausich RL, Brinkhaus FL, Mukharji I, Proffitt JH, Yarger JG and Yen HCB (1991) Biosynthesis of carotenoids with genetically engineered hosts. Patent application PCT/US91/01458.

Bartley GE and Scolnik PA (1995) Plant carotenoids: pigments for photoprotection, visual attraction and human health. *Plant Cell.,* **7** : 1027-1038.

Bartley GE, Scolnik PA and Giuliano G (1994) Molecular biology of carotenoid biosynthesis in plants. *Annu. Rev. Plant Physiol.,* **45** : 287-301.

Bartley GE, Scolnik PA and Beyer P (1999) Two *Arabidopsis thaliana* carotene desaturases, phytoene desaturase and ζ-carotene, expressed in *E. coli,* catalyze a poly-*cis* pathway to yield pro-lycopene. *Eur. J. Biochem.* **259** : 396-403.

Bird CR, Ray JA, Fletcher JD, Boniwell JM, Bird AS, Teulieres C, Blain I, Bramley PM and Schuch W (1991) Using antisense RNA to study gene function: inhibition of carotenoid biosynthesis in transgenic tomatoes. *BioTechnology* **9** : 635-639.

Bouvier F, Hugueney P, d'Harlingue A, Kuntz M and Camara B (1994) Xanthophyll biosynthesis in chromoplasts: isolation and molecular cloning of an enzyme catalysing the conversion of 5,6-epoxycarotenoids into ketocarotenoids. *Plant J.,* **6** : 4 5-54.

Bouvier F, d'Harlingue A, Hugueney P, Marin E, Marion-Poll A and Camara B (1996) Xanthophyll biosynthesis: cloning, expression, functional reconstitution and regulation of β-cyclohexenyl carotenoid epoxidase from pepper (*Capsicum annuum*) *J Biol. Chem.,* **271** : 28861-28867.

Bouvier F, Bachaus RA and Camara B (1998) Induction and control of chromoplast-specific carotenoid genes by oxidative stress. *J. Biol. Chem.,* **273** : 30651-30659.

Bramley PM (1992) Analysis of carotenoids by HPLC and diode array detection. *Phytochemical Anal.,* **3** : 97-104.

Bramley PM (1993) Inhibition of carotenoid biosynthesis. In: *Carotenoids in Photosynthesis* (Eds. Young AJ and Britton G), Chapman and Hall, London, pp. 127-159.

Bramley PM (1997a) Isoprenoid metabolism. In: *Plant Biochemistry* (Eds, Dey PM and Harbourne J), Academic Press, London, pp. 417-437.

Bramley PM (1997b) The regulation and genetic manipulation of carotenoid biosynthesis in tomato fruit. *Pure Appl. Chem.,* **69** : 2159-2162.

Bramley PM (2000) Is lycopene beneficial to human health? *Phytochemistry,* **54** : 233-236.

Bramley PM, Bird CR and Schuch W (1993) Carotenoid biosynthesis and manipulation. In: *Biosynthesis and Manipulation of Plant Products vol. 3* (Ed. Grierson G), Blackie, London, pp. 139-177.

Britton G (1991) Carotenoids. In: *Methods in Plant Biochemistry* (Eds. Charlwood BV and Banthorpe DV), Academic Press, London, pp. 437-518.

Britton G (1995) Structure and properties of carotenoids in relation to function. *FASEB J.,* **9** : 1551-1558.

Brown L, Rimm EB, Seddon JM, Giovannucci EL, ChasanTaber L, Spiegelman D, Willett WC and Hankison SE (1999) A prospective study of carotenoid intake and risk of cataract extraction in US men. *Am. J. Clin. Nutr.* **70** : 517-524.

Bugos RC and Yamamoto HY (1996) Molecular cloning of violaxanthin de-epoxidase from romaine lettuce and expression in *E. coli. Proc. Natl. Acad. Sci.,* **93** : 6320-6325.

Burkhardt PK, Beyer P, Wunn J, Kloti A, Armstrong GA, Schledz M, von Lintig J and Potrykus I (1997) Transgene rice (*Oryza sativa*) endosperm expressing daffodil (*Narcissus pseudonarcissus*) phytoene synthase accumulates phytoene, a key intermediate of provitamin A biosynthesis. *Plant J.,* **11** : 1071-1078.

Chappell J (1995) Biochemistry and molecular biology of the isoprenoid pathway in plants. *Annu. Rev. Plant Physiol. Plant Mol. Biol.,* **46** : 521-547.

Corona V, Aracci B, Kosturkova G, Bartley GE, Pitto L, Giorgetti L, Scolnik PA and Giuliano G (1996) Regulation of a carotenoid biosynthesis gene promoter during plant development. *Plant J.,* **9** : 505-512.

Cunningham FX Jr and Gantt E (1998) Genes and enzymes of carotenoid biosynthesis in plants. *Annu. Rev. Plant Physiol. Plant Mol. Biol.,* **49** : 557-583.

Cunningham FX Jr, Pogson B, Sun Z, McDonald KA, DellaPenna D and Gantt E (1996) Functional analysis of the β-and ε-lycopene cyclase enzymes of *Arabidopsis* reveals a mechanism for control of cyclic carotenoid formation. *Plant Cell,* **8** : 1613-1626.

Demmig-Adams B, Gilmore AM and Adams WW II (1996) *In vivo* function of carotenoids in higher plants. *FASEB J.,* **10** : 403-412.

Deruere J, Romer S, d'Harlingue A, Backhaus RA, Kuntz M and Camara B (1994) Fibril assembly and carotenoid overaccumulation in chromoplasts: a model for supramolecular lipoprotein structures. *Plant Cell,* **6** : 119-133.

Frank H and Cogdell RJ (1993) The photochemistry and functions of carotenoids in photosynthesis. In: *Carotenoids in Photosynthesis* (Eds. Young A J and Britton G), Chapman and Hall, London, pp. 252-326.

Frank H, Chynwat V, Desamero RZB, Farhoosh R, Erickson J and Bautista J (1997) On the photophysics and photochemical properties of carotenoids and their role as light harvesting pigments in photosynthesis. *Pure Appl. Chem.,* **69** : 2117-2124.

Fraser PD, Misawa N, Linden H, Yamano S, Kobayashi K and Sandmann G (1992) Expression in *E. coli,* purification and reactivation of recombinant *Erwinia uredovora* phytoene desaturase. *J. Biol. Chem.,* **267** : 19891-19895.

Fraser PD, Truesdale MR, Bird CR, Schuch W and Bramley PM (1994) Carotenoid boisynthesis during tomato fruit development. *Plant Physiol.,* **105** : 405-413.

Fraser PD, Kiano J, Truesdale MR, Schuch W and Bramley PM (1999) *Psy-2* enzyme activity does not contribute to carotenoid synthesis in ripe fruit. *Plant Mol. Biol.,* **40** : 687-698.

Fraser PD, Schuch W and Bramley PM (2000) Phytoene synthase from tomato (*Lycopersicon esculentum*) chloroplasts – partial purification and biochemical properties. *Planta,* **211** : 361-369.

Fray RG, Wallace A, Fraser PD, Valero D, Hedden P, Bramley PM and Grierson D (1995) Constitutive expression of a fruit phytoene synthase in transgenic tomatoes causes dwarfism by redirecting metabolites from the gibberellin pathway. *Plant J.,* **8** : 693-701.

Gann PH, Ma J, Giovannucci E, Willett W, Sacks FM, Hennekens CH and Stampfer MJ (1999) Lower prostate cancer risk in men with elevated plasma lycopene levels: results of a prospective study. *Cancer Res.,* **59** : 1225-1330.

Giovannucci E (1999) Tomatoes, tomato-based products, lycopene and cancer: review of the epidemiologic literature. *J. Natl. Cancer Inst.,* **91** : 317-331.

Giovannucci E, Ascherio A, Rimm EB, Stampfer MJ, Colditz GA and Willett WC (1995) Intake of carotenoids and retinol in relation to risk of prostate cancer. *J. Natl. Cancer Inst.* **87** : 1767-1776.

Giuliano G, Bartley GE and Scolnik PA (1993) Regulation of carotenoid biosynthesis during tomato fruit development. *Plant Cell,* **5** : 379-387.

Goodwin TW (1980) *Biochemistry of Carotenoids,* vol. 1. Chapman and Hall, London.

Gray JG (1987) Control of isoprenoid biosynthesis in higher plants. *Adv. Bot. Res.,* **14** : 25-91.

Gross J (1991) *Pigments in Vegetables. Chlorophylls and Carotenoids.* Van Nostrand Reinhold, USA.

Grusak MA and DellaPenna D (1999) Improving the nutrient composition of plants to enhance human nutrition and health. *Annu. Rev. Plant Physiol. Plant Biol.,* **50** : 133-161.

Harker M and Bramley PM (1999) Expression of 1-deoxy-D-xylulose-5-phosphatases in *E. coli* increases carotenoid and ubiquinone biosynthesis. *FEBS Lett,* **448** : 115-119.

Harker M and Hirschberg J (1998) Molecular biology of carotenoid biosynthesis in photosynthetic organisms. *Methods Enzymol.,* **297** : 244-263.

Hauptmann R, Eschenfeldt WH, English J and Brinkhaus FL (1997) Enhanced carotenoid accumulation in storage organs of genetically engineered plants *US patent,* **561** : 89-88.

Hirschberg J (1999) Production of high value compounds: carotenoids and vitamin E. *Curr. Opinion in Biotech.,* **10** : 186-191.

Hirschberg J, Cohen M, Harker M, Lotan T, Mann V and Pecker I (1997) Molecular genetics of the carotenoid biosynthetic pathway in plants and algae. *Pure Appl. Chem.,* **69** : 2151-2158.

Rice-Evans CA, Sampson J, Bramley PM and Holloway DE (1997) Why do we expect carotenoids to be antioxidants *in vivo*? *Free Rad. Res.,* **26** : 381-398.

Hugueney P, Badillo A, Chen HC, Klein A, Hirschberg J, Camara B and Kuntz M (1995) Metabolism of cyclic carotenoids: a model for the alteration of this biosynthetic pathway in *Capsicum annuum* chromoplasts. *Plant J.,* **8** : 417-424.

IUPAC Commission on Nomenclature of Organic Chemistry and IUPAC-IUB Commission on Biochemical Nomenclature (1975) Nomenclature of carotenoids. *Pure Appl. Chem.* **41** : 407-431.

Kajiwara S, Kakizono T, Saito T, Kondo K, Ohtani T, Nishio N, Nagai S and Misawa N (1995) Isolation and functional identification of a novel cDNA for astaxanthin biosynthesis form *Haematococcus pluvialis* and astaxanthin biosynthesis in *E. coli. Plant Mol. Biol.,* **29** : 343-352.

Kajiwara S, Fraser PD, Knondon K and Misawa N (1997) Expression of an exogenous IPP isomerase gene enhances isoprenoid biosynthesis in *E. coli. Biochem. J.,* **324** : 421-426.

Karvouni Z, John I, Taylor JE, Watson CF, Turner AJ and Grierson D (1995) Isolation and characterisation of a melon cDNA clone encoding phytoene synthase. *Plant Mol. Biol.,* **27** : 1153-1162.

Khachik F, Beecher GR, Goli MB, and Lusby WR (1991) Separation, identification and quantitation of carotenoids in fruits, vegetables and human plasma by HPLC. *Pure Appl. Chem.,* **63** : 71-80.

Kumagai MH, Donson J, Della-Cioppa G, Harvey D, Hanley K and Grill LK (1995) Cytoplasmic inhibition of carotenoid biosynthesis with virus-derived RNA. *Proc. Natl. Acad, Sci.,* **92** : 1679-1683.

Kumagai MH, Keller Y, Bouvier F, Clary D and Camara B (1998) Functional integration of non-native carotenoids into chloroplasts by viral-derived expression of CCS in *Nicotiana benthamiana. Plant J.,* **14** : 305-315.

Kuntz M, Romer S, Suire C, Hugueney P, Weil JH, Schantz R and Camara B (1992) Identification of a cDNA for the plastid-located geranylgeranyl diphosphate synthase from *Capsicum annuum*: correlative increase in enzyme activity and transcript level during fruit ripening. *Plant J.,* **2** : 25-34.

Landrum JT, Bone RA, Joa A, Kilburn MD, Moore LL and Sprague KE (1997) A one year study of the macular pigment: the effect of 140 days of a lutein supplement *Exp. Eye Res.,* **65** : 57-62.

Lange BM and Croteau R (1999) Isoprenoid biosynthesis via a MVA independent pathway in plants: cloning and heterologous expression of 1-deoxy-D-xylulose-5-phosphate reductoisomerase from peppermint. *Archiv. Biochem. Biophys.,* **365** : 170-174.

Lichtenthaler K (1999) The 1-deoxy-D-xylulose-5-phosphate pathway of isoprenoid biosynthesis in plants. *Annu. Rev. Plant Physiol. Plant Mol. Biol.,* **50** : 47-65.

Linden H, Misawa N, Saito T and Sandmann G (1994) A novel carotenoid biosynthesis gene coding for ζ-carotene desaturase-functional expression, sequence and phylogeneic origin. *Plant Mol. Biol.,* **24** : 369-379.

Lotan T and Hirschberg J (1995) Cloning and expression in *E. coli* of the gene encoding β-carotene -4-oxygenase that converts β-carotene to the ketocarotenoid canthaxanthin in *Haematococcus pluvialis. FEBS Lett.,* **364** : 125-128.

Marin E, Nussaume L, Quesada A, Gonneau M, Sotta B, Hugueney P, Frey A and Marion-Poll A (1996) Molecular identification of zeaxanthin epoxidase of *Nicotiana plumbaginifolia,* a gene involved in abscisic acid biosynthesis and corresponding to the ABA locus of *Arabidopsis thaliana. EMBO J.,* **15** : 2331-2342.

Marrs BL (1981) Mobilisation of the genes for photosynthesis from *Rhodopseudomonas capsulata* by a promiscuous plasmid. *J. Bacterial.* **140** : 1003-1012.

Math SK, Hearst JE and Poulter CC (1992) The *crtE* gene in *Erwinia herbicola* encodes GGPP synthase. *Proc. Natl. Acad. Sci.,* **89** : 6761-6764.

Mayne ST (1996) β-Carotene, carotenoids and disease prevention in humans. *FASEB J.,* **10** : 690-701.

McCaskill D and Croteau R (1996) Some caveats for bioengineering terpenoid metabolism in plants. *Trends in Biotech.*, **16** : 349-355.

Miller NJ, Sampson J, Candeias LP, Bramley PM and Rice-Evans CA (1996) Antioxidant activity of carotenes and xanthophylls. *FEBS Lett.* **384** : 240-242.

Misawa N and Shimada H (1998) Metabolic engineering for the production of carotenoids in non-carotenogenic bacteria and yeasts. *J. Biotechnol.* **59** : 169-181.

Misawa N, Truesdale MR, Sandmann G, Fraser PD, Bird C, Schuch W and Bramley PM (1994a) Expression of a tomato cDNA coding for phytoene synthase in *E. coli*, phytoene formation *in vivo* and *in vitro* and functional analysis of the various truncated gene products. *J. Biochem.*, **116** : 980-985.

Misawa N, Masamoto K, Hori T, Ohtani T, Boger P and Sandmann G (1994b) Expression of an *Erwinia* phytoene desaturase gene not only confers multiple resistance to herbicides interfering with carotenoid biosynthesis but also alters xanthophyll metabolism in transgenic plants. *Plant J.*, **8** : 481-489.

Misawa N, Kajiwara S, Kondon K, Yokoyama A, Satomi Y (1995) Canthaxanthin biosynthesis by the conversion of methylene to keto groups in a hydrocarbon β-carotene by a single gene. *Biochem. Biophys. Res. Commun.*, **209** : 867-876.

Ninu L, Ahmad M, Miarelli C, Cashmore AR and Giuliano G (1999) Cryptochrome 1 controls tomato development in response to blue light. *Plant J.*, **18** : 551-556.

Pecker I, Chamovitz D, Linden H, Sandmann G and Hirschberg J (1992) A single polypeptide catalysing the conversion of phytoene to ζ-carotene is transcriptionally regulated during tomato fruit development. *Proc. Natl. Acad. Sci.*, **89** : 4962-4966.

Pecker I, Gubbay R, Cunnigham FX Jr and Hirschberg J (1996) Cloning and characterisation of the cDNA for lycopene β-cyclase from tomato reveals a decrease in its expression during fruit ripening. *Plant Mol. Biol.*, **30** : 807-819.

Pfander H (1987) *Key to Carotenoids.* 2nd edn. Birkhauser Verlag, Basel.

Pogson B, McDonald KA, Troong M, Britton G and DellaPenna D (1996) *Arabidopsis* carotenoid mutants demonstrate that lutein is not essential for photosynthesis in higher plants. *Plant Cell,* **8** : 16327-1639.

Rock CD and Zeevart JAD (1991) The *aba* mutant of *Arabidopsis thaliana* is impaired in epoxy-carotenoid biosynthesis. *Proc Natl. Acad. Sci.*, **88** : 7496-7499.

Rockholm DC and Yamamoto YY (1996) Violaxanthin de-epoxidase. *Plant Physiol.*, **110** : 697-703.

Romer S, Hugueney P, Bouvier F, Camara B and Kuntz M (1993) Expression of the genes encoding the early carotenoid biosynthetic enzymes in *Capsicum annuum. Biochem. Biophys. Res. Commun.* **196** : 1414-1421.

Ronen G, Cohen M, Zamir D and Hirschberg J (1999) Regulation of carotenoid biosynthesis during tomato fruit development: expression of a gene for lycopene epsilon cyclase is down regulated during ripening and is elevated in mutant *delta. Plant J.*, **17** : 341-351.

Schledz M, Al-Babili S, von Lintig J, Haubruck H, Rabbani S, Kleinig H and Beyer P (1996) Phytoene synthase from *Narcissus pseudonarcissus*: functional expression, galactolipid requirement, topological distribution in chloroplasts and induction during flowering. *Plant J.*, **10** : 781-792.

Schulz A, Ort O, Beyer P and Kleinig H (1993) SC-0051, a 2-benzoyl cyclohexane, 1,3 dione bleaching herbicide, is a potent inhibitor of p-hydroxyphenylpyruvate dioxygenase. *FEBS Lett.,* **318** : 162-166.

Schwartz SH, Tan BC, Gage DA, Zeevart JAD and McCarty DR (1997) Specific oxidative cleavage of carotenoids by VP14 of maize. *Science,* **276** : 1872-1874.

Schwender J, Muller C, Zeidler J and Lichtenthaler K (1999) Cloning and heterologous expression of a cDNA encoding 1-deoxy-D-xylulose-5-phosphate reductoisomerase from *Arabidopsis thaliana. FEBS Lett.,* **455** : 140-144.

Scolnik PA and Bartley GE (1995) Nucleotide sequence of lycopene cyclase from *Arabidopsis. Plant Physiol.,* **108** : 1343.

Scott KJ and Hart DJ (1994) *The Carotenoid Composition of Vegetables and Fruit Commonly Consumed in the UK.* Institute of Food Research, Norwich.

Shewmaker CK, Sheehy JA, Daley M, Colburn S and Ke DY (1999) Seed-specific overexpression of phytoene synthase : increase in carotenoids and other metabolic effects. *Plant J.,* **20** : 401-412.

Spurgeon SL and Porter JW (1983) Biosynthesis of carotenoids. In: *Biosynthesis of Isoprenoids, vol. 2* (Eds. Porter J W and Spurgeon S L). Wiley, New York, pp. 1-122.

Sun ZR, Gantt E and Cunningham FX Jr (1996) Cloning and functional analysis of β-carotene hydroxylase of *Arabidopsis thaliana. J Biol. Chem.,* **271** : 24349-24352.

Thompson AJ, Tor M, Barry CS, Vrebalov J, Orfila C, Jarvis MC, Giovannoni JJ, Grierson D and Seymour GB (1999) Molecular and genetic characterisation of a novel pleiotropic tomato-ripening mutant. *Plant Physiol.,* **120** : 383-389.

Truesdale MR (1993) Carotenoid biosynthesis in tomato. PhD thesis, University of London.

Vecchi M, Englert G, Maurere R and Meduna V (1981) Separation and characterisation of the *cis* isomers of β, β-carotene. *Helv. Chim. Acta,* **64** : 2746-2758.

Vishnevetsky M, Ovadis M, Zuker A and Vainstein A (1999) Molecular mechanisms underlying carotenogenesis in the chromoplast: multilevel regulation of carotenoid associated genes. *Plant J.,* **20** : 423-431.

Weedon BCL and Moss G (1995) Structure, stereochemistry and nomenclature. In: *Carotenoids, vol. 1a* (Eds. Britton G, Pfander H and Liaaen-Jensen S), Birkhauser Verlag, Basel, pp. 27-44.

Wetzel CM and Roderwel SR (1998) Regulation of phytoene desaturase expression is independent of leaf pigment content in *Arabidopsis thaliana. Plant Mol. Biol.,* **37** : 1045-1053.

Ye X, Al-Babili S, Kloti A, Zhang J, Lucca P, Beyer P and Potrykus I (2000) Engineering the provitamin A (β-carotene) biosynthetic pathway into (carotenoid-free) rice endosperm. *Science,* **287** : 303-305.

Young AJ (1993) Occurrence and distribution of carotenoids in photosynthetic systems. In: *Carotenoids in Photosynthesis* (Eds. Young A J and Britton G), Chapman and Hall, London, pp. 16-71.

Chapter 11

PRODUCTION OF ANTIBODIES USING TRANSGENIC PLANTS

Kevin C Gough★ and Garry C Whitelam

Department of Biology, University of Leicester, University Road, Leicester, LE1 7 RH, UK

Summary

The ability of antibodies to interact with their cognate antigens with high specificity and affinity is exploited in a wide range of applications including immunotherapy, immunodiagnostics and affinity separations. In recent years, plants have been shown to be capable of synthesising and assembling numerous antibody molecules, ranging from small antigen-binding antibody fragments to large, multimeric antibody complexes. The use of plants as a vehicle for antibody expression opens the way for the relatively inexpensive, large-scale production of antibodies and offers the prospect of the development of plant-expressed antibodies ('plantibodies') in several passive immunotherapy applications. Plantibodies are also proving to be useful tools for several in situ applications where they exert their effects in the host plant cells. This has allowed the development of antibody-mediated disease resistance strategies and the use of plantibodies for the sequestration or inactivation of specific cellular target molecules.

Keywords : Antibodies; antibody fragments; scFv; passive immunotherapy

---

★Corresponding author : E-mail : kcg3@le.ac.uk

1. INTRODUCTION

Antibodies (immunoglobulins) are complex, multimeric proteins synthesised in animals in response to the presence of a foreign substance, or antigen. Typical antibodies, such as those belonging to the immunoglobling class, are composed of two heavy-chain polypeptides and two light-chain polypeptides, in a structure that is stabilised by the presence of intra- and inter-molecular disulphide bonds (Figure 1).

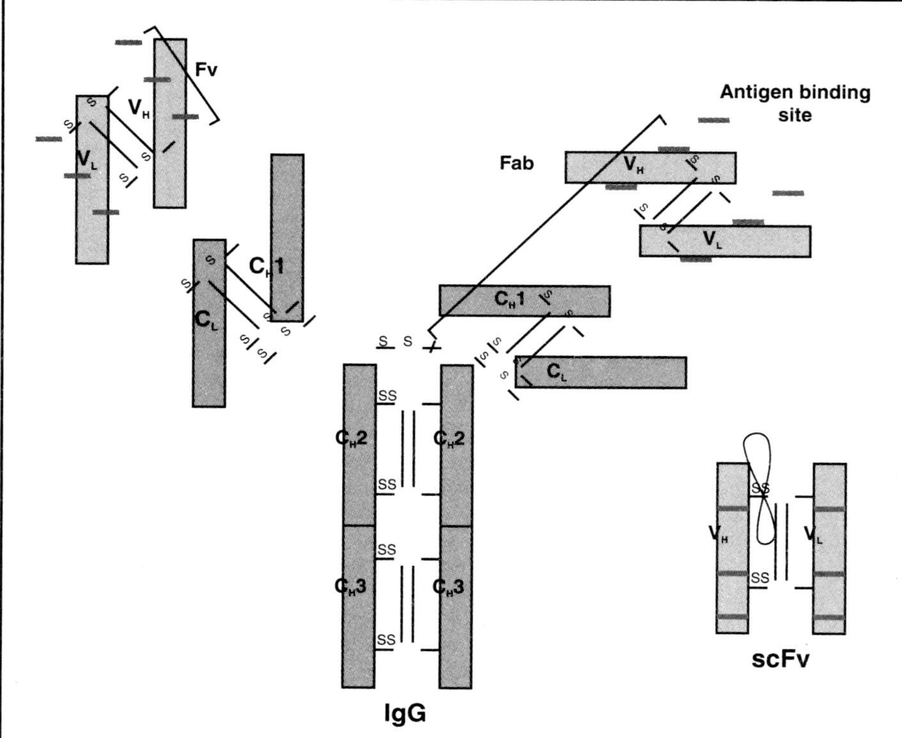

Figure 1 : The domain structure of an IgG molecule and antibody fragments. The IgG molecule comprises of two heavy (H) and two light (L) polypeptide chains that are covalently linked by intra- and inter-molecular disulphide bonds. Antigen binding is a function of the variable (V) domains, each of which possess three 'hypervariable' peptide loops (shaded in dark grey) that are the complementarity determining regions. The constant (C) domains are involved in secondary effector functions. Fab fragments consist of the V_H, C_H1, V_L and C_L regions and are stabilized by inter- and intra-molecular disulphide bonds. Fv fragments consist of the V_H and V_L regions and are stabilized by intra-molecular disulphide bonds only. When producing recombinant single chain Fv (scFv) fragments these V_H and V_L domains are linked by a synthetic polypeptide.

Immunoglobulins contain two antigen-binding regions, formed by the N-terminal variable domains of the heavy- and light chains, referred to as V_H and V_L, respectively (Figure 1). Within the variable domain of each of the chains are three hypervariable peptide loops that vary in amino acid sequence and length. These six regions, known as complementarity determining regions (CDRs), convey the antigen-binding activity of the molecule (Figure 1). As well as the V_H and V_L domains, immunoglobulins also contain a number of constant domains that do not affect antigen-binding activity but play other important roles in the immune system by mediating effector functions such as complement fixation. Also, the constant region of each of the heavy chains carries one or more N-linked, complex glycans.

The full antigen-binding activity of the immunoglobulin molecule is retained by several antibody fragments that contain the variable regions. Such fragments include Fab and single-chain Fv (scFv) antibodies (Figure 1) which can be produced either by proteolytic cleavage of intact antibodies or by recombinant DNA methods. Within the last few years the development of technologies for the creation of recombinant antibodies and antibody fragments has become a major route to the generation of new antibody specificities for use in a wide range of clinical and non-clinical applications.

A variety of heterologous expression systems are now available for the production of recombinant antibodies and antibody fragments. These include *Escherichia coli* (Pluckthun, 1991), yeast (Wood *et al.*, 1985; Horwitz *et al.*, 1988), insect cells (Putlitz *et al.*, 1990), mammalian cells (Biocca *et al.*, 1990), fungi (Nyyssonen *et al.*, 1993) and plants (Owen *et al.*, 1992; Fiedler and Conrad, 1995). Here we focus on the production of recombinant antibodies and antibody fragments in plants.

2. SOURCES OF ANTIBODY GENES

Hybridoma cells, secreting monoclonal antibodies, are the usual source of genes encoding antibodies with particular, desired specificities that can be subsequently expressed in plants. However, the recent advent of phage display technology has dramatically increased the accessibility of antibody genes for recombinant expression. This technology hinges on the ability to produce vast libraries of functional scFv or Fab antibody libraries displayed on the surface of M13 bacteriophage. The bacteriophage not only displays the antibody fragment on its surface but also harbours the antibody-encoding genes. Thus, selection of a particular antibody specificity, via bio-panning against the target antigen, leads to selection of the desired gene. Phage-display libraries have been produced from various sources including human (Nissim *et al.*, 1994), rodent (Clackson *et al.*, 1991), chicken (Davies *et al.*, 1995), rabbit (Lang *et al.*, 1996; Li *et al.*, 1999) and sheep (Li *et al.*, 2000). These libraries, which can contain up to 10^{10} scFv (Vaughan *et al.*, 1996) or Fab (Griffiths *et al.*, 1994) fragments; are produced from mRNA extracted from lymphocytes of immunised (Clackson *et al.*, 1991; Li *et al.*, 1999) and non-immunised (Marks *et al.*, 1991; Griffiths *et al.*, 1994; Vaughan *et al.*, 1996) donors. Phage display antibody libraries allow the rapid isolation of monoclonal antibody fragments against a wide range of antigens including self, toxic or non-immunogenic targets and offer a convenient method for the isolation of genes encoding specific antibody fragments for plant expression.

3. EXPRESSION OF WHOLE ANTIBODIES IN PLANTS

The most common strategy for synthesis of intact immunoglobulins in plants involves the separate expression of genes encoding heavy- and light-chains, each under the control of a strong promoter. In the first description of the stable expression on an intact antibody in plants, separate tobacco plants were transformed with expression cassettes for the heavy chain cDNA and the light chain cDNA (Hiatt *et al.*, 1989). In the F_1 progeny derived from a cross of these two lines, about 25% were found to express both cDNAs and to synthesise an intact antibody. The antibody was derived from a mouse monoclonal hybridoma cell line and was directed against a phosphonate ester. The levels of functional, assembled antibody in the F_1 plants were found to be significantly higher than the levels of the individual chains produced in the parent plants, suggesting that assembly of the chains had enhanced their stability. Yields of the antibody were in the range of 1% of total soluble protein.

Other expression strategies have been used to produce functional whole antibodies in plants. De Neve *et al.* (1993) used a double transformation technique to introduce both the heavy- and light-chain genes into the same plant cell simultaneously. Using genes for a mouse anti-human creatine-kinase IgG antibody, this procedure resulted in antibody yields of about 0.05% of total soluble protein. Several laboratories have adopted an approach in which the heavy- and light-chain genes are incorporated into a single T-DNA. Using this strategy van Engelen *et al.* (1994) observed a yield of up to 1.1% of the total soluble protein for the expression in tobacco of an intact mouse IgG raised against fungal cutinase and Voss *et al.* (1995) reported a maximum antibody accumulation of 0.16% of total soluble protein for the expression of a mouse IgG2b antibody against tobacco mosaic virus (TMV) in tobacco.

An elegant extension of the 'crossing' strategy for the creation of plants synthesising assembled antibodies, has also been developed by Ma *et al.* (1995) in order to produce tobacco plants synthesising a polymeric, secretory antibody. The antibody, Guys13, was raised against the cariogenic bacterium *Streptococcus mutans* and had been shown to be effective in preventing dental caries following topical application. The plant-produced, polymeric, secretory version of Guys13 was composed of two IgA units linked by a joining (J) chain and associated with a further polypeptide, the secretory component (SC). In order to generate a single transgenic tobacco plant synthesising this antibody complex, four separate transgenic lines were produced, expressing cDNAs encoding (1) the heavy chain, (2) the light chain, (3) the J chain and (4) a rabbit SC. By successive cross-pollinations, it was possible to select from among the progeny, individuals in which all four cDNAs were being expressed. These plants were shown to assemble the four polypeptides into a functional, complex, high molecular weight antibody comprising up to 0.05% of the fresh weight (Ma *et al.*, 1995). Furthermore, in human trials the plant-derived antibody was found to be more effective and more long-lived than the parental Guys13 antibody (Ma *et al.*, 1998).

The production and yield of whole antibodies in plants is largely dependent upon their assembly and processing within the plant cell. Correct processing and assembly requires the translocation of the nascent immunoglobulin polypeptides into the endoplasmic reticulum (ER). Thus, Hiatt el al. (1989) observed that transgenic tobacco plants expressing genes for truncated heavy- and light-chains, lacking the native signal sequences, did not accumulate significant amounts of transgenic protein. Further studies showed that signal sequences from a variety of sources, both plant and non-plant, could facilitate ER translocation of immunoglobulin polypeptides (Hiate *et al.*, 1989; Düring *et al.* 1990; Hein *et al.*, 1991). The correct folding and assembly of antibody chains within the ER lumen of plant cells may reflect the presence in that compartment of protein disulphide isomerase (PDI) and homologues of the stress proteins BiP (heavy chain binding protein) and GRP94 that may function as molecular chaperones (Fontes el al., 1991; Walther-Larsen *et al.*, 1993). In several studies, it has been shown that after targeting to the ER lumen, the assembled antibodies enter the default pathway and are secreted into intercellular spaces (Hiatt el al., 1989; De

Wilde *et al.*, 1996). De Wilde *et al.* (1996) have shown that upon secretion of plant-made intact antibodies or Fab fragments, a large proportion are transported within the apoplast, possibly by the water flow in the transpiration stream, and accumulate at sites where water passes on its radial pathway towards and within the vascular bundle.

In contrast to the secretion of plantibodies via the default pathway, Düring *et al.* (1990) found that expression of chimaeric genes made up of the barley α-amylase signal sequence fused to cDNAs coding for heavy- and light-chains of a mouse IgM antibody led to accumulation of assembled, functional antibody in the ER lumen and the chloroplast, but not in the intercellular spaces. This unexpected result may indicate that the barley α-amylase signal sequence provides more than simple ER-targeting information.

Despite the absence of BiP and GRP94-like chaperonins, and the presence of a reducing environment, the assembly of an intact antibody in the cytoplasm of plant cells has been reported. Steiger *et al.* (1991) transiently expressed truncated heavy- and light-chain cDNAs of an IgM antibody in the giant alga *Acetabularia*. The accumulation of assembled antibody within the cytoplasm suggests that targeting to the ER may not be an absolute requirement for the correct assembly of all plantibodies.

Mature, whole antibodies contain an N-linked, complex glycan and plantibodies are no exception. These complex glycans have been shown to differ in plantibodies and native mammalian antibodies (Faye *et al.*, 1989; Kornfeld and Kornfeld, 1985). The differences in glycosylation patterns for plantibodies have no effect on antibody binding or specificity (Hein *et al.* 1991; Cabanes Macheteau *et al.*, 1999).

4. EXPRESSION OF ANTIBODY FRAGMENTS IN PLANTS

A range of different antibody fragments have been stably expressed in transgenic plants including those with affinities for phytochrome (Owen *et al.*, 1992), abscisic acid (ABA; Artsaenko *et al.*, 1995), fungal cutinase (Schoutten *et al.* 1996), artichoke mottled crinkle virus (AMCV) coat protein (Tavladoraki *et al.*, 1993), oxazalone (Fiedler and Conrad, 1995), creatine kinase (Bruyns *et al.*, 1996), III nematode salivary secretions (Rosso *et al.*, 1996) and TMV coat protein (Zimmermann *et al.*, 1998). Benvenuto *et al.* (1991) also described the expression of a non-antigen binding single V_H domain from a rat anti-tachykinin neuropeptide antibody. This was shown to accumulate to up to 1% of the soluble protein and although this particular single domain did not display binding activity other functional single domain antibodies have been described (Ward *et al.*, 1989).

Many important targets reside within the cytosol and consequently the expression of antibody fragments in this cellular compartment has been of particular interest. Owen *et al.* (1992) expressed an anti-phytochrome scFv, derived from a mouse hybridoma cell line, in the cytosol of transgenic tobacco and demonstrated the accumulation of functional scFv protein up to around 0.05% of the total soluble protein. Similar low levels of protein accumulation were observed following expression of genes encoding cytosolic scFv proteins directed against the coat proteins of the plant viruses AMCV and TMV (Tavladoraki *et al.*, 1993; Zimmerman *et al.*, 1998; Schillberg *et al.*, 1999). However, in both of these cases, the low expression levels were sufficient to disrupt antigen activity and confer virus resistance to the transgenic plants.

As is the case with whole antibody expression, the targeting of antibody fragments to the apoplast has proven to be more conducive to higher levels of accumulation of functional antibody fragment. Firek *et al.* (1993) demonstrated this for the anti-phytochrome scFv, which accumulated up to 1.0% of total soluble protein when targeted to the ER using a signal sequence derived from a tobacco pathogen related protein. Likewise, Fiedler and Conrad (1995), studying the expression of antibody fragments in tobacco seeds, observed that scFv protein accumulation was only detected when the scFv was targeted to the ER using a signal peptide derived from the seed storage protein, legumin. In transgenic seeds, functional scFv accumulated up to 0.67% of the total soluble protein and in dry seeds, the scFv was stable for up to one year following storage at room temperature.

Several reports have shown that the additional incorporation of the retention tetrapeptide sequence, KDEL, to the C-terminus of an ER-targeted scFv can further significantly increase protein expression levels. Artsaenko *et al.* (1995) transformed tobacco with a gene encoding an anti-ABA scFv fused to a signal sequence and KDEL and detected protein accumulation to up to 4.5% of the total soluble protein. Significantly lower levels of the same antibody were detected in plants transformed with a gene encoding the scFv lacking the KDEL sequence. Similarly, Schouten *et al.* (1996) observed that addition of a KDEL sequence increased the accumulation of an ER-targeted anti-cutinase scFv in tobacco from about 0.01% of total soluble protein to 1% of the total soluble protein. This latter study also showed that inclusion of a C-terminal KDEL sequence also increased the accumulation of a cytosolic scFv. Further investigation (Shouten *et al.*, 1997) showed that for several cytosolic scFv proteins, C-terminal KDEL sequences increased protein accumulation without causing mis-translocation to other subcellular compartments. This has led to speculation that the KDEL sequences reduce proteolysis of the scFv fragments within the cytoplasm. The accumulation of cytosolic scFv protein can be increased by other fusion approaches. For example, Spiegel *et al.* (1999) showed that fusion of either glutathione S-transferase or the coat protein of TMV to the N-terminus of a TMV-specific scFv led to increased antibody accumulation in the cytoplasm of transgenic tobacco.

Antibody fragments such as the scFv possess only a single antigen-binding domain and so have a reduced avidity compared with whole antibodies. This may result in the binding energy of an scFv being reduced by up to one order of magnitude (Huston *et al.*, 1998; Schodin and Kranz, 1993). To overcome this potential limitation, two scFv proteins can be fused to form bivalent antibodies. Diabodies are one type of bivalent antibody and are produced by the non-covalent coupling of two scFv proteins that possess the same discrete binding activity (Holliger *et al.*, 1993). A functional diabody, directed against the coat protein of potato virus V (PVV), has been transiently expressed in tobacco plants using a potato virus X (PVX) episomal vector (Hendy *et al.*, 2000). A second type of bivalent antibody is the bispecific scFv that comprises of two independent binding sites for two discrete epitopes in a single polypeptide. Bispecific scFv proteins have the ability to cross-link different antigens. Fischer *et al.* (1999b) have described the production of a bispecific scFv in transgenic tobacco suspension culture cells and intact plants. In that study, two discrete anti-TMV scFv units were linked by the *Trichoderma reesi* cellobiohydrolase I linker and expressed in a secretory form to up to 1.65% of total soluble protein. The bispecific diabody retained the binding characteristics of both parental scFvs.

5. *EX SITU* APPLICATIONS OF PLANTIBODIES

5.1. Large scale production of plantibodies

Commercial monoclonal antibody production is based on the purification of antibodies from spent hybridoma culture medium or from mouse ascites fluid. In recent years several heterologous expression systems suitable for antibody production have become available including plants. Present day agricultural systems offer a commercially attractive proposition for the large-scale production of plantibodies and it is envisaged that agriculture could provide virtually unlimited quantities of recombinant antibodies for use as diagnostic and therapeutic tools in both health-care and life science applications (Ma *et al.*, 1995; Smith, 1996; Fischer *et al.*, 1999a). One of the attractions of producing antibodies in plants is the potential to exploit their natural storage capabilities. Fiedler and Conrad (1995) demonstrated that seeds are a suitable location for the expression and storage of scFvs. Furthermore, plants have been shown to be suitable hosts for the combination of genetic material in the production of functional sIgA molecules (Ma *et al.*, 1995).

In applications that require the use of purified recombinant antibody, downstream processing is likely to account for the major costs in the production of plantibodies. The use of edible transgenic plant tissues to both produce and deliver topically active antibodies to the mouth or gastro-intestinal tract is an attractive proposition. The use of such antibodies would remove the need for expensive purification steps and would facilitate the transport, storage and topical delivery of the antibody. Alternatively, the likely high costs of extraction and purification of proteins from biochemically complex plant tissues may be substantially reduced by using a secretion-based system utilising transgenic plant cells or plant organs cultivated aseptically *in vitro* (Firek *et al.*, 1993; Wongsamuth and Doran, 1997; Fischer *et al.*, 1999c). secretion is a basic function of plant cells and organs in planta and is especially well developed in plant roots (Shepherd and Davies, 1993). In a recent report by Borisjuk *et al.* (1999) an expression system was described which exploits root secretion in a simple hydroponic medium for the continuous production of recombinant proteins in a process termed "rhizosecretion". It could be envisaged that such a system could be utilised to continuously produce recombinant antibodies that would require relatively simple purification.

5.2. Passive Immunotherapy

Topically delivered plantibodies could be applied to the passive immunotherapy of a variety of diseases of mucosal surfaces, as well as of the oral cavity and the gastro-intestinal tract. A number of reports have shown that the use of topically applied antibodies can prevent the colonisation of the oral cavity by pathogenic bacteria and modify the resident bacterial flora in a highly specific manner (Ma and Hiatt, 1996; Bessen and Fischetti, 1988; Ma *et al.*, 1990). One such example used an antibody raised against a cell surface adhesin of Streptococcus mutans which when topically applied to the oral cavity prevented bacterial infection and reduced the levels of dental caries (Lehner *et al.*, 1985; Ma *et al.*, 1990). In a recent study, Ma *et al.* (1998) produced a recombinant, secretory, polymeric form of one such antibody in transgenic tobacco and showed that following its isolation, it afforded specific protection against oral streptococcal colonisation in human volunteers for over 4 months. The production of an apoplastically-targeted form of this antibody in edible plant tissues could provide a simple and convenient means of antibody delivery.

A further demonstration of the successful use of plantibodies for passive immunotherapy has been presented by Zeitlin *et al.* (1998) who described the expression of a humanized monoclonal antibody in soybean that is specific for glycoprotein D of herpes simplex virus-2 (HSV-2). When applied topically, the isolated plantibody successfully protected mice against vaginal HSV-2 infection.

6. IN SITU APPLICATIONS OF PLANTIBODIES

6.1. Intracellular immunisation

It is possible to use plantibodies to modify the metabolic and phenotypic characteristics of transgenic plants. Owen *et al.* (1992) expressed a functional anti-phytochrome A scFv in transgenic tobacco and showed that a number of phytochrome-mediated developmental responses, including light-dependent seed germination, were perturbed in the transgenic plants (Whitelam *et al.*, 1994). A very striking demonstration of intracellular immunisation has been provided by Artsaenko *et al.* (1995) using an intracellularly-produced anti-ABA scFv. *In situ* binding of this scFv to ABA resulted in a wilty phenotype that copies that of a hormone-deficient or hormone-insensitive mutant. Furthermore, compared with untransformed plants, or transgenic plants expressing a control scFv, those transgenic plants accumulating anti-ABA scFv to over 1% of the total soluble protein grew more slowly and were hypersensitive to water stress. In a further refinement of this approach tobacco plants were transformed with the anti-ABA scFv under the control of a seed-specific promoter (Philips *et al.*, 1997). The transformants were phenotypically similar to wild-type plants apart from their seeds. The transgenic seeds displayed aberrant embryo development, with green cotyledons and reduced levels of storage compounds. In addition, transgenic seeds displayed precocious germination following removal from seed capsules.

These observations clearly indicate the potential of plantibodies as tools for the selective modification of intracellular antigens.

6.2. Disease and pest resistance

Plantibodies are being used to develop novel disease and pest resistance strategies. Tavladoraki *et al.* (1993) expressed a cytosolic scFv against the coat protein of AMCV in tobacco and demonstrated a delay in viral symptom development and a lowering in the infection susceptibility of the transgenic plants. Although the mechanism of protection afforded by scFv expression is unknown, it was speculated that antibody-binding might interfere with uncoating or assembly of the virions. Subsequently, Voss *et al.* (1995) demonstrated virus resistance in tobacco plants following production of a secreted, whole antibody. The transgenic plants synthesised a mouse-derived monoclonal antibody against a surface epitope of intact TMV virions. Transgenic plants showed a reduction in the number of necrotic lesions that developed following their inoculation with TMV in inverse proportion to the level of antibody accumulation. Since the assembled, functional antibody was secreted to the apoplast it was speculated that the antibody might be able to 'decorate' the TMV particles in this compartment so neutralising them. However, in a further development of this work, tobacco plants were transformed with a gene encoding a cytosolic scFv derived from the same monoclonal anti-TMV antibody (Zimmermann *et al.*, 1998). Although the levels of scFv protein accumulation were very low, these transgenic plants showed increased resistance against infection with TMV. It was speculated that cytosolic anti-TMV scFv might have interfered with virus uncoating.

There are a few examples of the expression of plantibodies directed against pest-derived antigens. For example, Rosso *et al.* (1996) have detailed the transient expression of a functional scFv directed against salivary secretions of the root-knot nematode *Meloidogyne incognita* in tobacco leaf protoplasts. In similar studies, Baum *et al.* (1996) have expressed a functional whole plantibody specific to stylet secretions of the same nematode in tobacco. However, the transgenic plants expressing this antibody were found to be wild type with respect to root-knot parasitism.

7. CONCLUSIONS AND FUTURE PROSPECTS

The expression of antibodies in plants has both a commercial and scientific potential, including the large-scale production of antibodies, the production of antibodies for passive immunotherapy, plant disease resistance strategies, intracellular immunisation and bioremediation. As yet plantibodies have not been utilised in bioremediation but the potential is obvious. The sequestration of a wide range of environmental pollutants by plantibodies could form the basis for a natural bioremediation system. Such a system would require very little maintenance and would be applicable to pollution on any scale. Sequestration would be specific to a particular pollutant and could utilise very high affinity interactions allowing the extraction of very low levels of pollutants. Recent publications describe the expression of a number of antibody fragments specific for a range of environmental pollutants in *E. coli* (Graham *et al.*, 1995; Byrne *et al.*, 1996; Lie *et al.*, 1999) and plants (Longstaff *et al.*, 1998; Strachan *et al.*, 1998). Furthermore, antibodies from these expression systems have been used to remove trace levels of contaminants from aqueous samples (Graham *et al.*, 1995; Molloy *et al.*, 1995). Dhillon *et al.* (1999) described the *E. coli* cell surface display of an scFv specific for the herbicide and environmental pollutant, atrazine, and speculated on the use of these whole cells in bioremediation. Similarly, it could also be possible to anchor an antibody of interest within the apoplast of plants by fusing it to a membrane insertion sequence.

Other potential uses of plantibodies have been discussed in this chapter. The expression of antibodies directed against endogenous plant components allows phenotype modification and could include targets such as enzymes, plant growth regulators and other regulatory proteins or non-proteins. In principle almost any metabolic pathway could be manipulated in order to study individual components of such pathways or to alter their end products. Cytosolic antibodies can also be used to neutralise the activity of a non-gene or existing gene product and as an alternative to present methods of gene inactivation such as dominant negative mutants, antisense technology and targeted gene inactivation.

Antibody mediated plant protection against viruses has already been demonstrated and similar strategies can be envisaged against a wide range of plant pathogenic organisms. Where suitable epitopes are identified, the cognate antibodies expressed in plants could form the basis for novel disease resistant strategies. Such strategies against viral infection could include non-structural viral proteins as targets including those crucial for replication such as the replicases, proteases and helicases, or epitopes found on viral movement proteins. The use of a non-structural target in an antibody mediated viral disease resistance strategy has been demonstrated in cultured mammalian cells using a cytoplasmic antibody fragment directed against the HIV reverse transcriptase (Maciejewski *et al.*, 1995). Expression of this antibody fragment resulted in the host cells displaying HIV resistance.

A further potential for plantibodies is their use in topical passive immunotherapy. The availability of sIgA molecules in transgenic plants could open the way for plantibody mediated passive immunotherapy at all mucosal sites. Furthermore, the use of edible plants or targeting of antibodies to edible plant tissues such as seeds, tubers or fruits would allow for direct oral delivery. Such plantibody production could greatly reduce or even eliminate the need to purify plantibodies prior to treatment. Further developments of such treatments may include the design of antibodies resistant to proteolysis that could be used to deliver a high proportion of functional antibody locally within the gastrointestinal tract for the treatment of enteric infections or even to allow antibody absorption for the treatment or prevention of systemic conditions (Reilly *et al.*, 1997).

REFERENCES

Artsaenko O, Peisker M, zur Neiden U, Feidler U, Weiler EW, Muntz K and Conrad U (1995) Expression of a single chain antibody against abscisic acid creates a wilty phenotype in transgenic tobacco. *Plant J.,* **8** : 745-750.

Baum TJ, Hiatt A, Parrott WA, Pratt LH and Hussey RS (1996) Expression in tobacco of a functional monoclonal antibody specific to stylet secretions of the root-knot nematode. *Mol. Plant-Microbe Interact.,* **9** : 382-387.

Benvenuto E, Ordas RJ, Tavazza R, Ancora G, Biocca S, Cattaneo A and Galeffi P (1991) "Phytoantibodies": a general vector for the expression of immunoglobulin domains in transgenic plants. *Plant Mol. Biol.,* **17** : 865-874.

Bessen D and Fischett VA (1988) Passive acquired mucosal immunity to group A streptococci by secretory immunoglobulin A. *J. Exp. Med.,* **167** : 1945-1950.

Biocca S, Neuberger MS, and Cattaneo A (1990) Expression and targeting of intracellular antibodies in mammalian cells. *EMBO J.,* **9** : 101-108.

Borisjuk NV, Borisjuk LG, Logendra S, Petersen F, Gleba Y and Raskin I (1999) Production of recombinant proteins in plant root exudates. *Nature Biotechnol.,* **17** : 466-469.

Bruyns AM, De Jaeger G, De Neve M, De Wilde C, Van Montagu M and Depicker A (1996). Bacteria and plant-produced scFv proteins have similar antigen-binding properties. *FEBS Lett.,* **386** : 5-10.

Byrune FR, Grant SD, Porter AJ and Harris WJ (1996) Cloning, expression and characterisation of a single chain antibody specific for the herbicide atrazine. *Food Agric. Immunol.,* **8** : 19-29.

Cabenes Macheteau M, Fitchtte Laine AC, Loutelier Bourhis C, Lange C, Vine ND, Ma JKC, Lerouge P and Faye L (1999) N-glycosylation of a mouse IgG expressed in transgenic plants. *Glycobiol.,* **9** : 365-372.

Clackson T, Hoogenboom HR, Griffiths AD and Winter G (1991) Making antibody fragments using phage display libraries. *Nature,* **352** : 624-628.

Davies EL, Smith JS, Birkett CR, Manser JM, Anderson DD and Young JR (1995) Selection of specific phage-display antibodies using libraries derived from chicken immunoglobulin genes. *J. Immunol. Meth.*, **186** : 125-135.

De Neve M, De Loose M, Jacobs A, Van Houdt H, Kaluza B, Weidle U, Van Montagu M and Depicker A (1993) Assembly of an antibody and its derived antibody fragment in *Nicotiana* and *Arabidopsis. Transgenic Res.*, **2** : 227-237.

De Wilde C, De Neve M, De Rycke R, Bruyns AM, De Jaeger G, Van Montagu M, Depicker A and Engler G (1996) Intact antigen-binding MAK33 antibody and Fab fragment accumulate in intercellular spaces of *Arabidopsis thaliana. Plant Sci.*, **114** : 233-241.

Dhillon JK, Drew PD and Porter AJR (1999) Bacterial surface display of an anti-pollutant antibody fragment. *Lett. Appl, Microbiol.*, **28** : 350-354.

Düring K, Hippe S, Kreuzaler F and Schell J (1990) Synthesis and self-assembly of a functional monoclonal antibody in transgenic *Nicotiana tabacum. Plant Mol. Biol.*, **15** : 281-293.

Faye L, Johnson KD, Sturm A and Chrspeels MJ (1989) Structure, biosynthesis and function of asparagine-linked glycans on plant glycoproteins. *Physiol. Plant.*, **75** : 309-314.

Fiedler U and Conrad U (1995) High-level production and long term storage of engineered antibodies in transgenic tobacco seeds. *Biol/Tchnol.*, **13** : 1090-1093.

Firek S, Draper J, Owen MRL, Gandecha A, Cockburn B and Whitelam GC (1993) Secretion of a functional single-chain Fv protein in transgenic tobacco plants and cell suspension cultures. *Plant Mol. Biol.*, **23** : 861-870.

Fischer R, Liao YC, Hoffmann K, Schillberg S and Emans N (1999a) Molecular farming of recombinant antibodies in plants. *Biol. Chem.*, **380** : 825-839.

Fischer R, Schumann D, Zimmermann S, Drossard J, Sack M and Schillberg S (1999b) Expression and characterisation of bispecific single-chain fragments produced in transgenic plants. *Eur. J. Biochem.*, **262** : 810-816.

Fischer R, Liao YC and Drossard J (1999c) Affinity-purification of a TMV-specific recombinant full-size antibody from a transgenic tobacco suspension culture. *J. Immunol. Meth.*, **226** : 1-10.

Fontes EBP, Shank BB, Wrobel RL, Moose SP, Obrian GR, Wurtzel ET and Boston RS (1991) Characterisation of an immunoglobulin binding protein homolog in the maize *floury-2* endosperm mutant. *Plant Cell,* **3** : 483-496.

Graham BM, Porter AJR and Harris WJ (1995) Cloning, expression and characterisation of a single chain antibody fragment to the herbicide paraquat. *J. Chem. Technol. Biotechnol.*, **63** : 279-289.

Griffiths AD, Williams SC, Hartley O, Tomlinson IM, Waterhouse P, Crosby WL, Kontermann RE, Jones PT, Low NM, Allison TJ, Prospero TD, Hoogenboom HR, Nissim A, Cox JPL, Harrison JL, Zaccolo M, Gherardi E and Winter G (1994) Isolation of high affinity human antibodies directly from large synthetic repetoires. *EMBO J.*, **13** : 3245-3260.

Hein M, Tang Y, MeLeod DA, Janda KD and Hiatt A (1991) Evaluation of immunoglobulins from plant cells. *Biotechnol. Prog.*, **7** : 455-461.

Hendy S, Chen ZC, Barker H, Santa Cruz S, Chapman S, Torrance L, Cockburn W and Whitelam GC (2000) Rapid production of single-chain Fv fragments in plants using a potato virus X episomal vector. *J. Immuol. Meth.*, **231** : 137-146.

Hiatt A, Cafferkey R and Bowdish K (1989) Production of antibodies in transgenic plants. *Nature,* **342** : 76-78.

Horwitz AH, Chang CP, Better M, Hellstrom KE and Robinson RR (1988) Secretion of functional antibody and Fab fragment from yeast cells. *Proc. Natl. Acad. Sci. USA,* **85** : 8678-8682.

Holliger P, Prospero T and Winter G (1993) 'Diabodies': small bivalent and bispecific antibody fragments. *Proc. Natl. Acad. Sci. USA,* **90** : 6444-6448.

Huston JS, Levinson D, Mudgett-Hunter M, Tai MS, Novotny J, Margolies MN, Ridge RJ, Bruccoleri RE, Haber E, Crea R and Oppermann H (1988) Protein engineering of antibody binding sites: recovery of specific activity in an anti-digoxin single-chain Fv analogue produced in *Escherichia coli. Proc. Natl. Acad. Sci. USA,* **85** : 5879-5883.

Kornfeld R and Kornfeld S (1985) Assembly of asparagine-linked oligosaccharides. *Annu. Rev. Biochem.,* **54** : 631-664.

Lang IM, Barbas C III and Schleef RR (1996) Recombinant rabbit Fab with binding activity to type-1 plasminogen activator inhibitor derived from a phage-display library against human alpha-granules. *Gene,* **172** : 295-298.

Lehner T, Caldwell J and Smith R (1985) Local passive immunization by monoclonal antibodies against streptococcal antigen I/II in the prevention of dental caries. *Infect. Immun.,* **50** : 796-799.

Li Y, Cockburn W, Kilpatrick J and Whitelam GC (1999) Selection of rabbit single-chain Fv fragments against the herbicide atrazine using a new phage display system. *Food Agric. Immunol.,* **11** : 5-17.

Li Y, Kilpatrick J and Whitelam GC (2000) Sheep monoclonal antibody fragments generated using a phage display system. *J. Immunol. Meth.,* **236** : 133-146.

Langstaff M, Newell CA, Boonstra B, Strachan G, Learmonth D, Harris WJ, Porter AJ and Hamilton WDO (1998) Expression and characterisation of single-chain antibody fragments produced in transgenic plants against the organic herbicides atrazine and paraquat. *Biochim. Biophys. Acta,* **1381** : 147-160.

Ma JKC and Hiatt A (1996) Expressing antibodies in plants for immunotherapy. In: *Transgenic plants: A production system for industrial and pharmaceutical proteins* (Eds owen MRL and Pen J), J Wiley and Sons., Chichester pp. 229-243.

Ma JKC, Hiatt A, Hein M, Vine DN, Wang F, Stabila P, van Dollerweerd C, Mostov K and Lehner T (1995) Generation and assembly of secretory antibodies in plants: *Science,* **268** : 716-719.

Ma JKC, Hikman BY, Wycoff K, Vine ND, Chargelegue D, Yu L, Hein MB and Lehner T (1998) Characterization of a recombinant plant monoclonal secretory antibody and preventive immunotherapy in humans. *Nature Med.,* **4** : 601-606.

Ma JKC, Hunjan M, Smith R, Kelley C and Lehner T (1990) An investigation into the mechanism of protection by local passive immunisation with monoclonal antibodies against *Streptococcus mutans. Infect. Immun.,* **58** : 3407-3414.

Maciejewski JP, Weichold FF, Young NS, Cara A, Zella D, Retiz M S Jr and Gallo RC (1995) Intracellular expression of antibody fragments directed against HIV reverse transcriptase prevents HIV infection *in vitro. Nature Med.* **1** : 667-673.

Marks JD, Hoogenboom HR, Bonnert TP, McMafferty J, Griffiths AD and Winter G (1991) By-passing immunization: human antibodies from V-gene libraries displayed on phage. *J. Mol. Biol.,* **22** : 581-597.

Molloy P, Brydon L, Porter AJ and Harris WJ (1995) Separation and concentration of bacteria with immobilized antibody fragments. *J. App. Bacteriol.*, **78** : 359-365.

Nissim A, Hoogenboom HR, Tomlinson IM, Flynn G, Midgley C, Lane D and Winter G (1994) Antibody fragments from a 'single pot' phage display library as immunochemical reagents. *EMBO J.*, **13** : 692-698.

Nyyssonen E, Penttila M, Harkki A, Salheima A, Knowles JKC and Keranen S (1993) Efficient production of antibody fragments by the filamentous fungus *Trichoderma reesei. Bio/Techol.*, **11** : 591-595.

Owen M, Gandecha A, Cockburn W, Whitelam GC (1992) Synthesis of a functional anti-phytochrome single-chain Fv protein in transgenic tobacco. *Bio/Technol.*, **10** : 790-794.

Philips J, Artsaenko L, Fiedler U, Horstmann C, Mock HP, Muntz K and Conrad U (1997) Seed-specific immunomodulation of abscisic acid activity induces a developmental switch. *EMBO J.*, **16** : 4489-4496.

Pluckthun A (1991) Antibody engineering : Advances from the use of *Escherichia coli* expression systems. *Bio/Technol.*, **9** : 545-551.

Putlitz J, Kubasek WL, Duchene M, Marget M, Vonspecht BU and Domdey H (1990) Antibody production in Baculovirus-infected insect cells. *Bio/Technol.*, **8** : 651-654.

Reilly RM, Domingo R and Sandhu J (1997) Oral delivery of antibodies – Future pharmacokinetic trends. *Clin. Pharmacokin.*, **32** : 313-323.

Rosso MN, Schouten A, Roosien J, Borst-Vrenssen T, Hussey RS, Gommers FJ, Bakker J, Schots A and Abad P (1996) Expression and functional chracterisation of a single chain FV antibody directed against secretions involved in plant nematode infection process. *Biochem. Biophys. Res. Commun.*, **220** : 255-263.

Schillberg S, Zimmermann S, Voss A and Fischer R (1999) Apoplastic and cytosolic expression of full-size antibodies and antibody fragments in *Nicotiana tabacum. Transgenic Res.*, **8** : 255-263.

Schoding BA and Kranz JA (1993) Binding affinity and inhibitory properties of a single-chain anti-T cell receptor antibody. *J. Biol. Chem.*, **268** : 25722-25727.

Schouten A, Roosien J, deBoer JM, Wilmink A, Rosso MN and Bosch D (1997) Improving scFv antibody expression levels in the plant cytosol. *FEBS Letts.*, **425** : 235-241.

Schouten A, Roosien J, van Engelen FA, de Jong GAM, Borst-Vrenssen AWM, Zilverentant JF, Bosch D, Stiekma WJ, Gommers FJ, Schots A and Bakker J (1996) The C-terminal KDEL sequence increases the expression level of a single chain antibody designed to be targeted to both the cytosol and the secretory pathway in transgenic tobacco. *Plant Mol. Biol.*, **30** : 781-793.

Shepherd T and Davies HV (1993) Carbon loss from the roots of forage rape (*Brassica napus* L.) seedlings following pulse-labelling with CO_2. *Ann. Bot.*, **72** : 155-163.

Smith MD (1996) Antibody production in plants. *Biotech. Adv.*, **14** : 267-281.

Speigel H, Schillberg S, Sack M, Holzem A, Nahring J, Monecke M Liao YC and Fischer R (1999) Accumulation of antibody fusion proteins in the cytoplasm and ER of plant cells. *Plant Sci.* **149** : 63-71.

Stieger M, Neuhaus G, Momma T, Schell J and Kreuzaler F (1991) Self-assembly of immunoglobulins in the cytoplasm of the alga *Acetabularia mediterranea. Plant Sci.*, **73** : 181-190.

Strachan G, Grant SD, Learmonth D, Longstaff M, Porter AJ and Harris WJ (1998) Binding characteristics of anti-atrazine monoclonal antibodies and their fragments synthesised in bacteria and plants. *Biosens. Bioelect.,* **13** : 665-673.

Tavladoraki P, Benvenuto E, Trinca S, De Martinis D, Cattaneo A and Galeffi P (1993) Transgenic plants expressing a functional single-chain Fv antibody are specifically protected from virus attack. *Nature,* **366** : 469-472.

van Engelen FA, Schouten A, Molthoff JW, Roosien J, Dirske WG, Schots A, Bakker J, Gommers FJ, Jongsma MA, Bosch D and Stiekma WJ (1994) Coordinate expression of antibody subunit genes yields high levels of functional antibodies in roots of transgenic tobacco. *Plant Mol. Biol.,* **26** : 1707-1710.

Vaughan TJ, Williams AJ, Pritchard K, Osbourn JK, Pope AR, Earnshaw JC, McCafferty J, Hodits RA, Wilton J and Johnson KS (1996) Human antibodies with sub-nanomolar affinities isolated from a large non-immunised phage display library. *Nature Biotechnol.,* **14** : 309-314.

Voss A, Niersback M, Bain R, Hirsch JH, Liao YC, Kreuzaler F and Fischer R (1995) Reduced virus infectivity in *N. tabacum* secreting a TMV-specific full-size antibody. *Mol. Breeding.* **1** : 39-50.

Walther-Larsen H, Brandt J, Collinge DB and Thordal-Christensen A (1993) Pathogen-induced gene of barley encodes a hsp90 homologue showing striking similarity to vertebrate forms resident in the endoplasmic reticulum. *Plant Mol. Biol.,* **6** : 1097-1108.

Ward ES, Gussow D, Griffiths AD, Jones PT and Winter G (1989) Binding activities of a repertoire of single Ig variable domains secreted from *Escherichia coli. Nature,* **341** : 544-546.

Whitelam GC and Cockburn W (1996) Antibody expression in transgenic plants. *Trends Plant Sci.,* **1** : 268-272.

Whitelam GC, Cockburn W and Owen MRL (1994) Antibody production in transgenic plants. *Biochem. Soc. Trans.,* **22** : 940-944.

Wongsamuth R and Doran PM (1997) Production of monoclonal antibodies by tobacco hairy roots. *Biotechnol. Bioeng.,* **54** : 401-415.

Wood CR, Boss MA, Kenten JH, Calvert JE, Roberts NA and Emtage JS (1985) The synthesis and *in vitro* assembly of functional antibodies in yeast. *Nature,* **314** : 446-449.

Zeitlin L, Olmsted SS, Moench TR, Co MS, Martinell BJ, Paradkar VM, Russell DR, Queen C, Cone RA and Whaley KJ (1998) A humanized monoclonal antibody produced in transgenic plants for immunoprotection of the vagina against genital herpes. *Nature Biotechnol.,* **16** : 1361-1364.

Zimmermann S, Schillberg S, Liao, YC and Fischer R (1998) Intracellular expression of a TMV-specific single chain Fv fragment leads to improved virus resistance in *Nicotiana tabacum. Mol. Br.* **4** : 369-379.

Chapter 12

DEVELOPMENT OF TRANSGENIC PLANTS AS SOURCE OF EDIBLE VACCINES - A NEW APPROACH

G Lakshmi Sita★ and KJM Vally

Department of Microbiology and Cell Biology, Indian Institute of Science, Bangalore - 560 012, India

Summary

Production of transgenic plants as a source of edible vaccines against diseases is the most convenient and inexpensive system of immunization. Large amounts of antigens could be produced at a relatively low cost, using agriculture instead of sophisticated and expensive cell culture based expression system. By stimulating mucosal immunity, edible vaccines have the potential to stimulate both local and systematic immune responses. Immunological studies in model system like mice demonstrated that antigens such as Norwalk virus capsid protein (NVCP), Hepatitis-B and heat-labile enterotoxins (LT-B) delivered through transgenic plants trigger an immune response. Preliminary studies in humans with transgenic potato expressing LT-B showed detectable amounts of IgA and IgG in the peripheral blood. Among animal diseases, efforts to produce edible vaccines against Foot and Mouth disease in cattle are substantial. On similar lines, the expression of an antigen in plant systems against Rinderpest virus (RPV) and peste des petitis ruminant virus (PPRV), the causative agents of fatal contagious diseases of the small and large ruminants is being reported.

Keywords : Edible vaccine, genetic transformation, transgenic plants, immune response, antigen

---

★Corresponding author : E-mail : sitagl@mcbl.iisc.ernet.in

Abbreviations : FMDV - Foot and mouth disease virus; HbsAg - Hepatitis B virus surface antigen; i-rNVs - Norwalk virus like particles; LT-B - heat-labile enterotoxins; *nptII* - neomycin phosphotransferase; NVCP - Norwalk virus capsid protein; spaA - surface protein antigen A; *Ti* - tumour inducing

1. INTRODUCTION

Diseases are a major cause for concern all over the world, but more so in developing countries. About 16.5 million people died globally owing to infectious diseases and parasitic diseases in 1993 of which 16.3 million belonged to the developing countries. Ever since Edward Jenner developed small pox vaccine in 1798, the world has come to know that prevention is better than cure. A vaccine essentially is a small dose of the infectious agent (antigen) injected into the body of human/animal in a small dose responding in minor infection which triggers the immune response in the body and develops the antibodies to fight against the disease. An ideal vaccine should have 1) single dose administration, 2) life long protection, 3) safe from the risk of infections and side-effects, 4) heat stability, 5) simple and cost effective technology for mass production, 6) easy method of administration, such as oral instead of injectable preparation and 7) multipotent hybrid vaccines, where a single preparation can offer protection from a range of diseases.

Vaccines are of different kinds namely classical vaccines, recombinant vaccines and DNA vaccines. Classical vaccines are vaccines where the whole of the disease-causing organism is injected as the vaccine. Therefore, an allergy or anaphylaxis may develop in the individual. In recombinant vaccines only the protein that is responsible for causing the disease from the organism is genetically engineered in yeast cells and inoculated and there won't be any chance to g*et al*lergy. Recombinant vaccines are far more superior to classical vaccines. DNA vaccines on the other hand have the gene for the antigen, which is directly inoculated into the muscles of an individual. Although these vaccines were proved effective, as immunogens in eliciting specific immune response, the production is prohibitively expensive by fermentation based pilot plants. Spectacular developments in biotechnology, enabled the removal of the barriers in different areas of sciences, and made exciting discoveries possible. There are no longer boundaries between plant sciences and animal sciences. Genes cloned from different organisms can be transferred into each other. The discovery of 'plant genetic transformation technology' has facilitated the study of plant gene expression, and has resulted in great progress. During the last decade novel plants with enhanced production traits such as herbicide resistance, insect resistance, disease resistance were reported. In addition, transgenic plants are an attractive and cost effective alternative to microbial systems for the production of bio-molecules. The number of compounds that have successfully been produced in plants using this 'molecular farming' approach is steadily increasing. Molecular farming has already proved feasible for the production of carbohydrates, fatty acids, high value pharmaceuticals, polypeptides, industrial enzymes and biodegradable plastics.

The concept of vaccine production in transgenic plants popularly known as 'Edible Vaccines' was introduced by Chales J Artzen in 1992. According to him "This interest is stimulated by interest in evaluating the capacity of plants to produce different classes of proteins of pharmaceutical value and because of the practical need for new technology for the production and delivery of inexpensive vaccines especially in the developing world". They have also hypothesized that plants could be a useful system for producing vaccines, because large amounts of antigen could be produced at a relatively low cost, using agriculture instead of sophisticated and expensive cell-culture based expression systems. Even with technological advances, fermentation based sub-unit vaccine production may be a prohibitively expensive technology for developing countries, where novel oral vaccines are urgently required. Transgenic plants that express antigens in their edible tissues might be used as an inexpensive oral vaccine production and delivery system. Therefore, immunization might be possible simply through consumption of an edible plant part containing the 'vaccine'. In this review, we highlight the work carried in recent times along with brief discussion on development of transgenic plants and immune response in general.

2. IMMUNE RESPONSE AND ORAL VACCINES

The injection of an antigen into an animal/human body induces several important changes. The most important among them is the production of antibodies, which provides resistance to the animal body against infection. Usually the pathogens responsible for the disease make initial contact with the host at a mucosal site such as the respiratory tract, gastrointestinal tract, or genital tract. Stimulation of immune responses of suitable strength and quality to protect against illness is, therefore, particularly desirable at these mucosal sites. Although traditional parenteral (by injection) vaccines may be effective in preventing illness caused by pathogens, which enter at a mucosal site, parenteral immunizations rarely result in specific mucosal immune responses. On the other hand, mucosal immunization achieved by applying vaccine directly to the mucosal surfaces may induce systemic antibodies and elicit cellular immune response as well as local immune response.

The immune system of the body is monitored with the help of white blood cells called lymphocytes, which circulate between blood and the lymph system. The lymphocytes are of two types: B cells and T cells. The B cells synthesize antibodies causing agglutination of foreign antigen while T cells carry cellular immune response where cells having foreign antigens on their surface are attacked without any mediation by antibodies. The former response is called humoral immunity and the latter response is called cellular immunity. Mucosal vaccines have been sought after because of the potential for stronger immune response directed at the initial site of interaction of the pathogen and host. Attention has, therefore, been paid mainly to those antigens that stimulate mucosal immune system to produce secretory IgA (s-IgA) at mucosal surfaces, such as gut and respiratory epithelia. The oral and nasal routes have been most vigorously studied because of their easy accessibility in adults and children. These are also perceived as safer than parenteral routes and delivery of the vaccines orally or nasally may be more easier and more acceptable, since needles and syringes are not needed thus avoiding the dreaded diseases like AIDS and Hepatitis. In general, a mucosal response is achieved more effectively by oral instead of parenteral delivery of the antigen. Several particulate antigens have proven to be effective oral immunogens including live and killed microorganisms. An antigen produced in edible parts of a plant can serve as a vaccine against several infectious agents such as bacteria and viruses that are transmitted by contaminated water resulting in deadly diseases like diarrhea and whooping cough. Only negative point about oral immunization is, it is inefficient to stimulate immune response in small doses and requires larger doses (mg versus µg) of antigen. The choice of plant system for testing recombinant antigen production is initially driven by convenience and the need to evaluate different constructs quickly. For this reason, tobacco plants are generally utilized, however, because of the alkaloids present in the leaves; studies on the animal feeding are difficult and not practical. Therefore, potato has been selected as an alternate choice. Mice would accept raw potato as tubers in lieu of laboratory feed. Crops used for animal feed, such as alfalfa, grains and beans are obvious choices for animal vaccines.

3. PLANT GENETIC TRANSFORMATION SYSTEMS FOR EXPRESSING FOREIGN PROTEINS IN PLANTS

Two different strategies for transgene expression in plants for vaccine production have been well recognized. These involve either 1) stable genomic integration, with foreign gene introduced either by *Agrobacterium* T- DNA vectors or by direct means of microprojectile bombardment 2) transient expression using viral vectors. By using viral vectors inexpensive edible vaccine in greater yields can be produced in short period, however the production includes considerable risk in the dissemination of the virus from vaccine factories. However, transient expression is less easy to initiate, because the viral vectors must be inoculated into individual host plants. Thus, it is important to develop alternative approaches for producing effective and safe vaccines. Although several plant transformation systems are reported, widely used method is *Agrobacterium* mediated gene transfer. The transformation process inserts one or few copies of transgene, often into actively transcribed region of the plant genome and from a binary plasmid system 1-10 Kb DNA fragment can be introduced into the plant. Stable expression of introduced gene over the life of the plant and heritable expression in the progeny requires intregration of the gene and all of the regulatory sequences into the plant genome. Judicious use of genetic regulatory elements allows organ and tissue specific expression of foreign genes. This stable expression affords the advantage of the subsequent generation of a large number of transgenic plants by either vegetatively or sexual means.

3.1. Gene transfer using *Agrobacterium*

Agrobacterium tumefaciens and *A.rhizogenes* are soil bacteria which induce crown gall and hairy root disease respectively at wound sites on dicotyledonous plants. Only a few monocots are reported to be weakly susceptible to crown gall induction. These growth responses are as a result of natural genetic engineering event, and are associated with the presence within the bacteria of a mega plasmid , the *Ti* (tumour inducing) plasmid in the case of *A.tumefaciens* and the *Ri* (root inducing) plasmid in *A.rhizogenes*. Once induced, tumour tissue can grow axenically in tissue culture on media lacking exogenous supply of auxins and cytokinins. The T-DNA also encodes genes specifying enzymes involved in the production and secretion of opines *e.g.* nopaline, octopine, agricnopine, mannopine and agropine (derivatives of amino acids) depending on the strain. *Agrobacterium* utilizes these as a food source. *Ti* plasmid, found in virulent strains of *A.tumefaciens*, are around 200-250-kilobases (kb) in size and are stably maintained at temperatures below 30^0C. *Ti* plasmids found in different strain of *Agrobacterium* have four regions of homology as judged by DNA-DNA hybridization and heteroduplex mapping. Genetic analysis has shown that two regions are associated with tumour formation (*Ti* and *vir* regions) where as the other two are involved with conjugative transfer and the replicative maintenance of the plasmid within the *Agrobacterium*. During tumour formation a defined sequence of *Ti* plasmid, the T-DNA is transferred to the plant cell and integrated into the plant nuclear genome. The site of integration of T-DNA into plant DNA is apparently random. One or more copies of the T-DNA can be present, although multiple T-DNA copies can occur in tandem repeats they can also be separate and linked to different regions of plant DNA. It must be noted that the genes encoded by the T-DNA are all not required for the transfer of the T-DNA to the plant cell, nor its stable maintenance in the plant genome. The T-DNA regions in both *A. tumefaciens* and *A. rhizogenes* are flanked by 25 base pair (bp) direct repeats and the end points of integrated T-DNA in the plant genome are found close to these sequences. The natural ability of *Agrobacterium* to transfer defined sequences of DNA into the plant genome has been exploited in the development of variety of plant transformation vectors. During 1980s several research groups engineered Ti plasmids to remove all the T-DNA oncogenes (*onc* tumour proliferating).

The first important discovery was that the *onc* genes encoded by the Ti plasmid are neither required for the transfer of the T-DNA to the plant cells nor its integration into the nuclear genome. Hence these genes can be replaced, not only allowing the insertion of foreign DNA, but also removing the *onc* function in trans. Finally, non oncogenic T-DNAs present in regenerated whole plants are transmitted to progeny in a Mendelian fashion. Taking the characteristics of gene transfer mediated by *Agrobacterium* into account, any foreign DNA that has been cloned can be transferred into the genome of dicotyledonous plant cell. The foreign DNA to be transferred must be flanked by the T-DNA border sequences and stably maintained in an *Agrobacterium* strain harboring a full complement of *vir* genes either in *cis* or *trans* (located on a separate virulence helper plasmid). The removal of *onc* functions means that transformed tissues are no longer recognizable as neoplastic out growths that can be selected by their ability to grow vigorously without phytohormones. This led to the development of 'disarmed' Ti plasmids, in which the phytohormone biosynthetic genes are deleted from the T-DNA. To resolve the problem of identifying transformed cells, bacterial antibiotic-resistance genes have been placed under the control of T-DNA promoters and polyadenylation signals and inserted between the 25 bp repeat sequences. Such chimeric antibiotic- resistance genes are efficiently expressed in plant cells in a dominant fashion in any genetic background.

3.2. The process of transformation

The initial studies on transfer of foreign gene in plants involved the co-cultivation of plant protoplasts with *Agrobacterium.* A major technical advance was the demonstration that transgenic plants could be regenerated from leaf disc following co-cultivation with *Agrobacterium.* Subsequently transgenic plants have been produced from many families using this approach, or modification of it. Virtually every plant species explants have been co-cultivated with *Agrobacterium* and transgenic plants obtained. Such explants include, leaf bits, cotyledons, hypocotyls, petioles, stems, microspores and pre-embryos. The explant co-cultivation method involves dipping explant into a culture of modified *Agrobacterium* (any binary vector with desired gene, *e.g.* LB4404 harboring pBI121), blotting on sterile filter paper, and culturing on callussing or regeneration media. After 24-72 hours incubation, explants are transferred to similar medium containing an antibiotic *(e.g.* Cefotaxime and Carbenicillin) to suppress *Agrobacterium* growth. Selection for transgenic cells, by inclusion of the selective antibiotic in the culture medium, is usually initiated at this stage. In some instances, it is preferable to delay selection until 6-8 days after co-cultivation. The concentrations of the selection agent used vary widely depending on the sensitivity of the plant species and/or explant source. Generally 15-500 mg/l kanamycin is used. Cell colonies and/or shoots growing on the selection medium are transferred to fresh medium as required for the development of complete plants. The selective agent is usually maintained in all culture media throughout plant development. In some instances, it may inhibit shoot regeneration or root initiation and it may become necessary to omit or reduce the concentration of the selective agent.

To develop plants with vaccine genes, selected genes are introduced into binary vectors and mobilzed further into *Agrobacterium tumefaciens.* Such *bacterial* cultures are used for transformation of the selected plant, which can be eaten to get immunity. Frequencies at which transgenic plants have been selected and regenerated following *Agrobacterium* transformation vary considerably between plant species. Since plant tissues often decrease in their response under tissue culture conditions after co-cultivation cell/tissues must have high potential for growth and regeneration. An essential component of the *Agrobacterium* system is the induction of the *vir* operons. A number of methods have been used to induce three operons there by enhancing transformation. For example pre incubation with 20 μM acetosyringone increases the transformation frequency from 2-3% to 64% of explants producing transgenic plants.

4. DEVELOPMENTS IN "EDIBLE VACCINES" DURING THE LAST DECADE

The first step in demonstrating concept feasibility for plant based production came when Arntzen's group reported that the gene encoding HBsAg (Hepatitis B virus surface antigen) can be expressed in tobacco plants. The tobacco-derived rHBsAg self-assembles into subviral particles which are virtually indistinguishable from serum-derived HBsAg and yeast-derived rHBsAg with respect to size, density, sedimentation and antibody binding capacity. From their data they have demonstrated that plants can produce an immunonologically reactive antigen from an animal virus. Since this initial discovery major developments are reported in edible vaccine production for several diseases like Hepatitis, Rabies, *Escherichia coli* heat labile enterotoxin, cholera toxin B subunit, Norwalk virus, Foot and mouth disease etc.

The first report of an antigen being produced in plants is appeared as an European patent, where surface protein antigen A (spaA) from *Streptococcus mutans,* which causes tooth decay, was engineered into tobacco plants. The recombinant spaA protein represented 0.02 % of the total protein in leaf sample. Data from oral immunogenicity studies are also reported in which purified recombinant spaA protein made in *E.coli* was mixed with mouse chow and fed to mice. The mice developed sigA specific for spaA in their saliva. In addition, a study using lyophilized tobacco extracts containing spaA was described in which plant material was added to mouse chow, and no data was reported, so this may have been a theoretical study for patent purpose. Recent advances in developing edible vaccines against several human diseases and animal diseases are discussed in the following pages.

4.1. Hepatitis-B

The hepatitis B Virus (HBV) is a noncytopathic enveloped virus that causes acute and chronic necro inflammatory liver disease and hepatocellular carcinoma. The worldwide problem of hepatitis B virus infection and its association with chronic liver diseases has necessitated the development of an effective vaccine. The first vaccine consisted of Hepatitis B surface antigen (HBsAg) purified from the plasma of HBsAg chronic carriers. However, safety concerns of serum derived HBsAg resulted in the development of the first commercially available recombinant protein vaccine, derived from HBsAg producing yeast cells or chinese hamster ovary cells that contain the S or the S plus pre-S2 genes. Direct intramuscular injection of a plasmid vector encoding the HBsAg gave rise to secretion of the viral surface protein into the circulation, which led to the production of antibodies. A single injection of DNA was able to induce high levels of antibodies to HBsAg in mice that were sustained for at least 6 months. The first antibodies appeared within 1 or 2 weeks after infection of DNA and induced antibodies of the IgM isotype. Within few weeks helper T-lymphocyte activity switch over IgM to IgG. Hepatitis B virus gene induces a strong humoral response that is sustained up to 74 weeks without a booster dose. This can be further increased several fold by boosting with a second injection.

Arntzen's group have produced the surface antigen of hepatitis B in tobacco plants and found it similar in form and function to those from human serum or from recombinant yeast (Mason *et al.*, 1992). The tobacco-derived rHBsAg resembles yeast-derived rHBsAg in its size and immunogenicity and it is recovered as virus like particles of 22nm diameter. Subsequent to these findings they also demonstrated strong immune response in mice against rHBsAg. The response is qualitatively similar to that obtained by immunizing mice with yeast derived rHBsAg. All IgG subclasses and IgM against Hep B serological results are compared with standard antigen from *E. coli*. Both B- and T-cell epitopes of HBsAg are preserved when the antigen is expressed in a transgenic plant (Thanavala *et al.*, 1995). Hepatitis B is reported to accumulate to 0.01% of soluble protein level in transgenic tobacco.

4.2. Rabies

Rabies is one of the most fearsome diseases and one that certainly causes terrible death. Recent advances in DNA vaccines shown specific B and T cell mediated immune responses when a plasmid vector containing the full length rabies virus glycoprotein (G protein) under the control of SV 40 or cytomegalovirus promoter is injected into mice. Immune response resulted in the protection against the challenge with a virulent strain of the virus. Monkeys inoculated with DNA encoding the glycoprotein of rabies virus survived the virus challenge. Control of rabies among wildlife and feral dogs through conventional parenteral immunization is very difficult and thus efforts have focussed on oral immunization strategies. A recombinant vaccinia virus expressing rabies glycoprotein has been employed as an effective oral rabies vaccine in wildlife. Vaccinia virus engineered with rabies G-protein protected rabbits and mice completely from rabies, eliciting a high concentration of virus neutralizing antibodies. Raccoons fed with recombinant vaccinia virus vaccine orally are protected against rabies. Similarly Rabies virus glycoprotein expressed in baculovirus infected insect cells partially purified G-protein has given to raccoons orally in two doses elicited immune response. The ribonucleoprotein expressed in baculovirus is found to prime specific T-cells and mediated systemic immune response against rabies virus.

Michael's group (1995) from USA had developed transgenic tomato plants that express glycoprotein (G protein) gene which coats the outer surface of rabies virus including the signal peptide, under CaMV 35S promoter. Plants are transformed by *Agrobacterium* mediated transformation of cotyledons under culture conditions. The recombinant G- protein is immunoprecipitated from both leaf, fruit tissue and detected on western blots as two bands of 60 and 62.5 kD. The native G protein from denatured rabies virus is 66 kD. The difference in sizes may be due to altered glycosylation and /or specific enzymatic cleavage by plant tissue of sugar or aminoacid residues. They demonstrated the presence of important immunological epitopes in the plant produced G protein. The expression level of recombinant G protein was estimated to be between 1-10 ng/mg total soluble protein. Mice injected with extracts of fruits from these plants developed antibodies to rabies virus. Modelska *et al.*, 1998 demonstrated that two rabies virus epitopes stimulate virus neutralizing antibody synthesis in immunized mice. Tobacco and spinach leaves are used for oral administration of the vaccine. Mice immunized intraperitoneally or orally with engineered plant virus particles containing rabies antigen showed a local and systemic immune response. After the third dose of antigen, 40% of the mice are protected against challenge infection with a lethal dose of rabies virus. Oral administration of the antigen stimulated serum IgG and IgA synthesis and ameliorated the clinical signs caused by intranasal infection with an attenuated rabies virus strain.

4.3. Heat-labile enterotoxin and cholera toxin

Diarrhea is responsible for high mortality rate in developing countries, especially among children and travelers to these areas. *E. coli* cause diarrhea, which colonize in the small intestine and produce one or more toxins, one of which is called heat-labile enterotoxins (LT). The heat labile enterotoxin is a multimeric protein comprised of one A subunit (LT-A) of 27 kD and a pentamer of 3 subunits (LT-B) each with 11.6 kD. LT-A is an enzymatically active protein, which enters the epithelial cells of the gut, initiates cellular metabolic changes, and leads to the loss of water from the cells. LT-B is an enzymatically inactive protein, which forms a pentamer that binds to GM1 gangliosides in the membranes of epithelial cells. Attempts are made towards the production of edible vaccine by expressing heat-labile enterotoxin (LT-B) in tobacco and potato (Haq *et al.*, 1995). Antigen amounts in tobacco leaves and potato microtubers are quantified by ELISA. In both tobacco and potato, recombinant LT-B produced protein in microsomal vesicles. Purified protein from transgenic potato retained its immunogenecity. Both humoral and mucosal immune responses are elicited when potato tubers fed to the mice. Oral immunization with LT-B results in the production of anti LT-B immunoglobulins in serum (IgG and IgA) and in mucosal secretion (sIgA). Secretory antibodies in mucosal fluids prevent LT-B binding to the epithelial cells and thereby interfere with the toxic effect of LT. In human volunteers who ate 50-100 g raw transgenic potatoes, antibody-secreting cells are detected within 7-10 days. Both LT IgA and LT IgG are detected in their peripheral blood.

Cholera toxin (CT) is quite similar to heat-labile enterotoxin (LT) from *E. coli,* in its structure, function and antigenecity. LTB and CTB are both potent oral immunogens. CT-B protein derived from *V. cholerae* is recognised by mouse anti-CT-B antibody. An oral vaccine composed of the cholerae toxin B subunit (CT-B) with killed *V.cholerae* cells give protection against cholera and enterotoxigenic *E.coli.* However, the cost of vaccine for diarrheal disease is too high to afford. Cholera toxin has also been expressed in plants. Tobacco plants expresses CT-A and CT-B subunits of the toxin. CT-B has also been expressed in potatoes and shown immunogenicity (Arakawa *et al.*, 1998). In a Vero cell assay, sera from immunized mice neutralized the cytopathic effect of CT and mice are partially protected against the toxic effects of CT. When mice fed with 3 gm of potato tubers, shown comparable protection obtained with 30μg of purified bacterial CT-B. Transgenic potatoes retained 50% of the CT-B as pentameric GMI ganglioside binding form, even after boiling (Arakawa *et al.*, 1998). In addition co-administration or chemical binding of the cholera toxin B-subunit (CT-B) or LT-B to another antigen is capable enough to produce an adjuvant effect, *i.e.*, increasing the response to the second antigen compared to responses that occur when the antigen is given alone. Mason *et al.* (1998) increased the expression of *E.coli* enterotoxin B-subunit form by constructing a plant optimized synthetic gene. During construction of synthetic gene the sequences encoding inefficient processing or premature degradation of gene product are removed. Synthetic gene increased antigen accumulation in leaves and tubers by 3-14 fold. Antigenicity of the resulting vaccinogen was further demonstrated by immunization studies. Modelska *et al.* (1998) first detected a mucosal immune response after oral induction with a plant-virus-derived vaccinogen. Recombinant alfalfa mosaic virus was engineered to express two rabies virus epitopes. Mice vaccinated through diet produced double the level of anti-vaccinogen IgA. Brennan *et al.* (1999) further characterized the immune response induced by mucosal delivery of a plant derived vaccine. A chimeric CPMV expressing a vaccinogen from *Staphylococcus aureus* was delivered mucosally to mice. Vaccinogen specific IgA and IgG could be detected at distant mucosal sites like bronchia, intestine and vagina without the need of adjuvant.

4.4. Norwalk virus

Norwalk virus causes epidemics of acute gastroenteritis in human. Candidate vaccine has been developed by cloning the Norwalk virus genome and expressed the capsid protein in recombinant baculovirus infected insect cells. Norwalk virus capsid protein (NVCP) yields 58 kD protein which self assembles into Norwalk virus like particles (i-rNVs) lacking viral RNA and it reacts with Norwalk virus infected human sera. Electron cryomicroscopy revealed empty capsid of 38nm composed of 90 dimers of NVCP that form archlike capsomes. These particles are quite similar to authentic virus particles both in morphology and in antigenicity. Oral immunization of mice with 50 μg of i-rNVs per dose and 4 doses resulted in serum and mucosal antibody production against NVCP. i-rNVs is a nonreplicating vaccine and no adjuvant is needed to achieve immunization.

Mason *et al.* (1996) reported the expression of recombinant NVCP in transgenic tobacco leaves and potato tubers. Expression of recombinant NVCP with the patanin promoter in tissue cultured tobacco leaves was upto 0.23% total soluble protein and 0.37% in potato tubers or 34μg per gm fresh tuber. The tobacco leaves NVCP resembles i-rNVs both morphologically and physically. Mice fed directly potato tubers expressing NVCP or partially purified i-rNVs given orally stimulates the production of antibodies against NVCP. In these studies mice consumed four doses of 4g each of tuber containing 10-20μg NVCP per gram fresh tuber weight. Tuber slices were fed to mice with or without 10μg of the adjuvant CT being added to the potato. With adjuvant titer of antibody response is more. However, authors say it is not necessary to add adjuvant to the transgenic tubers to induce immune response. Transgenic tobacco leaf with NVCP elicited Norwalk virus specific serum IgG and sIgA responses. Potato tubers expressing NVCP stimulated serum IgG. The immune response elicited by the tubers was lower than tobacco induced response, where in both cases antigen quantity is remained same as 40 μg. It was estimated that only 50% of the NVCP were assembled into virus like particles in tuber material, whereas in leaf material all most all of the NVCP was in the form of virus like particle. Basing on this, it was assumed that practical form is more stable in the stomach and more readily taken up by the M-cells. It is important to note that NVCP like LT-B tubers is quite immunogenic molecule. Subcellular localization of NVCP in the cytoplasm may allow higher expression levels than the previously described LT-B and HBsAg which are directed into the endoplasmic reticulum. IgA titers were assayed in the mice orally immunized with plant material containing NVCP (Mason *et al.*, 1996). Although 32 out of 43 mice had significant serum titers against NVCP, only 12 out of 41 mice assayed were positive for fecal IgA specific for NVCP. To induce measurable IgA response larger oral doses of NVCP might be required.

4.5. Foot and mouth disease

Although substantial work is done towards development of vaccines against human diseases, limited work is done in the case of animal diseases. Efforts to develop edible vaccines against Foot and mouth disease in cattle are substantial. Foot and mouth disease virus (FMDV) is the causative agent, which affects meat and milk producing animals. The structural protein VPI carries critical epitopes responsible for the induction of neutralizing antibodies. Immunization with VPI or with synthetic peptides representing part of its amino acid sequence has been demonstrated to induce protection against challenge in experimental and natural hosts. Production of complete or fractional VPI in a diversity of expressing systems has been performed frequently in the search for an effective and inexpensive alternative, which would be highly immunogenic. Carrillo *et al.*, 1998 expressed structural protein VPI of foot and mouth disease virus in *Arabidopsis*. The mouse that was immunized intraperitonially with a leaf extract elicited immune response to synthetic peptides carrying various epitopes of VPI or to complete VPI. Further more all the mice immunized with the leaf extract were protected against challenge with virulent foot and mouth disease virus.

Wigdoroutz *et al.*, 1999 developed alfalfa transgenics expressing the structural protein VPI of foot and mouth disease virus (FMDV). They reported for the first time the complete open reading frame expression of an antigen in plants, using a recombinant plant virus permitting the use of the crude plant extracts as an experimental immunogen to protect against virus challenge. The mice orally or parenterally immunized with leaves obtained from transgenic plants developed a similar virus specific immune response. Immunized animals elicited a specific antibody response to a synthetic peptide representing amino acid residues 135-160 of VPI.

At the Indian institute of sciences, in the author's laboratory transgenics are being developed as source of edible vaccines against two animal diseases like Rinderpest and Peste des petitis ruminant. Rinderpest virus (RPV) and Peste des petitis ruminant virus (PPRV) are the causative agents of fatal contagious diseases of the small and large ruminants. Both RPV and PPRV belong to the family Paramyxoviridae and genus Morbillivirus. RPV has only one antigenic type (serotype) and an attenuated, live vaccine with high immunogenicity is available. Rinderpest has been eradicated in developed countries, but is still prevalent in parts of Africa, the Middle East and South Asia, where eradication campaigns are underway. The major draw back of the currently used Rinderpest vaccine is it's heat liability. In hot countries, the vaccine delivery is constrained by high costs, power failure, lack of refrigiration and cold chain to keep the potency of the vaccine. RPV carries two virus specific glycoproteins, the haemagglutanin (RPVH) and the fusion (F) glycoprotein. These glycoproteins are immunogenic and confer complete protective immunity against the disease. PPRV hemagglutinin protein "hemagglutinin nuraminidase" affects small ruminants by absorbing them through unknown biological activity. Recombinant Rinderpest vaccines have been developed using the Rinderpest H or F gene and creating recombinant vaccinia viruses carrying these genes, which are highly protective in experimental cattle. Effectiveness of the vaccinia recombinant H vaccine has been proved for at least one year after a single dose of vaccination in cattle. However, this was not field tested. A baculo virus recombinant carrying the RPVH gene expressing large quantities of H protein, which is antigenically authentic protein capable of eliciting protective immunity in cattle was reported by Scientists from Indian Institute of Science Bangalore (Naik and Shaila, 1997; Naik *et al.*, 1997). Even this one was not field tested and even, if, this works, well, the production costs would be high as it would involve cell-culture based production.

Peste des petits ruminants is a highly contagious disease of sheep, goats and wild ruminants, with a high mortality rate. The causative agent, peste des petits ruminant virus (PPRV) has only one antigenic type (serotype) and an attenuated, live vaccine has been made and used for immunization of small ruminants in Africa. The PPRV disease is still prevalent in parts of Africa, and the disease is rampant in India. This live attenuated vaccine has not been field-tested in India. The development of a safe, efficacious and inexpensive oral vaccine made in plants assumes strategic importance for a country like India. Attempts are being made to obtain a live attenuated vaccine using an Indian isolate. No vaccine is used for prevention at present. Plants could be a useful system for producing vaccines, because large amounts of antigen could be produced at a relatively low cost, using agriculture instead of sophisticated and expensive cell culture based, expression system.

Thus development of a safe efficacious and inexpensive oral vaccine made in plants assumes strategic importance for a country like India. Transgenic plants that express protective antigens of RPV and PPRV in their edible tissues might be used as an inexpensive oral vaccine production and delivery system.

RPVH and PPRHN are highly immunogenic and surviving animals are immune to the disease for the rest of their lifetime. Hence a programme was initiated in the author's laboratory to develop transgenic plants against these diseases with functional RPVH and PPRHN in tobacco, groundnut and pigeonpea. Transgenic plants with vaccine genes were produced in all the selected plants. Groundnut plants are usually fed to animals after harvest and could be used as host plants for the vaccine gene. Molecular analysis was done to confirm the integration of the vaccine gene into the plant genome by Southern hybridization. The expression levels of respective immunogenic proteins are analyzed and experiments to immunize mice orally with developed vaccine are currently in progress.

5. RECENT DEVELOPMENTS IN PLANT DERIVED VACCINES

Advanced research in plant biotechnology is concentrating on the application of plant-derived vaccines in human and animal health care. To commercialize edible vaccine, industries require information on transformation technology, quantification of immunogen, dosage, response type and best delivery method. The first human clinical trails for a transgenic plant derived antigen were reported in 1997 by Arntzen's group. It was approved by US Food and Drug Administration. Transgenic potatoes expressing a synthetic bacterial diahhrea vaccinogen were orally delivered to human volunteers in phase I/II clinical trails (Tacket *et al.*, 1998). Each participant received 50-100 g of raw transgenic potatoes displayed a significant rise in LT-B antibodies, whereas no specific antibodies were detected in control participant. Thus plant derived recombinant LT-B delivered in edible tissue was protected against digestion and proved capable of inducing an immune response in human. Orally delivered potato hepatitis B surface antigen used as a booster for the commercial hepatitis B vaccine. Phase I/II trails are currently in progress. Norwalk virus-like particles (VLPs) for a viral diarrhea vaccine taken from potato are also under clinical trails. In human, trials performed with edible vaccine showed that the vaccinogen is sufficient to induce systemic and mucosal immune responses without aid of adjuvant. Though the genetically modified materials are used as vaccines instead of natural source no adverse affects were observed (Thanavala and coworkers, unpublished data).

6. CONCLUSIONS AND FUTURE PROSPECTS

Genetic engineering in plants has opened up yet another successful application, in the production of edible vaccines and has now reached an exciting stage. Edible vaccines delivered in transgenic vegetable/fruit plants meet several desirable criteria for new vaccines. By manipulating the expression constructs, the level of antigen can be increased to that which is feasible for human testing of oral vaccine doses. Level of recombinant antigen comprising 1% of the total protein in edible tissues is a reasonable goal. The list of plant derived vaaccinogens continues to grow and includes viral, bacterial, enteric and non-enteric pathogen antigens. Production of edible vaccines may allow broader access to vaccination by reducing overall cost of the vaccination and promoting compliance by the user friendly oral administration in receiving vaccinations. In spite of successes reported, many challenges are to be met before the products are successfully marketed.

REFERENCES

Arakawa T, Chong DKX and Langridge WHR (1998) Efficacy of a food plant based oral cholera toxin B subunit vaccine. *Nature Biotech.,* **16** : 292-297.

Brennan FR, Bellaby T, Helliwell SM, Jones TD, Kamstrup S, Dalsgaard K, Flock JI and Hamilton WDO (1999) Chimeric plant virus particles administered nasally or orally induce systemic and mucosal immune responses in mice. *J. Virol.,* **73** : 930-938.

Carrillo C, Wigdorovitz A, Oliveros JC, Zamorano PI, Sadir AM, Gomez N, Salinas J, Escriban JM and Borca MV (1999) Protective immune response to foot and mouth disease virus with VP1 expressed in transgenic plants. *J. Virol.,* **72** : 1688-1690.

Chen I, Finn TM, Yanquing, Guoming Q, Rappuoli R and Pizza M (1998) A recombinant live attenuated strain of *Vibrio cholera* induces immunity against tetanus toxin and *Bordetella pertussin* trachael colonization factor. *Infect. Immuno.,* **66** : 1648-1653.

Dunwell JM (1999) Transgenic crops: The next generation, or an example of 2020 vision. *Ann. Bot.,* **84** : 269-277.

Haq TA, Mason HS, Clements JD and Arntzen CJ (1995) Oral immunization with a recombinant bacterial antigen produced in transgenic plant. *Science,* **268** : 714-716.

Koo M, Bendahmane M, Lettieri GA, Paoletti AD, Lane TE, Fichen JH, Buchmeier MU and Beachy RN (1999) Protective immunity against murine hapatitis virus (MHV) induced by intranasal or subcutaneous administration of hybrids of Tobacco mosaic virus that carries an MHV epitope. *Proc. Natl. Acad. Sci. USA,* **96** : 7774-7779.

Mason HS and Arntzen CJ (1995) Transgenic plants as vaccine production systems. *Trends Biotechnol.,* **13** : 388-392.

Mason HS, Haq TA, Clements JD and Arntzen CJ (1998) Edible vaccine protects mice against *Escheria coli* heat labile enterotoxin (LT): potato expressing a synthetic LT-B gene. *Vaccine,* **16** : 1336-1343.

Mason HS, Lam DMK and Arntzen CJ (1992) Expression of hepatitis B surface antigen in transgenic plants. *Proc. Natl. Acad. Sci.,* **89** : 11745-11749.

McGarvey PW, Hammond J, Dienelt MM, Hooper DC, Fu ZF, Dietzschold B, Koprowski H and Michaels FH (1995) Expression of the rabies virus glycoprotein in transgenic tomatoes. *Biotechnology,* **13** : 484-1487.

Modelska A, Dietzschold B, Sleyesh N, Fang Fu Z, Steplewski K, Hooper DC, Koprowski H and Yusibov V (1998) Immunization against rabies with plant derived antigen. *Proc. Natl. Acad. Sci. USA,* **95** : 2481-2485.

Naik S and Shaila MS (1997) Characterization of membrane -bound and membrane anchor-less forms of Hemagglutinin glycoprotein of Rinderpest virus expressed by baculovirus recombinants. *Virus Genes,* **14** : 95-104

Naik S, Renukaradhya G, Rajasekhar JJ and Shaila MS (1997) Immunogenic and protective prperties of haemgglutenin protein (H) of Rinderpest virus expressed by a recombinant Baculovirus. *Vaccine,* **15** : 603-607

Richter L and Kipp PB (1999) Transgenic plants as edible vaccines. *In CTMI* (Eds J Hammon, P.Mc Garvey and V. Yoshibov) pp 159-176.

Sharma AK, Mohanty A, Singh Y and Tyagi AK (1999) Transgenic plants for production of edible vaccine and antibiotics for immuno therapy. *Curr. Sci.,* **77** : 524-529.

Tacket CO, Mason HS, Losonsky G, Clements JD, Levine MM and Arntzen CJ (1998) Immunogenicity in humans of a recombinant bacterial antigen delivered in a transgenic potato. *Nat. Med.,* **4** : 607-609.

Tacket CO and Mason HS (1999) A review of oral vaccination with transgenic vegetables. *Microbes and infection,* **1** : 777-783

Thanavala Y, Yang YF, Lyons P, Mason HS and Arntzen CJ (1998) Immunogenecity of transgenic plant derived hepatitis B surface antigen. *Proc. Natl. Acad. Sci. USA,* **92** : 3358-3361

Walmsley AM and Arntzen CJ (2000) Plants for delivery of edible vaccines. *Current Opinion in Biotechnology,* **11** : 126-129

Wigdorovitz A, Carrillo C, DusSantos M, Trono JK, Peralta A, Gomez MC, Rios RD, Franze PM, Sadir AM, Escribano JM and Borca MV (1999) Induction of protective antibody response to foot and mouth disease virus in mice either orally or parentally immunized with alfalfa transgenic plants expressing the viral structural protein VP1. *Virology,* **255** : 347-353

Wigdorovitz A, Filgueira DMP, Robertson N, Carrillo C, Sadir AM, Morris JJ and Borca MV (1999) Protection of mice against challenge with foot and mouth disease virus (FMDV) by immunization with foliar extracts from plants infected with recombinant Tobacco mosaic virus expressing the FMDV structural protein VPI. *Virology,* **264** : 85-91.

Chapter 13

HAIRY ROOT CULTURES OF *CATHARANTHUS ROSEUS* : A MODEL FOR PRIMARY AND SECONDARY METABOLIC STUDIES

Victor M Loyola-Vargas★ and SM Teresa Hernandez-Sotomayor

Unidad de Biología Experimental, Centro de Investigación Científica de Yucatán, Calle 43 No. 130, Col. Chuburná de Hidalgo, Mérida Yuc. 97200, México.

Summary

Due to the interest in the production of alkaloids of pharmacological importance, Catharanthus roseus has been the focus of several research groups for more than 40 years. Hairy root cultures obtained by infection with Agrobacterium rhizogenes have advantages over other in vitro cultures due to their rapid auxin-independent growth, genetic stability and high-level production of secondary metabolites. Furthermore, they are an excellent model system for plant biochemistry and molecular studies. A review of the literature in this field has been presented and discussed.

Keywords : *Catharanthus roseus*, hairy roots, tissue culture, indole alkaloids

1. INTRODUCTION

Catharanthus spp., and in particular *Catharanthus roseus* (L.) G. Don, the Madagascar periwinkle, is one of the most extensively investigated medicinal plants. Investigations with *C. roseus* have focused especially on the isolation and semi-synthesis of its alkaloids. More than 100 alkaloids have been isolated from different parts of the plant, many

★Corresponding author : E-mail : vmloyola@cicy.mx

of them possessing remarkable pharmacological activity (Taylor and Farnsworth, 1975). The most important of these are the antileukemic alkaloids vinblastine and vincristine, the antihypertensive alkaloid ajmalicine and serpentine, which has a sedative effect (Figure 1). The low yields of the antileukemic alkaloids in the plant, combined with their high market price has encouraged intense research for alternative methods for the production of these alkaloids.

One alternative is the use of different type of cultures, such as hairy roots. These cultures produce the same alkaloid spectra as the roots of the mother plant and are obtained by genetic transformation by *Agrobacterium rhizogenes*. They have advantages over cell suspension cultures in the production of useful secondary metabolites due to their genetic and biochemical stability and are characterized by rapid auxin-independent growth, genetic stability and stable, high-level production of secondary metabolites (Flores *et al.*, 1999).

The aim of this review is to present and summarize the research made on *C. roseus* hairy roots, focused not only on those that refer to

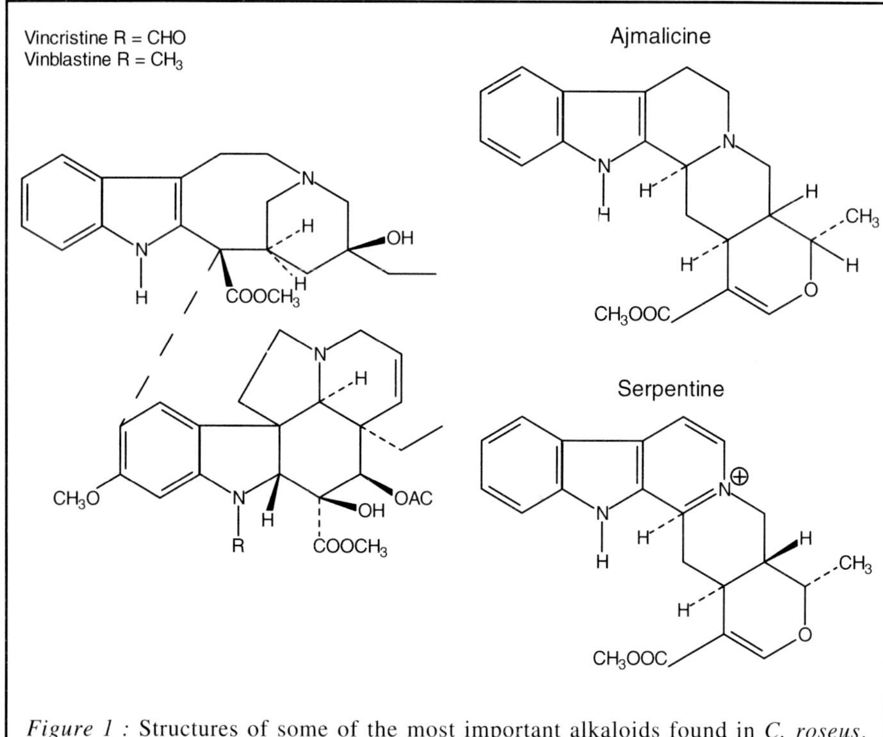

Figure 1 : Structures of some of the most important alkaloids found in *C. roseus*.

the production of secondary metabolites, but also to those on other biochemical studies.

2. SECONDARY METABOLITES FROM *C. ROSEUS* HAIRY ROOTS

Two species of *Catharanthus* spp. have been transformed with more than 20 Agrobacterium rhizogenes strains and several groups have reported the generation of *Catharanthus* spp. hairy root cultures (Table I, Fig. 2). Amongst the most important indole alkaloids found in these hairy root cultures are ajmalicine, catharanthine, serpentine, vindolina and yohimbine. Our group has also discovered a new alkaloid in hairy roots, the 19(S) epimiciline, which has not been found in the plant (Peraza-Sánchez *et al.*, 1998).

Each different hairy root line obtained shows particular characteristics. These depend on the explant, the *A. rhizogenes* strain, the conditions

Figure 2 : a) Stem of *C. roseus* infected with the *A. rhizogenes* strain 18555. b) Leaves of *C. roseus* infected with the *A. rhizogenes* strain 13333 36 days after infection. c) Bottom of a flask with an established culture of *C. roseus* hairy roots. d) Flask with an established culture of *C. roseus* hairy roots. All the pictures are from the laboratory of the authors.

Table 1. Reports of *Catharanthus* spp. hairy root cultures and the alkaloids they produce.

Species and cultivars	Explant	*A. rhizogenes* strain	Medium	Indole alkaloids found	Reference
C. roseus	Leaves	LBA 9402	Gamborg´s B_5	ajmalicine, serpentine, catharanthine, vindoline and vinblastine	(Parr *et al.*, 1988)
C. trichophyllus	Plantlets	15834	Mo medium	ajmalicine, vindolinine, tabersonine, lochnericine and another 12	(Davioud *et al.*, 1989a)
C. trichophyllus	Plantlets	15834	Mo medium	anthraserpine, dimethoxyanthraserpine and another 3	(Davioud *et al.*, 1989b)
C. roseus cv. Little Delicata and cv. Flower-less Variety	Leaves	15834, A2, A2-83, A47-83, R1000 and TR7	½ B_5	ajmalicine, vindolinine, catharanthine, yohimbine, tabersonine and another 8	(Toivonen *et al.*, 1989)
C. roseus cv. Little Delicata and cv. Little Linda	Hypoco-tyls and plantlets	15834	1/3 SH	ajmalicine and catharanthine	(Jung *et al.*, 1992; Sim *et al.*, 1994)

Table 1. Continue

Species and cultivars	Explant	*A. rhizogenes* strain	Medium	Indole alkaloids found	Reference
C. roseus cv. Little Bright Eye, cv. Rose Carpet, cv. Little Linda, cv. Snow Carpet	Plantlets	15834	½ B_5	ajmalicine, serpentine, catharanthine and vindoline	(Bhadra *et al.*, 1993)
C. roseus	Leaves, stems	15834, A41027, R1000, A4, AR10, A4PC, A4T, 2659T, 1855T, 1855TD, R1000CbR, R1200rolb, R1236, 2659, LBA1855, NcPPB2659, and 1855 pBI	½ B_5	Ajmalicine, serpentine, catharanthine, vincamine, ajmaline and another 10	(Ciau-Uitz *et al.*, 1994; Vázquez-Flota *et al.*, 1994; Moreno-Valenzuela *et al.*, 1999)
C. roseus	Leaves	LBA9402	½ B_5	19(S)-epimisiline	(Peraza-Sánchez *et al.*, 1998)

used for the transformation and the number of copies of the plasmid inserted into the plant genome. This is because every root obtained is the result of an independent event in a single cell of the explant. This characteristic has allowed the selection of hairy root lines that over-produce some particular alkaloids .

As *C. roseus* has been the center of attention for the synthesis of indole alkaloids, most of the work on *C. roseus* hairy roots has been focused on the production of greater amounts and on the specific production of those alkaloids. The production of alkaloids in hairy roots is regulated by genetic factors as well as by different external conditions such as the medium composition, nutritional factors, temperature, pH and abiotic factors.

3. NUTRITIONAL FACTORS

The effects of sucrose, phosphate, nitrogen and many other nutrients are known to be important in the regulation of both growth and indole alkaloid production. However, there is no universal recipe that can warrant either. The optimization of the medium for each particular hairy root line must be evaluated.

Toivonen *et al.*, (1991) studied the effect of sucrose, phosphate, nitrate and ammonia concentrations on growth and indole alkaloid production on *C. roseus* hairy roots. Contradictory effects of the nutrients were observed. The maximal growth was obtained with 77.8 mg/L $NaH_2PO_4.H_2O$ and 1.311 g/L KNO_3 in the medium, whereas the specific production of alkaloids was highest at the lowest levels of all the nutrients studied. In this study, when the cultures had enough nutrients to support growth, it was difficult to affect the specific alkaloid production rates. Sucrose, phosphate, nitrate, and ammonia concentrations in the culture medium were all found to have statistically significant effects on the growth rate and indole alkaloid production. Optimum concentrations for maximal growth were obtained for phosphate and nitrate. The effect of sucrose was more complicated high concentrations of sucrose induced more biomass but caused the callusing of roots. Of all the nutrients studied, ammonia had the best effect on growth.

The effects of the concentrations of inorganic salts in Schenk and Hilldebrandt (SH) medium on catharanthine production was also investigated by Jung *et al.*, (1994). Nitrate supported both the growth and catharanthine production. Ammonium and phosphate gave contradictory effects in respect to growth and alkaloid production. Borate and molibdate were also tested, but produced an inhibitory effect on both growth and alkaloid production.

In most cases, the optimization of the media composition to induce the accumulation of indole alkaloids, involves a two-stage batch process. In the first stage, optimum conditions for biomass accumulation are needed; in the second stage, alkaloid production is induced.

By optimizing the composition and concentration of inorganic salts in the culture medium, a two stage process for culturing *C. roseus* hairy roots was obtained (Jung *et al*., 1994). This resulted in an enhancement of catharanthine productivity (4.1-fold) and total alkaloid productivity (5.4-fold) compared with those for one stage culture in the original SH medium The carbohydrate source is also a key parameter that needs to be taken into account. Sucrose or glucose have been found to be effective in supporting high growth rates and biomass yields in plant cell cultures. In the case of hairy root cultures, sucrose is generally used as a carbon source. Sucrose, glucose and fructose, as carbon sources in culture medium, have been assayed for growth and alkaloid production in *C. roseus* hairy roots (Jung *et al*., 1992). The cultures preferentially consumed sucrose, resulting in about 40% higher growth rate. However, fructose enhanced the catharanthine yield about two-fold. A two stage culture using sucrose and fructose improved volumetric yields of catharanthine by two-fold. It has been suggested (Jung *et al*., 1992) that substitution of fructose for sucrose alters carbohydrate metabolism in hairy roots, which subsequently stimulates their secondary metabolism.

Since growth of cultured tissues depends on media composition, the consumption patterns of sugars and nutrients are important factors in the development of a continuous system. Moreno-Valenzuela *et al*., (1999a) investigated the utilization of the B_5 media component and its relationship to alkaloid production and growth. A hairy root line from *C. roseus* was cultured in a 14 L bioreactor. Nitrate and phosphate uptakes were similar to the same root line cultured in a 250 ml Erlenmeyer flask. However, sucrose consumption rate was slower in roots cultured in the bioreactor. These results suggested that it was feasible to upgrade that particular hairy root line to bioreactor level. Ajmalicine and catharanthine were produced and retained within the biomass tissue (Moreno-Valenzuela *et al*., 1999a).

Several reports on *C. roseus* hairy roots have focused on the accumulation of ajmalicine, serpentine, and catharnathine in a variety of short- and medium-term periods (Ciau-Uitz *et al*., 1994; Islas-Flores *et al.*, 1994; Vázquez-Flota *et al.*, 1994). The kinetics of growth, the uptake of macronutrients and the accumulation of indole alkaloids were investigated in long-term, heterotrophically cultured transgenic *C. roseus* hairy roots (Bhadra and Shanks, 1997). Tabersonine, ajmalicine and serpentine were monitored over a 70-day period. Tabersonine accumulation was distinctly growth-associated with maximum specific and total yields in the late-exponential phase of growth. Serpentine accumulation was non-growth associated with increasing specific and total levels in the stationary growth phase. The accumulation of ajmalicine also appeared to be growth associated.

4. pH OF THE CULTURE MEDIA AND TEMPERATURE

The alkaloid productivity and the storage capacity can be influenced by the size of the pH gradient between the medium and vacuoles. It has been shown, at least in suspension cultured cells of *C. roseus*, that shifting the medium pH between low and high values releases intracellular alkaloids into the culture medium.

The release of alkaloids from *C. roseus* hairy roots increases when the pH of the culture medium was reduced from 5.7 to 3.5. Total alkaloids increased almost 16- fold, ajmalicine and catharanthine also increased 380- and 30-fold respectively. This increase was also observed in cultures growing in a 14-L fermentor, when the medium pH was reduced. Reduction of the pH did not affect growth of the root cultures (Sáenz-Carbonell *et al.*, 1993).

The initial medium pH of the cultures, extracellular pH, cytoplasmic pH, vacuolar pH, uptake of inorganic phosphate, carbon utilization and growth of *C. roseus* hairy roots were studied using in vivo ^{31}P nuclear magnetic resonance spectroscopy to monitor subcellular compartmental pH and inorganic phosphate levels and ^{13}C to monitor carbon utilization (Ho and Shanks, 1992). Maximum growth rate was found in cultures at the initial pH values of 4.2 and 7.3. All cultures maintained a cytoplasmic pH of 7.4 throughout the growth cycle. However, vacuolar pH was 5.1-5.2 in cultures with an initial pH of 4.2, as opposed to 5.4-5.5 for other cultures. Sucrose was hydrolyzed completely to glucose and fructose by day 26 except for cultures with an initial pH of 7.3. Glucose was the preferred substrate after day 20 and 26 for cultures with initial pH values of 7.3 and 6.5 respectively. The best growth was observed for cultures grown at pH 6.5. The more acidic vacuolar pH observed at external pH of 4.2 may be related to these phenomena. Utilization of inorganic phosphate was unaffected by pH and, in contrast to the cultures at an initial medium pH of 4.2, the reduced growth observed at an external pH of 7.3 may be due to reduced sucrose hydrolysis (Ho and Shanks, 1992).

C. roseus hairy roots were cultured at different temperatures and were found to affect growth rate and indole alkaloid content (Toivonen *et al.*, 1992). When the temperature was lowered, no effect was detected on the distribution of indole alkaloids between the roots and the medium. Indeed, the level of alkaloid accumulation showed a clear increase with lowered temperature. An increase in the alkaloid content of hairy roots with decreased temperature is not necessarily due to enhanced biosynthesis as a response to environmental stress. However, temperature-induced changes in alkaloid transport, accumulation and degradation rates may affect the n*et al*kaloid content of the cultures (Schulze-Lefert *et al.*, 1989)

5. ABIOTIC AND BIOTIC STRESSES

The use of hairy roots for secondary metabolite production is inhibited because the products of interest are most often stored within the cells. To overcome this, *in situ* extraction by organic solvents have been studied. The use of solid adsorbents in combination with a bioconversion process is also expected to improve progress. The presence of a solid adsorbent can be expected to prevent the products from being degraded and to stimulate *de novo* synthesis by removal of the products. To enhance alkaloid production and secretion, a permeabilizing agent (dimethyl sulfoxide, DMSO) and a fungal elicitor to provide physical and biochemical stress, respectively, were used in *C. roseus* (cvs. Little Linda and Little Delicata) hairy roots (Sim *et al.*, 1994). The influence of the addition of XAD-7, an amberlite resin, and DMSO, a permeabilizing agent and a fungal elicitor, *Penicillium* sp. as physical and biochemical stress on the production and secretion of alkaloids was tested. The amberlite resin, greatly enhanced the release of catharanthine and ajmalicine. Combining *in situ* adsorption by XAD-7 sequentially with DMSO and fungal elicitation enhances both alkaloid production and secretion. The release ratio of catharanthine and ajmalicine was enhanced up to 20 and 70% respectively. These results indicated that *in situ* adsorption sequentially applied with permeabilization and fungal elicitation have a synergistic effect on the production and secretion of indole alkaloids (Sim *et al.*, 1994).

Other strategies have also been used to increase the production of indole alkaloids. One example is the addition of hydrolytic enzymes and biotic elicitors to *C. roseus* hairy root cultures (Vázquez-Flota *et al.*, 1994). It was shown that the nitrogen source induced differential responses in the individual alkaloid yields. By contrast no changes were observed when the concentration of vitamins or macro- and micronutrients were modified. *Aspergillus* treatment and the use of macerozyme increased the accumulation of ajmalcine selectively, while the addition of methyl jasmonate increased the yield of both alkaloids.

Late exponential phase hairy root cultures of *C. roseus* were elicited with pectinase and jasmonic acid. The effects of elicitor concentration and exposure time on growth and levels of several compounds in the indole alkaloid biosynthetic pathway were analyzed (Rijhwani and Shanks, 1998). Pectinase decreased the fresh weight to dry weight ratio of the roots, while the addition of jasmonic acid had no significant effect. An increase of 150% in the specific yield of tabersonine was observed upon addition of 72 units of pectinase. Transient studies at the same level demonstrated possible catabolism as serpentine, tabersonine, and lochnericine levels decreased immediately after elicitation. Jasmonic acid was found to be a unique elicitor leading to an enhancement of flux in several branches of the indole alkaloid pathway. Jasmonic acid addition caused an increase in the specific yields of ajmalicine (80%), serpentine (60%), and lochnericine (150%). Transient studies showed that lochnericine and tabersonine levels increase to a maximum, then decrease back to control levels and finally fall below the control levels. The yields of ajmalicine and serpentine increased continuously after the addition of jasmonic acid (Rijhwani and Shanks, 1998).

6. LIGHT

The potential of light-adapted hairy roots to accumulate increased levels of some alkaloids has been demonstrated in other plant species. Cultures of *C. roseus* hairy roots, were given periodic daily illumination to investigate the effect of light on growth and nutrient utilization and the accumulation of the indole alkaloids. Light-adapted roots appeared green and had a radically thickened morphology compared with the dark-grown controls. Their growth rates were higher than dark-grown controls. The specific and total levels of the indole alkaloid serpentine were enhanced and those of tabersonine were lowered, while the overall trends of growth and non-growth association of tabersonine and serpentine respectively, remained unaltered by light adaptation (Bhadra *et al.*, 1998).

7. DEVELOPMENTAL CONTROL

Hairy roots have been used as alternatives to cell suspensions for the production of secondary metabolites due to greater genetic and biochemical stability. The long and interlacing structure of hairy roots, however, causes problems with transfer to large scale culture facilities and inoculation.

It has been demonstrated that indole alkaloid biosynthesis is under developmental control. An excellent model to use when studying the role of differentiation in the biosynthesis of secondary metabolites is that of undifferentiated-dedifferentiated hairy root cultures. Moreno-Valenzuela *et al.*, (Moreno-Valenzuela *et al.*, 1998) showed that higher levels of indole alkaloids were accumulated in *in vitro* cultures of hairy roots derived from *C. roseus* than in cell suspension cultures. In that work, hairy roots were interconnected to undifferentiated cells by manipulation of the culture medium. That was achieved when the concentration of micronutrients in the culture medium was five times the concentration of Philips and Collins medium. Under those conditions, cell suspensions were obtained from hairy roots. However, the alkaloid content was five times lower in the cell suspensions that in the original hairy roots, but upon regeneration of the roots the alkaloid content regained its original level.

Jung *et al.*, (1995) reported an interchangeable culture system of hairy root and cell suspension of *C. roseus.* A rhizogenic cell suspension culture was established from callus. When cultured in SH medium with growth regulators, the rhizogenic callus produced catharanthine at a level of 41% of the initial level in the hairy roots. Upon transfer to SH basal medium, regenerated hairy roots produced this alkaloid at the original level of 1.5 mg/g dry weight. Using this cell/hairy root interchange method a new management system for hairy root culture in bioreactors has been devised and tested. It involves the production of biomass in the form of a cell suspension in medium supplemented with growth regulators and catharanthine production by hairy roots regenerated from these cells in medium without growth regulators (Jung *et al.*, 1995). Because it is more convenient to inoculate or transfer suspensions rather than hairy root cultures, the interchangeable cell/hairy root system represents a simplified process in the large scale production of catharanthine. Moreover, the secondary metabolite production of regenerated hairy roots from rhizogenic callus is comparable to the level of initial hairy roots. Thus, the interchangeable cell/hairy root culture system can be used for secondary metabolite production which maximizes both biomass, in the form of cells, and secondary metabolite production, in the form of hairy roots.

The bisindole alkaloids vinblastine and vincristine, originally isolated from *C. roseus*, have been the subject of a significant number of studies due to their importance as antineoplastic drugs in clinical medicine. The limited availability of these metabolites from plant materials has led many research groups to investigate plant tissue cultures as an alternative source. As a part of the studies on the metabolites produced by hairy roots, three known alkaloids, lochnericine, 21-hydroxycyclolochnerine and 19(*S*)-epimisiline, were isolated from hairy roots of *C. roseus* (Peraza-Sánchez *et al.*, 1998). It is important to point out that although over 100 indole alkaloids are reported as being produced by *C. roseus*, with approximately 60 of them having been isolated from plant tissue culture systems, this is the first report of 19(S)-epimisiline produced in *C. roseus* hairy roots.

8. BIOCHEMICAL STUDIES

Metabolic engineering of plant cell and tissue cultures for enhanced production of pharmaceutically valuable alkaloids has promise due to recent advances in plant molecular techniques. However, rational genetic manipulation of an alkaloid-producing pathway requires determination of the points in the pathway at which flux is most severely restricted. For identification of these points, a sufficient understanding of the biochemistry and regulation of the pathway under consideration is necessary. Several enzymes involved in the alkaloid biosynthesis pathway have been studied in *C. roseus* hairy roots. In addition to the studies of the metabolic pathways involved in alkaloid biosynthesis, it is also fundamental to know which are the metabolic pathways involved in the perception of signals that may alter secondary metabolism and, in consequence, alkaloid production.

9. METABOLIC PATHWAYS INVOLVED IN ALKALOID PRODUCTION

Tryptophan decarboxylase (TDC), the enzyme that catalyzes the decarboxylation of tryptophan to tryptamine, was studied in a *C. roseus* transformed root culture. Its activity was evaluated throughout the culture cycle. An increase in the specific activity of this enzyme was observed after 18 days of growth without significant changes in total alkaloid and tryptamine profiles (Islas-Flores *et al.*, 1994).

The activity of TDC is highly regulated by developmental processes, at least in *C. roseus* seedlings (De Luca *et al.*, 1988). TDC activity decreases by at least one order of magnitude when the hairy roots dedifferentiated into suspension cells. This effect was also shown for another key enzyme, strictosidine synthase (Moreno-Valenzuela *et al.*, 1998).

The conversion of tabersonine to lochnericine was investigated in *C. roseus* hairy roots cultures (Morgan and Shanks, 1999). The accumulation of lochnericine and horhammericine, like tabersonine, was associated with growth. Details of the metabolic pathway around tabersonine in hairy roots of *C. roseus* were elucidated by the use of oxygenase inhibitors, 1-aminobenzotriazole, clotimazole, and 2, 5-pyridinedicarboxylic acid. Using jasmonic acid in combination with the inhibitors suggests the existence of an inducible P-450 enzyme responsible for the formation of horhammericine. The inhibitor study also revealed that both lochnericine and horhammericine are turned over in hairy root cultures.

The behavior of the enzyme geraniol 10-hydroxylase (G10H), and its redox partner the NADPH: Cyt C (P450) reductase (CPR), was studied in *C. roseus* hairy roots subjected to different treatments applied on 24-day-old cultures. CPR activity was enhanced by all treatments, except zeatine, which had no effect. With fungus homogenate, macerozyme and P-free medium, modest increments of 50-75% over the control, were obtained. Higher increments (enhanced to 250%) were produced by ABA or 8% sucrose. Interestingly, no identical behavior was found with G10H activity, which increased either on P-free medium or on 8% sucrose. Although CPR activity increased 157% by adding ABA and 72% with the fungal homogenate, these treatments had no effect on G10H activity. Moreover, while zeatine produced no change in CPR activity, it inhibited G10H activity by 50%. However, as with CPR, the highest increment in G10H activity (200%) was observed with the 8% sucrose medium (Canto-Canché and Loyola-Vargas, 2000).

For several of the treatments, no concerted responses of either enzyme were observed, but in general, reductase was more responsive than G10H suggesting that in *C. roseus* CPR may have physiological functions other than the reduction of G10H. An alternative explanation for the "excess" of CPR activity would be its interaction with other classical P450s probably occurring in *C. roseus* roots.

10. METABOLIC PATHWAYS INVOLVED IN SIGNAL TRANSDUCTION

Plant cells are constantly receiving information from the external environment, to which they must react and respond. Signal transduction mechanisms have evolved to respond to these signals. Signal transduction research today contributes to all aspects of plant science, linking many fields of study in much the same way that signal transduction pathways mirror cellular processes.

Through their life cycle, plants and plant cells continually respond to signals and alter their physiology, morphology and development. Among the stimuli are, light, mineral nutrients, organic metabolites, gravity, water status, mechanical tension, heat, cold, growth regulators and hormones. Developmental age, as well as previous environmental experience, and the circadian clock modulate plant responses to stimuli. For growing cells, the response to these signals can be morphological and developmental.

Calcium is the second messenger par excellence in plant cells, regulating many different processes including development. One of the mechanisms that regulates calcium concentration involves the enzyme phospholipase C and the phosphoinositide pathway. Protein kinases also play crucial roles in the regulation of development. Genes encoding these enzymes represent close to 3 or 4% of the genome. At any time, a cell will be using hundreds of different protein kinases. The signaling pathways that utilize protein kinases and calcium together constitute a fine network.

Specific PLC enzyme activity in *C. roseus* hairy roots was 10 times higher in a membrane fraction than in cytosol. During a culture cycle, PLC activity in the cytosol reached a maximum that preceded the higher activity in the membranes. Both activities, soluble and membranal, are calcium-dependent for full activity. (De los Santos-Briones *et al.*, 1997) The effect of neomycin and several divalent cations was also analyzed (Piña-Chablé *et al.*, 1998).

Neomycin, an aminoglycoside antibiotic, inhibited PLC in a concentration-dependent fashion. The effect of different divalent cations such as Ni $^{2+}$, Cu $^{2+}$, and Zn $^{2+}$ were also studied. To test the effect of these cations, two conditions were selected: a) in the presence of and b) in the absence of calcium. In the presence of calcium, these three divalent cations were able to inhibit PLC activity in both fractions in a concentration-dependent manner.

Cells contains a group of compounds such as spermine, spermidine, and putrescine that posses several amine groups. These compounds are present in all eukaryotic cells, often in the millimolar concentration range. The increase in the concentrations observed in rapidly growing tissues has stimulated many investigations of these compounds. The specific functions of polyamines in plant systems remain unclear, but substantial amounts of experimental evidence suggest their involvement in a broad range of physiological processes including cell division, embryogenesis, stress phenomena and senescence. In *C. roseus* hairy roots (Piña-Chablé *et al.*, 1998), neomycin inhibited PLC in a concentration-dependent fashion. The authors propose that internal or external signals may regulate polyamine and micronutrient concentrations and this will be reflected in the regulation of PLC. How the primary metabolism can be regulating the levels of secondary metabolites and influence the production of alkaloids is something that needs further investigations.

Attempts have been made to link these two fields In *C. roseus* hairy roots cultures. The effect of macerozyme on accumulation of indole alkaloid and coumarine and on the induction of TDC, PAL and PLC activities on *C. roseus* hairy roots was investigated. Increasing concentrations of macerozyme induced an increase in indole alkaloid and coumarine accumulation. There was a 10-fold increase in TDC and a 4-fold increase in PAL activities in hairy roots treated with 1% of macerozyme. In a dose-response experiment PLC activity decreased 38% with 0.5% of macerozime and increased 40% with 1% macerozyme when compared to untreated roots. The results demonstrate that macerozyme treatment can affect the production of secondary metabolities and PLC activity in *C. roseus* hairy roots (Moreno-Valenzuela *et al.*, 1999b).

The superfamily of GTP-binding proteins consist of several members, including traslational factors, tubulins, and signal-transducing GTP-binding proteins. The GTP-binding proteins have the ability to efficiently bind and subsequently hydrolyze guanine nucleotides. In several cases, it was shown that GTP-binding proteins act as molecular signal transducers, whose active/inactive state depends on the binding of GTP or GDP, respectively. As stated before, plant cells respond to a variety of signals, both from the environment and from other cells. However, little is known about the mechanisms of signal transduction in plants. Among all the information cited above, very little is known regarding the presence and the role of the GTP-binding proteins in plant tissue culture and, more specifically, in transformed tissues. Biochemical analysis revealed the presence of GTP-binding proteins (G-proteins) in *C. roseus* hairy roots. In a microsomal fraction, several proteins, with the characteristics of G proteins were identified using antisera against the Ga-subunit (Suárez-Solis *et al.*, 1999). Regulators of G-proteins such as GTP-gS and sodium fluoride are able to regulate PLC activity.

Phosphorylation and dephosphorylation of proteins is the most important mechanism regulating cellular function in eukaryotic cells and integrating multicellular systems. A number of protein kinases and phosphatases have been identified and characterized in higher plants and several plant proteins have been found to be regulated by reversible phosphorylation. Most of the phosphorylation in both animal and plant cells occurs on serine on threonine residues and a smaller proportion on tyrosine residues (less than 0.1% of total phosphorylations). Phosphotyrosine is a minor component of animal cells, but cells stimulated to divide by certain factors may contain five- to ten-fold more phosphotyrosine than resting cells. Protein-tyrosine phosphorylation is important for the regulation of fundamental cellular processes such as proliferation, differentiation and oncogenic transformation. In *C. roseus* hairy roots, homogenate fractions (soluble and particulate) from transformed roots of *C. roseus* showed several phosphorylate proteins that gave positive signals with monoclonal antiphosphotyrosine antibodies (Rodríguez-Zapata and Hernández-Sotomayor, 1998b). Tyrosine kinase activity was also detected using an exogenous substrate specific to these enzymes. The presence of protein tyrosine phosphatase activity in the same system was also detected (Rodríguez-Zapata and Hernández-Sotomayor, 1998a). This is the first time that the presence of protein-tyrosine kinase and tyrosine-phosphatase activities have been reported in transformed plant tissues. The presence of protein tyrosine phosphatase activity was also detected (Rodríguez-Zapata and Hernández-Sotomayor, 1998a).

11. CONCLUSIONS AND FUTURE PROSPECTS

The application of plant biotechnology has some advantages. For instance, in the past few years the culture of transformed roots emerged as an alternative method to obtain different natural products. Different *C. roseus* hairy root lines have been obtained and a number of different strategies were followed to achieve high yields of indole alkaloids. Among these strategies are changes in nutritional factors, light, and elicitors. All of them can usually regulate growth as well as alkaloid productivity. The same factors regulate key enzymes and pathways involved in signal transduction. A challenge for future investigations is to find the link between the perception of the external signals and the induction of secondary metabolites. The complete metabolic pathway must be found to take advantage of metabolic engineering techniques for the biosynthesis of the monoterpen indole alkaloids.

REFERENCES

Bhadra R, Morgan JA and Shanks JV (1998) Transient studies of light-adapted cultures of hairy roots of *Catharanthus roseus*: growth and indole alkaloid accumulation. *Biotechnol. Bioeng.* **60** : 670-678

Bhadra R and Shanks JV (1997) Transient studies of nutrient uptake, growth, and indole alkaloid accumulation in heterotrophic cultures of hairy roots of *Catharanthus roseus*. *Biotechnol. Bioeng.* **55** : 527-534

Bhadra R, Vani S and Shanks JV (1993) Production of indole alkaloids by selected hairy root lines of *Catharanthus roseus*. *Biotechnol. Bioeng.* **41** : 581-592

Canto-Canché B and Loyola-Vargas VM (2000) Non-coordinated response of geraniol 10-hydroxylase and NADP:cyt C (P-450) reductase in *Catharanthus roseus* hairy roots under different conditions. *Phyton* **66** : 183-190

Ciau-Uitz R, Miranda-Ham ML, Coello-Coello J, Chí B, Pacheco LM and Loyola-Vargas VM (1994) Indole alkaloid production by transformed and non-transformed root cultures of *Catharanthus roseus*. *In Vitro Cell. Dev. Biol.* **30P** : 84-88

Davioud E, Kan C, Hamon J, Tempé J and Husson HP (1989a) Production of indole alkaloids by *in vitro* root cultures from *Catharanthus trichophyllus*. *Phytochem.* **28** : 2675-2680

Davioud E, Kan C, Quirion JC, Das BC and Husson HP (1989b) Epiallo-yohimbine derivatives isolated from *in vitro* hairy-root cultures of *Catharanthus trichophyllus*. *Phytochem.* **28** : 1383-1387

De los Santos-Briones C, Muñoz-Sánchez JA, Chín-Vera J, Loyola-Vargas VM and Hernández-Sotomayor SMT (1997) Phosphatidylinositol 4, 5-biphosphate-phospholipase C activity during the growing phase of *Catharanthus roseus* transformed roots. *J. Plant Physiol.* **150** : 707-713

De Luca V, Fernández AJ, Campbell D and Kurz WGW (1988) Developmental regulation of enzymes of indole alkaloid biosynthesis in *Catharanthus roseus*. *Plant Physiol.* **86** : 447-450

Flores HE, Vivanco JM and Loyola-Vargas VM (1999) "Radicle" biochemistry: the biology of root-specific metabolism. *Trends Plant Sci.* **4** : 220-226

Ho CH and Shanks JV (1992) Effects of initial medium pH on growth and metabolism of *Catharanthus roseus* hairy root cultures. A study with ^{31}P and ^{13}C NMR spectroscopy. *Biotechnol. Lett.* **14** : 959-964

Islas-Flores IR, Loyola-Vargas VM and Miranda-Ham ML (1994) Tryptophan decarboxylase activity in transformed roots from Catharanthus roseus and its relationship to tryptamine, ajmalicine, and catharanthine accumulation during the culture cycle. *In Vitro Cell. Dev. Biol.* **30P** : 81-83

Jung KH, Kwak SS, Choi CY and Liu JR (1994) Development of two stage culture process by optimization of inorganic salts for improving catharanthine production in hairy root cultures of *Catharanthus roseus*. *J. Ferment. Bioeng.* **77** : 57-61

Jung KH, Kwak SS, Choi CY and Liu JR (1995) An interchangeable system of hairy root and cell suspension cultures of *Catharanthus roseus* for indole alkaloid production. *Plant Cell Rep.* **15** : 51-54

Jung KH, Kwak SS, Kim SW, Lee H, Choi CY and Liu JR (1992) Improvement of catharanthine productivity in hairy root cultures of *C. roseus* by using monosaccharides as carbon source. *Biotechnol. Lett.* **14** : 695-700

Moreno-Valenzuela O, Coello-Coello J, Loyola-Vargas VM and Vázquez-Flota F (1999a) Nutrient consumption and alkaloid accumulation in a hairy root line of *Catharanthus roseus. Biotechnol. Lett.* **21** : 1017-1021

Moreno-Valenzuela OA, Monforte-González M, Muñoz-Sánchez JA, Méndez-Zeel M, Loyola-Vargas VM and Hernández-Sotomayor SMT (1999b) Effect of macerozyme on secondary metabolism plant product production and phospholipase C activity in *Catharanthus roseus* hairy roots. *J. Plant Physiol.* **155** : 447-452

Moreno-Valenzuela OA, Galaz-Avalos RM, Minero-García Y and Loyola-Vargas VM (1998) Role of differentiation on the regulation of indole alkaloid production in *Catharanthus roseus* hairy root. *Plant Cell Rep.* **18** : 99-104

Morgan JA and Shanks JV (1999) Inhibitor studies of tabersonine metabolism in *C-roseus* hairy roots. *Phytochem.* **51** : 61-68

Parr AJ, Peerless ACJ, Hamill JD, Walton NJ, Robins RJ and Rhodes MJC (1988) Alkaloid production by transformed root cultures of *Catharanthus roseus. Plant Cell Rep.* **7** : 309-312

Peraza-Sánchez SR, Gamboa-Angulo MM, Erosa-López C, Ramírez-Erosa I, Escalante-Erosa F, Peña-Rodríguez LM and Loyola-Vargas VM (1998) Production of 19(*S*)-epimisiline by hairy root cultures of *Catharanthus roseus. Nat. Prod. Lett.* **11** : 217-224

Piña-Chablé ML, De los Santos-Briones C, Muñoz-Sánchez JA, Echevarría-Machado I and Hernández-Sotomayor SMT (1998) Effect of different inhibitors on phospholipase C activity in *Catharanthus roseus* transformed roots. *Prostaglandins & other Lipid Mediators* **56** : 19-31

Rijhwani SK and Shanks JV (1998) Effect of elicitor dosage and exposure time on biosynthesis of indole alkaloids by *Catharanthus roseus* hairy root cultures. *Biotechnol. Progress* **14** : 442-449

Rodríguez-Zapata LC and Hernández-Sotomayor SMT (1998a) Detection of tyrosine phosphatase activity in *Catharanthus roseus* hairy roots. *Plant Physiol. Biochem.* **36** : 731-735

Rodríguez-Zapata LC and Hernández-Sotomayor SMT (1998b) Evidence of protein-tyrosine kinase activity in *Catharanthus roseus* roots transformed by *Agrobacterium rhizogenes. Planta* **204** : 77-84

Sáenz-Carbonell L, Maldonado-Mendoza IE, Moreno V, Ciau-Uitz R, López-Meyer M, Oropeza C and Loyola-Vargas VM (1993) Effect of the medium pH on the release of secondary metabolites from roots of *Datura stramonium, Catharanthus roseus* and *Tagetes Patula* cultured *in vitro. Appl. Biochem. Biotechnol.* **38** : 257-267

Schulze-Lefert P, Dangl JL, Becker-André M, Hahlbrock K and Schulz W (1989) Inducible *in vivo* DNA footprints define sequences necessary for UV light activation of the parsley chalcone synthase gene. *EMBO J.* **8** : 651-656

Sim SJ, Chang HN, Liu JR and Jung KH (1994) Production and secretion of indole alkaloids in hairy root cultures of *Catharanthus roseus*: effects of *in situ* adsorption, cell permeabilization. *J. Ferment. Bioeng.* **78** : 229-234

Suárez-Solis VM, Carrillo-Pech MR, Muñoz-Sánchez JA, Coria-Ortega R and Hernández Sotomayor SM (1999) Presence of guanine nucleotide-binding proteins in *Catharanthus roseus* transformed roots. *Physiol. Plant.* **105** : 593-599

Taylor WI and Farnsworth NR (1975) The *Catharanthus* Alkaloids. Marcel Dekker Inc., New York.

Toivonen L, Balsevich J and Kurz WGW (1989) Indole alkaloid production by hairy root cultures of *Catharanthus roseus. Plant Cell Tissue Org. Culture* **18** : 79-93.

Toivonen L, Laakso S and Rosenqvist H (1992) The effect of temperature on hairy root cultures of *Catharanthus roseus*: Growth, indole alkaloid accumulation and membrane lipid composition. *Plant Cell Rep.* **11** : 395-399

Toivonen L, Ojala M and Kauppinen V (1991) Studies on the optimization of growth and indole alkaloid production by hairy root cultures of *Catharanthus roseus. Biotechnol. Bioeng.* **37** : 673-680

Vázquez-Flota F, Moreno-Valenzuela OA, Miranda-Ham ML, Coello-Coello J and Loyola-Vargas VM (1994) Catharanthine and ajmalicine synthesis in *Catharanthus roseus* hairy root cultures. Medium optimization and elicitation. *Plant Cell Tissue Org. Culture* **38** : 273-279.

Chapter 14

GENETIC ENGINEERING FOR MODIFICATIONS OF LIGNIN

Smita Rastogi and UN Dwivedi*

Department of Biochemistry, Lucknow University, Lucknow – 226 007, India

Summary

Lignin is a complex phenolic heteropolymer resulting from the dehydrogenative polymerization of three types of cinnamyl alcohols, the monolignols (p-coumaryl, coniferyl, sinapyl alcohols) synthesized through phenylpropanoid pathway. Lignin is deposited in the secondary walls of specific cells as xylem-vessels, tracheids and fibers, at the last stage of their differentiation, where it reinforces and water-proofs the wall of these specialized cells. Lignin provides rigidity, resistance to pathogen attack, resistance to compressive stresses and also water impermeability to conductive tissues. Thus, though lignin is useful to plants, it is one of the greatest obstacles for the optimal utilization of biomass for pulp production in paper industry, nutrition of live-stock and new technologies concerned with biodegradation of ligno-cellulosic materials. Lignin with reduced levels or with high syringyl subunit composition can be efficiently extracted during the kraft pulping process, while forage digestibility and biodegradation processes are aided by reduced lignin content. Stimulation of lignification is advantageous for the use of plant material as fuel and as defence mechanism against microbial attack. Therefore, biotechnological alteration in lignin content and composition is desirable in both woody tree species and crop plants. During the period spanning around a decade, a number of gene clones pertaining to monolignol biosynthesis, its transport

*Corresponding author : E-mail : upendradwivedi@hotmail.com

and polymerization, have been characterized and used to alter lignin content and composition, through genetic engineering tools, in model plants as well as forest trees. Most of these studies, involving down-regulation or over-expression of genes pertaining to lignin biosynthesis, have resulted in alterations in corresponding enzymes and lignin content and composition.

The results already obtained through genetic manipulation strategies, offer promising perspectives for a range of biotechnological applications and provide new insights on key steps and regulation mechanisms in the lignification process. New strategies, to modulate lignin content or composition, relative to various biotechnological applications have to be defined in future to optimize plant biomass utilization.

Keywords : Lignin, monolignol biosynthesis, PAL, C4H, C3H, OMT, CCoAOMT, F5H, 4CL, CCR, CAD, CBG, POD, LAC, PO, genetic manipulation of lignin, Antisense and sense expression, transgenic plants, pulp and paper industry, forage digestibility.

Abbreviations : C4H : Cinnamate 4-hydroxylase; C3H : Cinnamate 3-hydroxylase; CCoAOMT : Caffeoyl coenzyme A 3-O-methyltransferase; 4CL : 4-coumarate: coenzyme A ligase; CCR : Cinnamoyl coenzyme A reductase; CAD : Cinnamyl alcohol dehydrogenase; CBG : Coniferin β-glucosidase; CAO : Coniferin alcohol oxidase; F5H : Ferulate 5-hydroxylase; G : Guaiacyl unit; H : Hydroxyphenyl unit; LAC : Laccase; OMT : O-methyltransferase; PAL : Phenylalanine ammonia-lyase; POD : Peroxidase; PO : Phenoloxidase; S : Syringyl unit; S/G : Syringyl : Guaiacyl; UDPG-GT : Uridine diphosphate D-glucose coniferin glucosyltransferase.

1. INTRODUCTION

Lignin forms an integral cell wall component of all vascular plants and constitutes the second most abundant organic constituent on earth after cellulose, representing on an average of 25% of the terrestrial plant biomass (Sarkanen *et al.*, 1967). Chemically, it is a complex phenolic heteropolymer synthesized from the dehydrogenative polymerization of monolignols, namely coumaryl, coniferyl and sinapyl alcohols. It plays a fundamental role in conferring rigidity, strength, resistance to pathogen attack and water impermeability to the polysaccharide-protein matrix of the cell wall allowing solute conductance (Brown, 1985). From the functional point of view, lignin is predominantly synthesized and deposited in the secondary cell wall of certain specialized cells such as xylem vessels, tracheids and fibers. Lignin is also deposited in minor amounts in the periderm where in association with suberin, it provides a protective role. Despite its biological importance in plants, lignin composition, quantity and distribution is known to affect the agro-industrial utilization of plant biomass. It is one of the greatest obstacles for the optimal utilization of biomass for pulp production in paper industry, nutrition of live-stock and biodegradation of ligno-cellulosic materials. The presence of residual lignin results in pulp and paper with inferior performance characteristics and poor brightness stability or yellowing with age. During paper pulping, lignin is removed from wood by chemical- and energy- intensive processes, which are not eco-friendly as it releases toxic pollutants. The quantity of lignin in forage crop is also negatively correlated to digestibility and appetance. Furthermore, the resistance of lignin to microbial degradation enhances persistence of plant biomass in soil leading to decreased biodegradation. On the other hand, in parallel with the obtention of plant materials with reduced amounts of lignin, hyper-lignified woody materials are also desirable for their use as fuel, as a renewable energy source, or to enhance defence reaction of plant against pathogens. For these industrial and agricultural perspectives, it is desirable to develop improved plant varieties with altered lignin content and composition (Whetten and Sederoff, 1991; Sederoff *et al.*, 1994; Boudet *et al.*, 1995; Boerjan *et al.*, 1996; Boudet and Grima-Pettenati, 1996; Boudet *et al.*, 1997; Baucher *et al.*, 1998; Taochim *et al.*, 1998; Grima-Pettenati and Goffner, 1999; Boudet, 1998, 2000).

Plant materials exhibit a great heterogeneity with regards to lignin content and composition. Lignin subunit composition and overall quantity in a cell may vary depending on the location in the wall, cytological origin, conditions of growth, developmental state of the cell and tissue and the influence of environmental stresses. Natural variations due to taxonomic differences, developmental changes, mutation, responses to pathogen attack, herbicides, light and hormones are quite common (Grand *et al.*, 1982; Campbell and Sederoff, 1996). This variation in lignin content and composition suggests that human induced changes in lignin content and composition through genetic engineering approaches is a realistic possibility. The biotechnological manipulation of plant metabolism related to lignin biosynthesis, requires the identification of rate determining steps as targets for modulation by gene transfer, exemplified by recent attempts to suppress lignification in transgenic plants. With the availability of a number of cDNA clones pertaining to lignin biosynthesis (Table 1), extensive research efforts have been done in the last 10 years or so towards genetic engineering of lignin which has mainly focused on the reduction of lignin content and/or alteration of its composition.

In this chapter, we have summarized the results and discussed the implications of targeting various genes for altering lignin content and composition, especially in the areas of pulp and paper industry and forage digestibility. We have also discussed the future approaches for genetic engineering of lignin in the light of these findings.

Table 1. c-DNA clones available for genetic manipulation of enzymes involved in lignin biosynthesis.

Enzyme	Source
PAL	*Arabidopsis*, Bean, Loblolly pine, Parsley, Pea, Poplar, Rice, Sweet Potato, Tobacco.
C4H	Alfalfa, Jerusalem artichoke, Mung-bean.
OMT	Alfalfa, Aspen, Barley, Eucalyptus, Ice plant, Maize, Poplar, Rye grass, *Stylosanthes humilis*, Sugarcane, Tobacco, *Vanilla planifolia*, Zinnia.
CCoAOMT	Aspen, Loblolly pine, Parsley, *Stellaria longipes*, Tobacco, *Vitis vinifera*, Zinnia.
F5H	*Arabidopsis*.
4-CL	*Arabidopsis*, *Lithospermum*, Loblolly pine, Parsley, Poplar, Potato, Rice, Soyabean, Tobacco, *Vanilla planifolia*.
CCR	Eucalyptus, Maize, Poplar, Sugarcane, Tall fescue, Tobacco.
CAD	Alfalfa, *Aralia cordata*, Bean, Eucalyptus, Loblolly pine, Lucerne, Maize, Poplar, Spruce, Sugarcane, Tobacco.
CBG	Pinus.
POD	Barley, Cucumber, Maize, Peanut, Potato, Rice, Sweet potato, Tobacco, Tomato, Wheat, Zucchini.
LAC	Loblolly pine, Sycamore maple, Tobacco, Yellow poplar.

2. LIGNIN BIOSYNTHESIS PATHWAY

Lignin is a polymer of three alcohol monomers or monolignols: p-coumaryl alcohol, coniferyl alcohol and sinapyl alcohol. These monolignols are synthesized through the phenylpropanoid pathway as shown in Fig. 1. This pathway converts phenylalanine (phe) to one of the three monolignols. The first step is deamination of phe by phenylalanine ammonia-lyase (PAL). The product of PAL activity on phe is trans-cinnamic acid, which is the substrate for a P450 hydroxylase, cinnamate 4-hydroxylase (C4H). C4H hydroxylates the aromatic ring of cinnamate to produce p-coumaric acid. In the next step p-coumaric acid is hydroxylated at the 3-position by p-coumarate 3-hydroxylase (C3H). According to an alternative hypothesis, p-coumaric acid is converted to p-coumaryl CoA by 4-coumarate: coenzyme A ligase (4CL) and hydroxylated at the level of the CoA thioester, to produce caffeoyl CoA. The traditional pathway holds that caffeic acid is methylated in the next step, at 3-hydroxyl to produce ferulic acid by caffeic acid O-methyltransferase (OMT). More recently, a caffeoyl coenzyme A 3-O-methyltransferase (CCoAOMT) enzyme has been hypothesized to play a role in monolignol biosynthesis. Ferulic acid produced, can be activated by 4-coumarate: coenzyme A ligase (4CL) through conjugation to CoA, as can p-coumarate, and these CoA thioesters are presumed intermediates in the synthesis of coniferyl alcohol and p-coumaryl alcohol, respectively. An alternative fate for ferulic acid is hydroxylation at the 5-position by another P 450 hydroxylase, ferulate 5-hydroxylase (F5H). 5-hydroxyl ferulic acid can be methylated by OMT to produce sinapic acid. The CoA thioesters of p-coumarate, ferulate and sinapate are considered to be substrates for cinnamoyl coenzyme A reductase (CCR) which reduces these thioesters to yield the corresponding aldehydes. The aldehydes, in turn, are substrates for cinnamyl alcohol dehydrogenase (CAD) which reduces the aldehyde end group to an alcohol to yield the three monolignols. After their synthesis in the cytoplasm these monolignols are transported outside the cell as glucoside derivatives where they are hydrolysed by β-glucosidase and finally polymerized through oxidative coupling to give rise to lignin polymer. Peroxidases, laccases and phenoloxidases are reported to carry out these polymerization reactions. Monolignols can also form bonds with other cell wall polymers in addition to lignin, cross-linking polysaccharides and proteins. Lignin biosynthesis has been extensively reviewed in the past few years (Higuchi, 1985; Lewis and Yamamoto, 1990; Whetten and Sederoff, 1995; Douglas, 1996; Whetten *et al.*, 1998; Weisshar and Jenkins, 1998).

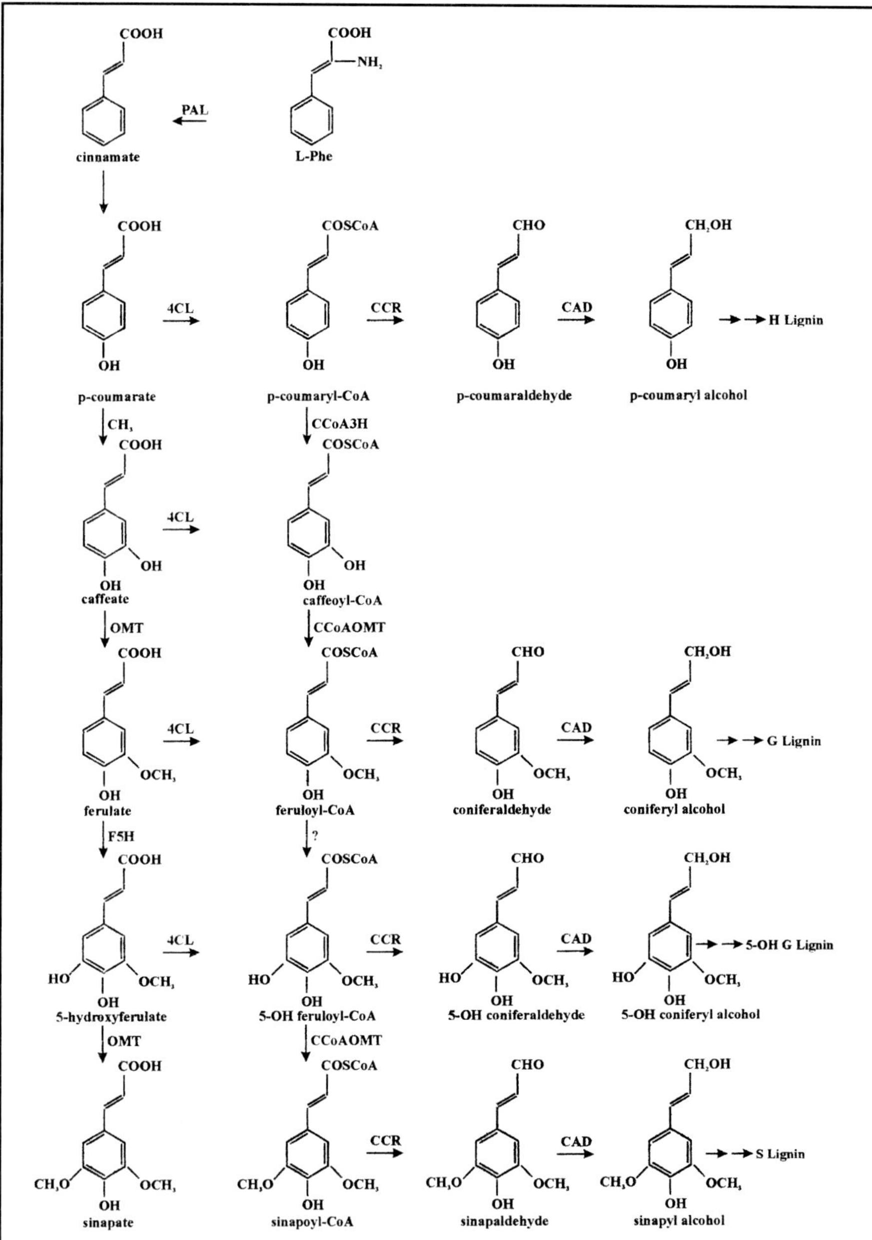

Figure 1 : An outline of monolignol biosynthesis. PAL, phenylalanine ammonia-lyase; C4H, cinnamate 4-hydroxylase; OMT, O-methyl transferase; CCoAOMT, caffeoyl-coenzyme A O-methyltransferase; F5H, ferulate 5-hydroxylase; 4CL, 4-coumarate: coenzyme A ligase; CCR, cinnamoyl coenzyme A reductase; CAD, cinnamyl alcohol dehydrogenase; H, hydroxy phenyl; G, guaiacyl; 5-OH G, 5-hydroxy guaiacyl; S, sinapyl.

Lignin content and composition have long been known to vary between the major groups of higher plants and between species. The two major classes of lignin are (a) gymnosperm lignin, which primarily contains guaiacyl subunits (G units) polymerized from coniferyl alcohol, and a small proportion of p-hydroxyphenyl units (H units) polymerized from p-coumaryl alcohol; (b) angiosperm lignin, which contains syringyl units (S units) polymerized from sinapyl alcohol, and G units with small proportions of H units. Exceptions to this basic classification are found in both gymnosperms (*e.g., Podocarpus* sp., certain species of Gnetales as *Ephedra trifurca*) (Sarkanen, 1971) and angiosperms (*e.g., Erythrina cristagalli*) (Kutsuki and Higuchi, 1978). Some non-conventional lignin subunits (novel lignin subunits) as substituted cinnamaldehyde, benzaldehyde, 2-methoxy benzaldehyde, dihydro-cinnamyl alcohol (amyl propanols) and cinnamyl alcohol ester have also been reported in grasses, kenaf, pine and tobacco (Whetten *et al.*, 1998). Lignin with high proportion of syringyl (S) units is more efficiently extracted during the kraft pulping process. Though, the structure and biosynthesis of lignin have been studied for more than 100 years, the structure of lignin is not yet completely established.

3. GENETIC ENGINEERING OF LIGNIN

Since the overall rate of lignin deposition is regulated not only by the monolignol biosynthesis but also by the co-ordinated transport, storage, mobilization and polymerization of monolignol precursors to the wall, conceptually, the genetic engineering of lignin can be accomplished at three levels of control, namely, synthesis of monolignols and its transport and polymerization at the site of deposition, as depicted in Fig. 2.

Out of these three levels of control at which lignin content can be manipulated, so far most of the studies have been at the level of monolignol biosynthesis and polymerization while the monolignol transport is least touched.

3.1. Genetic engineering of lignin at the level of monolignol biosynthesis

3.1.1. Phenylalanine ammonia-lyase (PAL) (EC 4.3.1.5)

PAL catalyzes the first step in the synthesis of a diverse array of plant natural products based on the phenylpropane skeleton (Jones 1984). PAL mRNA and enzyme levels are highly regulated spatially and temporally, associated with the tissue specific accumulation of

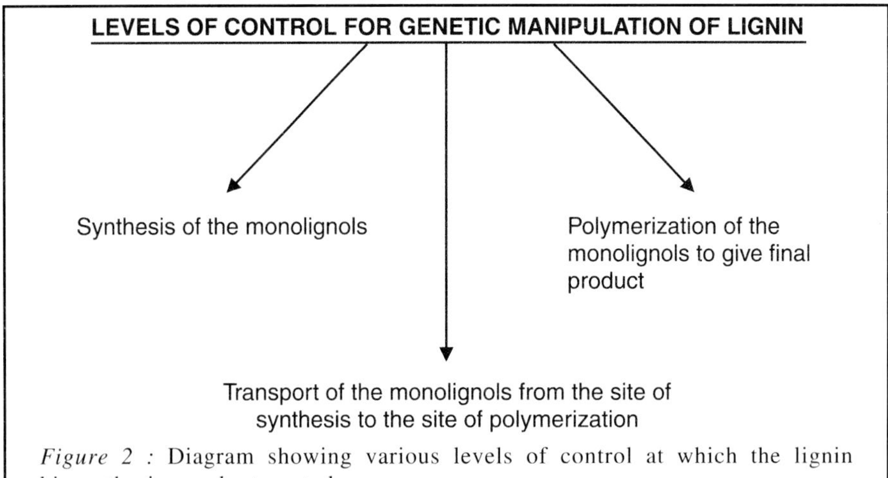

Figure 2 : Diagram showing various levels of control at which the lignin biosynthesis can be targeted.

phenylpropanoid products, exemplified by selective expression in differentiating xylem cells for production of lignin monomers at the onset of secondary wall deposition. Distinct isoforms of the PAL enzyme subunit are encoded by a small family of differentially regulated genes (Boudet *et al.*, 1995). cDNAs encoding for PAL have been characterized from sweet potato (Tanaka *et al.*, 1989), bean (Cramer *et al.*, 1989), rice (Minami *et al.*, 1989), parsley (Lois *et al.*, 1989), loblolly pine (Whetten and Sederoff, 1992), pea (Yamada *et al.*, 1992), poplar (Subramaniam *et al.*, 1993), tobacco (Nagai *et al.*, 1994) and *Arabidopsis* (Ohl *et al.*, 1990; Wanner *et al.*, 1995). The PAL cDNA sequence from gymnosperms shows considerable similarity to pal genes isolated from herbaceous angiosperms.

Transgenic tobacco transformed with a heterologous PAL 2 gene from bean exhibited a significant reduction of PAL activity concomitant with slight reduction in lignin content corresponding to co-suppression events (Elkind *et al.*, 1990; Bate *et al.*, 1994). These were the first genetically engineered plants to exhibit a reduction in lignin content, though unusual phenotypes with localized fluorescent lesions, altered leaf shape and texture, stunted growth, reduced pollen viability and altered flower morphology and pigmentation were also observed (Elkind *et al.*, 1990). In a separate study on the subsequent generations of these transgenic plants involving isogenic lines, a reduction in lignin content of <10% (thioglycolate extractable lignin) as compared to that of wild type was shown with severe suppression of PAL activity (Bate *et al.*, 1994). Because PAL is at the entry point of the phenylpropanoid metabolism, the suppression of PAL alters biosynthesis of diverse phenolic compounds, which are important for plant growth and survival, and hence leads to severe phenotypic changes. The tobacco plants with suppressed PAL activities have also been shown to exhibit increased disease susceptibility owing to correspondingly lower levels of chlorogenic acid (Maher *et al.*, 1994).

The paradoxical suppression of PAL activity, however, progressively reversed in succeeding generations homozygous for the PAL 2 transgene. Transgenic tobacco plants raised from seeds of a primary transformant containing the bean PAL 2 gene, which had shown homology dependent silencing of the endogenous tobacco PAL genes (in primary transformants), exhibited over-expression of the PAL (Howles *et al.*, 1996). Further analysis of these transgenic plants revealed a high level of expression of the bean PAL 2 transcript with absence of corresponding proteins. A rapid post-translational degradation or defects in polysome recruitment or translational elongation/initiation (Seymour *et al.*, 1993) has been postulated for these observations. Recently, it has been shown that in transgenic tobacco, a reduction in PAL activity to approximately 15% of wild type levels gave an approximately two-fold decrease in Klason lignin as a percentage of dry matter. Analysis of lignin methoxyl content and pyrolysis suggested an increased S/G ratio indicative of a reduction in G lignin (Sewalt *et al.*, 1997).

3.1.2. Cinnamate 4-hydroxylase (C4H) (EC 1.14.13.11)

C4H is a cytochrome P450 dependent enzyme. cDNA clones encoding C4H have been reported from mung bean (Mizutani *et al.*, 1993), Jerusalem artichoke (Teutsch *et al.*, 1993) and alfalfa (Fahrendorf and Dixon, 1993). The antisense and sense C4H in transgenic tobacco employing heterologous C4H gene from alfalfa led to reduced levels of Klason lignin accompanied by decreased S/G monomer ratio (Sewalt *et al.*, 1997). This observation was opposite to that obtained in case of PAL down-regulation, where an increase in S/G ratio was observed (Sewalt *et al.*, 1997). Recently, Blount *et al.* (2000) generated transgenic tobacco lines with altered level of C4H by sense or antisense expression of an alfalfa C4H cDNA. The altered expression of C4H led to reduction in PAL activity in leaves and stems of plants which implicated that trans-cinnamic acid is a feedback modulator of PAL.

3.1.3. Cinnamate 3-hydroxylase (C3H)

C3H catalyzes the hydroxylation of p-coumaric acid to caffeic acid. cDNA clone for this enzyme has not yet been reported.

3.1.4. O-Methyltransferase (OMT) (EC 2.1.1.68)

Angiosperm dicot OMTs are bispecific enzymes catalyzing the methylation of caffeic acid and 5- hydroxy ferulic acid to form ferulic acid and sinapic acid, respectively. Gymnosperm OMTs are monospecific and catalyze the methylation of caffeic acid only. The lack of efficient methylation of 5-hydroxy ferulic acid catalyzed by gymnosperm OMTs suggests one reason for the absence of S lignin in gymnosperms. Tobacco has been shown to possess three classes of OMTs differing in substrate specificities and expression patterns. Class I OMT is highly expressed in lignified tissues whereas class II and III OMTs are barely detectable in healthy plants but are strongly induced upon infection (Boudet *et al.*, 1995). cDNA clones for OMT have been isolated from aspen (Bugos *et al.*, 1991), alfalfa (Gowri *et al.*, 1991), poplar (Dumas *et al.*, 1992), maize (Collazo *et al.*, 1992), ice plant (Vernon and Bohnert, 1992), baerly (Gregersen *et al.*, 1994), eucalyptus (Poedomenge *et al.*, 1994), *Stylosanthes humilis* (McIntyre *et al.*, 1995), zinnia (Ye and Varner, 1995), tobacco (Jaeck *et al.*, 1996), ryegrass (Heath *et al.*, 1998), Vanilla planifolia (Xue and Brodelius, 1998) and sugarcane (Selman-Housein *et al.*, 1999). A high degree of homology has been observed in the sequence of all dicot OMTs, though the homology was lower for monocot and dicot OMTs. The high homology of cDNA of angiosperm dicot OMTs with each other at the nucleotide and amino acid levels has raised the possibility of using heterologous antisense constructs to alter lignin precursor biosynthesis. Thus, an OMT antisense gene construct was stably integrated into the tobacco genome and inherited in transgenic plants with a normal phenotype (Dwivedi *et al.*, 1994). BiOMT activity of the stems was decreased by an average of 29% in the four transgenic plants analyzed. Chemical analyses of woody tissue of stems indicated a reduced content of S unit while G unit remained unchanged in most of the transgenic plants, leading to decreased S/G ratio. However, the lignin content decreased only slightly (12%). Transgenic tobacco with a heterologous OMT construct from alfalfa in antisense orientation has also been obtained (Ni *et al.*, 1994). The constitutive expression of the transgene led to a 20% residual OMT activity and a significant reduction in the lignin content without apparent alterations in lignin monomer composition. Antisense inhibition and co-suppression of OMT along with reduction in lignin content has also been demonstrated in *Arabidopsis* using a heterologous gene construct for aspen bi-OMT (Dwivedi and Campbell, 1995).

Sense or antisense expression of sequences encoding homologous OMT has also been shown to modulate enzyme activity in tobacco (Atanassova *et al.*, 1995). The full-length sequence was found to be active in both orientations. The sense construct gave rise to plants with an increased OMT activity ranging from 170 to 370% of the mean value with no change in lignin content and composition. On the other hand, the expression of full-length sequence in antisense orientation led to a significant reduction in OMT activity. In the other transformants, with only a fragment of OMT gene, expressed in either orientation, a depletion in OMT activity was achieved. The inhibition in OMT activity, however, sustained through the adult stage only in plants transformed with the complete cDNA. Not much variation in lignin content was achieved, however, the S content of lignin was markedly decreased in proportion to decrease in OMT activity. G content either remained unchanged or showed a slight increase in few cases. Thus, a spectacular decrease in S/G ratio (upto 0.07) was achieved in the most inhibited transgenics, which was one-tenth of the control as well as that achieved by Dwivedi *et al.* (1994). Moreover, the thioacidolysis method demonstrated the accumulation of 5-hydroxy guaiacyl (5-OH G) units while this component was not detected with the alkaline nitrobenzene oxidation procedure used in earlier studies (Dwivedi *et al.*, 1994; Ni *et al.*, 1994). The presence of this unusual monomeric compound (5-OH G) was also determined in the OMT down-regulated poplar (*P. tremula* × *P. alba*) with an antisense construct encoding OMT from poplar (*P. trichocarpa* × *P. deltoides*) (van Doorssalaere *et al.*, 1995). A strong inhibition of OMT activity (less than 5% residual activity) was attained with an antisense fragment (Van Doorssalaere *et al.*, 1995) while a similar reduction of OMT in tobacco was achieved by a full-length antisense construct (Atanassova *et al.*, 1995). In both the studies, the total amount of lignin assessed by different methods was found to be unchanged even in transgenic plants where OMT was dramatically depleted. In these transgenic poplar plants, the analysis of lignin monomeric composition revealed a decrease in the S/G ratio, which was mainly due to lower S content as compared to that in control. These data suggest that caffeic and 5-hydroxy ferulic acids are methylated to varying degree in angiosperms by a bifunctional OMT. This was the first report on modulation of OMT activity leading to discoloration of xylem in transgenic plants (Van Doorssalaere *et al.*, 1995). Homologous sense suppression of OMT in the xylem of quaking aspen has been demonstrated recently in transgenic plants exhibiting novel phenotypes with either mottled or complete red-brown coloration in their woody stem (Tsai *et al.*, 1998). The analysis of OMT activity and transcript level in xylem, leaves and sclerenchyma suggested that the introduction of a homologous sense transgene could result in an ectopic expression of the transgene in tissues with insignificant endogene expression (*e.g.*, leaves, sclerenchyma) and a co-suppression of endogene and transgene in tissues where the endogene is highly expressed (xylem). Besides Tsai *et al.* (1998), none of the studies on OMT has yielded unequivocal evidence of red coloration in lignified tissues, characteristic of maize bm3 mutants (Vignols *et al.*, 1995). This was found to be attributable to the accumulation of coniferyl aldehyde residues into the lignin.

OMT has been suggested to play an important role in mediating the synthesis of S lignin (Douglas, 1996; Li *et al.*, 1997) and this enzyme together with CAD is found to be responsible for the low lignin content associated with maize and *Sorghum* brown midrib (bm) mutants that exhibit red coloration in their lignified tissues (Cherney *et al.*, 1991; Vignols *et al.*, 1995). Possibly this coloration was due to the incorporation of non-conventional moieties into the lignin polymer (*e.g.*, 5-OH G) or the formation of unusual bonding patterns resulting either directly or indirectly from a reduced OMT activity or due to accumulation of coniferyl aldehyde residues into the lignin.

Recently, transgenic tobacco with antisense suppression of OMT has been produced (Zhong *et al.*, 1998). The results were in accordance with that of Atanassova *et al.* (1995) and Van Doorssalaere *et al.* (1995) depicting a decrease in S/G ratio owing to decreased S only and no change in lignin content. A reduction of 90-95% OMT activity leading to increased G lignin and no change in lignin content has been demonstrated recently in antisense suppressed poplar (*P. tremula* × *P. alba*) (Lapierre *et al.*, 1999). Transgenic tobacco plants with an antisense expression of a heterologous OMT gene construct from *Stylosanthes humilis* have been recovered recently (Talas *et al.*, 2000). The amount of lignin in one of the transgenic lines decreased by 62% compared with untransformed control plant. Recently, transgenic poplars (*P. tremula* × *P. alba*) were obtained by introduction of sense homologous OMT transgene under the control of CaMV double 35S or eucalyptus CAD promoters (Jouanin *et al.*, 2000). Moderate over-expression of OMT was found in some lines, while negligible (less than 3%) OMT was achieved in a transgenic line with CaMV 35S[2] promoter. Lignin content in this line was also substantially reduced (17% decrease), which was in contrast to the earlier reports on transgenic poplar trees (Lapierre *et al.*, 1999). Lignin composition was also severely altered with an almost absence of S units, incorporation of 5OH-G units and higher content in condensed bonds. Xylem showed brownish coloration similar to that in transgenic aspen (Tsai *et al.*, 1998).

All OMT related studies, so far, have revealed that the expression of an antisense OMT transgene could influence the lignin composition with a concomitant appearance of 5-OH G lignin moieties in few cases, with not much decrease in lignin content. The differences in results, however, were a function of the species, gene construct size, sense or antisense suppression, homologous or heterologous source of OMT, degree of down-regulation of OMT activity, degree of sequence homology and endogene expression and the method of analysis. Despite these discrepancies, OMT may be considered a good target for changing the lignin composition to increase the calorific value of wood for its use as fuel.

3.1.5. Caffeoyl coenzyme A O-methyltransferase (CCoAOMT) (EC 2.1.1.104)

CCoAOMT catalyzes the methylation of caffeoyl-CoA to feruloyl-CoA and 5-hydroxy feruloyl-CoA to sinapoyl-CoA (Ye *et al.*, 1994; Ye, 1997; Zhong *et al.*, 1998) and has been shown recently to be involved in an alternative methylation pathway of lignin biosynthesis. Recently, it has been shown that the expression of CCoAOMT is closely associated with lignifying tissues in a number of dicot species, including zinnia, forsythia, tobacco, alfalfa, soybean and tomato (Zhong *et al.*, 1998). Since the first cloning of CCoAOMT cDNA from parsley (Schmitt *et al.*, 1991), CCoAOMT cDNAs have also been isolated from a number of species such as Zinnia (Ye *et al.*, 1994), aspen (Meng and Campbell, 1995), *Stellaria longipes* (Zhang *et al.*, 1995), *Vitis vinifera* (Busam *et al.*, 1997), tobacco (Martz *et al.*, 1998) and loblolly pine (Li *et al.*, 1999). Recently, a cDNA clone encoding a bifunctional methyltransferase, named AEOMT has been characterized from a gymnosperm, which shares the common features with OMT and CCoAOMT (Li *et al.*, 1997). Genomic DNA encoding CCoAOMT of citrus has also been cloned recently (Grycuik and Tsuyumu, 2000). Transgenic tobacco plants with antisense gene construct of CCoAOMT have been developed (Zhang *et al.*, 1995). Transgenic plants with a simultaneous reduction in both CCoAOMT and OMT activities have also been generated (Zhang *et al.*,

1995). It seems that in transgenic plants with suppression of OMT, CCoAOMT efficiently compensates for the loss of OMT to synthesize normal levels of Klason lignin so that overall lignin content remained unaltered. Furthermore, the antisense suppression of OMT alone resulted in nearly complete blockage of S lignin production with no change in G lignin. On the other hand, in case of CCoAOMT suppression, a reduction in lignin content as well as reduction in both G and S lignin monomers was observed (Zhang *et al.*, 1995). The magnitude of decrease of lignin content was found to increase when OMT and CCoAOMT were expressed simultaneously. Thus, antisense suppression of CCoAOMT alone led to decrease in lignin content to 53-67% of wild type, increased S/G ratio with a preferential decrease in G, whereas simultaneous suppression of CCoAOMT and OMT led to reduction upto 34-42% of that of the wild type and a decreased S/G ratio. These results were an unequivocal demonstration for the involvement of both CCoAOMT and OMT in methylation reactions. The transgenic plants resulting from antisense suppression of either CCoAOMT and OMT or CCoAOMT alone bore normal flowers and grew to heights similar to wild type plants though showed deformed vessel walls. The phenotype exhibited by CCoAOMT antisense plants was quite different from PAL suppressed plants (Elkind *et al.*, 1990). Because PAL is at the entry point of phenylpropanoid pathway, biosynthesis of diverse phenolic compounds is also altered upon suppression of PAL. In contrast, the block of CCoAOMT does not affect the biosynthesis of important compounds involved in defence, such as flavanoid, chlorogenic acid and salicylic acid, thus indicating that block of CCoAOMT may be more specific to monolignol biosynthesis. The residual level of monolignols produced in transgenic plants may be sufficient to sustain normal plant growth. Because no abnormal visible phenotype was associated with substantial reduction in lignin in the antisense plants, it is anticipated that CCoAOMT alone as well as a combination of CCoAOMT and OMT may be ideal targets for manipulation of lignin content and composition in trees and forage crops. Furthermore, the preferential reduction of G lignin over S lignin, leading to higher S/G ratio of CCoAOMT antisense plants, presents additional benefits to the pulping process because lignin with higher S/G ratio requires low chemicals for the kraft pulping process (Chiang *et al.*, 1988). Recently, repression of CCoAOMT expression upto 70% has been achieved by the antisense approach in transgenic poplar (Zhong *et al.*, 2000). The lignin content was also reduced to 60% of that of the wild type. The reduction in lignin content was the result of a decrease in both G and S units leading to less condensed and less cross-linked lignin structure in wood. This decrease in lignin content, however, led to no significant effect on plant growth and morphology, except for minor changes in vessel wall shapes. Wood displayed a light orange color and an elevation of p-hydroxybenzoic acid. In another study, a down-regulation of CCoAOMT by 90% in poplar (*P. tremula* × *P. alba*) xylem suppressed the amount of lignin by 12% (Meyermans *et al.*, 2000). This depletion in lignin was due to reduced synthesis of both G and S monomers with an increased (11%) S/G ratio. p-hydroxy benzoate level was also increased. Stem xylem of these CCoAOMT down-regulated lines showed a pink red coloration. These two studies indicate a depletion of both G and S monomers which contradicts the idea that the methylation step accomplished by CCoAOMT is specific for G monomer synthesis as suggested earlier (Ye *et al.*, 1994; Van Doorssalaere *et al.*, 1995).

3.1.6. Ferulate 5-hydroxylase (F5H)

An *Arabidopsis thaliana fah1* mutant having a lesion in the F5H enzyme, leading to a deficiency in the synthesis of sinapate esters or sinapic acid derived lignin has been characterized (Chapple *et al.*, 1992). This mutant was phenotypically normal and was unable to produce sinapyl alcohol and thus contained only G lignin (Meyer *et al.*, 1996). This suggested that feruloyl-CoA and its reduced derivatives are not efficiently diverted into sinapyl alcohol biosynthesis. The F5H cDNA has been cloned via an allelic TDNA tagged mutant. A genetically induced methoxylation at C-5 would reduce the resistant lignin interunit bonds and would thereby facilitate the pulping of gymnosperm wood. In angiosperms, it might be possible to produce high syringyl lignin, through over-expression of F5H and of a specific 5-OH ferulic acid OMT. Transgenic *Arabidopsis* plants with over-expressed F5H have been produced recently (Meyer *et al.*, 1997; Meyer *et al.*, 1998). This over-expression of F5H has led to predominant production of S lignin with little G lignin accumulating. Moreover, the tissue specificity of S lignin deposition was abolished and S lignin was no longer restricted to interfascicular, sclerified parenchyma cells but were also associated with vascular bundles. Thus, F5H expression determines lignin monomeric composition as well as tissue specificity of S lignin deposition. To determine whether F5H has a similar regulatory role in plants that undergo secondary growth, Rochus *et al.* (2000) over-expressed F5H gene in tobacco and poplar under the control of CaMV 35S and C4H promoters. An increase in lignin S monomer content in petioles with no detectable effect on lignification in stems was observed with CaMV 35S promoter, while a significant increase in S monomer content in stem was observed with C4H promoter. In C4H-F5H lines, a reduced deposition of lignin and an alteration of hydroxy cinnamyl aldehydes content were observed as compared to wild type. The results indicated that hydroxylation of G substituted lignin precursors control lignin monomer composition in woody plants. Taken together, these studies suggest that F5H could provide an excellent control point for genetically engineered major changes in lignin quality through introduction of 5-OH groups to lignin monomers especially in softwoods (gymnosperm), which are lacking in F5H.

3.1.7. 4-Coumarate : coenzyme A ligase (4CL) (EC 6.2.1.12)

The coenzyme-A esters of 4-coumarate and other cinnamate derivatives formed as a result of 4CL action are central intermediates in the synthesis of many phenylpropanoid derivatives in higher plants. Several isoforms of 4CL have been characterized in plants (Boudet *et al.*, 1995). The genes for 4CL have been isolated from potato (Lozoya *et al.*, 1989), rice (Zhao *et al.*, 1990), parsley (Becker-Andre *et al.*, 1991), tobacco (Hauffe *et al.*, 1991), soybean (Uhlmann and Ebel, 1993), poplar (Allina and Douglas, 1994), *Arabidopsis* (Lee *et al.*, 1995), *Lithospermum* (Yazaki *et al.*, 1995), loblolly pine (Zhang *et al.*, 1997) and *Vanilla planifolia* (Brodelius and Xue, 1997). The first transgenic tobacco plants with reduced 4CL activity were reported by Kajita *et al.* (1996) using chimeric sense and antisense gene constructs for 4CL. In the transgenic plants, the cell walls of the xylem tissue in stems were brown rather than pale white as in control. Analysis of the lignin content and composition revealed that the level of total aldehydes from brown tissue was one-third of that of normal tissue, level of lignin was 65% of that of control, S/V ratio was greater, while cinnamaldehyde residues, S and G, showed depressed levels than that of normal colored tissue. The reduction in the level of total aldehydes detected by oxidation analysis suggested the presence of a relatively high level of condensed units, rich in C-C linkages in brown tissue. According to Monties (1989), a low proportion of S units in lignin leads to a condensed lignin polymer. Thus, the low level of total aldehydes detected, suggest a reduced proportion of S units in the tissue. In an extension of the above study, structural characterization of transgenic tobacco with depressed 4CL revealed the presence of a novel lignin in the xylem (Kajita *et al.*, 1997). In the brownish tissue, the levels of 3-hydroxy cinnamic acids, p-coumaric, ferulic and sinapic acids bound to cell walls were apparently increased as a result of down-regulation of the expression of the gene for 4CL which led to an increase in the level of condensed units. Transgenic *Arabidopsis* lines having as low as 8% residual 4CL activity have also been generated (Lee *et al.*, 1997). These plants with reduced 4CL activity were morphologically normal but had decreased thioglycolic acid extractable lignin (upto 50%). Statistically significant decreases in lignin content were not attained until residual activity of 4CL decreased to >20% of wild type levels. This was quite similar to reduction in PAL activity which became rate limiting for lignin accumulation when enzyme activity was decreased by 75-80% relative to wild type (Bate *et al.*, 1994). There was a significant decrease in G units in 4CL suppressed *Arabidopsis*, however the levels of S units remained unchanged leading to an increase in S/G ratio in these plants as determined by nitrobenzene oxidation (NBO) method. Thus, the results of the study in tobacco are consistent with many of the observations in *Arabidopsis*. A little discrepancy between the results of nitrobenzene oxidation and thioglycolic acid extraction method has, however, been observed which reflects the differences in mechanisms of extraction procedures in the two processes. The results of these studies suggest that in transgenic *Arabidopsis*, 4CL is encoded by a single gene and is not required for sinapyl alcohol biosynthesis and thus S units are formed via 4CL independent pathway, while in transgenic tobacco, 4CL is necessary for the biosynthesis of S type monomers. Thus, the suppression of 4CL activity may be an effective way both to modestly decrease the lignin content of wood and to shift the S/G ratio of this lignin in favour of S units so as to produce crops with improved pulping efficiency or digestibility. Recently, transgenic aspen trees (*P. tremuloides* Michx) with down-regulated 4CL have been produced (Hu *et al.*, 1998; Hu *et al.*, 1999). The transgenic aspen trees that accumulated structurally normal lignin at substantially reduced levels (55-97%) were developed. However, the reduced lignin was compensated by a concomitant increase in cellulose (15%) leading to a nearly doubled cellulose/lignin ratio. Thioglycolysis analysis of stem woody tissue revealed that the S/G lignin ratio was essentially the same in transgenic (1.8-2.2) and control (2.3) plants and no incorporation of unusual monomeric units into lignin was observed. Along with the cellulose increase, the enhanced growth of transgenic trees was a surprising observation not reported in transgenic tobacco or *Arabidopsis* with any of the down-regulated lignin pathway enzymes (Sederoff, 1999). It was thus demonstrated that cellulose and hemicellulose deposition and perhaps less directly growth are sensitive to changes in carbon flow directed into lignin biosynthesis and thus suggested the involvement of some other pathways besides lignin biosynthesis. Thus, the potential for improving plant utilization with 4CL down-regulation is enormous.

3.1.8. Cinnamoyl coenzyme A reductase (CCR) (EC 1.2.1.44)

CCR catalyzes the first committed step in the monolignol specific/ branch pathway. Consequently, this enzymatic step is an important control point regulating the flux of phenylpropanoid metabolites towards the biosynthesis of lignin. A full-length cDNA clone encoding CCR was isolated for the first time from *Eucalyptus gunnii* (Lacombe *et al.*, 1994; Lacombe *et al.*, 1997). This cDNA was further used to clone the corresponding cDNAs in tobacco, poplar and tall fescue (Boudet and Grima-Pettenati, 1996), maize (Pichon *et al.*, 1998) and more recently in sugarcane (Selman-Housein *et al.*, 1999). A CCR genomic clone has also been isolated from *Eucalyptus gunnii* (Van Doorssalaere *et al.*, 1994). After the characterization of CCR cDNA clones, down-regulation of CCR activity was aimed at by some workers in tobacco (Boudet and Grima-Pettenati, 1996; Piquemal *et al.*, 1998; Ralph *et al.*, 1998). Thus, tobacco plants transformed with antisense CCR cDNA fragments (homologous as well as heterologous), under the control of either constitutive promoter (CaMV 35S promoter or CaMV 35S with a double enhancer) or lignification specific promoter (CAD promoter), have been produced (Boudet and Grima-Pettenati, 1996; Piquemal *et al.*, 1998; Ralph *et al.*, 1998). Significant down-regulation of CCR activity was obtained only by ectopic expression of the homologous tobacco antisense gene. A decrease in lignin content, upto 75%, has been reported in transgenic tobacco with antisense suppression of CCR activity (Boudet and Grima-Pettenati, 1996). These transformants exhibited an increased S/G ratio. Later, Piquemal *et al.* (1998) also suggested an increased S/ G ratio in CCR down-regulated tobacco plants which was mainly due to a decrease in G units suggesting a potential improvement of lignin profiles in terms of pulp making. The less severely depressed lines exhibited a normal phenotype and a very slight reduction of thioacidolysis yield. On the other hand, severely depressed CCR lines exhibited a reduction in lignin content, presence of unusual cell wall bound phenolics and an abnormal heritable phenotype as reduced size, abnormal leaf morphology, collapsed vessels and orange brown coloration. In these CCR down-regulated plants, the hypolignified xylem vessels were unable to withstand the compressive forces generated during transpiration and had a tendency to collapse inwards which in turn had a negative effect on water supply required for normal growth and development. A concomitant decrease of other monolignol derived compounds as lignans and/or dehydrodiconiferyl glucosides (DCGs) was also observed with a decrease in CCR activity. Studies for semi-*in-vivo* incorporation of ferulic acid and sinapic acid suggested that the orange brown coloration was in part due to an increase in ferulic acid deposition in the walls. The incorporation of ferulate tyramine has also been detected in the walls of CCR antisense lines by ^{13}C NMR (Ralph *et al.*, 1998). The data suggest that CCR controls the carbon flux into the lignin pathway, however, a large decrease in lignin content is incompatible with normal vascular development. Thus, CCR silencing induced a marked reorientation of phenolic metabolism, resulting in a new partitioning of phenolic precursors into lignin and other related phenolic sinks. Contradictory to earlier findings, a decrease in S/G ratio has been shown recently, in sense suppressed tobacco lines (O'Connell *et al.*, 1998).

3.1.9. Cinnamyl alcohol dehydrogenase (CAD) (EC 1.1.1.195)

CAD catalyzes the reduction of hydroxy-cinnamaldehydes (sinapyl, p-coumaryl, coniferyl aldehydes) to the corresponding alcohols, which are the direct monomeric precursors of lignin. Two isoforms of CAD have been characterized in angiosperms (Wyrambic and Grisebach, 1975). Walter *et al.* (1988) reported the cloning of the cDNA for a putative bean CAD. However, they later expressed doubts about the identity of their cDNA clone (Walter *et al.*, 1990). They found extensive similarity between the CAD clone and that of NADP + malic enzyme from maize (Rothermal and Nelson, 1989). Since then CAD cDNA clones have been characterized in higher plants (Deborah *et al.*, 1998) such as loblolly pine (O'Malley *et al.*, 1992), tobacco (Knight *et al.*, 1992), spruce (Galliano *et al.*, 1993), *Eucalyptus gunnii* (Grima-Pettenati *et al.*, 1993; Grima-Pettenati *et al.*, 1994), *Aralia cordata* (Hibino *et al.*, 1993), alfalfa (Van Doorssalaere *et al.*, 1995), poplar (Van Doorssalaere *et al.*, 1995), lucerne (Brill *et al.*, 1999) and sugarcane (Selman-Housein *et al.*, 1999). The *cad* gene appears to be well conserved and exhibits 80% identity and 10% similarity. Lower but still important homologies are observed between dicot and monocot CADs, angiosperm and gymnosperm CADs. Gymnosperm CAD exhibits very low affinity for sinapaldehyde whereas angiosperm CADs uses all three substrates with roughly identical affinities, at least *in vitro*. Mutants with lower CAD activity and reduced lignin content have been isolated from maize, soybean, sorghum, pearl millet and pine. They are designated as bmr and *cad-n1* mutants (Jorgensen, 1931; Kuc and Nelson, 1964; Bucholtz *et al.*, 1980; Cherney *et al.*, 1990; Pillonel *et al.*, 1991; Barriere and Argillier, 1993; Mackay *et al.*, 1997; Provan *et al.*, 1997; Ralph *et al.*, 1997; Halpin *et al.*, 1998; Wu *et al.*, 1999). These lines of evidence imply a possible way to reduce lignification in plants by manipulating expression of CAD involved in lignin biosynthesis. The first CAD down-regulated plants were obtained through ectopic expression of homologous CAD cDNA antisense construct in tobacco (Halpin *et al.*, 1994). The backcross of two of the selected transformants with wild type plants provided a population of azygous (control) and hemizygous (antisense) plants in a uniform genetic background (line 40 and 50). Although exhibiting a very low residual CAD activity (20% and 7% for line 40 and 50, respectively), no change in growth, development and lignin content of these transgenic plants as compared to control was observed. In several primary transformants and in line 50, the xylem colored red brown. Pyrolysis mass spectrometry revealed an increase, within the polymer, in both coniferyl and sinapyl aldehydes. The shift in the aldehyde- to alcohol- ratio was greater for sinapyl than coniferyl moieties, when CAD activity was limiting, leading to decreased S/G ratio in transgenic plants. These observations were confirmed and extended by other studies on tobacco CAD down-regulated plants (Myton *et al.*, 1995; Tollier *et al.*, 1995). The red brown coloration of transgenic tobacco plants with reduced CAD activity could be ascribed to "abnormal lignins" with increased cinnamaldehyde groups (Higuchi *et al.*, 1994). S/G ratio of antisense CAD tobacco plants was slightly increased with 55% residual CAD activity. An antisense gene for *Aralia cordata* CAD was introduced into tobacco plants which was effective in suppressing activity of heterologous tobacco CAD in the transgenic plants (Hibino *et al.*, 1995). A reduction upto 20% in CAD activity was observed. However, lignin content showed no significant difference. The content of p-hydroxy cinnamaldehyde groups in lignin was higher in the transgenic plants and was inversely correlated to levels of CAD activity. Down-regulation of CAD activity was further studied in transgenic poplar (*P. tremula* × *P. alba*) by both antisense and co-suppression strategies (Baucher *et al.*, 1996). Approximately, 70% reduction in CAD activity with no change in lignin content has been achieved. Moreover, the lignin monomeric composition (S/G) was not significantly modified. In addition, no increased amounts of cinnamaldehyde derivatives were identified. These results were in contradiction to that obtained with transgenic tobacco plants. However, the red coloration of xylem tissue was common among transgenic tobacco (Higuchi *et al.*, 1994; Myton *et al.*, 1995; Tollier *et al.*, 1995; Hibino *et al.*, 1995), poplar (Baucher *et al.*, 1996), bm mutants of maize (Jorgensen, 1931; Kuc and Nelson, 1964; Barriere and Argillier, 1993), Sorghum (Bucholtz *et al.*, 1980; Pillonel *et al.*, 1991) and pine mutants (Mackay *et al.*, 1995; Ralph *et al.*, 1997). The red color of the stem was unstable. In the bm mutants, the disappearance of color was observed upon maturation. In transgenic tobacco, the red color of the stem was visible during the growth of the plants but disappeared at the stage of flowering. In transgenic poplars, the red color was intense during the first growing season but disappeared or was much paler during the winter. In the second year of growth, the red color of the CAD down-regulated plants might be due to the presence of conjugated aldehydes in the lignin and the degree of conjugation may be reduced during aging resulting in disappearance of red color. Recently, transgenic tobacco plants have been produced with CAD activity ranging between 8-11% of the control (Yahiaoui *et al.*, 1998). The reduction in lignin content was, however, insignificant but a decreased S/G ratio was achieved. A significant increase in S/G ratio in transgenic poplar plants with upto 90% reduction in CAD activity has also been achieved recently (Lapierre *et al.*, 1999). This was attributable to increased S units and unchanged G units. And unlike transgenic tobacco with reduced CAD activity (Halpin *et al.*, 1994; Higuchi *et al.*, 1994), the red coloration of xylem tissues could not be assigned to any increased incorporation of cinnamaldehyde units into lignin. In the same study (Lapierre *et al.*, 1999), a transgenic line with CAD down-regulation was re-transformed with a construct containing the OMT antisense gene. This double transformant showed a reduction of OMT activity in the xylem tissue of about 90% as observed in case of other OMT down-regulated poplars. A similar reduction of CAD activity was also observed (Lapierre *et al.*, 1999). Incorporation of dihydroconiferyl alcohol into lignin of transgenic tobacco plants has also been suggested (Ralph *et al.*, 1998). More recently, down-regulation of CAD activity (upto 30%) has been achieved in alfalfa transformed with the full-length or 3' fragment of CAD cDNA in antisense orientation (Baucher *et al.*, 1999). These lines revealed an altered lignin composition, being more condensed and having decreased S+G yields (upto 25%) owing to decreased amount of S units. The lignin quantity, however, remained unchanged. The findings were in concordance with that obtained for transgenic tobacco (Yahiaoui *et al.*, 1998), however, the reduction in S+G was much higher in transgenic tobacco (50%). The transgenic alfalfa lines with most reduced CAD activity had enhanced alkali solubility and improved *in situ* degradability and hence digestibility.

These data suggest that a reduction in CAD activity is associated with increased aldehyde content in lignin as determined by various methods, such as phloroglucinol staining (Pillonel *et al.*, 1991; Mackay *et al.*, 1997), thioacidolysis (Higuchi *et al.*, 1994), pyrolysis mass spectrometry (Halpin *et al.*, 1994), Fourier transform infra red and Raman spectroscopy (Stewart *et al.*, 1997). The difference in S/G ratio was ascribed to differences in residual CAD activity, methods of analysis and lignification patterns between annual and perennial plants. Reduction or suppression of CAD activity in these diverse plant species has either no detectable effect or a small effect on lignin content, indicating that CAD is normally not rate limiting and has a minor role in regulating metabolic flux through the lignin biosynthetic pathway. However, some minimal level of CAD enzyme is needed to achieve wild type lignin content.

3.2. Genetic Engineering of Lignin at the Level of Monolignol Transport

The enzymes regulating the transport and mobilization of monolignol precursors are UDPG-coniferin glucosyl transferase (UDPG-GT) and β-glucosidase. The transport or storage forms of monolignols include 4-O-β-glucosides of monolignols that accumulate in the cambial sap of gymnosperms and some angiosperms (Freudenberg and Harkin, 1963; Terezawa *et al.*, 1984; Savidge, 1989). These glucosides are formed by transfer of a glucose residue from UDP-glucose to the phenolic hydroxyl of the monolignols, catalyzed by a specific UDPG-glucosyl transferase (Ibrahim and Grisebach, 1976). Hydrolysis of coniferin, a glucoside of coniferyl-alcohol, by a β-glucosidase is essential before the polymerization in the cell wall may proceed. Till date, no reports are available on the down-regulation of any gene of this category. However, a great potential exists for genetic manipulation of lignin at this level of control in future.

3.2.1. Coniferin β-Glucosidase (CBG)

During lignin formation, the monolignol precursors are transported to the cell wall as glucosides and that specific β-glucosidase within the cell wall would regenerate the monolignol aglycones for *in situ* polymerization (Freudenberg and Harkin, 1963). A coniferin β-glucosidase (CBG) has been purified and characterized from pine (Leinhos *et al.*, 1994; Dharamawardhana *et al.*, 1995). Significant headway has been made on the molecular characterization of monolignol specific β-glucosidases. After purification of CBG from developing xylem in *Pinus contorta* (Dharamawardhana *et al.*, 1995), a cDNA clone was subsequently isolated from xylem specific *P. contorta* library (Dharamawardhana *et al.*, 1995). A coniferin β-glucosidase has also been described in *P. banksiana* (Leinhos *et al.*, 1994). Functional data, based on genetic approaches for these enzymes, is still lacking. As the enzyme is apparently very closely associated with lignification, its manipulation should avoid undesirable pleiotropic effects. Moreover, the presence of a similar monolignol specific glucosidases in angiosperm is yet to be determined and the application of genetic engineering to raise transgenic plants has yet to be explored.

3.3. Genetic engineering of lignin at the level of polymerization

Peroxidases, laccases, H_2O_2 independent phenol oxidases including coniferyl alcohol oxidase have all been proposed to participate in lignification by forming higher order polymers from monolignols. However, no unequivocal proof has ever been obtained for the involvement of any particular oxidase in lignification through loss-of-function experiments in transformed plants. A dehydrogenative mechanism for the production of free radical intermediate is involved, that could polymerize into lignin as shown in Fig. 3.

3.3.1. Peroxidase (POD) (EC 1.11.1.7)

The POD plays an integral role in secondary cell wall biosynthesis by the polymerization of cinnamyl-alcohols into lignin and by forming rigid cross-links between cellulose, pectin, hydroxyproline rich glycoproteins and lignin. In the tobacco plant, there are at least 12 distinguishable isoforms (Boudet *et al.*, 1995). The molecular analyses of POD began with the cloning and characterization of the anionic POD isozyme from tobacco (Lagrimini *et al.*, 1987). Till date, cDNA clones for anionic POD have been characterized from wheat (Rebmann *et al.*, 1991; Baga *et al.*, 1995), tobacco (Lagrimini *et al.*, 1987), barley (Johansson *et al.*, 1992; Thordal-Christensen *et al.*, 1992; Kristensen *et al.*, 1999), cucumber (Rasmussen, 1995), maize (Teichmann *et al.*, 1997), potato (Roberts and Kolattukudy, 1993), rice (Ito *et al.*, 1994), zucchini (Carpin *et al.*, 1999), tomato (Sherf *et al.*, 1993), peanut (Lige *et al.*, 1998), sweet potato (Kim *et al.*, 2000). The genetic manipulation of an anionic POD gene expression using antisense/sense constructs has been studied extensively in tomato (Sherf *et al.*, 1993) and tobacco (McIntyre *et al.*, 1996; Kolattukudy *et al.*, 1992; Liu *et al.*, 1994; Lagrimini *et al.*, 1997) plants. A significant reduction in the level of specific POD has been achieved in transgenic tomato plants experiencing constitutive expression of homologous antisense POD gene, however, the lignin content remained unaltered. The phenotype and developmental progression of transgenic lines were indistinguishable from those of non-transformed control plants. However, in transgenic tobacco, an observable change in plant growth

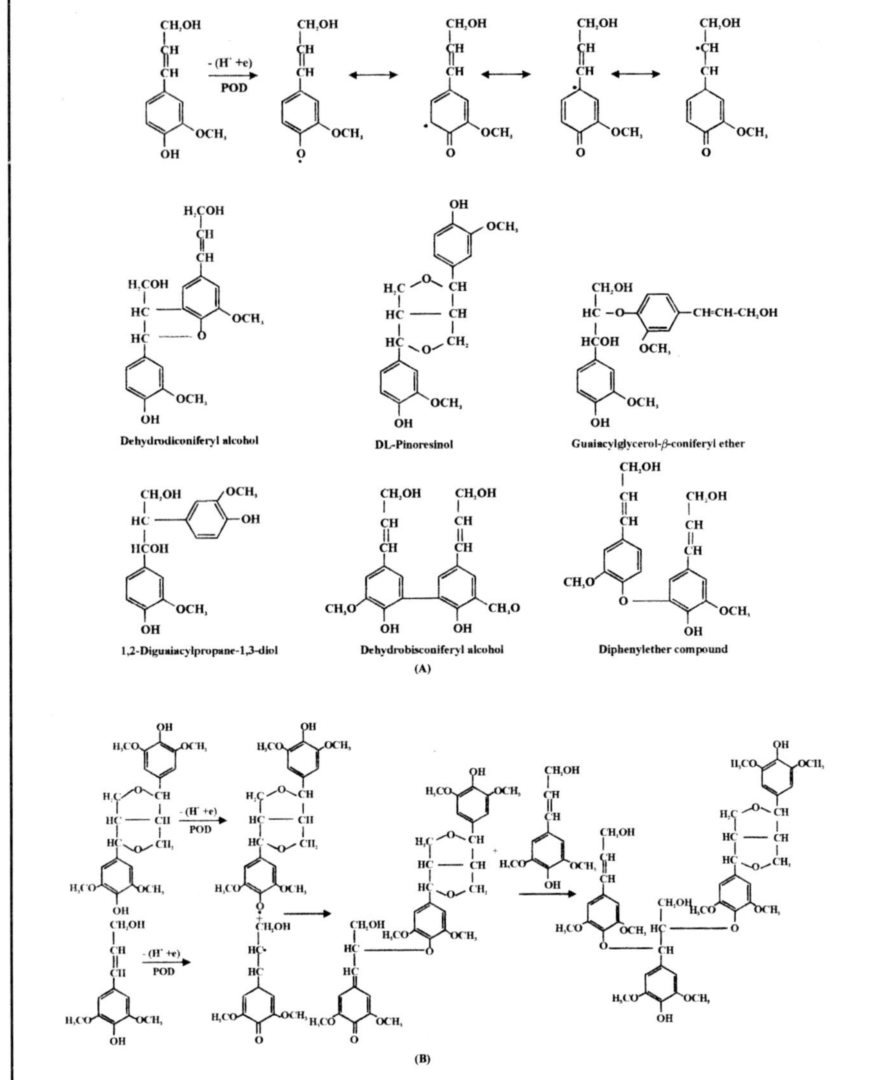

Figure 3 : Polymerization of monolignols, coniferyl alcohol (A) and sinapyl alcohol (B) through oxidative coupling by peroxidase. It may be noted that the phenyl-phenyl linkage (C_5-C_5), encountered in G-lignin, which shows resistance in pulping, can not be obtained in case of sinapyl lignin.

and development has been reported (Kolattukudy *et al.*, 1992; Liu *et al.*, 1994; McIntyre *et al.*, 1996; Lagrimini *et al.*, 1997). McIntyre *et al.* (1996) studied the expression of POD antisense construct (As) with CaMV 35S and NOS promoters and were the first to report the use of ribozyme constructs (Rz) for the suppression of endogenous plant genes in transgenic plants. POD levels were found to be significantly reduced in transgenic plants carrying 35S-As, 35S-Rz and tRNA-Rz constructs without significant difference in lignin levels. The results of this study indicated 35S-As construct to be more effective than that of NOS-As construct and a tRNA vector could also be used effectively for gene suppression in transgenic plants. Transgenic tobacco plants that under-express anionic POD to as much as 1600 fold were also generated (McIntyre *et al.*, 1996), using antisense anionic POD cDNA, but even at this tremendous level of reduction, the lignin content in leaf, stem and root was unchanged. It was, however, predicted that antisense RNA suppression of tobacco anionic POD would result in a measurable decrease in lignification, based on previous studies of the purified enzyme and the results of over-expression in transformed tobacco and tomato plants (Lagrimini, 1991; Lagrimini *et al.*, 1993; Lagrimini, 1993). Antisense transformed plants appeared normal at initial observation, however, growth studies showed that plants with reduced POD activity were taller, flowered sooner and bore more developed axillary buds as compared to control plants. In contrast, transformed plants over-expressing anionic POD with 10 fold higher activity were shorter, bore less developed axillary buds and flowered later than controls (Lagrimini, 1990). Transformed plants with over-expressed POD showed a unique phenotype of chronic severe wilting through loss of turgor in leaves, which was initiated at the time of flowering, although having functional stomata and normal vascular anatomy and physiology (Lagrimini *et al.*, 1997). The POD induced wilting was shown to be an effect of diminished water uptake through the roots, decreased conductance of water through the xylem, or increased water loss through the leaf surface or stomata. Close examination of root systems revealed that at flowering, root growth rate and total root mass in transformed plants were less than 50% of those of control plants although shoot mass and growth rate were unchanged. The results related to POD suppression are ambiguous and recently promoter expression analysis indicated that this anionic POD is not correlated to lignifying tissues in tobacco (Klotz *et al.*, 1998).

3.3.2. Laccase (LAC) (EC 1.10.3.2)

Laccase is a copper containing metallozyme and catalyzes the $4e^-$ reduction of O_2 to H_2O. Laccase are somewhat promiscuous with respect to substrate specificity oxidizing a variety of diphenol, dinaphthol, phenylene diamine and anthranilate derived metabolites. Several reports suggest that plant laccases in conjuction with POD catalyze the formation of lignin (Freudenberg, 1968; Kolattukudy *et al.*, 1992; Hoopes *et al.*,

1992; Droiuich *et al.*, 1992; Savidge and Udagama-Randeniya, 1992; Sterjiades *et al.*, 1993; Bao *et al.*, 1993). Four closely related cDNAs encoding laccase isozymes from xylem tissues of yellow poplar have been identified and sequenced recently (Ranocha *et al.*, 1999; LaFayette *et al.*, 1999). Two other reports suggest the cloning of a low pI laccase from the suspension cultured sycamore maple cells (LaFayette *et al.*, 1995) and high pI laccase from tobacco stems (Keifer-Meyer *et al.*, 1996). A laccase cDNA has also been isolated from loblolly pine. Laccase genes from different species have a relatively low level of sequence homology. Antisense suppression of laccase activity has been reported for few plant species (LaFayette *et al.*, 1994; Eriksson *et al.*, 1995; Dean *et al.*, 1996, 1998). Recently, antisense laccase plants have been obtained for poplar (Boudet, 2000). In these plant species, no significant differences in growth or development and even in lignin content and composition were observed as compared to control plants. Yellow poplar transgenic lines over-expressing the *Acer pseudoplatanus* laccase gene (Dean *et al.*, 1998), have been shown to exhibit an increased fluorescence of cell walls, suggesting an increased deposition of phenolic material or a change in cell wall phenolic components. As no change either in lignin content or composition was obtained, the role of laccase in lignification remains uncertain.

3.3.3. Phenol oxidase (PO) (EC 1.10.3.1)

Phenol oxidases capable of oxidizing polyphenols even in the absence of H_2O_2 have been suggested to participate in the polymerization of lignin monomers along with POD (Sterjiades *et al.*, 1993; O'Malley *et al.*, 1993; Dean and Eriksson, 1994). However, cloning and subsequent down-regulation of phenol oxidase are yet to be achieved. Coniferyl alcohol oxidase (CAO), a phenol oxidase, constitutes a potential target for genetic modification of lignin (Savidge and Udagama-Randeniya, 1992; Chabanet *et al.*, 1994; Hoopes *et al.*, 1995).

4. PULP AND PAPER MANUFACTURE *VIS-Á-VIS* LIGNIN MODIFICATION

Removal of lignin from the wood cell walls is the most energy intensive and environmentally damaging step in wood processing for pulp and paper, amounting for more than 1% of world's economy (WRI, 1994). During paper making, the pulping process is followed by bleaching to remove the residual lignin from the pulp, which results in a reduction of the pulp yield with increase in brightness. Equally, the cellulose DP value decreases because the bleaching agents attack the cellulose polymer. In pulp production, the efficiency of the delignification process depends both on the plant material and on the monomeric composition of the lignin, which determines the interunit linkages between monomers. Genetic modifications to reduce lignin content in wood or to alter its chemistry to enhance extractability offers new opportunities and attractive perspectives for improving woody species with improved quality and efficiency of pulping as well as decreased mill effluents. Any increase in efficiency that allows production of more wood products from less land helps conserve natural forests and reduces the environmental impact of processing wood into pulp and paper. In general, the efficiency of wood pulping is directly proportional to the amount of syringyl residues in lignin. An increase in S units is considered advantageous. It has been clearly demonstrated that gymnosperm lignin is very difficult to degrade owing to the availability of free 5-aromatic position for very strong C-C bonds (5-5 linkages). Thus, a genetically induced methoxylation at C-5 would reduce the resistant C_5-C_5 interunit bonds in lignin and would thereby facilitate the pulping of gymnosperm wood more efficiently. This could be achieved in transgenic conifers by introducing a bifunctional OMT gene as well as F5H gene. In angiosperms, coniferyl aldehyde 5-hydroxylation and methylation direct syringyl lignin biosynthesis (Osakabe *et al.*, 1999). Thus, it might be possible to produce "high S" lignin through over-expression of F5H and of a specific 5-hydroxy ferulic acid OMT. To improve delignification in paper manufacture, another strategy has been proposed based on the higher hydrophilicity of hydroxy groups compared to methoxy groups and their ability to ionize the aqueous media. The partial substitution of O-CH_3 groups with OH groups could improve lignin solubilization in the pulping liquor and/or would allow the use of less damaging chemicals to obtain an equivalent pulp grade. In angiosperm lignin, this could be potentially achieved through genetic reduction of OMT activity.

Lignin of the CAD down-regulated plants has an altered reactivity towards alkali when compared with the control plants. The higher extractability of lignin upon alkali treatment might occur for several reasons. The incorporation of aldehydes into the lignin polymer would (a) render the free hydroxyl group at the C4 position more susceptible to ionization in alkali and (b) render the α and β carboxyl more susceptible to chemical attack by alkali. This would not only explain the higher solubility of the lignin but also the higher amounts of the C_6-C_1 benzaldehydes, vanillin and syringaldehyde. The increased levels of these

C_6-C_1 compounds in the down-regulated CAD plants are thus probably derived from C_6-C_3 aldehydes. An additional reason why the lignin would be more extractable is that the γ-terminal alcohol groups might be involved in the formation of cross-links with polysaccharides. Replacing part of this alcohol group with aldehyde group may result in a reduced frequency of cross-links to polysaccharides. The kraft method of pulping is the major one used by the pulp industry. During this process, splitting and degradation of lignin occurs mostly of β-O-4 linkages between monomeric phenylpropane units. In cinnamaldehyde rich lignin, the β-O-4 linkages are thought to be more effectively cleaved than those of normal lignin because the aldehyde groups attract electron more strongly from the linkages than the alcohol groups in normal lignin. Therefore, transgenic plants, especially trees with increased amounts of cinnamaldehyde groups in lignin could be more amenable to pulping by kraft method. Thus, CAD or CCR down-regulation (Halpin *et al.*, 1994; O'Connell *et al.*, 1998; Boudet *et al.*, 1998; Lapierre *et al.*, 1999) has been thought to be suitable target to improve delignification during pulping process. The lignin of these plants has been shown to have a lower kappa number, improved bleachability and higher levels of brightness. In contrast, OMT down-regulation (Lapierre *et al.*, 1999) has recently been shown to have a negative impact on lignin extractability. During OMT down-regulation, a reduction in S units and labile ether bonds and an increase in resistant biphenyl bonds (C_5-C_5 linkages) is achieved that are characteristic of soft wood lignin. But this highly cross-linked lignin is more resistant to thermal degradation and can be employed as fuel (Table 2). More recently, transgenic poplars with OMT level close to zero have been shown to possess substantially reduced lignin (17% decrease) and a strongly altered structure with a complete lack of S units and a high condensation degree (Jouanin *et al.*, 2000). The impact of this transformation on kraft pulping performance of poplar trees was positive with respect to pulp yield (10% relative increase) but negative in terms of industrial degradation of lignin. Thus, a decrease in lignin content was beneficial to the pulp yield (kraft pulp), but the alteration of lignin structure made the polymer less amenable to degradation leading to decreased production of bleached pulp.

5. FORAGE DIGESTIBILITY *VIS-Á-VIS* LIGNIN MODIFICATION

Enormous lignification in the forage crop and their strong association with cellulose reduces digestibility by cattle (Jung and Vogel, 1986; Buxton and Russel, 1988; Jung and Deetz, 1993; Buxton and Casler, 1993; Argillier

Table 2. Technological properties (pulping characteristics) of transgenic lignin of CAD, CCR and OMT down-regulated transgenic plants.

1. CAD depleted tobacco plants	Improved extractability of lignin and slight increase in pulp yield, lower kappa number (indicating lower amounts of residual lignin and therefore less chemical treatment needed in bleaching process).
2. CAD antisense poplar plants	Improved extractability, slight increase in pulp yield and lower kappa number.
3. CCR down-regulation in tobacco	Improved extractability, lower kappa number, improved bleachability, higher levels of brightness, improved mechanical properties of transgenic pulp.
4. OMT down-regulation in poplar	Decreased extractability, increase in kappa number, decrease in S units with increased biphenyl bonds with highly cross-linked lignin, not good for pulp production but is suitable for use of wood as fuel.

CAD and CCR are suitable targets for improving delignification during pulp processing which is economic and eco-friendly. These studies provide new insight for tailoring lignin structure for improved pulping.

et al., 1996). Thus, a reduction in the amount of lignin is important. The p-coumarate : ferulate ratio (pCA: FA) has been accepted to be inversely correlated with digestibility. Previous work has shown that the amount of esterified p-coumaric acid was greatly reduced in bm1 mutants, although the soluble pool of p-coumarate remained unchanged (Kuc and Nelson, 1964). The data of Halpin *et al.* (1998) demonstrated a large reduction in the proportion of lignin in labile ether linked structures in bm1 which may affect this change in pCA:FA ratio by reducing the number of sites available for p-coumarate esterification onto the lignin core. The ultra-structure and distribution of lignified tissues within plant materials is also important in determining digestibility (Bernard Vailhe *et al.*, 1996, 1998; Halpin *et al.*, 1998; Baucher *et al.*, 1999). Changes in phenolic deposition in parenchyma cells may both speed the kinetics of digestion in these cells and aid the penetration of digestive enzymes to less digestible tissues.

6. CONCLUSIONS AND FUTURE PROSPECTS

Down-regulation of the genes at the level of monolignol biosynthesis and its polymerization has been extensively aimed at modifying the lignin profiles of plants through antisense and sense suppression. Though, studies on these transgenes revealed alterations in the lignin quantity as well as composition, the results are ambiguous suggesting that plants have a high level of metabolic plasticity in the formation of lignin. When lignin biosynthetic pathway is down-regulated, plants develop means by which they circumvent genetic manipulation. They compensate by incorporating other structurally related aromatic compounds including cinnamaldehydes, dihydro-coniferyl alcohol, ferulate esters and feruloyl tyramine into wall that may provide similar functions. Based on the results of these studies, it can be concluded that enzymes of general phenylpropanoid biosynthetic pathway such as PAL, C4H, 4CL are not good targets for specific manipulation of monolignol synthesis since many other biological functions could be adversely affected. OMT and CCoAOMT are good targets for changing the lignin composition. Lignins from OMT down-regulated plants have been shown to be good for increasing the calorific value of wood instead of improving the pulping characteristics. CCR and CAD are suitable targets for improving delignification during pulp processing as lignin of these CCR and CAD depleted plants have been shown to possess improved pulping characteristics such as increased lignin extractability, increased pulp yield, reduced kappa number and improved brightness. Enzymes involved in monolignol transport are not yet been seriously targeted. In the future, the exciting prospect can be the manipulation of UDPG-GT or CBG. The enzymes involved in the polymerization of monolignols in the wall are still a matter of debate. However, all the criteria for acceptance of an enzyme for its involvement in lignification, such as its location in wall of differentiating xylem, capability of *in vitro* oxidation of monolignols and an impact on lignin content or composition on the knocking out of a corresponding gene, have never been met simultaneously for any of the enzyme (POD, LAC, PO). The cloning of CAO is also necessary to ascertain its role in lignification and its universal occurrence through out the plant kingdom. More judicious strategy could be the genetic manipulation of more than one gene simultaneously (double constructs) (Lapierre *et al.*, 1999; Abbott *et al.*, 1998) and usage of xylem specific promoters (Feuillet *et al.*, 1995; Chen *et al.*, 1998) to prevent pleiotropic effects. The identification of novel target genes will also be instrumental in opening up new avenues of lignin research in the future. In this direction, screening of *Arabidopsis* mutants with altered lignification has been recently envisaged. These include *Arabidopsis* mutants as ifl 1 (Zhong *et al.*, 1997), rsw 1 (Aroili *et al.*, 1998), irx A (Jones and Turner, 1998), eli-1 (Cano *et al.*, 1998), lig 1 (Ye *et al.*, 1998), monopteros mutant (Przemeck *et al.*, 1996), lop 1 (Carland *et al.*, 1996), xtc 1, xtc 2, amp 1 (Conway *et al.*, 1997) and avb 1 (Zhong *et al.*, 1999). These mutants may represent invaluable tools for the identification of important regulatory genes controlling the lignification process. Other strategies deal with transcription factor genes as MYB (Tamagnone *et al.*, 1998), Nt lim 1 (Kawaoke and Ebinuma, 1999), dirigent protein from *Forsythia* (Davin *et al.*, 1997) involved in free radical coupling and xylem proteins for which the EST (expressed sequence tag) analysis of xylem is required (Bloksberg *et al.*, 1998). Impact of genetic engineering of lignin on cellulose metabolism and biomass yield needs in depth investigation in future (Hu *et al.*, 1999).

ACKNOWLEDGEMENTS

Financial supports in the form of a research project from the Department of Biotechnology, New Delhi and a senior research fellowship from Council of Scientific and Industrial Research, New Delhi to SR are gratefully acknowledged.

REFERENCES

Abbott J, Barakate A, Schuch W and Halpin C (1998) Strategies for simultaneous down-regulation of cinnnamyl alcohol dehydrogenase and O-methyltransferase in tobacco. In : *Proceedings of 8th Int. Cell Wall Meeting,* Norwich.

Allina SM and Douglas CJ (1994) Isolation and characterization of the 4-coumarate: coenzyme A ligase gene family in a poplar hybrid. *Plant Physiol.,* **105** : 154.

Argillier O, Barriere Y, Lila M, Jeanneteau F, Gelinet K and Menanteau V (1996) Genotypic variation in phenolic components of cell walls in relation to the digestibility of maize stalks. *Agronomie,* **16** : 123-130.

Arioli T, Peng L and Betzner AZS (1998) Molecular analysis of cellulose biosynthesis in *Arabidopsis. Science,* **279** : 717-720.

Atanassova R, Favet N, Martz F, Chabbert B, Tollier MT, Monties B, Fritig B and Legrand M (1995) Altered lignin composition in transgenic tobacco expressing O-methyltransferase sequence in sense and antisense orientations. *Plant J.,* **8** : 465-477.

Baga M, Chibbar RN and Kartha KK (1995) Molecular cloning and expression of peroxidase genes from wheat. *Plant Mol. Biol.,* **29** : 647-662.

Bao W, O'Malley DM, Whetten R and Sederoff RR (1993) A laccase associated with lignification in loblolly pine xylem. *Science,* **260** : 672-674.

Barriere Y and Argillier O (1993) Brown-midrib genes of maize, a review. *Agronomie,* **13** : 865-895.

Baucher M, Bernard-Vailhe MA, Chabbert B, Besle JM, Opsomer C, van Montagu M and Botterman J (1999) Down-regulation of cinnamyl alcohol dehydrogenase in transgenic alfalfa (*Medicago sativa* L.) and the effect on lignin composition and digestibility. *Plant Mol. Biol.,* **39** : 437-447.

Baucher M, Chabbert B, Pilate G, Van Doorssalaere J, Tollier MT, Petit-Conil M, Cornu D, Monties B, van Montagu M, Inze D, Jouanin L and Boerjan W (1996) Red xylem and higher lignin extractability by down-regulating a cinnamyl alcohol dehydrogenase in poplar (*Populus tremula* × *P. alba*). *Plant Physiol.,* **112** : 1479-1480.

Baucher M, Monties B, van Montagu M and Boerjan W (1998) Biosynthesis and genetic engineering of lignin. *Crit. Rev. Plant Sci.,* **17** : 125-197.

Bate NJ, Orr J, Ni W, Meromi A, Nadler-Hassar T, Doerner P, Dixon RA, Lamb CJ and Elkind Y (1994) Quantitative relationship between phenylalanine ammonia-lyase levels and phenylpropanoid accumulation in transgenic tobacco identifies a rate-determining step in natural product synthesis. *Proc. Natl. Acad. Sci., U.S.A.,* **91** : 7608-7612.

Becker-Andre M, Schulze-Lefert P and Hahlbrock K (1991) Structural comparison modes of expression and putative cis-acting elements of the two 4-coumarate: coenzyme A ligase genes in potato. *J. Biol. Chem.,* **266** : 8551-8559.

Bernard-Vailhe MA, Besle JM, Maillot MP, Cornu A, Halpin C and Knight M (1998) Effect of down-regulation of cinnamyl alcohol dehydrogenase on cell wall composition and on degradability of tobacco. *J. Sci. Food Agric.,* **76** : 505-514.

Bernard-Vailhe MA, Migne C, Cornu A, Maillot MP, Grenet E, Besle JM, Atanassova R, Martz F and Legrand M (1996) Effect of modification of the O-methyltransferase activity on cell wall composition, ultrastructure and degradability of transgenic tobacco. *J. Sci. Food Agric.,* **72** : 385-392.

Bloksberg LN, Nieuwenhiuzen NJ and Strabala TJ (1998) EST database-assisted cloning of the lignin biosynthesis pathway gene sense and antisense expression reduces lignin content in plants. In : *Proceedings of 8th Int. Cell Wall Meeting,* Norwich.

Blount JW, Korth KL, Masoud SA, Rasmussen S, Lamb C and Dixon RA (2000) Altering expression of cinnamate 4-hydroxylase in transgenic plants provides evidence for a feed back loop at the entry point into the phenylpropanoid pathway. *Plant Physiol.,* **122** : 107-116.

Boerjan W, Meyermans H, Chen C, Leple JC, Christensen JH, Van Doorssalaere J, Baucher M, Petit-Conil M, Chabbert B, Tollier MT, Monties B, Pilate G, Cornu D, Inze D, Jouanin L and van Montagu M (1996) Genetic Engineering of lignin biosynthesis in poplar. In : *Somatic cell genetics and molecular genetics of trees* (Eds. Ahuja M R *et al.*), Kluwer Academic Publishers, Dordrecht, pp. 81-88.

Boudet AM (1998) A new view of lignification, *Trends Plant Sci.,* **3** : 67-71.

Boudet AM (2000) Lignins and Lignification: Selected Issues. *Plant Physiol. Biochem.,* **38** : 81-96.

Boudet AM, Chabanes M and Goffner D (1998) Controlled down-regulation of genes involved in the last steps on lignin synthesis may significantly change the lignin profiles of plants. In : *Proceedings of Molecular Breeding of Woody Species,* Tokyo, Japan, pp. 1-21.

Boudet AM, Goffner D and Grima-Pettenati J (1997) Lignins and Lignification. In : *Recent biochemical and biotechnological advances,* C. R Acad. Sci., Paris, **319** : 317-331.

Boudet AM, Lapierre C and Grima-Pettenati J (1995) Biochemistry and Molecular Biology of Lignification. *New Phytol.,* **129** : 203-236.

Boudet AM and Grima-Pettenati J (1996) Lignin Genetic Engineering. *Mol. Breeding,* **2** : 25-39.

Boudet AM and Grima-Pettenati J (1996) Lignification genes: structure, function and regulation. In : *Proceedings of Keystone Symp. Extracell. Matrix Plants: Mol. Cell Dev. Biol.,* Tamarron, Colo, Wiley-Liss, New York, p. 7.

Brill EM, Abrahams S, Hayes CM, Jenkins CLD and Watson JM (1999) Molecular characterization and expression of a wound-inducible cDNA encoding a novel cinnamyl alcohol dehydrogenase in lucerne (*Medicago sativa* L.). *Plant Mol. Biol.,* **41** : 279-291.

Brodelius PE and Xue ZT (1997) Isolation and characterization of a cDNA from cell suspension cultures of *Vanilla planifolia* encoding 4-coumarate : coenzyme A ligase. *Plant Physiol. Biochem.,* **35** : 497-506.

Brown A (1985) Review of lignin in biomass. *J. Appl. Biochem.,* **7** : 371-387.

Bucholtz Y, Cantrell RP, Axtell JD and Lechtenberg VL (1980) Lignin biochemistry of normal and bm mutant *Sorghum. J. Agric. Food Chem.,* **28** : 1239-1241.

Bugos RC, Chiang VLC and Campbell WH (1991) cDNA cloning, sequence analysis and seasonal expression of lignin bi-specific caffeic acid/5-hydroxyferulic acid O-methyltransferase of aspen. *Plant Mol. Biol.,* **17** : 1203-1215.

Busam G, Junghanns KT, Kneusel RE, Kassemeyer HH, and Matern U (1997) Characterization and expression of caffeoyl coenzyme A 3-O-methyltransferase proposed for the induced resistance response of *Vitis vinifera* L. *Plant Physiol.,* **115** : 1039-1048.

Buxton DR and Casler MD (1993) Environmental and genetic effects on cell wall composition and digestibility. In : *Forage cell wall structure and digestibility* (Eds. Jung H *et al.*), *Am. Soc. Agron.,* Madison, WI., pp. 685-714.

Buxton DR and Russel JR (1988) Lignin constituents and cell wall digestibility of grass and legume stems. *Crop Sci.,* **28** : 553-558.

Campbell MM and Sederoff RR (1996) Variation in lignin content and composition. *Plant Physiol.,* **110** : 3-13.

Cano AI, Metzlaff K and Bevan M (1998) Xylem developmental mutants in *Arabidopsis thaliana.* In : *Proceedings of 8th Int. Cell Wall Meeting,* Norwich.

Carland FM and MacHale NA (1996) LOP 1: a gene involved in auxin transport and vascular patterning in *Arabidopsis. Development,* **122** : 1811-1819.

Carpin S, Crevecoeur M, Greppin H and Penel C (1999) Molecular cloning and tissue specific expression of an anionic peroxidase in *Zucchini. Plant Physiol.,* **120** : 799-810.

Chabanet A, Goldberg R, Catesson AM, Quinet-Szeyl M, Delaunay AM and Faye L (1994) Characterization of a phenoloxidase in mung-bean hypocotyl cell walls. *Plant Physiol.,* **106** : 1095-1102.

Chapple CCS, Vogt T, Ellis BE and Somerville CR (1992) An *Arabidopsis* mutant defective in the general phenylpropanoid pathway. *Plant Cell,* **4** : 1413-1424.

Chen C, Meyermans H, van Montagu M and Boerjan W (1998) Cell specific expression of chimeric caffeoyl coenzyme A 3-O-methyltransferase genes in transgenic poplar. In : *Proceedings of 8th Int. Cell Wall Meeting,* Norwich.

Cherney JH, Cherney DJR, Akin DE and Axtell JD (1991) Potential of bm, low lignin mutants for improving forage quality. *Adv. Agron.,* **46** : 157-198.

Cherney DJR, Patterson JA and Johnson KD (1990) Digestibility and feeding value of pearl millet as influenced by the brown-midrib, low lignin trait. *J. Anim. Sci.,* **68** : 4345-4351.

Chiang VL, Puumala RJ, Takeuchi H and Eckert RE (1988) Comparison of soft wood and hard wood draft pulping. *TAPPI J,* **71** : 173-176.

Collazo P, Montoliu L, Puigdomenech P and Rigau J (1992) Structure and expression of the lignin O-methyltransferase gene from *Zea mays* L. *Plant Mol. Biol.,* **20** : 857-867.

Conway LJ and Poethig RS (1997) Mutation of *Arabidopsis thaliana* that transform leaves into cotyledons. *Proc. Natl. Acad. Aci., U.S.A.,* **94** : 10209-10214.

Cramer CL, Edwards K, Dron M, Liang X, Dildine SL, Bolwell GP, Dixon RA, Lamb CJ and Schuch W (1989) Phenylalanine ammonia-lyase gene organisation and structure. *Plant Mol. Biol.,* **12** : 367-383.

Davin LB, Wang HB, Crowell AL, Bedgar DL, Martin DM, Sarkanen S and Lewis NG (1997) Stereoselective bimolecular phenoxy radical by an auxillary (Dirigent) protein without an active center. *Science,* **275** : 362-366.

Dean JFD and Eriksson KEL (1994) Laccase and the decomposition of lignin in vascular plants. *Holzforschung,* **48** : 21-33.

Dean JFD, LaFayette PR, Rugh CL, Merkle SA and Eriksson KEL (1996) The role of laccase in lignin biosynthesis. In : *Proceedings of Keystone Symp. Extracell. Matrix Plants : Mol. Cell Dev. Biol.,* Wiley-Liss, Tamarron, Colo, New York, p. 6.

Dean JFD, LaFayette PR, Rugh C, Tristram AH, Hoopes JT, Eriksson KEL and Merkle SA (1998) Laccases associated with lignifying vascular tissues. In : *Lignin and lignan biosynthesis* (Eds. Lewis N G and Sarkanen S), American Chemical Society, pp. 96-108.

Deborah G, Van Doorssalaere J, Yahaioui N, Samaj J, Grima-Pettenati J and Boudet A M (1998) A novel aromatic alcohol dehydrogenase in higher plants: Molecular cloning and expression. *Plant Mol. Biol.,* **36** : 755-765.

Dharmawardhana DP and Ellis BE (1998) β-glucosidases and glucosyl transferases in lignifying tissues, In : *ACS Symp. Series : Lignin and Lignan Biosynthesis.* Washington DC, **697** : 76-83.

Dharmawardhana DP, Ellis BE and Carlson JE (1995) β-glucosidases from lodgepole pine xylem specific for the lignin precursor coniferin. *Plant Physiol.,* **107** : 331-339.

Douglas CJ (1996) Phenylpropanoid metabolism and lignin biosynthesis: From weeds to trees. *Trends Plant Sci.,* **1** : 171-178.

Dumas B, Van Doorssalaere J, Gielen J, Legrand M, Fritig B, van Montagu M and Inze D (1992) Nucleotide sequence of a cDNA sequence encoding O-methyltransferase from poplar. *Plant Physiol.,* **98** : 796-797.

Driouich A, Laine AC, Vian B and Faye L (1992) Characterization and localization of laccase formation in stem and cell cultures of Sycamore. *Plant J.,* **2** : 13-24.

Dwivedi UN, Campbell WH, Datla RSS, Bugos RC, Chiang VL and Podila GK (1994) Manipulation of lignin biosynthesis in transgenic tobacco through expression of an antisense O-methyltransferase gene from quaking aspen. *Plant Mol. Biol.,* **26** : 61-71.

Dwivedi UN and Campbell WH (1995) Antisense inhibition and co-suppression of lignin biosynthesis in transgenic *Arabidopsis* harboring O-methyltransferase gene constructs from aspen. *Plant Physiol.* (*Life Sci. Adv.*), **14** : 45-54.

Elkind Y, Edwards R, Marandad M, Hedrick SA, Ribak O, Dixon RA and Lamb CJ (1990) Abnormal plant development and down-regulation of phenylpropanoid biosynthesis in transgenic tobacco containing a heterologous phenylalanine ammonia-lyase gene. *Proc. Natl. Acad. Sci., U.S.A.,* **87** : 9057-9061.

Eriksson KE, LaFayette PR, Merkle SA and Dean JFD (1995) Laccase as a target for decreasing lignin content in transgenic trees through antisense genetic engineering.

In : *Proceedings of 6th Int. Conf. on Biotechnol. in Pulp and Paper Industry* (Eds. Srebotnik E and Messner K.), Vienna, pp. 310-314.

Fahrendorf T and Dixon RA (1993) Stress responses in alfalfa (*Medicago sativa*) XVIII: Molecular cloning and expression of the elicitor-inducible cinnamic acid 4-hydroxylase cytochrome P450. *Arch. Biochem. Biophys.*, **305** : 509-515.

Feuillet C, Lauvergeat V, Deswarte C, Pilate G, Boudet AM and Grima-Pettenati J (1995) Tissue- and cell- specific expression of a cinnamyl alcohol dehydrogenase promoter in transgenic poplar plants. *Plant Mol. Biol.,* **27** : 651-667.

Freudenberg K (1968) The constitution and biosynthesis of lignin. In : *Constitution and biosynthesis of lignin* (Eds. Freudenberg K and Neish AC), Springer Verlag, Heidelberg, pp. 47-116.

Freudenberg K and Harkin JM (1963) The glucosides of cambial sap of spruce. *Phytochem,* **2** : 189-193.

Galliano H, Cabane M, Eckerskorn C, Lottspeich F, Sandermann H Jr. and Ernst D (1993) Molecular cloning, sequence analysis and elicitor/ ozone induced accumulation of cinnamyl alcohol dehydrogenase from Norway spruce (*Picea abies* L). *Plant Mol. Biol.,* **23** : 145-156.

Gowri G, Bugos RC, Campbell WH, Maxwell CA and Dixon RA (1991) Stress responses in alfalfa X: Molecular cloning and expression of S-Adenosyl-L-methionine: caffeic acid 3-O-methyltransferase, a key enzyme of lignin biosynthesis. *Plant Physiol.,* **97** : 7-14.

Grand C, Boudet AM and Ranjeva R (1982) Natural variations and controlled changes in lignification process. *Holzforschung,* **36** : 217-223.

Gregersen PL, Christensen AB, Sommer-Knudsen J and Collinge DB (1994) A putative O-methyltransferase from barley is induced by fungal pathogens and ultra violet light. *Plant Mol. Biol.,* **26** : 1797-1806.

Grima-Pettenati J, Campargue C, Boudet A and Boudet AM (1994) Purification and characterization of cinnamyl alcohol dehydrogenase isoforms from *Phaseolus vulgaris, Phytochem.,* **37** : 941-947.

Grima-Pettenati J, Feuillet C, Goffner D, Borderies G and Boudet AM (1993) Molecular cloning and expression of a *Eucalyptus gunnii* cDNA clone encoding cinnamyl alcohol dehydrogenase. *Plant Mol. Biol.,* **21** : 1085-1095.

Grima-Pettenati J and Goffner D (1999) Lignin genetic engineering revisited. *Plant Sci.,* **145** : 51-65.

Grycuik AA and Tsuyumu S (2000) Cloning of a genomic DNA encoding caffeoyl coenzyme A 3-O-methyltransferase of citrus. *J. Gen. Plant Pathol.,* **66** : 144-148.

Halpin C, Holt K, Chojecki J, Oliver D, Chabbert B, Monties B, Edwards K, Barakate A and Foxon GA (1998) Brown-midrib maize (bm 1)- a mutation affecting the cinnamyl alcohol dehydrogenase gene. *Plant J.,* **14** : 545-553.

Halpin C, Knight ME, Foxon GA, Campbell MM, Boudet AM, Boon JJ, Chabbert B, Tollier MT and Schuch W (1994) Manipulation of lignin quality by down-regulation of cinnamyl alcohol dehydrogenase. *Plant J.,* **6** : 339-350.

Hauffe KD, Paszkowski U, Schulzelefert P, Hahlbrock K, Dangl JL and Douglas CJ (1991) A parsley 4CL-1 promoter fragment specifies complex expression patterns in transgenic tobacco. *Plant Cell,* **3** : 435-443.

Heath R, Huxley H, Stone B and Spangenberg G (1998) cDNA cloning and differential expression of three caffeic acid O-methyltransferase homologues from perennial rye grass (*Lolium perenne*). *J. Plant Physiol.,* **153** : 649-657.

Hibino T, Shibata D, Chen JQ and Higuchi T (1993) Cinnamyl alcohol dehydrogenase from *Aralia cordata*: cloning of the cDNA and expression of the gene in lignified tissues. *Plant Cell Physiol.,* **34** : 659-665.

Hibino T, Takabe K, Kawazu T, Shibata D and Higuchi T (1995) Increase of cinnamaldehyde groups in lignin of transgenic tobacco plants carrying an antisense gene for cinnamyl alochol dehydrogenase. *Biosci. Biotechnol. Biochem.,* **59** : 929-931.

Higuchi T (1985) Biosynthesis of lignin. In : *Biosynthesis and Biodegradation of Wood Components* (Ed. Higuchi T), Academic Press, Orlando, FL, pp. 141-160.

Higuchi T, Ito T, Umezawa T, Hibino T and Shibata D (1994) Red brown color of transgenic plants with antisense cinnamyl alcohol dehydrogenase gene: wine red lignin from coniferyl aldehyde. *J. Biotechnol.,* **37** : 151-158.

Hoopes JT, Tristram AH, LaFayette PR, Eriksson KEL and Dean JFD (1995) Isolation of laccases and their encoding genes from *Zinnia elegans. Plant Physiol.,* **108** : 73.

Howles PA, Sewalt VJH, Paiva NL, Elkind Y, Bate NJ, Lamb C and Dixon RA (1996) Over-expression of L-phenylalanine ammonia-lyase in transgenic tobacco plants reveals control points for flux into phenylpropanoid biosynthesis. *Plant Physiol.,* **112** : 1617-1624.

Hu WJ, Lung L, Popko JL, Harding SA, Kawaoka A, Kao YY, Hideki S, Stokke DD, Rinaldi PL, Tsai CJ and Chiang VL (1998) Reduced lignin quality, increased cellulose content and enhanced growth in transgenic aspen trees. In : *8th Meeting of Conifer Biotechnol. Working Group,* Rutgers Univ., U.S.A.

Hu WJ, Harding SA, Lung J, Chiang VL, Ralph J, Douglas DS, Tsai CJ and Chiang VL (1999) Repression of lignin biosynthesis promotes cellulose accumulation and growth in transgenic trees. *Nat. Biotechnol.,* **17** : 808-812.

Ibrahim R and Grisebach H (1976) Purification and properties of UDP-G coniferyl alcohol glucosyl transferase from suspension cultures of Paul's scarlet rose. *Arch. Biochem. Biophys.,* **176** : 700-708.

Ito H, Kimizuka F, Ohbayashi A, Matsui H, Henma M, Shinmyo A, Ohashi Y, Caplan AB and Rodriguez RL (1994) Molecular cloning and characterization of two cDNA encoding putative peroxidases from rice shoots. *Plant Cell Rep.,* **13** : 361-366.

Jaeck E, Martz F, Stiefe, V, Fritig B and Legrand M (1996) Expression of class I O-methyltransferase in healthy and tobacco mosaic virus infected tobacco. *Mol. Plant Microbe Interact.,* **9** : 681-688.

Johansson A, Rasmussen SK, Harthill JE and Welinder KG (1992) cDNA, amino acid and carbohydrate sequence of barley seed-specific peroxidase BP-1. *Plant Mol. Biol.,* **18** : 1151-1161.

Jones DH (1984) Phenylalanine ammonia-lyase: regulation of its induction and its role in plant development. *Phytochem.,* **23** : 1349-1359.

Jones L and Turner S (1998) Irx-4a novel mutation affecting lignin biosynthesis. In : *Proceedings of 8th Int. Cell Wall Meet.,* Norwich.

Jorgensen LR (1931) Brown midrib in maize and its linkage relations. *J. Am. Soc. Agron.,* **23** : 549-557.

Jouanin L, Goujon T, deNadai V, Martin MT, Mila I, Vollet C, Pollet B, Yoshinaga A, Chabbert B, Petit-Conil M and Lapierre C (2000) Lignification in transgenic poplars with extremely reduced caffeic acid O-methyltransferase activity. *Plant Physiol.,* **123** : 1363-1373.

Jung HG and Deetz DA (1993) Cell wall lignification and degradability. In : *Forage cell wall structure and digestibility* (Eds. Jung H *et al.*), *Am. Soc. Agron.,* Madison, WI, pp. 315-340.

Jung HG and Vogel KP (1986) Influence of lignin on digestibility of forage cell wall material. *J. Anim. Sci.*, **62** : 1703-1712.

Kajita S, Hishiyama S, Tomimura Y, Katayama Y and Omori S (1997) Structural characterization of modified lignin in transgenic tobacco plants in which the activity of 4-coumarate: coenzyme A ligase is depressed. *Plant Physiol.*, **114** : 871-879.

Kajita S, Katayama Y and Omori S (1996) Alterations in the biosynthesis of lignin in transgenic plants with chimaeric genes for 4-coumarate: coenzyme A ligase. *Plant Cell Physiol.*, **37** : 957-965.

Kawaoka A and Ebinuma H (1999) Cloning and characterization of transcription factor Ntliml involved in lignin biosynthesis. In : *Proceedings of 10th Int. Symp. on Wood and Pulping Chemistry,* Yokohama, pp. 510-515.

Keifer-Meyer MC, Gomord V, O'Connell A, Halpin C and Faye L (1996) Cloning and sequence analysis of laccase encoding cDNA clones from tobacco. *Gene,* **178** : 205-207.

Kim KY, Kwon HK, Kwon SY, Lee HS, Hur Y, Bang JW, Choi KS and Kwak SS (2000) Differential expression of four sweet potato peroxidase genes in response to abscissic acid and ethephon. *Phytochem.*, **54** : 19-22.

Klotz KL, Liu TTY and Lagrimini LM (1998) Expression of the tobacco anionic peroxidase gene is tissue specific and developmentally regulated. *Plant Mol. Biol.*, **36** : 509-520.

Knight ME, Halpin C and Schuch W (1992) Identification and characterization of cDNA clones encoding cinnamyl alcohol dehydrogenase from tobacco. *Plant Mol. Biol.*, **19** : 793-801.

Kolattukudy PE, Mohan R, Bajar MA and Sherf BA (1992) Plant oxygenases, peroxidases and oxidases. *Biochem. Soc. Trans.*, **2** : 333-337.

Kristensen B, Bloch H and Rasmussen SK (1999) Barley coleoptile peroxidases: purification, molecular cloning and induction by pathogens. *Plant Physiol.*, **120** : 501-512.

Kuc J and Nelson OE (1964) The abnormal lignins produced by the bm mutants of maize. *Arch. Biochem. Biophys.*, **105** : 103-113.

Kutsuki H. and Higuchi T (1978) The formation of lignin in *Erythrina cristagalli. Mokuzai Gakkaishi.*, **24** : 625-631.

Lacombe E, Hawkins S, Van Doorssalaere J, Piquemal J, Goffner D, Poedomenge O, Boudet AM and Grima-Pettenati J (1997) Cinnamoyl coenzyme A reductase, the first committed enzyme of lignin branch biosynthetic pathway: cloning, expression and phylogenic relationships. *Plant J.*, **11** : 429-442.

Lacombe E, Poedomenge O, Van Doorssalaere J, Piquemal J, Goffner D, Boudet AM and Grima-Pettenati J (1994) Cloning and characterization of an *Eucalyptus gunnii* cDNA encoding cinnamoyl coenzyme A reductase, a key enzyme in lignification. In : *22nd Aharon Katzir-Katchalsky Conf. on Plant Mol. Biol. Biotechnol. Environ.*, Koln, Germany.

LaFayette PR, Eriksson KEL and Dean JFD (1995) Nucleotide sequence of a cDNA clone encoding an acidic laccase from Sycamore maple (*Acer pseudoplatanus* L.) *Plant Physiol.*, **107** : 667-668.

LaFayette PR, Eriksson KEL and Dean JFD (1999) Characterization and heterologous expression of laccase cDNAs from xylem tissues of yellow-poplar (*Liriodendron tulipifera*). *Plant Mol. Biol.*, **40** : 23-35.

LaFayette PR, Merkle SA Eriksson KEL and Dean JFD (1994) Molecular characterization of hardwood laccase genes and their application in lignin reduction *via* antisense technology. In : 2nd *Int. Symp. on Applications of Biotechnol. to Tree Culture, Protection & Utilization,* Minnesota, U.S.A.

Lagrimini LM (1991) Wound induced deposition of polyphenols in transgenic plants overexpressing peroxidase. *Plant Physiol.,* **96** : 577-583.

Lagrimini LM (1993) Analysis of peroxidase function in transgenic plants. In : *Plant Peroxidase: Biochem. Physiol.* (Eds. Welinder K G *et al.*), Switzerland, pp. 301-306.

Lagrimini LM, Bradford S and Rothstein S (1990) Peroxidase induced wilting in transgenic tobacco plants. *Plant Cell,* **2** : 7-18.

Lagrimini LM, Burkhart W, Moyer M and Rothstein S (1987) Molecular cloning of cDNA encoding the lignin forming peroxidase from tobacco: molecular analysis and tissue specific expression. *Proc. Natl. Acad, Sci., U.S.A.,* **84** : 7542-7546.

Lagrimini LM, Gingas V, Finger F, Rothstein S and Liu TTY (1997) Characterization of antisense transformed plants deficient in the tobacco anionic peroxidase. *Plant Physiol.,* **114** : 1187-1196.

Lagrimini LM, Joly RL, Dunlap JR and Liu TY (1997) The consequence of peroxidase over-expression in transgenic plants on root growth and development. *Plant Mol. Biol.,* **33** : 887-895.

Lagrimini LM, Vaughn J, Erb W and Miller SA (1993) Peroxidase over production in tomato: wound-induced polyphenol deposition and disease resistance. *Hort. Sci.,* **28** : 218-221.

Lapierre C, Pollet B, Petit-Conil M, Toval G, Romero J, Pilate G, Leple JC, Boerjan W, Ferrat V, Nadai VD and Jouanin L (1999) Structural alterations in lignin in transgenic poplars with depressed cinnamyl alcohol dehydrogenase or caffeic acid O-methyltransferase activity have an opposite impact on the efficiency of industrial kraft pulping. *Plant Physiol.,* **119** : 153-163.

Lee D, Ellard M, Wanner LA, Savis KR and Douglas CJ (1995) The *Arabidopsis thaliana* 4-coumarate: coenzyme A ligase gene: stress and developmentally regulated expression and sequence of its cDNA. *Plant Mol. Biol.,* **28** : 871-884.

Lee D, Meyer K, Chapple C and Douglas CJ (1997) Antisense suppression of 4-coumarate: conenzyme A ligase activity in *Arabidopsis* leads to altered lignin subunit composition. *Plant Cell,* **9** : 1985-1998.

Leinhos V, Udagama-Randeniya PV and Savidge RA (1994) Purification of an acid coniferin-hydrolysing β-glucosidase from developing xylem of *Pinus banksiana. Phytochem.,* **37** : 311-315.

Lewis NG and Yamamoto E (1990) Lignin: occurrence, biogenesis and biodegradation. *Annu. Rev. Plant Physiol. Plant Mol. Biol.,* **41** : 455-496.

Li L, Osakabe Y, Joshi CP and Chiang VL (1999) Secondary xylem-specific expression of caffeoyl-coenzyme A 3-O-methyltransferase plays an important role in the methylation pathway associated with lignin biosynthesis in loblolly pine. *Plant Mol. Biol.,* **40** : 555-565.

Li L, Popko JL, Zhang XH, Osakabe K, Tsai CJ, Joshi CP and Chaing VL (1997) A novel multifunctional O-methyltransferase implicated in a dual methylation pathway associated with lignin biosynthesis in loblolly pine. *Proc. Natl. Acad. Sci., U.S.A.* **94** : 5461-5466.

Lige B, Ma S and van Huystee RB (1998) Expression of the peanut peroxidase cDNA in *E. coli* and purification of protein by nickel affinity. *Plant Physiol. Biochem.,* **36** : 335-338.

Liu L, Dean JFD, Friedman WE and Eriksson KEL (1994) A laccase-like phenoloxidase is correlated with lignin biosynthesis in *Zinnia elegans* stem tissues. *Plant J.,* **6** :

213-224.

Lois R, Dietrich A, Hahlbrock K and Schuch W (1989) A phenylalanine ammonia-lyase gene from parsley: structure, regulation and identification of elicitor and light responsive cis acting elements. *EMBO J.,* **8** : 1641-1648.

Lozoya E, Hoffmann H, Douglas CS, Scheel D and Hahlbrock AK (1989) Primary structure and catalytic properties of isoenzymes encoded by the two 4-coumarate: coenzyme A ligase genes in parsley. *Eur. J. Biochem.,* **176** : 661-667.

Mackay JJ, Liu W, Whetten RW, Sederoff RR and O'Malley DM (1995) Genetic analysis of cinnamyl alcohol dehydrogenase in loblolly pine: single gene inheritance, molecular characterization and evolution. *Mol. Gen. Genet.,* **247** : 537-545.

Mackay JJ, O'Malley DM, Presnell T, Fitzgerald LB, Campbell MM, Whetten RW and Sederoff RR (1997) Inheritance, gene expression and lignin characterization in a mutant pine deficient in cinnamyl alcohol dehydrogenase. *Proc. Natl. Acad. Sci., U.S.A.,* **94** : 8255-8260.

Maher EA, Bate NJ, Ni W, Elkind Y, Dixon RA and Lamb CJ (1994) Increased disease susceptibility of transgenic tobacco plants with suppressed levels of preformed phenylpropanoid products. *Proc. Natl. Acad. Sci., U.S.A.,* **91** : 7802-7806.

Martz F, Maury S, Pincon G and Legrand M (1998) cDNA cloning, substrate specificity and expression study of tobacco CCoAOMT, a lignin biosynthetic enzyme. *Plant Mol. Biol.,* **36** : 427-437.

McIntyre CL, Bettenay HM and Manners JM (1996) Strategies for the suppression of peroxidase gene expression in tobacco II: *In vivo* suppression of peroxidase activity in transgenic tobacco using ribozyme and antisense constructs. *Transg. Res.,* **5** : 263-270.

McIntyre CL, Rae AL, Curtis MD and Manners JM (1995) Sequence and expression of a caffeic acid O-methyltransferase cDNA homologue in the tropical forage legume *Stylosanthes humilis. Aust. J. Plant Physiol.,* **22** : 471-478.

Meng H and Campbell WH (1995) Cloning of aspen (*Populus tremuloides*) xylem caffeoyl coenzyme A 3-O methyltransferase. *Plant Physiol.* **108** : 1749.

Meyer K, Cusumano JC, Ruegger M, Bell-Lelong DA and Chapple CCS (1997) Regulation and manipulation of lignin monomer composition by expression of F5H, a cytochrome P450 dependent monooxygenase required for S lignin biosynthesis. *Int. Wood Biotechnol. Symp. 2nd* Canberra, p 9.

Meyer K, Cusumano JC, Sommerville C and Chapple CCS (1996) F5H from *Arabidopsis thaliana* defines a new family of cytochrome P450 dependent monooxygenases. *Proc. Natl. Acad. Sci., U.S.A.,* **93** : 6869-6874.

Meyer K, Shirley AM, Cusumano JC, Bell-Lelong D and Chapple C (1998) Lignin monomer composition is determined by the expression of a cytochrome P450 dependent monooxygenase. *Proc. Natl. Acad. Sci., U.S.A.,* **95** : 6619-6623.

Meyermans H, Moreel K, Lapierre C, Pollet B, DeBruyn A, Busson R, Herdewijn P, Devreese B, van Beeuman J, Marita JM, Ralph C, Chen C, Burggraeve B, van Montagu M, Messens E and Boerjan W (2000) Modification in lignin and accumulation of phenolic glucosides in poplar xylem upon down-regulation of caffeoyl-coenzyme A 3-O-methyltransferase, an enzyme involved in lignin biosynthesis, *J. Biol. Chem.,* **275** : 36899-36909.

Minami E, Ozeki Y, Matsuoka M, Koizuka N and Tanaka Y (1989) Structure and some characterization of the gene for phenylalanine ammonia-lyase from rice plants. *Eur. J. Biochem.,* **185** : 19-25.

Mizutani M, Ward E, DiMaio J, Ohta D, Ryals J and Sato R (1993) Molecular cloning and sequencing of a cDNA encoding mung-bean cytochrome P450 (P450C4H) possessing cinnamate 4-hydroxylase activity. *Biochem. Biophys. Res. Commun.,* **190** : 875-880.

Monties B (1989) Molecular structure and biochemical properties of lignins in relation with possible self organisation of lignin networks. *Annales des Sciences Forestieres,* **46** : 846-855.

Myton K, Stewart D, Yahiaoui N, McDougall G, Marque C and Boudet AM (1995) Lignin characterization in cinnamyl alcohol dehydrogenase down-regulated tobacco. In : *Cell Wall Meeting CW95,* Santiago de Compostela.

Nagai N, Kitauchi F, Shimosaka M and Okazaki M (1994) Cloning and sequencing of a full-length cDNA coding for phenylalanine ammonia-lyase from tobacco cell cultures. *Plant Physiol.,* **104** : 1091-1092.

Ni W, Paiva NL and Dixon RA (1994) Reduced lignin in transgenic plants containing a caffeic acid O-methyltransferase antisense gene. *Transg. Res.* **3** : 120-126.

O'Connell A, Lapierre C and Pollet B (1998) Reduction of lignin content and significant changes to lignin profiles through inhibition of cinnamoyl coenzyme A reductase. In : *Proceedings of 8th Int. Cell Wall Meeting,* Norwich.

Ohl S, Hedrick SA, Chory J and Lamb CJ (1990) Functional properties of a phenylalanine ammonia-lyase promoter from *Arabidopsis. Plant Cell,* **2** : 837-848.

O'Malley DM, Porter S and Sederoff RR (1992) Purification, characterization and cloning of cinnamyl alcohol dehydrogenase in loblolly pine (*Pinus taeda* L). *Plant Physiol.,* **98** : 1364-1371.

O'Malley DM, Whetten RW, Bao W, Chen CL and Sederoff RR (1993) The role of laccase in lignification. *Plant J.,* **4** : 751-757.

Osakabe K, Tsao CC, Li L, Popko JL, Umezawa T, Carraway DT, Smeltzer RH, Joshi CP and Chiang VL (1999) Coniferyl aldehyde 5-hydroxylation and methylation direct syringyl lignin biosynthesis in angiosperms. *Proc. Natl. Acad. Sci., U.S.A.,* **96** : 8955-8960.

Pichon M, Courbou J, Beckert M and Boudet AM (1998) Cloning and characterization of two maize cDNAs encoding cinnamoyl coenzyme A reductase and differential expression of the corresponding genes. *Plant Mol. Biol.,* **38** : 671-676.

Pillonel C, Mulder MM, Boon JJ, Forster B and Binder A (1991) Involvement of cinnamyl alcohol dehydrogenase in the control of lignin formation in *Sorghum bicolor* L. Moench. *Planta,* **185** : 538-544.

Piquemal J, Lapierre C, Myton K, O'Connell A, Schuch W, Grima-Pettenati J and Boudet AM (1998) Down-regulation of cinnamoyl coenzyme A reductase induces significant change of lignin profiles in transgenic tobacco plants. *Plant J.,* **13** : 71-83.

Poedomenge O, Boudet AM and Grima-Pettenati J (1994) A cDNA encoding S-Adenosyl-methionine: caffeic acid 3-O-methyltransferase from *Eucalyptus. Plant Physiol.,* **105** : 749-750.

Provan GJ, Scobbie L and Chesson A (1997) Characterization of lignin from cinnamyl alcohol dehydrogenase and O-methyltransferase deficient mutants of maize. *J. Sci. Food Agric.,* **73** : 133-142.

Przemeck GKH, Mattson J, Hardtke CS, Sung ZR and Berleth T (1996) Studies on the role of the *Arabidopsis* gene monopteros in vascular development and plant cell axialization. *Planta,* **200** : 229-237.

Ralph J, Hatfield RD, Piquemal J, Yahiaoui N, Pean M, Lapierre C and Boudet AM (1998) NMR characterization of altered lignins extracted from tobacco plants down-regulated for lignification enzymes cinnamyl alcohol dehydrogenase and cinnamoyl coenzyme A reductase. *Proc. Natl. Acad. Sci., U.S.A.,* **95** : 12803-12808.

Ralph J, Mackay JJ, Hatfield RD, O'Malley DM, Whetten RW and Sederoff RR (1997) Abnormal lignin in loblolly pine mutant. *Science,* **277** : 235-239.

Ranocha P, McDougall G, Hawkins S, Sterjiades R, Borderies G, Stewart D, Cabanes-Macheteau M, Boudet AM and Goffner D (1999) Biochemical characterization, molecular cloning and expression of laccases, a divergent gene family – in poplar. *Eur. J. Biochem.,* **259** : 485-495.

Rasmussen R (1995) cDNA cloning and systemic expression of acidic peroxidases associated with systemic acquired resistance to disease in cucumber. *Physiol. Mol. Plant Pathol.,* **46** : 389-400.

Rebmann G, Hertig G, Bull J, Mauch F and Dudler R (1991) Cloning and sequencing of cDNAs encoding a pathogen induced putative peroxidase of wheat. *Plant Mol. Biol.,* **16** : 329-331.

Roberts E and Kolattukudy PE (1988) Cloning and sequencing of cDNAs for a highly anionic peroxidase from potato and the induction of its mRNA in suberizing potato tubers and tomato fruits. *Plant Mol. Biol.* **11** : 15-26.

Rochus F, McMichael CM, Meyers K, Shirley K, Cusumano JC and Chapple C (2000) Modified lignin in tobacco and poplar plants over-expressing the *Arabidopsis* gene encoding ferulate 5-hydroxylase. *Plant J.,* **22** : 223-224.

Rothermal BA and Nelson T (1989) Primary structure for the maize NADP-dependent malic enzyme. *J. Biol. Chem.,* **264** : 19587-19592.

Sarkanen KV (1971) Precursors and their polymerization. In *Lignins: occurrence, formation, structure and reactions* (Eds. Sarkanen K V and Ludwig C G), Wiley-Interscience, New York, pp. 95-163.

Sarkanen KV, Chang HM and Allan GG (1967) Species variation in lignins III Hardwood lignins. *TAPPI J.,* **50** : 587-590.

Savidge RA (1989) Coniferin, a biochemical indicator of commitment to trachied differentiation in conifers. *Can. J. Bot.,* **67** : 2663-2668.

Savidge R and Udagama-Randeniya P (1992) Cell wall bound coniferyl alcohol oxidase associated with lignification in conifers. *Phytochem.,* **31** : 2959-2966.

Schmitt D, Pakusch AE and Matern U (1991) Molecular cloning, induction and taxonomic distribution of caffeoyl coenzyme A 3-O methyltransferase, an enzyme involved in disease resistance. *J. Biol. Chem.,* **266** : 17416-17423.

Sederoff R (1999) Building better trees with antisense. *Nat. Biotechnol.,* **17** : 750-751.

Sederoff RR, Campbell MM, O'Malley D and Whetten R (1994) Genetic regulation of lignin biosynthesis and potential modification of wood by genetic engineering in loblolly pine. In : *Genetic Engineering of Plant Secondary Metabolism* (Eds. Ellis B E *et al.*), Plenum Press, New York.

Selman-Housein G, Lopez MA, Hernandez D, Civardi L, Miranda F, Rigau J and Puigdomenech P (1999) Molecular cloning of cDNAs coding for three sugarcane enzymes involved in lignification. *Plant Sci.,* **143** : 163-171.

Sewalt VJH, Ni W, Blount JW, Jung HG, Masoud SA, Howles PA, Lamb CJ and Dixon RA (1997) Reduced lignin content and altered lignin composition in transgenic tobacco down-regulated in expression of L-phenylalanine ammonia-lyase or cinnamate 4-hydroxylase. *Plant Physiol.,* **115** : 41-50.

Seymour GB, Fray RG, Hill P and Tucker GA (1993) Down-regulation of two non-homologous endogenous tomato genes with a single chimaeric sense gene construct. *Plant Mol. Biol.,* **23** : 1-9.

Sherf BA, Bajar AM and Kolattukudy PE (1993) Abolition of an inducible highly anionic peroxidase activity in transgenic tomato. *Plant Physiol.,* **101** : 201-208.

Sterjiades R, Dean JFD and Eriksson KEL (1992) Laccase from Sycamore maple (*Acer pseudoplatanus*) polymerizes monolignols. *Plant Physiol.,* **99** : 1162-1168.

Sterjiades R, Dean JFD, Gamble G, Himmelsbach DS and Eriksson KEL (1993) Extracellular laccases and peroxidases from Sycamore maple (*Acer pseudoplatanus*) cell suspension culture: reactions with monolignols and lignin model compounds. *Planta,* **190** : 75-87.

Stewart D, Yahiaoui N, McDougall GJ, Myton K, Marque C, Boudet AM and Haigh J (1997) Fourier Transform Infrared and Raman spectroscopic evidence for the incorporation of cinnamaldehydes into the lignin of transgenic tobacco (*N. tabacum*) plants with reduced expression of cinnamyl alcohol dehydrogenase. *Planta,* **201** : 311-318.

Subramaniam R, Reinolds S, Molitor EK and Douglas CJ (1993) Structure, inheritance and expression of hybrid poplar (*Populus trichocarpa* × *Populus deltoides*) phenylalanine ammonia-lyase genes. *Plant Physiol.,* **102** : 71-83.

Talas OT, Kazan K and Gozukirmiri V (2000) Reduction of lignin in tobacco through the expression of an antisense caffeic acid O-methyltransferase. *Turkish J. Bot.,* **24** : 221-226.

Tamagnone L, Merida A, Parr A, Mackay S, Culianez-Macia FA, Roberts K and Martin C (1998) The AmMYB308 and AmMYB330 transcription factors from *Antirrhinum* regulate phenylpropanoid and lignin biosynthesis in transgenic tobacco. *Plant Cell,* **10** : 135-154.

Tanaka Y, Matsuoka M, Yamamoto N, Ohashi Y, Kano-Murakami Y and Ozaki Y (1989) Structure and characterization of a cDNA clone for phenylalanine ammonia-lyase from cut-injured roots of sweet potato. *Plant Physiol.,* **90** : 1403-1407.

Taochim GJH and Ni W (1998) Lignification of plant cell walls: Impact of genetic manipulation. *Proc. Natl. Acad. Sci., U.S.A.,* **95** : 12742-12743.

Teichmann T, Guan C, Kristoffersen P, Muster G, Teitz O and Palme K (1997) Cloning and biochemical characterization of an anionic peroxidase from *Zea mays. Eur. J. Biochem.,* **247** : 826-832.

Terezawa M, Okuyama H and Miyake M (1984) Seasonal variations in phenolic glycosides in the cambial sap of wood. *Mokuzai Gakkaishi,* **30** : 329-334.

Teutsch HG, Hasenfratz MP, Lesot A, Stoltz C, Garnier JM, Jeltsch JM, Durst F and Werck-Reichhart D (1993) Isolation and sequencing of a cDNA encoding the Jerusalem artichoke cinnamate 4-hydroxylase, a major plant cytochrome P450 involved in the general phenylpropanoid pathway. *Proc. Natl. Acad. Sci., U.S.A.,* **90** : 4102-4106.

Thordal-Christensen H, Brandt J, Cho BH, Rasmussen SK, Gregerson PL, Smedegaard-Peterson V and Collinge DB (1992) cDNA cloning and characterization of two barley peroxidase transcripts induced differentially by the powdery mildew fungus *Erysiphe graminis. Phys. Mol. Plant Path.,* **40** : 395-409.

Tollier MT, Chabbert B, Lapierre C, Monties B, Francesch C, Rolando C, Jouanin L, Pilate G, Cornu D, Baucher M and Inze D (1995) Lignin composition in transgenic poplar plants with modified cinnamyl alcohol dehydrogenase activity with respect to dehydropolymer models of lignin. In : *Polyphenols 94* (Eds. Brouillard R *et al.*), INRA Editions, Paris, pp. 339-340.

Tsai CJ, Popko JL, Mielke MR, Hu WJ, Podila GK and Chiang VL (1998) Suppression of O-methyltransferase gene by homologous sense transgene in quaking aspen causes red brown wood phenotypes. *Plant Physiol.,* **117** : 101-112.

Uhlmann A and Ebel J (1993) Molecular cloning and expression of 4-coumarate: coenzyme A ligase, an enzyme involved in the resistance of soybean against pathogen infection. *Plant Physiol.,* **102** : 1147-1156.

Van Doorssalaere J, Baucher M, Chognot E, Chabbert B, Tollier M T, Petti-Conil M, Leple JC, Pilate G, Cornu D, Monties B, van Montagu M, Inze D, Boerjan W and Jouanin L (1995) A novel lignin in poplar trees with a reduced caffeic acid/5-hydroxyferulic acid O-methyl transferase activity. *Plant J.,* **8** : 855-864.

Van Doorssalaere J, Baucher M, Feuillet C, Boudet AM, van Montagu M and Inze D (1995) Isolation of cinnamyl alcohol dehydrogenase cDNAs from two important economic species: alfalfa and poplar. Demonstration of high homology of the gene within angiosperms. *Plant Physiol. Biochem.,* **33** : 105-109.

Van Doorssalaere J, Cabrillae D, Grima-Pettenati J and Boudet AM (1994) Isolation of *Eucalyptus gunnii* clones coding for cinnamoyl coenzyme A reducatse, a key enzyme in lignin synthesis. In : *22nd Aharon Katzir-Katchalsky Conf. on Plant Mol. Biol. Biotechnol. Environ.,* Koln, Germany.

Vernon DM and Bohnert HJ (1992) A novel methyltransferase induced by osmotic stress in the facultative halophyte *Mesenbryanthemum crystallinum. EMBO J.,* **11** : 2077-2085.

Vignols F, Rigau J, Torres MA, Capellades M and Puigdomenech P (1995) The brown-midrib 3 (bm3) mutation in maize occurs in the gene encoding caffeic acid O-methyltransferase. *Plant Cell,* **7** : 407-416.

Walter MH, Grima-Pettenati J, Grand C, Boudet AM and Lamb CJ (1988) Cinnamyl alcohol dehydrogenase, a molecular marker specific for lignin synthesis: cDNA cloning and mRNA induction by fungal elicitor. *Proc. Natl. Acad. Sci., U.S.A.,* **85** : 5546-5550.

Walter MH, Grima-Pettenati J, Grand C, Boudet AM and Lamb CJ (1990) Extensive sequence similarity of the bean CAD 4 to a maize malic enzyme. *Plant Mol. Biol.,* **15** : 525-526.

Wanner LE, Li G, Ware DH, Somssich I and Davis KR (1995) the phenylalanine ammonia-lyase gene family in *Arabidopsis thaliana. Plant Mol. Biol.,* **27** : 327-338.

Weisshar B and Jenkins GI (1998) Phenylpropanoid biosynthesis and its regulation. *Curr. Opin. Plant Biol.,* **1** : 251-257.

Whetten RW, Mackay JJ and Sederoff RR (1998) Recent advances in understanding lignin biosynthesis. *Annu. Rev. Plant Physiol. Plant Mol. Biol.,* **49** : 585-609.

Whetten RW and Sederoff RR (1991) Genetic Engineering of Wood. *J. Forest Ecol. Mange.,* **43** : 301-316.

Whetten RW and Sederoff RR (1992) Phenylalanine ammonia-lyase from loblolly pine: purification of the enzyme and the isolation of cDNA clones. *Plant Physiol.,* **98** : 380-386.

Whetten RW and Sederoff RR (1995) Lignin Biosynthesis. *Plant Cell,* **7** : 1001-1013.

World Resources Institute, World Resources 1994-1995, Oxford Univesity Press, 1994.

Wu RL, Remington DL, Mackay JJ, Mckeand SE and O'Malley DM (1999) Average effect of a mutation in lignin biosynthesis in loblolly pine. *Theor. Appl. Genet.*, **99** : 705-710.

Wyrambic D and Grisebach H (1975) Purification and properties of isoenzymes of cinnamyl alcohol dehydrogenase from soybean cell suspension cultures. *Eur. J. Biochem.*, **59** : 9-15.

Xue ZT and Brodelius PE (1998) Kinetin induced caffeic acid O-methyltransferase in cell suspension cultures of *Vanilla planifolia* Andr. and isolation of caffeic acid O-methyltransferase cDNAs. *Plant Physiol. Biochem.*, **36** : 779-788.

Yahiaoui N, Marque C, Myton K, Negrel J and Boudet AM (1998) Impact of different levels of cinnamyl alcohol dehydrogenase down-regulation on lignins of transgenic tobacco plants. *Planta,* **204** : 8-15.

Yamada T, Tanaka Y, Sriprasertsak P, Kato H, Hashimoto T, Kawamata S, Ichinose Y, Kato H, Shiraishi T and Oku H (1992) Phenylalanine ammonia-lyase genes from *Pisum sativum* : structure organ specific expression and regulation by fungal elicitor and suppressor. *Plant Cell Physiol.*, **33** : 715-725.

Yazaki K, Ogawa A and Tabata MC (1995) Isolation and characterization of two cDNAs encoding 4-coumarate : coenzyme A ligase in *Lithospermum* cell cultures. *Plant Cell Physiol.*, **36** : 1319-1329.

Ye ZH (1997) Association of caffeoyl coenzyme A 3-O methyltransferase expression with lignifying tissues in several dicot plants. *Plant Physiol.*, **115** : 1341-1350.

Ye ZH, Kneusel RE, Matern U and Varner JE (1994) An alternative methylation pathway in lignin biosynthesis in *Zinnia elegans. Plant Cell,* **6** : 1427-1439.

Ye ZH and Varner JE (1995) Differential expression of two O-methyltransferase in lignin biosynthesis in *Zinnia elegans. Plant Physiol.*, **108** : 459-467.

Ye ZH, Zhong R and Ripperger A (1998) Ectopic deposition of lignin in pith of *Arabidopsis* mutants. In : *9th Int. Conf. on Arabidopsis Research,* Univ. Wisconsin, Madison.

Zhang XH and Chiang VL (1997) Molecular cloning of 4-coumarate : coenzyme A ligase in loblolly pine and the roles of this enzyme in the biosynthesis of lignin in compression wood. *Plant Physiol.*, **113** : 65-74.

Zhang XH, Dickson EE and Chinnappa CC (1995) Nucleotide sequence of a cDNA clone encoding caffeoyl coenzyme A 3-O methyltransferase of *Stellaria longipes* (Caryophyllaceae). *Plant Physiol.*, **108** : 429-430.

Zhao Y, Kung SD and Dube SK (1990) Nucleodide sequence of rice 4-coumarate : coenzyme A ligase gene 4CL-1. *Nucl. Acids Res.*, **18** : 6144.

Zhong R, Morrison WH, Himmelsbach DS, Poole FL and Ye ZH (2000) Essential role of caffeoyl-coenzyme A 3-O-methyltransferase in lignin biosynthesis in woody poplar plants. *Plant Physiol.*, **124** : 563-577.

Zhong R, Morrison WH, Negrel J and Ye ZH (1998) Dual methylation pathways in lignin biosynthesis. *Plant Cell,* **10** : 2033-2046.

Zhong R, Taylor JJ and Ye ZH (1997) Disruption of interfascicular fiber differentiation in an *Arabidopsis* mutant. *Plant Cell,* **9** : 2159-2170.

Zhong R, Taylor JJ and Ye ZH (1999) Transformation of the collateral vascular bundles into amphivasal vascular bundles in an *Arabidopsis* mutant. *Plant Physiol.*, **120** : 53-64.

Chapter 15

MANIPULATION OF FLOWER COLOUR BY GENETIC ENGINEERING

Yoshikazu Tanaka★ and John Mason[1]

★*Institute for Fundamental Research, Suntory Ltd., 1-1-1 Wakayamadai, Shimamoto-cho, Mishima-gun, Osaka, 618-8503, Japan*
[1]*Florigene Ltd., 16 Gipps St. Colligwood, Victoria 3066, Australia*

Summary

Flower colour is one of the most important characters used by consumers in selecting flowers at the point of purchase. The colours observed in nature are due to the presence of a range of compounds classed as flavonoids, carotenoids or betalains. Anthocyanins, a coloured class of flavonoids, are the most widely distributed and have a broad colour spectrum. The biosynthetic pathways leading to simple anthocyanins, anthocyanidin 3-glucosides, have been well characterized and are conserved among flowering plants. The genes encoding enzymes involved in biosynthesis of flavonoids that affect flower colour and transcriptional factors regulating the flavonoid biosynthetic pathway have recently been studied in some depth. Furthermore, plant transformation techniques are continually being improved both in the efficiency within a given variety and by extension of the capability of additional varieties or species. Such progress has enabled manipulation of flower colour by genetic engineering. An example of this is the violet Moon™ series carnations expressing heterologous flovonoid biosynthetic pathway genes that have been successfully commercialized in a number of countries including Japan, Australia and USA.

Keywords : Anthocyanins, colour, flavonoid, flower, genetic engineering

---

★Corresponding author : E-mail : Yoshikazu_Tanaka@suntory.co.jp

Abbreviation : AAT; aromatic acyltransferase, ANS; anthocyanidin synthase, CHI; chalcone isomerase, CHS; chalcone synthase, DFR; dihydroflavonol 4-reducatse, F3H; flavanone 3-hydroxylase, F3'H; flavonoid 3'-hydroxylase, F3'5'H; flavonoid 3',5'-hydroxylase, FLS; flavonol synthase; FNS; flavone synthase, 3GT: flavonoid 3-glucosyl transferase, 5GT; flavonoid 5-glucosyl transferase

1. INTRODUCTION

Traditional plant breeding has been used for more than a thousand years to improve food production and has enriched our lives through the production of hundreds of varieties of ornamental flower species. Although until recently these methods have produced all available commercial varieties of flowers, the plant breeder remains constrained by the limits of a given species' genes pool and a lack of precision.

The aesthetic appeal of a flower is a combination of many factors including colour, scent and form; modification of one character through classic hybridization can often be at the expense of another equally valuable feature. Recombinant DNA techniques provide an opportunity to produce novel and elite flower varieties that will command a high premium in the market place. Genetic engineering techniques can be incorporated into breeding programs to produce novel characters by the introduction of genes from unrelated species. In addition, antisense, co-suppression or RNAi technology can be used to inactivate the expression of endogenous genes so leading to the development of novel colours in a particular species or cultivar. An added advantage of such techniques is that traits can be added, singly or in combination, to a given variety leaving other characters unaltered. Development times for new varieties can in this way often be significantly reduced adding to the attraction of such methodology.

These methods are already being used to modify flower colour and have been reviewed on a number of occasions (Davies and Schwinn, 1997; Mol *et al.*, 1998; Tanaka *et al.*, 1998; Tanaka *et al.*, 1999; Mol *et al.*, 1999). In this chapter we focus on recent progress in understanding flavonoid biosynthesis and the application of this knowledge to the production of novel coloured flowers which are directed at the marketplace.

2. FLOWER COLOUR CHEMISTRY

Flavonoids, carotenoids and betalains are the major determinants of flower colour. Carotenoids are largely responsible for the production of yellow and orange flowers such as sunflower and daffodil (Bartley and Scolnik, 1995). Betalains are yellow to red nitrogenous compounds derived from tyrosine and their distribution is confined to the Caryophyllales (Stafford, 1994). Flavonoids have a wide range of colours from pale yellow to red, purple and blue (Goto and Kondo, 1991). Flavonoids are frequently present with carotenoids leading to dark red and bronze colours often seen in chrysanthemums for example. The flavonoid biosynthetic pathway has been the primary target of efforts to engineer novel flower colour. In this article we will confine discussion to the flavonoids.

Anthocyanins are a coloured class of flavonoids and accumulate in vacuoles usually of petal peidermal cells. Anthocyanidins are aglycon forms and chromophores of anthocyanins. There are six major anthocyanidins; pelarogonidin, cyanidin, peonidin (3' *O*-methyl cyanidin), delphinidin, petunidin (3' *O*-methyl delphinidin) and malvidin (3', 5' *O*-dimethyl delphinidin). As the number of hydroxyl groups on the B-ring is increased, the visible absorption maximum is shifted to longer absorption maximum. Anthocyanin colour depends on pH. They are in flavylium ion (protonated) forms in acidic condition and the colour is redder and stable. They are bluer in neutral to alkaline condition but unstable unless they are stabilized with proper modification and complexation with copigments or metal ions for example (Goto and Kondo, 1991; Brouillard and Dangles, 1994).

Hundreds of anthocyanins have been purified and their structures determined (Strack and Wray, 1994). Modification of anthocyanidins through glycosylation and acylation results in a wide range of anthocyanins. A hydroxyl group of 3 position is always glycosylated (usually glucosylated), that of 5 position is often glycosylated. Anthocyanidin 3, 5-diglucosides are commonly found in flowers. Hydroxyl groups of 7, 3' and 5' positions are sometimes glycosylated. These glycosyl groups are often further modified with aliphatic and aromatic acyl groups. Aliphatic acylation increases solubility of anthocyanins. Aromatic acylation contrinbutes to stabilization and bluing of anthocyanins. Such modifications contribute to the observed colour variation in anthocyanin-based flower pigments. When anthocyanins complex with polyphenols, their absorbance wavelength shifts longer by upto 35 nm (bathocromic shift, Goto and Kondo, 1991). This effect is called copigmentation and also results in a large increase in absorptivity. Intermolecular stacking (self-association of anthocyanin and copigmentation), intramolecular stacking of anthocyanins modified with poly aromatic acids and metal complexation increase stability of anthocyanins by preventing water molecules from hydration of the chromophores (Goto and Kondo, 1991; Brouillard and Dangles, 1994). Variation in epidermal cell shape also leads to flower colour changes (Mol *et al.*, 1998).

3. FLAVONOID BIOSYNTHESIS

Since the flavonoid biosynthetic pathway shown in Figure 1 has been extensively reviewed (Holton and Cornish, 1995; Davies, 1997; Mol *et al.*, 1998; Tanaka *et al.*, 1998), recent progress only is the focus of this section. The pathway leading to anthocyanidin 3-glucosides is generally conserved among plant species and well understood.

Anthocyanidin 3-glucosides are further modified with sugars and both aliphatic and aromatic acids. There are both species and variety differences in the extent of modification and the types of glycosyl and acyl groups attached (Figure 2).

The first enzyme committed to flavonoid production is chalcone synthase (CHS) that catalyzes the stepwise condensation of three acetate units starting from malonyl-CoA with *p*-coumaroyl-CoA to yield 4,2',4'6'-tetrahydroxychalcone. CHS is a kind of polyketide synthases. The three dimensional structure of CHS has been determined, which will enable the engineering of CHS-like enzymes to produce new products (Ferrer *et al.*, 1999). Isomerization of 4,2',4', 6'-tetrahydroxychalcone to naringenin is catalyzed by chalcone isomerase (CHI). The isomerization can also proceed non-enzymatically. Flavanone 3-hydroxylase (F3H), catalyzes hydroxylation of naringenin to dihydrokaempferol (DHK). Hydroxylation of DHK to dihydroquercetin (DHQ) and to dihydromyricetin (DHM) is carried out by flavonoid 3'-hydroxylase (F3'H) and flavonoid 3',5'-hydroxylase (F3'5'H), respectively (Figure 1). These two hydroxylases belong to the same cytochrome P-450 family. Both are key enzymes in determining flower colour because they determine the hydroxylation patterns of the B-ring of the dihydroflavonols and this hydroxylation eventually determines the structure of the anthocyanidin leading to flower colour. They have broad substrate specificity and catalyse hydroxylation of flavanones, flavones and flavonols as well as dihydroflavonol. A metabolic grid has been proposed as a model for

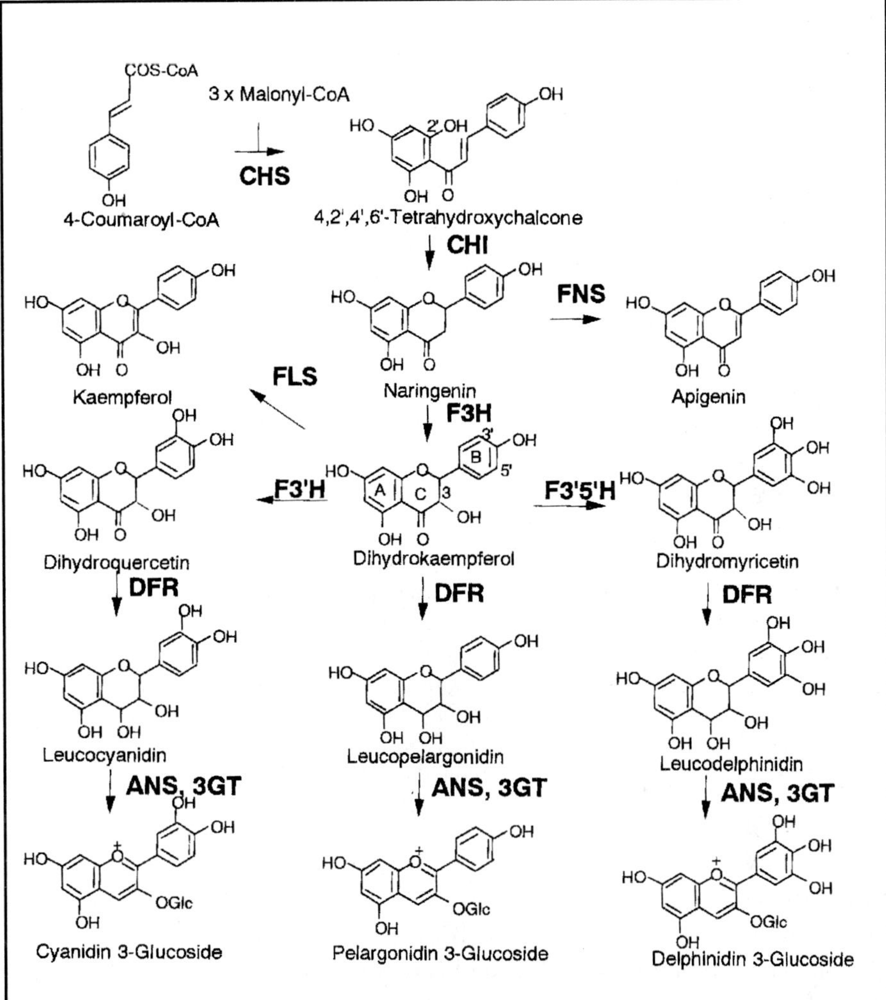

Figure 1 : Flavonoid biosynthesis pathway relevant to flower colour. CHS; chalcone synthase, CHI; chalcone isomerase, F3H; flavanone 3-hydroxylase, F3'H; flavonoid 3'-hydroxylase, F3'5'H; flavonoid 3', 5'-hydroxylase, DFR; dihydroflavonol 4-reductase, ANS; anthocyanidin synthase, 3GT; flavonoid 3-glucosyltransferase, FNS; flavone synthase, FLS; flavonol synthase, Glc; glucose.

pathway operation on the basis of the broad substrate specificity (Stafford, 1990). Dihydroflavonol 4-reducatase (DFR) catalyzes the reduction of dihydroflavonols to leucoanthocyanidins. The conversion of leucoanthocyanidins to anthocyanidins is catalyzed by anthocyanidin synthase (ANS). The reaction was recently characterized biochemically (Saito *et al.*, 1999). Anthocyanidin 3-glucosides, the first anthocyanins,

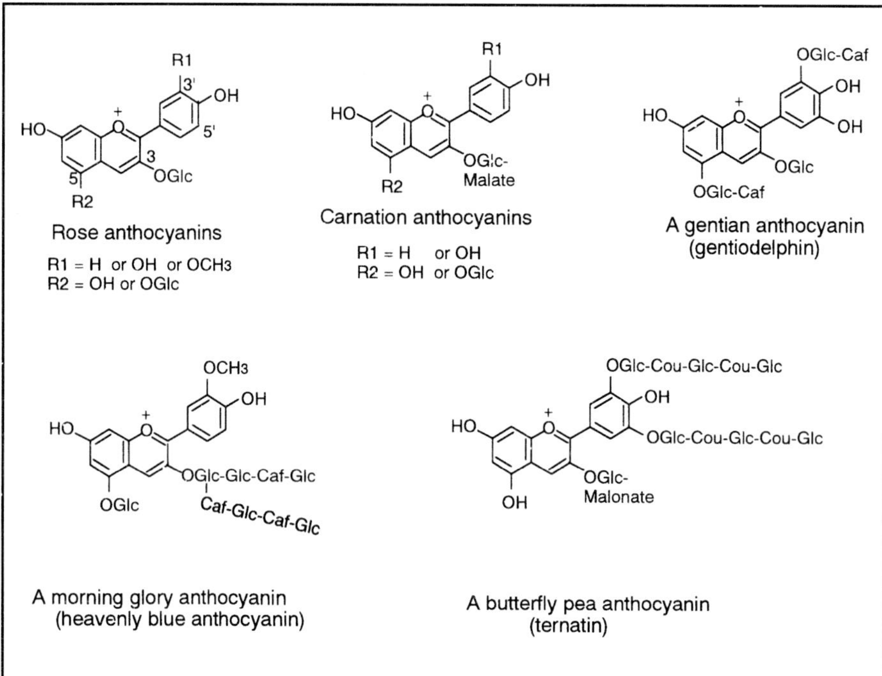

Figure 2 : Anthocyanins in rose, carnation, gentian, morning glory and butterfly pea. Blue flowers tend to contain highiy modified delphinidin type anthocyanins. Addition of glucose and aromatic acyl groups are catalyzed by specific enzymes at least in rose, carnation and gentian. Glc: glucose, Caf:caffeic acid, Cou: coumaric acid.

are synthesized by UDP-glucose: flavonoid 3-*O*-glucosyltransferase (3GT). A petunia 3GT cDNA has been isolated which reveal that it can glucosylate both anthocyanidins and flavonols (Yamazaki *et al.*, In press). Peonidin type anthocyanins are synthesized from cyanidin types and petunidin and malvidin type anthocyanins are synthesized from delphinidin types by the transfer of a methyl groups from S-adenosylmethionine catalyzed by anthocyanin methyltransferases.

Anthocyanins are modified with glutathione (Marrs *et al.*, 1995; Alfenito *et al.*, 1998) and postulated to be transported into vacuoles by ATP-binding cassette transporters (Lu *et al.*, 1998). The pH environment in the cytoplasm is unfavourable for anthocyanin stability leading to speculation on how the biosynthetic process might be isolated from degradation during synthesis and prior to transport across the tonoplast. Evidence for a biosynthetic complex associated with the endoplasmic reticulum is longstanding and has received recent support by Burbulis and Winkel-Shirley (1999). Arabidopsis CHS, CHI and DFR were shown to interact in an orientation-dependent manner on the basis of a yeast two-hybrid system. Affinity chromatography and immunoprecipitation assays demonstrated interactions between CHS, CHI and F3H (Burbulis and Winkel-Shirely, 1999). These results support the hypothesis that flavonoid enzymes assemble as a macromolecular complex to contribute to metabolic channelling. Such chanelling may also help protect pigment molecules from degradation en route to the vacuole. How such complexes operate and contribute to overall regulation of anthocyanin biosynthesis merits further study.

Flavonols and flavones are common copigments. Flavonols derived from dihydroflavonols by flavonol synthae (FLS). Flavones are synthesized from flavanones by flavone synthase (FNS). Interestingly there are two kinds of FNS, a cytochrome P-450 and a dioxygenase.

Some new molecular tools for the manipulation of flower colour have been recently isolated. An F3'H cDNA encoded by the *Ht1* locus was cloned from petunia (Brugliera *et al.*, 1999). A petunia lacking F3'H and F3'5'H activities was transformed with the cDNA under a constitutive promoter. The transgenic plants had dark pink flower colour and elevated peonidin type anthocyanins (Brugiera *et al.*, 1999). F3'H cDNAs from rose, carnation and crysanthemum have also been successfully cloned using the petunia sequences as a selection probe (unpublished results).

A few new flavonoid glycosyltransferase genes have been reported in recent years. UDP-galactose: flavonoid 3-O-galactosyltransfersase cDNA was isolated from *Vigna mungo* (black gram) seedlings (Mato *et al.*, 1998). Flavonol 3-*O*-galactosyltransferase that is specific to flavonols and UDP-galactose has been purified from petunia pollen and its cDNA has been cloned (Miler *et al.*, 1999).

An anthocyanin 5-*O*-glucosyltransferase (5GT) cDNAs have been isolated from some plants. Perilla and verbena 5GT have broad substrate specificity and catalyse glucosylation of anthocyanidin 3-glucosides, cyanidin 3-(6-cafferoyl) glucoside and cyanidin 3-(6-*p*-coumaryol) glucoside (Yamazaki *et al.*, 1999). Petunia 5GT is specific to anthocyanidin 3-acylrutinosides (Yamazaki *et al.*, In press). Such differences in substrate specificity must be considered when engineering the anthocyanin biosynthesis pathway to modify flower colour.

A cDNA clone encoding betanidin 5-*O*-glucosyltransferase was isolated from a *Dorotheanthus bellidiformis* library. Betanidin and its conjugates are flower and fruit pigments in most of the Caryophyllales that do not synthesize anthocyanins. Interestingly, the enzymes can also exhibit the glucosyltransferase activity towards 4'-and 7-hydroxyl groups of flavonoids (Vogt *et al.*, 1999). A few cDNA encoding flavonoid 7-*O*-glucosyltransferase activity have been isolated from scutellaria and tobacco (Hirotani *et al.*, 2000; Vogt *et al.*, 1999).

The phylogenetic tree based on flavonoid glucosyltransferases indicates that 3GT (including flavonoid 3-*O*-galactosyltransferase), 5GT and 7GT (including betanidin 5-*O*-glucosyltransfase) from distinctive subfamilies within the glycosltransferase family.

Genes of aromatic acyl transferase catalyzing the transfer of an aromatic group to the 3 or 5-glucose of anthocyanins (3AAT and 5AAT, respectively) have been cloned from perilla (Yonekura-Sakakibara *et al.*, 2000) and gentian (Fujiwara *et al.*, 1998), respectively. Both of these enzymes transfer hydroxycinnamoic acid (*P*-coumaric acid or cinnamic acid) to specific glucose moieties of anthocyanins. The enzymes have broad substrate specificity for anthocyanins. Anthocyanidin 3-glucosides followed by anthocyanidin 3,5-diglucoside are the preferable substrate for 3AAT. Whereas, anthocyanidin 3,5-glucosides and anthocyanidin 3-(aliphatic acyl)-glucoside 5-glucoside are not preferred. A petunia gene encoding an aromatic acyl transferase catalyzing the transfer of an aromatic group to the 3-rutinoside of an anthocyanin has also been isolated (unpublished results).

P450 type flavone synthase (flavone synthase II) cDNAs have been isolated from snapdragon and torenia (Akashi *et al.*, 1999) and gerbera (Martens and Forkmann, 1999). The enzyme catalyzes apigenin and luteolin biosynthesis from naringenin and eriodyctiol, respectively.

4. REGULATION OF FLAVONOID GENES

Expression of structural genes or enzymes of flavonoid biosynthesis in flowers, leaves and seedlings has been well studied in many plants such as petunia (Brugliera *et al.*, 1994), snapdragon (Jackson *et al.*, 1992), gerbera (Helariutta *et al.*, 1993), carnation (Stich *et al.*, 1992), rose (Tanaka *et al.*, 1995), eggplant (Toguri *et al.*, 1993), *Arabidopsis* (Pelletir *et al.*, 1997), grape (Boss *et al.*, 1996), perilla (Gong *et al.*, 1997) and lisianthus (Nielsen and Podivinsky, 1997). Expression of these genes if both spatially and developmentally regulated in a coordinated way that parallels flavonoid biosynthesis. Flavonoid biosynthesis thus appears to be transcriptionally regulated at the very least. This is supported by comparative Nothern analysis of high and low anthocyanin producing flowers.

Two gene families encoding *Myc*- and *Myb*-type transcriptional factors regulate the expression of the structural genes in the pathway in maize, snapdragon and petunia (Holton and Cornish, 1995; Mol *et al.*, 1998). Regulatory genes controlling anthocyanin biosynthesis are functionally conserved among plant species. However, they have distinct sets of target genes (Quattroccio *et al.*, 1993; Martin, 1996), which allows regulatory diversity in the pathway depending on the species. In petunia, *Anl* (a *Myc* gene), *An2* (a *Myb* gene) and *An11* are regulatory genes of anthocyanin biosynthesis in the flower. *An11* encodes a novel Trp-Asp repeat protein and is a component of a signal transduction cascade that modulates *an2* function (de Vetten *et al.*, 1997). Molecular events at the *an2* locus that occur during *Petunia* species evolution have been analyzed and the loss of *an2* function is proposed to reinforce genetic isolation and complete speciation of petunia species (Quattrocchio *et al.*, 1999). In *Perilla frutescens*, red forms accumulate anthocyanins in the leaves while green forms do not. A Myb and a Myc gene that are specifically expressed in the red form have been found (Gong *et al.*, 1999a; Gong *et al.* 1999b). A homeobox gene (*ANTHOCYANINLESS*2) controlling the anthocyanin pigmentation of the leaf subepidermal layer and cellular organization of the primary root has been cloned in *Arabidopsis* (Kubo *et al.*, 1999).

Increased synthesis of anthocyanins has been achieved via overexpresseion of genes encoding transcription factors. For example the flower colour of tobacco was changed from pink to intense red by expressing the maize *Lc* allele gene (a *Myc*) with CaMV35S. *C1* (a *Myb*) driven by the same promoter had no effect. *Arabidopsis* root, petal and stamen accumulate anthocyanins by constitutive expression of both *C1* and *Lc* (Lloyd *et al.*, 1992). Transgenic petunia (cv. Mitchell) plants that had increased anthocyanins in floral and vegetative tissues were obtained by constitutive expression of *Lc*. Leaves of this transgenic petunia are purple due to accumulation of anthocynanins (Bradley *et al.*, 1998). *Delila* (an R homolog controlling flavonoid synthesis in snapdragon) increased anthocyanins in tobacco flower and tomato flowers and vegetative tissues (Mooney *et al.*, 1994). The expression of a perilla *Myc*-like gene resulted in an increase in the anthocyanin content of flowers of tobacco and vegetative tissues and flowers of tomato (Gong *et al.*, 1999b). However, there are apparent limitations to the broad application of this strategy to generate novelty as similar attempts to enhance anthocyanin biosyntheis in carnation and rose using the same genes either failed to produce flowers with significantly enhanced anthocyanin biosynthesis or resulted in reduced anthocyanin biosynthesis (unpublished results). The constitutive expression of *Lc* in *Petunia hybrida* (cv. Surfina Purple) did not give any color changes in petals and leaves (unpublished results). Such findings are likely a reflection of the complexity of gene regulation and the fact that transcription factors are known to act in concert via protein-protein and protein-DNA interactions. Thus the degree of conservation between transcription factors across species or even varieties will lead to variability in the efficacy of heterologous expression strategies aimed at modifying anthocyanin levels.

5. ENGINEERING ACYANIC FLOWERS BY GENE SUPPRESSION

Stable suppression of endogenous gene expression is more challenging than expression of an introduced gene in transgenic plants. Plant gene expression can be suppressed by introducing either an antisense or a sense gene of interest or its close homologue. The mechanism of sense suppression (cosuppression) is not yet fully understood (Bruening, 1998; Gallie, 1998). The phenomenon is explained by a number of models invoking transcriptional and/or post transcriptional events. Double-stranded RNA is proposed to be a mediator in sequence specific genetic silencing and cosuppression (Montgomery and Fire, 1998; Fire, 1999). The frequency and degree of cosuppression depend on transgenic promoter strength and an intact full length cDNA may be preferable for efficient suppression (Que *et al.*, 1997) although it is not essential to use a full length cDNA to obtain phenotypes of antisense or sense suppression (Gutterson, 1995). The frequency and degree of antisense and sense suppression also depend on the targeted gene and species. The frequency varies from less than 1% to 40% in our laboratories. More recent reports indicate changes to constructs engineered for gene silencing can lead to suppression efficiencies of upto 96% (Hamilton *et al.*, 1998; Waterhouse *et al.*, 1998).

White is not a novel flower colour because of its abundance in nature. Nevertheless, molecular breeding of a white variety is commercially significant because as only flower colour can be modified without sacrificing other desirable characteristics with molecular breeding, there are often opportunities for engineering such targets. It should be possible to obtain white flowers from anthocyanin producing flowers by suppressing the expression of one of many structural or regulatory genes in the pathway. Down-regulation of anthocyanin biosynthesis has been reported in petunia (van der Krol, 1988), gerbera (Elomaa *et al.*, 1993), chrysanthemum (Courtney-Gutterson *et al.*, 1994), rose (Gutterson, 1995), carnation (Gutterson, 1995), lisianthus (Deroles *et al.*, 1998) and torenia (Suzuki *et al.*, 2000; Aida *et al.*, 2000). It is interesting that novel colour patterns were obtained in petunia and lisianthus while only uniform suppression was observed in rose, gerbera, and chrysanthemum. This may be relevant to the fact that petunia and lisianthus have natural white-sectored patterns (Deroles *et al.*, 1998). Furthermore, incomplete suppression of anthocyanin biosynthesis has been observed in experiments aimed at complete suppression in carnation, rose and chrysanthemum leading to the production of pale pink flowers. Growth conditions, promoter performance and the degree of sequence divergence between introduced and target sequences (often members of multigene families) are all possible contributors to these results. Recent improvements in sense suppression technology have yet to be evaluated in colour modification (Hamiltion *et al.*, 1998; Waterhouse *et al.*, 1998).

Considerable success in cosuppression of ethylene production through the disruption of 1-aminocycloropane-1-carboxylic acid synthase and oxidase expression in carnation has been obtained in a wide range of varieties in Florigene Ltd. These enzymes are involved in biosynthesis of ethylene that triggers senescence of the flower. At issue from a commercial standpoint is the low efficiency with which stable cosuppression is observed. Thus, recent reports of improved efficiency have the potential for high impact in the modification of flower colour. However an area of concern relates to the impact of viral infection on maintaining such suppression (Voinnet *et al.*, 1999). Little work aimed at transcriptional silencing has been reported relating to modification of flower colour.

Suntory Ltd. markets Summerwave™ (a torenia) that are superior to conventional torenias with their creeping and vigorous characters. White and blue/white (two out of four petals are white and the remaining two petals are blue) transgenic torenia plants were successfully generated from blue variety, Summerwave Blue, by sense suppression of CHS or DFR genes (Suzuki *et al.*, 2000). In both cases the transgenic plants generally retain the superior characteristics of the host and only flower colour has been modified. Permission for general release has been granted for a white line (CHS suppression) and a blue and white line (DFR suppression) in Japan after having cleared the reglatory procedures for

Figure 3 : Transgenic ornamental plants. Left. *Torenia hybrida* cv. Summerwave (From left, Blue and White made by DFR cosuppression, Violet (a native plant), White made by CHS cosuppression, Blue (the native host plant). Right. Transgenic carnation producing delphinidin (Moonshadow™). They have blue hue due to expressing F3'5'H of other plant species.

genetically modified organisms. Both new varieties grow normally outdoors (Figure 3).

When *Torenia fournieri* was transformed with sense or antisense CHS or DFR constructs, the patterns in modification of colour among the transformants fell into three groups, (i) the same colour as the target plant (i.e. 'wild-type'), (ii) the entire corolla changed to a uniformly light colour (i.e. reduced pigmentation) and (iii) similar to (ii) but with a further lightening in the tube relative to the lip (i.e. even less pigmentation in the flower tube). Transformants incorporating antisense transgene(s) tended to fall into group (ii), with no plants observed in group (iii). Transformants harbouring senses transgenes) tended to fall into group (iii). Sense genes and antisense genes thus seem to have a different potential for changing flower colour. The same approach has also produced transformants with novel flower colours in torenia. For example: lines with pastel flowers, wavy patterned flowers and partly-coloured flowers (Aida *et al.*, 2000).

6. ENGINEERING BLUE FLOWERS

Most blue flowers contain aromatically acylated delphinidin derivatives (Figure 2). The absorbance of anthocynanin shifts towards longer wavelength (bluer) by about 10 nm with each hydroxylation of the B ring and 4 nm following an aromatic acylation. Rose, chrysanthemum and carnation only have pelargonidin and cyanidin derivatives unmodified with aromatic acyl goups. Rose, chrysanthemum and carnation only produce pelargonidin and cyanidin derivatives none of which are modified with aromatic acyl groups. These three flowers represent over 50% of the world cutflower market and thus have become targets for attempts at engineering the synthesis of delphinidin derivatives with the hope of eventually generating blue flowers.

The key enzyme in biosynthesis delphinidin is F3'5'H as previously described. Genes encoding F3'5'H have been isolated from petunia (Holton *et al.*, 1993a), gentian (Tanaka *et al.*, 1996), lisianthus (Nielsen and Podivinsky, 1997), periwinkle (Kaltenbach *et al.*, 1999) and many others. Petunia F3'5'H cDNAs (petunia has two F3'5'H loci, *Hf1* and *Hf2*) could complement their deficiency in petunia that was also deficient in F3'H and a low pH line (see below). The flower colour changed from pale pink to reddish purple (Holton *et al.*, 1993a). F3'5'H genes from petunia and prairie gentian (lisianthus) change the color of petunia lacking F3'5H from pink to magenta due to an increase in 3'5'-hydroxylated anthocyanins. Transgenic tobacco expressing these genes produced delphinidin but gave minimal color changes (Shimada *et al.*, 1999). A cytochrome b5 that specifically transfers electrons to petunia *Hf1* F3'5'H and so directs efficient biosynthesis of 3'5'-hydroxylated anthocyanins has recently been isolated (de Vetten *et al.*, 1999).

Florigene Ltd. and Suntory Ltd. successfully developed violet carnations by introduction of petunia F3'5'H and DFR genes into a DFR deficient white carnation (Holton *et al.*, unpublished results). The petals of the carnations predominantly contain delphinidin that native carnations never produce. Delphinidin is glucosylated and mallylated as are pelargonidin and cyanidin in native carnations (unpublished results). Such a bluish hue of the transgenic flowers has not been achieved by traditional breeding of carnation (Figure 3). The transgenic violet carnations named Moon™ series such as and Moonshadow™ have been marketed in Australia, Japan and USA after being granted general release permissions for genetically modified plants. The transgenic carnations have been shown to be "substantially equivalent" to native carnations except for flower colour and not to have any altered environmental effects. The same strategy has been utilized to generate similar and even darker violet-purple colours in different carnation varieties bearing different flower forms. These are currently being trialed preparatory to seeking approval for release by regulatory authorities in key production and market countries. Now that the genes of aromatic acyltransferases are available, coexpression of F3'5'H and one of the acyltransferase may yield flowers with a bluer hue. Although intramolecular stacking of anthocyanins (Figure 4A) may be achieved by introducing multiple AAT and GT genes, not all genes to synthesize highly modified anthocyanins have been cloned. Furthermore, the stable introduction and expression of multiple genes in transgenic plants is still a considerable challenge.

Copigmentation (Figure 4C) also plays an important role in flower colour as mentioned above. For example, anthocyanin-flavone copigmentation was observed in bluish purple flowers of Japanese garden iris (*Iris ensata* Thunb). Their flowers contain malvidin 3-(acyl)-rutinoside 5-glucoside, petunidin 3-(acyl)-rutinoside 5-glucoside and delphinidin 3-(acyl)-rutinoside 5-glucoside as anthocyanins and isovitexin as the flavone. These anthocyanins brought about copigmentation (bluing effect) of visible λ max due to increased concentrations of isovitexin (Yabuya *et al.*, 1997).

Flavonol levels can be modified by genetic engineering because a number of FLS genes are available (Holton *et al.*, 1993b). Because dihydroflavonols are the common precursors of anthocyanins and colourless flavonols, DFR and FLS compete with each other for dihydroflavonols. Antisense suppression of the FLS gene in petunia and tobacco resulted in increase of anthocyanin and more intense flower colour (Holton *et al.*, 1993b). Overexpression of a petunia FLS gene in

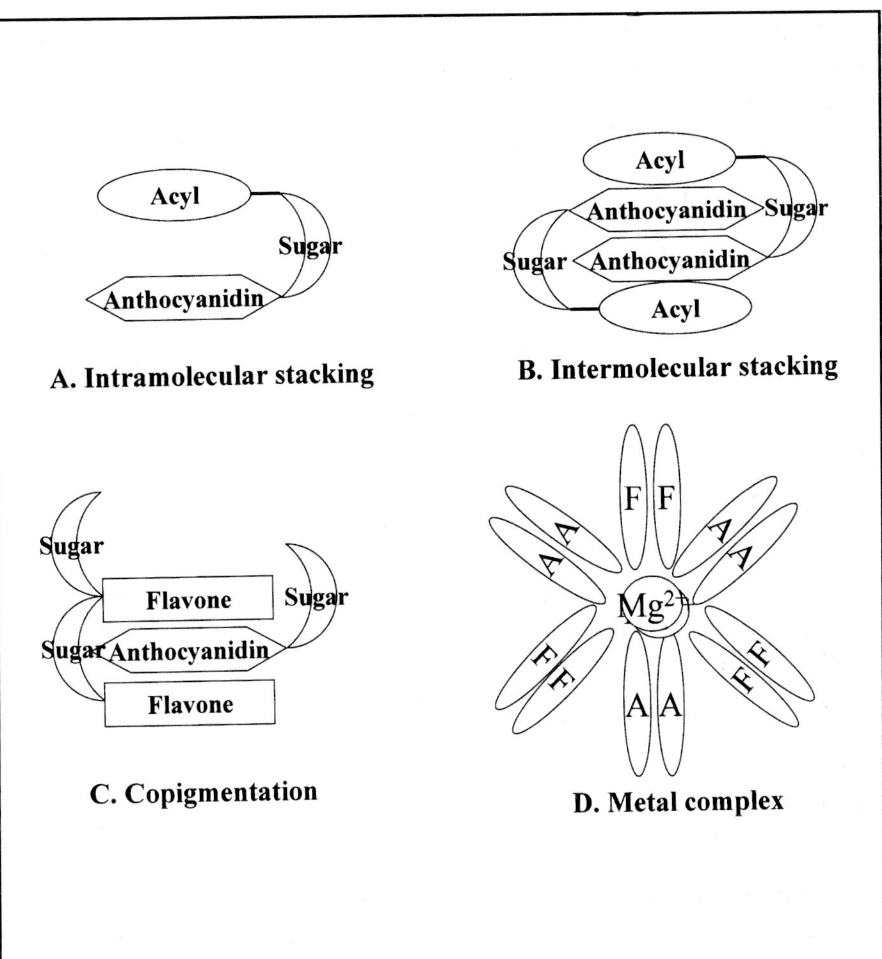

Figure 4 : Anthocyanins are stabilized by stakcings and complexations (Goto and Kondo 1991, Brouillard and Dangles (1994). A. Intramolecular stacking, B. Intermolcular stacking, C. Copigmentation (intermolecular stacking) and D. Metal complexation. Metal complexation also stabilizes anthocyanins (Kondo *et al.* 1992). Such stackings and metal complexations are important for blue colour.

petunia has not been successfully due to cosuppression. A rose FLS gene was isolated and introduced into petunia (cv. Surfinia violet). About one third of the transgenic plants exhibited paler flower colour due to a decrease in anthocyanins (Manuscript in preparation), presumably through competition for the same substrate as mentioned above.

Even a peonidin derivative can make flowers blue as in the blue petals of morning glory it is highly modified with six glucose and three caffeic acid molecules (Figure 2) and its vacuole has a high pH, 7.7

(Yoshida *et al.*, 1995). Because non-blue flowers, however, tend to have lower vacuolar pH and elevating pH is important to the engineering of blue flowers. Petunia is known to have seven loci controlling vacuolar pH (*ph1-ph7*). When one of them is homozygous recessive, the pH of the petal vacuole increases by upto 1 pH unit as estimated from petal extracts. Only *Ph6* has been cloned (Chuck *et al.*, 1993) and *An1* and *Ph6* turned out to be alleles of the same locus (Mol *et al.*, 1998). Mutations in *An2* and *An11* also affect pH (Mol *et al.*, 1998). Cloning of *Ph* genes and their functional analysis may be useful to elevate flower vacuolar pH (Mol *et al.*, 1998). The high pH of morning glory has been proven to be achieved by a Na^+-H^+ antiporter (Fukuda-Tanaka *et al.*, 2000). The relevant gene is encoded by the *Purple* locus because its mutation changes the flower colour from blue to purple. The antiporter is homologous to an arabidopsis antiporter that is involved in salt resistance and assumed to export H^+ out of the vacuoles across the tonoplast. This antiporter may also be useful in attempts to elevate vacuolar pH of other species. Some species such as cyclamen and pelargonium produce non-blue flowers with delphinidin (Davis *et al.*, 1997). Elevation of vacuolar pH is especially important for these species.

Attempts at manipulating vacuolar pH in plants have been confined to disruption of vacuolar H^+ ATPase function or assembly with no success reported to date. Compensation for down regulation of ATPase activity through increased vacuolar H^+ pyrophosphatase activity may make the strategy ineffective even if the down regulation can be engineered at a tissue-specific level. Although mechanisms for the maintenance of cytoplasmic pH would appear likely to override many attempts to significantly alter vacuolar pH there are examples in nature of where such changes do appear to occur thus providing a major challenge to the genetic engineer in devising strategies aimed at increasing petal vacuolar pH.

It is important to note that almost all measurements of vacuolar pH are crude estimates based on petal crush or epidermal peel crush methods. However, estimates do appear reproducible given the knowledge that there are developmental, diurnal and seasonal variations in the values obtained. The assumption is made that cytoplasmic pH is closely regulated around 7.5-8.0 and most of the epidermal cell volume is comprised of the vacuole. Epidermal peels can thus deliver reasonable vacuolar pH estimates. Studies of vacuolar content and development in petals are at an early stage and will provide valuable additional information in understanding how pH is modulated and what other factors, such as organic acids and metal ions, might contribute to flower colour.

7. ENGINEERING BRICK-RED FLOWER

Petunia flowers rarely contain pelargonidin type anthocyanins because DFR of petunia cannot reduce DHK and thus do not produce so-called brick-red colours. Transgenic brick-red petunias accumulating pelargonidin type anthocyanins have been obtained from DHK accumulating petunia (deficient in F3'5'H, F3'H and FLS) by using a maize DFR cDNA (Meyer *et al.*, 1987). The stability of the expression of maize DFR has been examined in a field test for a line of transgenic petunias containing a single copy of the DFR gene. White or variegated phenotypes and plants with flowers showing weak pigmentation were observed. A section of the maize DFR gene was deleted in four plants and methylation of the CaMV 35S promoter was observed in white and variegated flowering plants. More than 90% of plants produced red flowers in mid-June but less that 40% produced red flowers in late August. It was concluded that this variation depended on environmental factors and endogenous factors (Meter and Heidmann, 1994). Methylation of the maize DFR gene was also reported (Meyer and Heidmann, 1994).

They newly engineered flower colour was pale brick-red and commercially unattractive. In addition to plant had other negative characters, possibly due to somaclonal variation, which made it unsuitable for commercialization. Following introduction of the DFR gene into a breeding program by sexual hybridization F_4 plants exhibited strong and stable orange colour with a good general performance (Oud *et al.*, 1995). Such combination of genetic engineering and traditional breeding will be a powerful way to generate new plants with high marketability.

The pelargonidin producing petunias have been made using gerbera and roseDFR genes (Helariutta *et al.*, 1993; Tanaka *et al.*, 1995). Petunias containing maize and gerbera DFR genes were compared (Elomaa *et al.*, 1995). The flower of the gerbera DFR showed intense and stable colouration. It was concluded that the properties of the transgene itself influence the inactivation process. Gerbera DFR had a lower GC content and fewer methylation sites than maize DFR gene (Elomaa *et al.*, 1995). These results indicate that choice of a gene source is a factor to be considered engineering plants even if the gene encodes the same enzyme activity. Furthermore the same enzyme activity, modulated by numerous factors and conditions, may vary significantly when introduced into different species.

The orchid genus *Cymbidium* lacks orange-coloured flowers, and only cyanidin-type anthocyanins have been identified. Its DFR does not reduce DHK and this is the reason why *Cymbidium* does not produce pelargonidin. Developing an orange *Cymbidium* requires the transformation with a DFR gene that can reduce DHK. Phylogenetic analysis of DFR sequences indicates that the inability to reduce DHK has occurred at least twice during angiosperm evolution (Johnson *et al.*, 1999).

We assume that many other flowers other than petunia and *Cymbidium* do not produce pelargonidin due to the poor ability of DFR to reduce DHK.

8. ENGINEERING YELLOW FLOWERS

Yellow flower colour is sometimes due to the presence of flavonoids although typically it is due to carotenoids. Absorption maxima of yellow flavonoids (flavonols, flavones, chalcones and aurones) ranges from 380 nm to 440 nm, which, at least for human eyes, is not an intense colour. Accumulation of flavonols and flavones only produces very pale yellow flowers. Chalcones glycosylated at the 2' position generate yellow flower colour in some varieties of carnation and peony. In yellow varieties of some *Asteraceae* plants such as cosmos and dehlia 6'-deoxychalcones are the main pigments (Davies *et al.*, 1997). Aurones that produce the brightest yellow among flavonoids are found in some yellow flowers incuding dahlia, limonium, snapdragon and zinnia. Reports of the engineering of yellow flower colour have been sparse due to the absence of the isolated genes encoding the enzymes responsible for the synthesis of these compounds.

The only significant report of molecular breeding of yellow colour is a transgenic petunia containing a chalcone reductase gene. The flavonoid pathway (Figure 1) contains 4, 2',4'6'-tetrahydroxychalcone that is isomerized through the action of CHI. The isomerization readily occurs spontaneously. The deoxylation at 6'-position that is catalyzed by chalcone reductase stabilizes the chalcone. By introducing a chalcone reductase cDNA from *Medicago sativa* into a white petunia, chalcones such as butein 3-*O*-glucoside and butein 4-*O*-glucoside were accumulated and pale yellow flowers were obtained. The colour was, unfortunately, not intense enough to represent a new yellow variety (Davies *et al.*, 1998).

CONCLUSIONS AND FUTURE PROSPECTS

The manipulation of flower has progressed considerably since the early experiments of Meyer *et al.* (1987) and van der Krol *et al.* (1988) in petunia using maize DFR and petunia CHS (antisense) respectively. As more genes are isolated and with a better understanding of the many factors contributing anthocyanin-based flower colour the scope for manipulation now appears greater than ever. Two major limitations remain from a genetic engineering perspective (i) Despite the increased understanding of factors essential to flower colour only a limited number of these can currently be manipulated. This is either due to an absence of appropriate gene isolations or that effects of single introduced genes are either insufficient alone or compensated for by other endogenous gene product activities (in whole or in part) in what remains a complex biosynthetic pathway from the first committed step in biosynthesis to deposition of pigment in the epidermal vacuole of petals (ii) As much of the recent effort has had a commercial focus limited progress has been made due to the high cost of engineering any changes in commercial flower varieties as against what are often modest returns. Further progress will rest on more traditional research funding to maintain progress in understanding some of the fundamentals in anthocyanin biosynthesis and regulation, a significant reduction in the costs of genetic engineering and more reliance on more traditional breeding to transfer introduced genes into new varieties with the resultant opportunities for novel colours.

Genetic engineering remains primarily, though not exclusively, a means of breaking the species gene pool barrier. Thus the biochemistry and physiology of a target still often overrides the current capabilities of the genetic engineer. Genetic engineering can be further limited by the poor efficiency in successful transformation of target varieties and difficulties with the coordinated performance of an increasing number of introduced genes. Such introductions are at present primarily *Agrobacterium*-mediated and carried out utilizing a single T-DNA which stretches the limits of the technology. This strategy is necessary in many crops such as carnation and rose as propagation is vegetative and sequential additions using a combination of plant transformation and traditional crossing is not practical. Nevertheless ongoing progress in transformation technology means that more examples of colour modified flowers are appearing in the lab if not in the marketplace.

The regulatory climate, particularly in Europe, may continue to impact on the release of new flower varieties for some time. Although flowers are not generally classified as food the opposition to genetic modification is in many cases a broad one and new guidelines and regulations will quite possibly include flowers. The high cost of labelling may also make commercialization of many new varieties unlikely.

These issues aside future progress will come with an increased understanding of the key factors in biosynthesis, stabilization and deposition of floral pigments and a more sophisticated means of delivering multiple genes to a plant such that their performance is controlled in a predictable fashion. This is work enough for researchers worldwide in a rich mix of disciplines and with a range of exciting goals in mind driven by both curiosity and commercial gain.

REFERENCES

Aida R, Kishimoto S, Tanaka Y and Shibata M (2000) Modification of flower color in torenia (*Torenia fournieri* Lind). by genetic transformation. *Plant Sci.,* **153** : 33-42.

Akashi T, Fukuchi-Mizutani M, Aoki T, Ueyama Y, Yonekura-Sakakibara K, Tanaka Y, Kusumi T and Ayabe S (1999) Molecular cloning and biochemical characterization of a novel cytochrome P450, flavone synthase II, that catalyzes direct conversion of flavanones to flavones. *Plant Cell Physiol.,* **40** : 1182-1187

Alfenito MR, Souer E, Goodman CD, Buell R, Mol J, Koes R and Walbot V (1998) Functional complementation of anthocyanin sequestration in the vacuole by widely divergent glutathion S-transferases. *Plant Cell,* **10** : 1135-1149.

Bartley GE and Scolnik PA (1995) Plant carotenoids:pigments for photoprotection, visual attraction, and human health. *Plant Cell,* **7** : 1027-1038.

Boss PK, Davies C and Robinson, SP (1996) Expressin of anthocyanin biosynthesis pathway genes in red and white grapes. *Plant Mol. Biol.,* **21** : 565-569.

Bradley JM, Davies KM, Deroles SC, Bloor SJ and Lewis DH (1998) The maize *Lc* regulatory gene up-regulates the flavonoid biosynthetic pathway of Petunia. *Plant J.,* **13** : 381-392.

Brouillard R and Dangles O (1994) Flavonoids and flower colour. *In The Flavonoids: Advaces in Research since 1986* (Ed, Harborne, JB) Chapman and Hall, London pp 565-588.

Bruening G (1998) Plant gene silencing regularized. *Proc. Natl. Acad. Sci. USA,* **95** : 13349-13351.

Brugliera F, Barri-Rewell G, Holton TA and Mason JG. (1999) Isolation and characterization of a flavonoid 3'-hydroxylase cDNA clone corresponding to the *Ht1* locus of *Petunia hybrida. Plant J.,* **19** : 596-602.

Brugliera F, Holton TA, Stevenson TW, Farcy E, Lu CY and Cornish E (1994) Isolation and characterization of a cDNA clone corresponding to the *Rt* locus of *Petunia hybrida. Plant J.,* **5** : 81-92.

Burbulis IE and Winkel-Shirley B (1999) Interactions among enzymes of the Arabidopsis flavonoid biosynthetic pathway. *Proc. Natl. Acad. Sci. USA,* **96** : 12929-12934.

Chuck G, Robbins T, Nijjar C, Ralston E and Courtney-Gutterson N (1993) Tagging and cloning of a petunia flower color gene with the maize transposable element Activator. *Plant Cell,* **5** : 371-378.

Courney-Gutteson N, Napoli C, Lemieux C, Morgan A, Firoozababy E and Robison KEP (1994) Modification of flower color in Florist' *Chrysanthemum* : Production of a white-flowering variety through molecular genetics. *Bio/ Technology,* **12** : 268-271.

Davies KM and Schwinn KE (1997) In *Biotechnology of Ornamental Plants.* (Eds, Geneve RL, Preece JE and Merkle SA), CAB International, Wallingford, pp 259-294.

Davies KM, Bloor SJ, Spiller GB and Deroles SC (1998) Production of yellow colour in flowers: redirection of flavonoid biosynthesis in *Petunia. Plant J.,* **13** : 259-266.

de Vetten N, Quattrocchio F, Mol J and Koes R (1997) The an11 locus controlling flower pigmentation in petunia encodes a novel WD-repeat protein conserved in yeast, plants and animals. *Gene and Development,* **11** : 1422-1434.

de Vetten N, ter Host J, van Schaik HP, de Boer A, Mol J and Koes R (1999) A cytochrome *b5* is required for full activity of flavonoid 3',5' hydroxylase, a cytochrome P450 involved in the formation of blue flower colours. *Proc. Natl. Acad. Sci. USA,* **96** : 778-784.

Deroles S, Bradley JM, Schwinn KE, Markham KR, Bloor S, Manson DG and Davies KM (1998) An antisense chalcone synthase cDNA leads to novel colour patterns in lisianthus (*Eustoma grandiflorum*) flowers. *Mol. Breed.,* **4** : 59-66.

Elomaa P, Hondanen J, Puska R, Seppanen P, Helariutta Y, Mehto M, Kotilainen M, Navalainen L and Teeri TH (1993). *Agrobacterium*-mediated transfer of antisense chalcone sythase cDNA to *Gerbera hybrida* inhibits flower pigmentation. *Bio/Technology,* **11** : 508-511.

Elommaa P, Helariutta Y, Griesbach RJ, Kotilainen M, Seppanen P and Teeri TH (1995) Transgenic inactivation in *Petunia hybrida* is influenced by the properties of the foreign gene. *Mol. Gen. Genet.,* **248** : 649-65.

Ferrer JL, Jez JM, Bowman ME, Dixon RA and Noel JP (1999) Structure of chalcone synthase and the molecular basis of plant polyketide biosynthesis. *Nature Structural Biol.,* **6** : 775-784.

Fire A (1999) RNA-triggered gene silencing. *Trends in Genetics* **15** : 358-363.

Fujiwara H, Tanaka Y, Yonekura-Sakakibara K, Fukuchi-Mizutani M, Nakao M, Fukui Y, Yamaguchi M, Ashikari T and Kusumi T (1998) cDNA cloning gene expression and subcellular of anthocyanin 5-aromatic acyltransferaase from *Gentiana triflora. Plant J.,* **16** : 421-431.

Fukuda-Tanada S, Inagaki Y, Yamaguchi T, Saito N and Iida S (2000). Colour-enhancing protein in blue petals. *Nature,* **407** : 581.

Gallie DR (1998) Controlling gene expression in transgenics. *Curr. Opinion in Plant Biology,* **1** : 166-172.

Gong Z, Yamazaki M, Sugiyama M, Tanaka Y and Saito K (1997) Cloning and molecular analysis of structural genes involved in anthocyanin biosynthesis and expression in a forma-specific manner in *Perilla frutescens. Plant. Mol. Biol.,* **35** : 915-927.

Gong ZZ, Yamazaki M, Saito K (1999a) A light-inducible *Myb*-like gene that is specifically expressed in red *Perilla frutescens* and presumably acts as a determining factor of the anthocyanin forma. *Mol. Gen. Gent.,* **262** : 65-72.

Gong ZZ, Yamagishi E, Yamazaki M and Saito K (1999b) A consitutively expressed Myc-like gene involved in anthocyanin biosynthesis from *Perilla frutescens*: molecular characterization, heterologous expression in transgenic plant and transactivation in yeast cells. *Plant Mol. Biol.,* **41** : 33-44.

Goto T and Kondo T (1991) Structure and molecular stacking of anthocyanin-flower color variation. *Angrew. Chem. Int. Ed. Engl.,* **30** : 17-33.

Gutterson N (1995) Anthocyanin biosynthetic genes and their application to flower colour modification through sense suppression. *Hort. Sci.,* **30** : 964-966.

Hamilton AJ, Broen S, Yuanhai H, Ishizuka M, Lowe A, Solis AGA and Grierson D (1998) A transgenic with repeated DNA causes high frequency, post-transcriptional suppression of ACC-oxidase gene expression in tomato. *Plant J.,* **15** : 737-746.

Helariutta Y, Elomaa P, Kotilainen M, Seppanen P and Teeri TH (1993) Cloning of cDNA for dihydroflavonol 4-reducatse (DFR) and characterization of dfr expression in the corallas of *Gerbera hybrida* var. Regina (Compositae). *Plant Mol. Biol.,* **22** : 183-193.

Hirotani M, Kuroda R, Suzuki H and Yoshikawa T (2000) Cloning and expression of UDP-glucose: flavonoid 7-*O*-glucosyltransferase from hairy root cultures of *Scutellaria baicalensis", Planta,* **210** : 1006-1013.

Holton TA, Brugliera F, Lester DR, Tanaka Y, Hyland CD, Menting JGT, Lu CY, Farcy E, Stevenson TW and Cornish EC (1993a) Cloning and expression of cytochrome P450 genes controlling flower colour. *Nature,* **366** : 276-279.

Holton TA, Brugliera F and Tanaka Y (1993b) Cloning and expression of flavonol synthase from *Petunia hybrida. Plant J.,* **4** : 1003-1010.

Holton TA and Cornish EC (1995) Genetics and biochemistry of anthocyanin biosynthesis. *Plant Cell,* **7** : 1071-1083.

Jackson D, Roberts K and Martin C (1992) Temporal and spatial control of expression of anthocyanin biosynthetic genes in developing flowers of *Antirrhinum majus. Plant J.,* **2** : 425-434.

Johnson ET, Yi H, Shin B, Oh BJ, Cleong H and Choi G (1999) *Cymbidium hybrida* dihydroflavonol 4-reductase does not efficiently reduce dihydrokaempferol to produce orange pelargonidin-type anthocyanins. *Plant J.,* **19** : 81-85.

Kaltenbach M, Schroder G, Schmelzer E, Lutz V and Schroder J (1999) Flavonoid hydroxylase from *Catharanthus reseus*: cDNA heterologous expression, enzyme properties and cell-type specific expression in plants. *Plant J.,* **19** : 183-193.

Kondo T, Yoshida Y, Nakagawa A, Kawai T, Tamura H and Goto T (1992) Structural basis of blue-colour development in flower petals from *Commelina communis. Nature,* **358** : 515-518.

Kubo H, Peeters AJM, Aarts MGM, Pereira A and Koornneef M (1999) ANTHOCYANINLESS 2, a homeobox gene affecting anthocyanin distribution and root development in *Arabidopsis. Plant Cell,* **11** : 1217-1226.

Lloyd AM, Walbot V and Davis RW (1992) *Arabidopsis* and *Nicotiana* anthocyanin production activated by maize regulators R and C1. *Science,* **258** : 1773-1775.

Lu YP, LiZ S, Drozdowicz YM, Hortensteiner S, Martinoia E, Rea PA (1998) *AtMRP2*, an *Arabidopsis* ATP binding cassette transporter able to transport glutathoine S-conjugates abd chlorophyll catabolites: functional comparisons with AtMRP. *Plant Cell,* **10** : 267-282.

Marrs KA, Alfenito MR, Lloyd AM and Walbot V (1995) A glutathione S-transferase involved in vacuolar transfer encoded by the maize gene *Bronz-2. Nature,* **375** : 397-400.

Mato M, Ozeki Y, Itoh Y, Higeta D, Yoshitama K, Teramoto S, Aida R, Ishikura N and Shibata M (1998) Isolation and characterization of a cDNA clone of UDP-galactose: flavonoid 3-O-galactosyltransferase (UF3GaT) expressed in *Vigna mungo* seedlings. *Plant Cell Physiol.,* **39** : 1145-1155.

Martin C (1996) Transcrptional factors and the manipulation of plant traits *Curr. Opinions in Biotechnology,* **7** : 130-138.

Martens S and Forkmann G (1999) Cloning and expression of flavone synthase II from *Gerbera* hybrids. *Plant J.,* **20** : 611-618.

Meyer P, Heidemann I, Forkamann G and Saedler H (1987) A new petunia flower colour generated by transformation of a mutant with a maize gene. *Nature,* **330** : 677-678.

Meyer P and Heidmann I (1994) Epigenic variants of a transgenic petunia line show hypermethylation in trangene DNA: an indication for specific recognition of foreign DNA in transgenic plants. *Mol. Gen. Genet.,* **243** : 390-399.

Miller KD, Guyon V, Evans JNS, Shuttleworth WA and Taylor LP (1999) Purification, cloning, and heterologous expression of a catalytically efficient flavonol 3-*O*-galactosyltransferase expressed in the male gametophyte of *Petuina hybrida. J. Biol. Chem.,* **274** : 34011-34019.

Mol J, Gtotewold E and Koes R (1998) How gene paint flowers and seeds. *Trends Plant Sci.,* **3** : 212-217.

Mol J, Cornish E, Mason J and Koes R (1999) Novel coloured flowers. *Curr. Opinion Biotech.,* **10** : 198-201.

Montgomery MK and Fire A (1998) Double-stranded RNA as a mediator in sequence-specific genetic silencing and co-suppression. *Trends Genet.,* **14** : 255-258.

Mooney M, Desnos T, Harrison K, Jones J, Carpenter R and Coen E (1995) Altered regulation of tomato and tobacco pigmentation genes caused by the *delila* gene of *Antirrhinum. Plant J.,* **7** : 333-339.

Nielson KM and Podivinsky E (1997) cDNA cloning and endogenous expression of a flavonoid 3'5'-hydroxylase from petals of lisianthus (*Eustoma grandiflorum*) *Plant Sci.,* **129** : 167-174.

Oud JSN, Schneiders H, Kool AJ and van Grinsven MQJM (1995) Breeding of transgenic orange *Petunia hybrida* varieties. *Euphytica,* **84** : 175-181.

Pelletier MK, Murrell JR and Shirley BW (1997) Characterization of flavonol synthase and leucoanthocyanidin dioxygenase genes in *Arabidopsis. Plant Physiol.,* **113** : 1437-1445.

Quattroccio F, Wing JF, Leppen HTC, Mol JNM and Koes R (1993) Regulatory genes controllong anthocyanin pigmentation are functionally conserved among plant species and have distinct sets of target genes. *Plant Cell,* **5** : 1497-1512.

Quattroccio F, Wing JF, van der Woude K, Souer E, de Vetten N, Mol JNM and Koes R (1999) Molecular analysis of the *anthocyanin2* gene of petunia and its role in the evolution of flower color. *Plant Cell,* **11** : 1433-1444.

Que Q, Wang HY, English JJ and Jorgensen RA (1997) The frequency and degree of cosuppression by sense chalcone synthase transgenes are dependent on transgene promoter strength and are reduced by premature nonsense codons in the transgene coding sequence. *Plant Cell,* **9** : 1257-1368.

Saito K, Kobayashi M, Gong Z, Tanaka Y and Yamazaki M (1999) Direct evidence for anthocyanindin synthase as a 2-oxoglutarate-dependent oxygenase: molecular cloning and functional expression of cDNA from red forma of *Perilla frutescens. Plant J.,* **17** : 181-189.

Shimada Y, Nakano-Shimada R, Ohbayashi M, Okinaka Y, Kiyokawa S and Kikuchi Y (1999) Expression of chimeric P450 genes encoding flavonoid-3',5'-hydroxylase in transgenic tobacco and petunia plants. *FEBS Lett.*, **461** : 241-245.

Stafford HA (1990) Flavonoid Metabolism. CRC Press, Boca Raton.

Stafford HA (1994) Anthocyanins and betalains:evoluation of the mutually exclusive pathway. *Plant Sci.*, **101** : 91-98.

Stich K, Eidenberger T, Wurst F and Forkmann G (1992) flavonol synthase activity and the regulation of flavonol and anthocyanin biosynthesis during flower development in *Dianthus caryophyllus* L. (carnation). *Z. Naturforsch.*, **47C** : 553-560.

Strack D and Wray V (1994) The anthocyanins In: *The Flavonoid: Advances in Research since 1986,* (Ed Harborne, JB) Chapman and Hall, London pp1-22,

Suzuki K, Zue H, Tanaka Y, Fukui Y, Fukuchi-Mizutani M, Murakami Y, Katsumoto Y, Tsuda S and Kusumi T (2000) Flower color modifications of *Torenia hybrida* by cosuppression of anthocyanin biosynthesis genes. *Mol. Breed.*, **6** : 239-246.

Tanaka Y, Fukui Y, Fukuchi-Mizutani M, Holton TA, Higgins E and Kusumi T (1995) Molecular cloning and characterization of *Rosa hybrida* dihydroflavonol 4-reducatse gene. *Plant Cell Physiol.*, **36** : 1023-1031.

Tanaka Y, Tsuda S and Kusumi T (1998) Metabolic engineering to modify flower colour. *Plant Cell Physiol.*, **30** : 1119-1126.

Tanaka Y, Tsuda S and Kusumi T (1999) Application of recombinant DNA to floriculture. In: *Applied Plant Biotechnology*, (Eds Chopra VL, Malik VS and Bhat SR), Oxford & IBH, New Delhi. pp 177-231.

Tanaka Y, Yonekura K, Fukuchi-Mizutani M, Fukui Y, Fujiwara H, Ashikari T and Kusumi T (1996) Molecular and biochemical characterization of three anthocyanin biosynthetic enzymes from *Gentiana triflora. Plant Cell Physiol.,* **37** : 711-716.

Toguri T, Umemoto N, Kobayashi O and Ohtani T (1993). Activation of anthocyanin synthesis genes by white light in eggplant hypocotyl tissues, and identification of an inducible P-450 cDNA. *Plant Mol. Biol.,* **23** : 933-946.

van der krol AR, Lenting PE, Veenstra J, van der Meer IM, Koes RE, Gerats AGM, Mol JNM and Stuitje AR (1988). An antisense chalcone synthase gene in transgenic plants inhibits flower pigmentation. *Nature,* **3333** : 866-869.

Vogt T, Grimm R and Strack D (1999) Cloning and expression of a cDNA encoding betanidin-5-O-glucosyltransferase, a betanidin-and flavonoid-species enzyme with high homology to inducible glucosyltransferases from the Solanaceas. *Plant J.,* **19** : 509-519.

Voinnet O, Pinto YM and Baulcombe DC (1999) Suppression of gene silencing: A general strategy by diverse DNA and RNA viruses of plants. *Proc. Natl. Acad. Sci. USA,* **96** : 14147-14152.

Waterhouse P, Braham MW and Wang MB (1998) Virus resistance and gene silencing in plants can be induced by simulatneous expression of sense and antisense RNA. *Proc. Natl. Acad. Sci. USA,* **95** : 13959-13964.

Yabuya T, Nakamura M, Iwashina T, Yamagichi M and Takehara T (1997) Anthocyanin-flavone copigmentation in bluish purple flowers of Japanese garden irsi (*Iris ensata* Thunb). *Euphytica,* **98** : 163-167.

Yamazaki M, Gong Z, Fukuchi-Mizutani M, Fukui Y, Tanaka Y, Kusumi T and Saito K (1999) Molecular cloning and biochemical characterization of a novel anthocyanin-5-*O*-glucosyltransferase by mRNA differential display of plant forms regarding anthocyanin. *J. Biol. Chem.,* **274** : 7405-7411.

Yonekura-Sakakibara K, Tanaka Y, Fukuchi-Mizutani M, Fujiwara H, Fukui Y, Ashikari T, Murakami Y, Yamaguchi M and Kusumi T (2000) Molecular and biochemical characterization of hydroxycinnamol-CoA:anthocyanin 3-*O*-glucoside-6"-*O*-hydroxycinnamoltransferase from *Perilla frutescens Plant Cell Physiol.,* **41** : 495-502.

Yoshida K, Kondo T, Okazaki Y and Katou K (1995) Cause of blue petal colour. *Nature,* 373-291.

Chapter 16

TRANSGENIC PLANTS : ENVIRONMENTAL CONCERNS

HC Sharma★, N Seetharama, KK Sharma and Rodomiro Ortiz

Germplasm Resources and Enhancement Program, International Crops Research Institute for the Semi-Arid Tropics (ICRISAT), Patancheru 502 324, Andhra Pradesh, India

Summary

Significant progress has been made over the past two decades in handling and introduction of exotic genes into microorganisms and crop plants. The recombinant DNA technology has helped to extend the advantages of conventional plant breeding to meet the increasing need for food in the near future. Genetic engineering of crop plants to confer resistance to insect pests offers an environmental friendly method of crop protection. Development and deployment of transgenic crops for pest control need to address the issues related to: impact of the transgenic crops on insect density, development of resistance, effects on the nontarget organisms, potential for introgression of exotic genes into the closely related wild species, and biosafety of the genetically modified food to the human beings. There is a need for a more responsible public debate and better presentation of the benefits to the general public. Production and release of transgenic plants should be based on experience, and there is a need to streamline and harmonize the regulatory requirements for deployment of genetically engineered plants for sustainable crop production.

Keywords : Allergenic foods, biosafety, geneflow, genetically modified organisms, insecticidal genes

---

★Corresponding author : E-mail : h.sharma@cgiar.org.

1. INTRODUCTION

Human population is expected to exceed 8 billion by 2025, and most of this increase will occur in the developing countries. Food production capacity is quite substantial, and yet millions of people are too poor to meet the basic need for food. Agricultural growth is central to economic growth in the developing countries, and very few low-income group countries have achieved rapid growth in non-agricultural sector without a corresponding increase in agricultural production. Agriculture is the primary interface between people and the environment. Therefore, agricultural transformation will be essential to meet the global challenges of reducing poverty and protecting the environment. Socio-economic transformation will have to occur at the level of smallholder farmers so that their complex farming systems can be made more productive and efficient in the use of resources. An essential aspect of the response to this challenge is to harness all the technologies for a sustainable growth in agriculture.

The promise of biotechnology for increasing the production and productivity of crops for sustainable crop production has been dimmed by the intrinsic safety of the transgenic organisms (Miller and Flamm, 1993), and evolution of resistant strains of insects (Williamson *et al.*, 1990). In developed countries, the social and environmental groups have raised hue and cry about the real or conjectural effects on the nontarget organisms, while in the developing countries, the caution has given rise to fear because of lack of adequate information. In response to these concerns, a biosafety working group has been formed by Food and Agricultural Organization (FAO), United Nations Environment Program (UNEP), United Nations Industrial Organization (UNIDO), and World Health Organization (WHO); and guidelines for handling and release of genetically modified organisms have been published (Tzotzos, 1995). Recently, a convention on harmonizing the regulations for biotechnology has been adopted at a conference in Canada.

Genetically modified organisms have a better predictability of gene expression than the conventional breeding methods, and transgenes are not conceptually different than the use of native genes or organisms modified by conventional technologies. The focus of biosafety regulations needs to be on safety, quality, and efficacy (Levin, 1988; Wynaarden, 1990). The need and extent of safety evaluation may be based on the comparison of the new food and the analogous food, if any, and the interaction of the transgene with the environment. The potential of recombinant technologies to allow a greater modification than is possible with the conventional technologies may have a greater bearing on the environment (Tiedje *et al.*, 1989). The ability to insert genes into plants, and to synthesize genes in the laboratory has caused concern to the general public. Historically, crop plants have not been subjected to risk/safety analysis or risk management, and are improved by cross-pollination between plants with desirable traits or with species that are sexually compatible. The management, interpretation, and utilization of information will be an important component of risk assessment, and determine the effectiveness and reliability of this technology in a transparent manner.

2. THE PROMISE OF BIOTECHNOLOGY

The promise of biotechnology as an instrument of development lies in its capacity to improve the quantity and quality of plants quickly and effectively. The time required to combine favorable traits through traditional crop breeding is greatly reduced. Increased precision in plant breeding translates into improved predictability of the products. The application of biotechnology can create plants that are resistant to drought, insect pests, weeds, and diseases. Plant characteristics can also be altered for early maturity, increased transportability, reduced post-harvest losses, and improved nutritional quality.

Significant progress has been made over the past decades in handling and introduction of exotic genes into plants, and has provided opportunities to modify crops to increase yields, impart resistance to biotic and abiotic stress factors, and improve nutrition. Genes from bacteria such as *Bacillus thuringiensis* (Bt) and *B. sphearicus* have been used successfully for pest control through transgenic crops on a commercial scale (Hilder and Boulter, 1999; Sharma *et al.*, 2000). Insecticidal genes such as Bt, trypsin inhibitors, lectins, ribosome inactivating proteins, secondary plant metabolites, vegetative insecticidal proteins, and small RNA viruses can also be used alone or in combination with Bt genes (Sharma *et al.*, 2000). In addition to widening the pool of useful genes, genetic engineering also allows the use of several desirable genes in a single event, and thus reducing the time required to introgress novel genes into the elite background. Several processes have been employed to insert foreign genes into crop plants, and fall into four major groups (Potrykus, 1990). These include:

- *Agrobacterium*-mediated transformation,
- DNA uptake into protoplasts,
- particle bombardment, and
- partial digestion of cells in multicellular structures.

The Bt toxin gene was cloned in 1981, and the first transgenic plants were produced by mid-1980s (Barton *et al.*, 1987; Fischoff *et al.*, 1987; Vaeck *et al.*, 1987). Since then, several crop species have been genetically engineered to produce Bt toxins to control the target insect pests. Genes conferring resistance to insects have been inserted into maize, cotton, potato, tobacco, rice, broccoli, lettuce, walnuts, apples, alfalfa, and soybean (Federici, 1998; Griffiths, 1998). The first transgenic crop was grown in 1994, and large-scale cultivation was taken up in 1996 in USA (McLaren, 1998). Since then, there has been a rapid growth in the area under transgenic crops in USA, Australia, and China. The area planted to transgenic crops increased from 1.7 million ha in 1996 to 39.5 million ha in 1999. In 1997, transgenic crops were grown in 12 countries, and most of the area planted to genetically improved crops was in Australia, Canada, Argentina, China, and the United States of America. Among the crops produced included insect-resistant cotton and maize, herbicide-resistant soybean, and tomatoes with a long shelf-life (Federici, 1998; Griffiths, 1998).

2.1. Advantages

Biotechnology can help to achieve the productivity gains needed to feed the growing global population. Insect-resistant varieties and biocontrol agents may reduce the over dependence on pesticides, reduce the farmers' crop protection costs, and thus benefiting both the environment and public health. Biotechnology would also offer cost-effective solutions to micronutrient malnutrition, such as vitamin A and iron. Research in biotechnology on increasing the efficiency of utilising the farm inputs could also lead to development of crops that use water more efficiently and extract nutrients from the soil more effectively. The development of cereal plants capable of capturing nitrogen from the air could contribute greatly to plant nutrition, helping the poor farmers, who often cannot afford fertilizers.

Most of the early products of agricultural biotechnology focused on crop protection. Use of herbicide-tolerant varieties facilitated weed control using certain types of herbicides. It also enabled farmers to employ soil conservation practices such as minimum tillage, which reduces soil erosion. In 1998, an estimated 7.7 million hectares were planted to transgenic crops with resistance to insect pests. This has resulted in the reduced use of insecticides, a positive impact not only on farm income, but also on the environment, *e.g.*, transgenic maize containing Cry 1A(b) gene results in 76% reduction in leaf feeding by the stem borer (*Diatraea* spp.) (Bergvinson *et al.*, 1997). Transgenic cotton with Bt gene causes 94 and 91% reduction in damage to flowers and bolls, respectively. The reduction in insect damage resulted in 39% gain in cottonseed yield (Benedict *et al.*, 1996). In addition to the reduction in losses due to insect pests, the development and deployment of transgenic plants with insecticidal genes will also lead to:

- a major reduction in insecticide sprays,
- reduced exposure of farm labor and nontarget organisms to the pesticides,
- increased activity of natural enemies,
- reduced amounts of pesticide residues in the food and food products, and
- a safer environment to live.

The benefits to growers have been higher yields, lower costs, and ease of management. Additional impact of these varieties is the reduction in the number of pesticide applications. The primary benefit to growers of adopting Bt cotton varieties is controlling insect pests that have become resistant to commonly used insecticides and reduction in insecticide costs. Pesticide use data has shown a considerable reduction in the use of insecticides that are recommended for the control of these insects since the introduction of Bt cotton varieties in USA. It is expected that growers will also experience increased yields by using Bt cotton varieties compared to conventional varieties.

The overall aim of crop improvement is to increase the production and productivity of crop. A series of criteria can be established to determine the sustainability of cropping system in which the transgenic crop is to be deployed. Such a system can weigh the benefits against any potential hazards that may be foreseen. This may include economics, environmental concerns, impact on natural resource base, and acceptance. Whether or not a transgenic crop should be introduced in an area should depend on:

- reduction in pesticide use,
- durability of resistance,

- ability to grow the crop without soil tillage,
- improve soil quality and prevent surface runoff, and
- fertilizer use efficiency.

2.2. Limitations

The developments in plant biotechnology have both promise and problems. There are ethical and safety issues, which are further complicated by the intellectual property rights (IPRs). Many protests have been made by the nongovernmental organizations on ethical or ecological grounds. The dominance of the private sector, where the bulk of developments in agro-biotechnology have taken place, raises fears that this will create a new phase of comparative disadvantage and increased dependency in the developing countries. There is a fear that patenting and exercise of IPRs will lead to:

- a monopolization of knowledge,
- restricted access to germplasm,
- control over the research process,
- selectivity in the focus of research, and
- increased marginalization of the majority of the world's population.

These concerns need to be addressed properly. The critical issue is that every instrument of agricultural transformation should be mobilized in our efforts to feed the hungry, help the poor, and protect the environment. We must find ways of realizing the promise of biotechnology, while avoiding the pitfalls. Transgenics are not a panacea for solving all the pest problems. There are some genuine or perceived concerns. The major limitations of using transgenic plants are:

- secondary pests are not controlled in the absence of sprays for the major pests,
- need to control the secondary pests through chemical sprays will kill the natural enemies and thus offset one of the advantages of transgenics,
- proximity to sprayed fields and insect migration may reduce the benefits of transgenics, and
- development of resistance in insect populations may limit the usefulness of transgenics.

The evidence on these issues is still inconclusive and warrants careful monitoring before the transgenic crops are deployed on a large scale by the subsistence farmers. The major risk of modern biotechnology for developing countries is that technological development may bypass poor farmers because of a lack of enlightened adaptation. It is not that biotechnology is irrelevant, but research needs to focus on the problems of small farmers in developing countries. Private sector research is unlikely to take on such a focus, given the lack of future profits. Without a stronger public sector role, a form of scientific apartheid may develop, in which cutting edge science becomes oriented exclusively toward industrial countries and large-scale farming (Serageldin, 1999). A better synergy could be obtained if an effective public and private sector cooperation could be established.

3. TRANSGENIC CROPS AND ENVIRONMENT

A number of ecological and economic issues need to be addressed while considering the production and deployment of transgenic crops for insect control (Sharma and Ortiz, 2000). The most important consideration is the immediate reduction in the amount of pesticides applied for pest control. The number of pesticide applications on a crop such as cotton varies from 10 to 40, and most of the sprays are directed against the key pests such as cotton bollworm or legume pod borer, *Helicoverpa armigera*. In case the transgenic crops are introduced, number of pesticide applications are likely to be reduced by two-third to half. Reduction in number of pesticide applications would lead to an increased activity of the natural enemies, while some of the minor pests may tend to attain higher pest densities in the absence of sprays applied for the control of major pests.

Efficacy of transgenic crops for controlling the target and the nontarget pests need to be determined in each region. Some of the pests maintain high pest densities on alternate hosts. A reduction in the numbers of eggs and larvae of the natural hosts may also affect the activity of the natural enemies. The significance of such effects would depend on the importance of the immature stages of the target pest for maintaining the populations of the natural enemies. This may reduce the numbers of certain natural enemies in areas planted with transgenic crops, but their populations may be maintained on the other crops that serve as a host to the target pests. A few of the known predators are specialists on one insect, and hence, the populations of the predators and parasites with a wide host range would be maintained on the other insect species (Fitt *et al.*, 1994). Within field impact may be greater for parasitoids that are monophagous. The populations of such natural enemies can only be maintained on the nontransgenic crops or other hosts of the target pest.

The effect of transgenic crops on the abundance of natural enemies should be compared with the nontransgenic fields of the same crop where the natural enemies may be virtually absent because of heavy pesticide application.

In diverse agricultural systems such as those prevailing in the tropics, it would be important to understand the biology and behavior of all the insect species in an eco-system so that informed decisions can be made as to which crops to transform, and the toxins to be deployed. It is also important to consider the resistance management strategies, economic value, and environmental impact of the exotic genes in each crop, and whether a crop serves as a source or sink for the insect pests and their natural enemies (Gould, 1998). As in the past, the technologies that have been used in the developed countries may not be suitable for the developing countries with complex cropping systems and the species involved. Therefore, there is an urgent need to give a serious thought to these problems and develop appropriate strategies for production and deployment of insect-resistant transgenic crops. Major issues arising out the deployment of transgenic crops are:

- Environmental concerns,
- Biosafety of the transgenic food, and
- Public attitude.

4. ENVIRONMENTAL CONCERNS

The environmental issues that need to be addressed while introducing transgenic crops for pest control include: i) effect of transgenic plants on population dynamics of target and non-target insects, ii) performance limitations, iii) secondary pest problems, iv) insect sensitivity, v) evolution of insect biotypes, vi) environmental influence on gene expression, vii) development of resistance viii) gene escape into the environment, ix) effects on non-target organisms, x) social and ethical issues, and x) unanticipated consequences.

4.1. Effect of transgenic plants on population dynamics of target and non-target pests

Effects of transgenic crops on the population dynamics of the insects would be similar to the plants with conventional host plant resistance (Luginbill and Knipling, 1969), *e.g.*, continuous planting of the stem fly-resistant (*Cephus cinctus*) wheat cultivars would completely suppress the stem fly populations below the economic threshold levels within six years. The stem flies can also be kept under check by alternate planting of the resistant and susceptible cultivars. Similar models for the effect of insect-resistant cultivars on insect abundance have also been developed for sorghum shoot fly (*Atherigona soccata*), spotted stem borer (*Chilo partellus*), sorghum midge (*Stenodiplosis sorghicola*), and sorghum head bug (*Calocoris angustatus*) (Sharma, 1993; Sharma *et al.*, 1999). In wheat, no direct relationship has been observed between the planting of the resistant cultivars and the population of the Hessian fly (*Myetiola destructor*) (Foster *et al.*, 1991). Expression of resistance is not the same under different population densities, and under different environmental conditions. Activity of the natural enemies is density dependent, and this may reduce the parasitoid numbers over time. Such an interaction might result in population densities that are higher in magnitude than those predicted by the simulation models. Therefore, there is a need:

- to understand the natural population regulation of the target pest,
- field performance of insect-resistant cultivars under diverse environmental conditions,
- long-term effects of the resistant cultivars on insect populations, and
- level of adoption of the insect-resistant cultivars.

4.2. Performance limitations

The Bt toxins cannot produce the same dramatic effect on insect mortality as the synthetic insecticides. The farmers need to be educated about the efficacy and mode of action of transgenic crops. The expectations have to be real, and remedial measures should be taken as the situation warrants. The effects of the transgenic crops on insects will be relatively slower, but cumulative over time. Transgenic crops may not be able to withstand the pest density in some seasons. Therefore, careful monitoring of pest populations is an essential component of pest management involving the transgenic crops. The value of the transgenic crops can be best realized when deployed as a component of pest management for sustainable crop production (Sharma and Ortiz, 2000). However, enough information has not been generated involving transgenics in a genuine integrated pest management system. Such trials can demonstrate long-term benefits of the transgenic crops, especially if environmental and human health hazards are taken into account. Currently deployed transgenic crops produce only one Bt toxin protein, while the Bt strains used for commercial formulations produce several toxins in addition to other factors that increase insect mortality. The current CryIA(b) construct employs PEP-carboxylase promoter, which enables the expression in the green tissue, and as a result, the expression is greater in the young plants. Some insects such as stem borers and shoot fly migrate into the plant whorl or stem tissue with incomplete chlorophyll formation. If the toxin is expressed in insufficient amounts in such tissues, the insects can develop mechanisms to withstand low levels of toxins in the transgenic plants. Behavioral avoidance of the tissue expressing the toxin gene can be another component in insect resistance to the transgenic plants. Therefore, care should be taken to express the toxins in sufficient amounts at the site of damage/feeding by the insects.

4.3. Secondary pest problems

Most crops are not attacked by a single pest species, but a number of insect pests. In the absence of competition from the major pests, secondary pests may assume a major pest status (Hilder and Boulter, 1999). The Bt toxins may be ineffective against such pests, *e.g.,* leaf hoppers, mirid bugs, root feeders, mites, etc. This will offset some of the advantages expected of the cultivation of transgenic crops. Management of phytophagous stink bugs is necessary in transgenic Bt cotton (Greene *et al.*, 1997). Insecticide application for the control of stink bugs is necessary if more than 20% of the bolls are damaged in mid- to late-season. Parker and Huffman (1997) did not observe any difference between transgenic and nontransgenic cultivars in boll weevil or aphid damage, beneficial arthropods or fiber characteristics. Effective and timely control measures should be adopted for the control of secondary pests on transgenic crops. There is a need to identify genes that could be deployed to control pests not susceptible to the Bt. While there is a trend to develop target specific compounds for chemical control, it will be desirable to have genes with broad-spectrum of activity for use in genetic transformation of crops, provided this does not affect the beneficial organisms.

4.4. Insect sensitivity

There are many species of insects that are not susceptible to the currently available Bt proteins. There is a need to broaden the pool of genes, which can be effective against insects that are not sensitive to the currently available genes. Since first generation transgenics have only one Bt toxin gene, lack of control of less sensitive species may present another problem in pest management. This is not the same as development of resistance (which is a progressive decrease in sensitivity to a chemical by an insect population in response to the use of a product to kill the insects). If there is low or no sensitivity to a chemical in an insect species, it is not resistance. *Helicoverpa virescens* is less sensitive to Cry 1A(a), Cry 1C and Cry 1E, while *Spodoptera littoralis* is insensitive to most of the Bt toxins (Gill *et al.*, 1992). *Spodoptera litura* is less sensitive to toxins from *B. thuringiensis* var *kurstaki* than *H. armigera, Achoea janata, Plutella xylostella*, and *Spilosoma obliqua* (Meenakshisundram and Gujar, 1998). Bt toxins Cry IC and Cry1E, which are active against *H. virescens* (MacIntosh *et al.*, 1991), are ineffective against *H. armigera* (Chakrabarti *et al.*, 1998). Cry1B is slightly active against *H. armigera*, while it has been reported to be inactive against *H. virescens* (Hofte and Whiteley, 1989). Thus, there are considerable differences in the sensitivity of different insect species to various Bt toxins, and due care has to be taken to deploy Bt toxins in different crops/cropping systems.

4.5. Evolution of new insect biotypes

Experience from the conventional breeding has shown that there is no direct relationship between the deployment of insect resistant cultivars and the evolution of new insect biotypes. In case of the Hessian fly (*Myetiola destructor*) in wheat, no direct relationship has been observed between the planting of the resistant cultivars and the population of the Hessian fly. Planting of the Hessian fly-resistant cultivars did not lead to evolution of new biotypes. The time needed for adaptation to antibiosis resistant genes has been predicted to be 3 to 8 years. However, in case of greenbug (*Schizaphis graminum*), the breeding programs continue to struggle to keep pace with the evolution of new biotypes (Daniels, 1981; Wood, 1971). However, there is no relationship between the deployment of greenbug-resistant wheat cultivars and the development of new greenbug biotypes (Porter *et al.*, 1997). For sorghum, only 3 of the 11 biotypes of greenbug have shown a correlation between the use of resistant hybrids and the development of new biotypes. Even within the 3 biotypes, no clear cause-and-effect relationship has been established. Based on analysis of these specific insect-plant interactions, future plant resistance efforts should focus on the use of the most effective resistance genes; despite past predictions of what effect these genes may have on insect population genetics.

4.6. Environmental influence on gene expression

There have been some failures in insect control through the transgenic crops. Cotton bollworm (*Heliothis zea*) destroyed Bt cottons due to high tolerance to Bt toxin, Cry IA(c) in Texas, USA (Kaiser, 1996). Similarly, *H. armigera* and *H. punctigera* destroyed the cotton crop in the second half of the growing season in Australia because of reduced production of Bt toxins in the transgenic crops (Hilder and Boulter, 1999). Possible causes for the failure of insect control may be:

- inadequate production of the toxin genes,
- effect of environment on expression of transgene,
- locally resistant insect populations, and
- development of resistance due to inadequate management.

Cotton crop flooded with 3 to 4 cm deep water for 12 days lost resistance to insects significantly compared with the control plants irrigated normally (Wu *et al.*, 1997). Similar reaction has been observed in Bt cotton, which grew in overcast and rainy weather continuously for 21 days. When the waterlogging was over, the cotton plants recovered gradually and their insect resistance increased again to some extent. Under flooded conditions, the activity of superoxide dismutase increased considerably in Bt cotton plants at first, and then dropped continuously. Epistatic and environmental effects on foreign gene expression could influence the stability, efficacy, and durability of the foreign genes (Sachs *et al.*, 1998). CryIA gene expression is variable and is influenced by genetic and environmental factors. The CryIA phenotype segregates as a simple, dominant Mendelian trait. However, non-Mendelian segregation occurred in some lines derived from MON 249. Expression of transgene is influenced by:

- site of gene insertion,
- gene construct,
- epistasis,
- somaclonal mutations, and
- the physical environment.

Appropriate evaluation and selection procedures should be used in a breeding program to develop crop varieties with pest-resistant traits conferred by the foreign genes.

4.7. Development of resistance

There are concerns that the deployment of transgenics will lead to development of resistance in insect populations, to the herbicide genes in weeds, and to the antibiotic genes used as markers. While some of these concerns may be real, the others seem to be highly exaggerated. Careful thought should be given to while considering the production and release of transgenic crops in different agro-ecosystems.

4.7.1. Development of resistance in insect populations

Insect pest populations have shown a remarkable capacity to develop resistance to chemical pesticides. Over 500 species of insects have developed resistance to insecticides (Moberg, 1990). Most of the transgenic Bt crops express only one toxin gene and lack the complexity of the commercial Bt formulations. In addition, the plants continuously produce the toxins, and the insects are exposed to the Bt toxins throughout the feeding cycle/season, and this places the insect population under continuous and heavy selection pressure. With the development of resistance to Bt toxins, the value of microbial insecticides based on Bt proteins will diminish greatly due to lower sensitiveness of the target pest to the Bt formulations. One of the consequences of such a development would be that the farmers would have to return to broad-spectrum insecticides, which will lead to environmental hazards associated with the use of synthetic insecticides. The potential for development of resistance to Bt proteins is not only of concern to the farmers, but to the scientists, extension agencies, and the transgenic plant industry. The investment made in the past would be turned useless unless this issue is addressed on an urgent basis. Most of the transgenic plants produced so far have Bt genes under the control of cauliflower mosaic virus (CaMV35S) constitutive promoter, and this system may lead to development of resistance in the target insects as the toxins are expressed in all parts of the plant (Harris, 1991). However, several site or tissue specific promoters have been developed in the recent past. Toxin production may also decrease over the crop-growing season. Decreasing levels of toxin production may lead to devclopment of resistance to the toxin used, and to other related Bt toxins to which the insect populations may initially be quite sensitive. Low doses of the toxins eliminate the most sensitive individuals of a population, leaving a population, in which resistance can develop much faster. Since most Bt toxins have a similar mode of action, resistance developed against one toxin can also lead to development of cross-resistance to other toxins. However, there are reports that insects selected for resistance to one Bt toxin may not be resistant to other Bt toxins (Sharma and Ortiz, 2000).

The ability of insects to overcome host plant resistance is always a grave risk, and ways to delay the onset of resistance in insect populations will be an ongoing debate as transgenic crops are deployed. Extensive and intensive exposure of pests to Bt toxins through transgenic crop plants or other tactics may lead to widespread pest resistance to Bt. There are several reports on the development of resistance to Bt in different insect species. Laboratory screening has resulted in the development of Bt resistant populations in Lepidoptera, Coleoptera, and Diptera (Tabashnik, 1994). This highlights the fact that possibilities for resistance development are real. With the transgenic plants now being produced in both public and private sectors, the real challenge is to develop a strategy for deployment of transgenic plants for sustained protection of crops from insect pests. Different insect species react to Bt toxins differently. The resistance can develop quickly in *Diatraea saccharalis*, as considerable proportion of larvae have been observed to survive up to 8 days on the transgenic maize (Bergvinson *et al.*, 1997). However, *D. grandiosella* has shown a considerably lower frequency of surviving individuals. Diamond back moth, *Plutella xylostella* populations in several parts of the world have developed resistance to Bt (Tang *et al.*, 1999). The rapid response to laboratory selection shows genetic variation in populations of diamondback moth in their susceptibility to Bt and suggests that intense selection may produce much higher levels of resistance than those previously reported from the field (Tabashnik *et al.*, 1991). Soybean looper collected from soybean and Bt-cotton was less susceptible to Bt formulation *Condor XL* in dosage-mortality and discriminating concentration bioassays than the reference strain, and Bt-cotton strains were least susceptible to *Condor XL* (Mascarenhas *et al.*, 1998). These data indicated reduced susceptibility of field soybean looper strains compared to the reference strain.

However, in some insect species, the probability of development of resistance may be very low, *e.g.*, *Ostrinia nubilalis* has been observed to develop some tolerance to low levels of CryIA(b) in the diet, but it has not been possible to initiate or sustain the insect colonies at concentrations in the diet closer to the actual levels expressed in the transgenic maize plants (Lang *et al.*, 1996). After 13 generations of selection pressure, no colony survived on transgenic Bt maize hybrids in the greenhouse. Development of resistance to Bt may not be a serious issue since the Bt and the pests have co-evolved for million of years (Bauer, 1995; Tabashnik, 1994). Because of limited exposure and several toxins produced by Bt, the rate of development of resistance under natural conditions may not be high. In transgenic plants, the insects are continuously exposed to the exotic genes, and there are possibilitics of resistance development in the target pests.

4.7.2. Development of resistance to herbicides

Herbicide genes have been inserted into several crops to provide selectivity for herbicides that are degraded in the environment quickly. There is a need to know whether the herbicide-resistant plant can establish as a weed, and the possibility of gene transfer into the wild relatives of the crop plant. Genes from plants engineered for herbicide resistance could cross over to other plants, creating super weeds. Genes introduced into genetically transformed crops can spread into closely related native species (Chevre *et al.*, 1997). Studies in Norway and the United States have shown that the genes for herbicide resistance can move from cultivated canola to wild relatives. Genes from the conventionally bred *Brassica napus* have been moving to the wild turnip, *B. rapa* (Raybould and Gray, 1993). Genes from unrelated sources may change the fitness and population dynamics of the hybrids between native plants and the wild species.

4.7.3. Development of resistance to antibiotic genes

The antibiotic gene used as a marker to select for gene transfer may lead to resistance in pathogens infecting human beings. However, general scientific view is that the risk of compromising the therapeutic value of antibiotics is almost negligible. Most genetically engineered plants contain a gene for antibiotic resistance as an easily identifiable marker. Hypothetically, antibiotic resistance genes may move from a crop into bacteria in the environment. Since bacteria readily exchange antibiotic resistance genes, the antibiotic resistance genes may move into disease-causing bacteria. Gene transfer from plants to microorganisms is possible in laboratory studies (Gebhard and Samalla, 1998), and possibly has happened during evolution (Doolittle, 1999). The probability of movement of genes from plants to human pathogens (antibiotics) is negligible. Several studies have established that there is little chance that such a transfer would occur (Calgene, 1990), but there is a continuing debate whether such a gene should be present in the commercial varieties. Methods have been developed for removing selectable marker genes after selection of the transgenics (Yoder and Goldsbrough, 1994; Ebinuma *et al.*, 1997). There are alternatives to the antibiotic markers, and systems are also available to carry out the transformation without involving any markers.

The marker gene can also be excised after two lines are crossed (Dale and Ow, 1991).

4.8. Gene escape into the environment

The introgression of transgenes into the wild relatives is of potential concern (Gregorius and Steiner, 1993; Serratos *et al.*, 1997). Pollen dispersal from transgenic cotton is low, but increases with an increase in the size of the source plot (Llewellyn and Fitt, 1996). The results have shown that a 20-m buffer zone would serve to limit dispersal of transgenic pollen from small-scale field tests. Behavioral differences between resistant and susceptible Colorado potato beetles may affect gene flow between transgenic and the adjacent nontransgenic crops (Alyokhin and Ferro, 1999).

Probably no genetically engineered product has generated as much controversy as the so-called terminator technology, which makes plants produce sterile seeds. The terminator genes might cross over and cause plants to produce sterile seeds, resulting in a significant loss of food and diversity. There is no difference to the farmer between buying a hybrid seed with a terminator gene or buying a hybrid produced through conventional technology. However, it has implications for the developing countries where the farmers depend on the seed saved from the previous crops. In areas where farmers routinely save and replant seeds, the prospect of being forced to buy seeds each year has generated considerable unease among the farming communities.

Under laboratory conditions, plasmid transfer between *B. thuringiensis* subsp. *tenebrionis* and *B. thuringiensis* subsp. *kurstali* HD 1 (resistant to streptomycin) strains occurs at 10^{-2} (Thomas *et al.*, 1997). However, no plasmid transfer has been observed in soil release experiments, and in insects on leaf discs. The Bt toxins were detectable on the clay-particle-size fraction of nonsterile soil after 40 days. When the toxins bind on clay minerals, they become resistant to utilization by microorganisms. Binding of the Bt toxins to humic acids reduced their potential for microbial biodegradation (Crecchio and Stotzky, 1998). These results indicate that Bt toxins in transgenic plants and microbes could persist, accumulate, and remain insecticidal in the soil as a result of binding to humic acids, where they might pose an environmental hazard to non-target organisms.

The greatest risk of a transgenic plants released into the environment is its potential spread beyond the plant area to become a weed. However, there are no records of a plant becoming a weed as a result of plant breeding (Cook, 2000). This may be because of:

- low risk of crop plants to the environment,
- extensive testing of the crop varieties before release, and
- adequate management practices to mitigate any risks inherent in the crop plants.

Plant breeding efforts have been tended to decrease rather than increase the toxic substances, as a result, making the improved varieties more susceptible to insect pests. However, there is a feeling that genes introduced from outside the range of sexual compatibility might present new risks to the environment and humans. However, many of such apprehensions are not supported by data. A study conducted by the National Academy of Sciences, USA (NAS, 1987), has concluded that:

- there is no evidence of hazards associated with DNA techniques,
- the risks, if any, are similar to those with conventional breeding techniques,
- the risks involved are related to nature of the organism rather than the process, and
- there is a need for a planned introduction of the modified organisms into the environment.

One of the hazards in gene transfer from the transgenic plants to the wild relatives is of concern if the wild relatives are under selection pressure (biological control) from the pest. If the target pest does not play any role in population regulation of the wild hosts, the gene transfer will not constitute to any hazard. The build up of resistance in the wild relatives can also act as a component of pest management to the target pest.

Interspecific hybridization is a common process, but hybrids are rare, and most are sterile, and there is very low chance of gene introgression into the wild relatives (Fitter *et al.*, 1990). Transgenic plants may become weeds, except in the context of their normal agricultural environment. Gene escape may occur when the plant invades a semi-natural habitat or transferred into the wild relative, and persist in the uncultivated land. Its' spread can be checked by methods similar to any other single trait. There are differences among plant species to disperse from the environment other than the one in which they are released, and their ability to establish feral populations. Such an event has to be compared with that of the original plant.

Resistance to insects and diseases can also make the plant to be more persistent in the wild environment. It can also confer similar advantages on the wild relatives. The chances for development of resistance in the pest populations need to be assessed, and strategies devised to use the same gene in different crops in the same environment. Transgenic crops containing protease inhibitors may pose similar problems, and their use need to be carefully planned to avoid the evolution of pest populations capable of withstanding the transgene.

Genes conferring resistance to abiotic stress factors have also been identified, and plants with resistance may be released in the near future (Fraley, 1992). Resistance to abiotic stress factors may present additional challenge, as this would enable the plants to grow in environments where they were unable to do well earlier. This confers additional advantage to the transgenic plant, and there are chances for gene transfer through cross-pollination.

The exotic species model can be used to assess the risk of introducing transgenic plants with resistance to abiotic stress factors. The risk assessment in such cases requires more information, and the nature of competitive advantage conferred by the transgene under specific conditions. Assessment of realistic risk for gene transfer through pollen is available for many crops (Raybould and Gray, 1993), and good agriculturally sound procedures need to be developed for different regions (Boulter, 1995).

4.9. Effects on non-target organisms

One of the major concerns of transgenic crops is their effects on the non-target organisms, about which little is known at the moment. The Bt proteins are rapidly degraded by the stomach juices of the vertebrates. Most Bt toxins are specific to insects as they are activated in the alkaline medium of the insect gut. Bt proteins can have harmful effects on the beneficial insects, although such affects are much less severe than those of the broad-spectrum insecticides.

The secondary metabolites may be toxic to the non-target organisms. The information that use of genetically modified corn may have toxic effects on larvae of the monarch butterfly, *Papilio demoleus* has generated a huge amount of publicity and almost as much misinformation. Groups opposing biotechnology have used this preliminary data to argue against the production and development of genetically engineered crops. Scientists are now conducting follow-up studies to examine the effects of bio-engineered corn pollen on butterflies. A review of current research indicates that scientists have found some risk to monarch butterfly caterpillars from Bt corn pollen, but very few definitive conclusions can be made at this juncture. However, such effects would still be much less than the large-scale application of broad-spectrum insecticides. Possible effects of transgenes on the beneficial insects have been discussed below.

4.9.1. Honeybees

There are no significant effects of transgenic crops on the honeybees. Transgenic rape does not appear to have harmful effects on the lifespan and behavior of honeybees, but further tests may be necessary (Pham Delegue and Jouanin, 1997). Chitinase in genetically modified oil seed rape did not affect learning performance of honeybees; beta-1, 3 glucanase affected the level of conditioned responses (the extinction process occurring more rapidly as the concentration increased), and cowpea trypsin inhibitor (CpTI) induced marked effects in both conditioning and testing phases, especially at high concentrations (Picard-Nizou *et al.*, 1997). The decrease in learning performance induced by CpTI observed at the individual level has been confirmed at the colony level. Trypsin inhibitor and wheat germ agglutinin (WGA) did not show acute toxicity in honeybees. Serine proteinase inhibitor (PI) (from soybean), cysteine PI (OCI from rice), chicken egg white cystatin, and Bowman-Birk soybean inhibitor do not produce harmful effects on honeybees at the concentrations expressed in transgenic plants (Pham Delegue and Jouanin, 1997; Bottino *et al.*, 1988; Girard *et al.*, 1988). Consumption of high doses of protease inhibitors induces proteinase overproduction (Jouanin *et al.*, 1998). Trypsin endopeptidase inhibitor, bovine pancreatic trypsin inhibitor (BPTI), and soybean trypsin inhibitor (SBTI), have been found to be toxic to adult honeybees at 1% weight: volume in sugar solution (Malone *et al.*, 1995).

4.9.2. Predators

No major differences have been observed in the abundance of predators between the transgenic and non-transgenic crops (Hoffman *et al.*, 1992; Sims, 1995; Wang and Xia, 1997). Some observations have suggested that there may be a reduction in the fitness of the predatory chrysopid larvae directly attributable to caterpillars fed on Bt-maize (Hilbeck *et al.*, 1998; Hoffmann *et al.*, 1992). Laboratory studies have shown no adverse effects of the Bt based insecticide on the Colorado potato beetle predator, *Coleomegilla maculata* (Giroux *et al.*, 1994). Cry3A-intoxicated *L. decemlineata* can be eaten by *C. maculata* without any observable adverse effects on their survival or predation potential (Riddick and Barbosa, 1998). Its predatory activity can also decrease the rate at which *L. decemlineata* adapted to the Bt toxins if mixed plantings are used (Arpaia *et al.*, 1997). However, under choice conditions, the predator showed a distinct preference for the untreated eggs than those treated with Bt (Girard *et al.*, 1994). The predator activity was not affected by pure transgenic and mixed seed potato fields (Riddick *et al.*, 1998). No statistically significant effects on survival, aphid consumption, development, or reproduction have been observed in *Hippodamia convergens* fed on *Myzus persicae*, reared on potatoes expressing δ-endotoxin of *Bacillus thuringiensis* subsp. *tenebrionis* (Dogan *et al.*, 1996). Two spotted ladybirds (*Adalia bipunctata*) fed for 12 days on peach-potato aphids (*M. persicae*) colonizing transgenic potatoes expressing lectin from *Galanthus nivalis* in leaves have shown a decrease in fecundity, egg viability, and longevity (Birch *et al.*, 1999). No acute toxicity to the ladybird beetles was observed, although female longevity was reduced by up to 51%. Adverse effects on ladybird reproduction caused by eating peach-potato aphids from transgenic potatoes were reversed after switching the ladybirds to pea aphids from non-transgenic bean plants. Lozzia *et al.* (1998) did not observe any effect on pre-imaginal development or mortality of *Chrysoperla carnea* reared on *Rhopalosiphum padi* that had fed on Bt-maize. However, abundance of *Labia grandis* was lower in pure and mixed plants of transgenic potatoes than in pure non-transgenic potato plants (Riddick *et al.*, 1998). Thus, the effects of transgenic plants on the activity of predators vary across crops and the insect species involved.

4.9.3. Parasites

Increased levels of parasitism by *Campoletis sonorensis* have been observed on the transgenic plants compared to the nontransgenic plants, which may be due to fewer larvae on the transgenic plants. *Campoletis sonorensis* and transgenic plants act synergistically, decreasing the larval survival beyond the level expected for an additive interaction (Johnson and Gould, 1992). Synergistic increases in mortality and parasitism have been detected when development rates on toxic plants and control plants were equal, indicating existence of another type of interaction between natural enemies and transgenic crops (Johnson and Gould, 1992). *Bacillus thuringiensis* toxin-mediated partial resistance appeared compatible with natural enemies for the control of *H. virescens*.

The parasitoid, *Cardiochiles nigriceps* does not reduce the survival of the host larvae significantly, and its' activity is not influenced by the transgenic plants (Johnson, 1997; Johnson *et al.*, 1997). Egg parasitism of third-generation noctuids in the field of Bt-transgenic cotton has been observed to be lower than in the conventional cottons (Wang and Xia, 1997). In natural and integrated control plots, the parasitoids *Campoletis chlorideae* and *Microplitis* sp. density decreased by 79.2 and 87.5, and 88.9 and 90.7%, respectively, and the activity of *Lysiphlebia japonica* increased by 85.1 and 90.2%, respectively (Cui and Xia, 1998). Percentage of parasitism by the parasitoid *Diadegma insulare* was not significantly different between the mixed and non-mixed plots of transgenic crop (Riggin Bucci and Gould, 1997). There is no effect of transgenic corn on the parasitization of *O. nubilalis* by *Eriborus tenebrans* and *Macrocentrus grandii* (Orr and Landis, 1997). Intra-field mixtures could serve to decrease density of a target pest such as the diamondback moth, while not adversely affecting the activity of natural enemies. The effects of transgenic crops on the natural enemies vary across crops and the cropping systems. Some of the variation may be due to differences in pest abundance between the transgenic and the non-transgenic crops. Wherever the transgenic crops have shown adverse effects on the natural enemies, these effects may still be far lower those of the broad-spectrum pesticides.

4.9.4. Microflora

Under field conditions, the microflora of Bt transgenic potato plants has been observed to be minimally different from that of chemically and microbially treated commercial potato plants (Donegan *et al.*, 1996). It is unlikely that expression of Bt and any other genes in transgenic plants would have an adverse effect on the soil microflora.

5. SOCIAL AND ETHICAL ISSUES

Genetically engineered foods may be unacceptable for ethical or religious reasons. Rabbis, ministers, vegetarians, and others may be concerned about genes from animals or species that are proscribed by certain religions. Labeling allows those consumers to choose according to their conscience without imposing that view on others. Many consumers are suspicious of who is controlling a technology that promises to revolutionize agriculture. Biotechnology enables agricultural production to become more vertically integrated, consolidated, and centralized, largely in the hands of multinational corporations.

6. UNANTICIPATED CONSEQUENCES

No technology, no matter how much beneficial, is risk free. And with any new technology, there may be unanticipated consequences. Genetic engineering could also bite back by exacerbating the growing problem of antibiotic resistance. Genetically engineering a plant to have a particular trait can have unexpected effects on the ecosystem that cradles it. It is consumers' grasp of this fundamental phenomenon that underlines much of the concern over biotechnology.

Expression of virulence from a pathogen in a transgene may trigger an uncontrolled hypersensitive response, which is potentially lethal to the plants (de Wit *et al.*, 1998). But such a genetic disease can be eliminated early in research and development. Concerns have also been expressed about the virus-derived genes, which may lead to the development of new viruses with a potential to kill the plants.

7. BIOSAFETY OF TRANSGENIC FOOD

There is a need for the new technologies to be tested rigorously for potential allergenic, toxic, and antimetabolic effects in a transparent manner (Gillard *et al.*, 1999).

The Bt proteins are rapidly degraded by the stomach juices of vertebrates. Most Bt toxins are specific to insects as they are activated in the alkaline medium of the insect gut. No major changes have been observed in the composition of the transgenic tomatoes and potatoes. The levels of the antinutrients gossypol, cyclopropenoid fatty acids, and aflatoxins in the seed from the insect-protected lines were similar to or lower than the levels present in the parental variety and reported for other commercial varieties. The seed from the Bt transformed cotton lines is compositionally equivalent to, and as nutritious as seed from the parental and other commercial cotton varieties (Berberich *et al.*, 1996). Very little amounts of Bt toxins may remain in plant parts to be consumed by human beings or dairy cattle, *e.g.*, the raw seed of line 81 [with Cry IA(b) gene] showed 14.00 μg per g active protein, and line 531 [with CryIA(c) gene] contained 2.22 μg per g of active protein by ELISA method. Processing removed > 97% of the active proteins in the transgenic cottonseed (Sims and Berberich, 1996). CryIA(b) protein as a component of postharvest transgenic maize plants dissipates readily on the surface of, or cultivated into, soil (Sims and Holden, 1996), and has not been detected in silage prepared from transgenic plants (Fearing *et al.*, 1997).

There are no specific receptors for Bt protein in the gastrointestinal tract of mammals, including man (Kuiper and Noteborn, 1994). Histopathological effects have been observed in the gut mucosa, but no systemic adverse effects have been observed in mice and rabbits following oral administration. There are no major changes in composition of the transgenic tomatoes. Transgenic Bt tomatoes pose no additional risk to human and animal health. However, a number of aspects concerning the safety assessment of transgenic Bt tomatoes would require further study (Noteborn *et al.*, 1996). There were no differences in the survival and body weight of broilers reared on meshed or pelleted diets prepared with Bt transgenic maize and similar diets prepared using control maize (Brake and Vlachos, 1998).

Several protein families that contribute to the defense mechanisms of food plants are allergens or putative allergens, and some of these proteins have been used to confer resistance to insect pests. These include α-amylase and trypsin inhibitors, lectins, and pathogenesis-related proteins (Franck Oberaspach and Keller, 1997). Several of these secondary plant substances are toxic to mammals, including humans. This may result in a trade off between nature's pesticides produced by transgenic plants or varieties from traditional breeding programs, synthetic pesticides, and mycotoxins or other poisonous products of pests.

Rats fed on purified cowpea trypsin (EC 3.4.21.4) inhibitor in a semi-synthetic diet based on lactoalbumin (10 g inhibitor kg^{-1}) for 10 days showed a moderate reduction in weight gain in comparison with controls, despite an identical food intake (Pusztai *et al.*, 1992). Although most of the cowpea trypsin inhibitor (CpTI) was rapidly broken down in the digestive tract, its inclusion in the diet led to a slight, though significant, increase in the nitrogen content of faces, but not of urine. Accordingly, the net protein utilization in rats fed on inhibitor-containing diets was also slightly lower, while their energy expenditure was elevated. The slight anti-nutritional effects of CpTI were probably due mainly to the stimulation of growth and metabolism of pancreas. Thus, the nutritional penalty for increased insect-resistance after the transfer of the cowpea trypsin inhibitor gene into food plants is quite low in the short-term.

Level of *Galanthus nivalis* lectin expression that provides insecticidal protection for plants, but did not reduce the growth of young rats, showed a negligible effect on weight and length of the small intestine, even though there was a slight, but significant hypertrophy of this tissue (Pusztai *et al.*, 1996). However, the activity of brush border enzymes was affected. Sucrase-isomaltase activity was nearly halved, and those of alkaline phosphatase and aminopeptidase increased significantly. Incorporation of N-acetylglucosamine-specific agglutinins from wheat germ (WGA), thorn apple (*Datura stramonium*) or nettle (*Urtica dioica*) rhizomes in the diet at the level of 7 g kg^{-1} reduced the apparent digestibility and utilization of dietary proteins and the growth of rats, with WGA being the most damaging (Pusztai *et al.*, 1993). As a result of their binding and endocytosis by the epithelial cells of the small intestine, all three lectins interfered with its metabolism and function to varying degrees. WGA also induced the hypertrophic growth of the pancreas and caused thymus atrophy. Transfer of WGA genes into crop plants has been advocated to confer resistance to insect pests. However, the presence of this lectin in the diet may harm higher animals at concentrations required to be effective against most pests.

Expression of a new gene in a crop could also introduce new allergens, normally not present in the nontransformed plants (Lehrer, 2000). Some people fear allergenic reactions from GM food. Allergic reactions to foods are hard to predict, but they can be life-threatening. Genetic engineering can introduce unknown allergens into food. Virtually every gene transfer in crops results in some protein production, and proteins trigger the allergic reactions. Biotechnology can introduce new proteins into food crops, not only from the known sources (such as peanuts and shellfish), but also from plants, bacteria, and viruses whose allergenicity is unknown. This might lead to people needlessly avoiding foods that are actually safe. If the indigenous proteins or the new proteins are from the known sources of allergens, then assessing the allergens within the genetically modified plants is easier. If the source of the allergenic protein is known, and is related to the introduced gene from sources that have not been used as a human food, then one has to rely on criteria with which to assess their potential allergenicity. Eight commonly allergenic, and 160 less allergenic foods have been identified and scientists can avoid the transfer of genes with known allergenic effect (Lehrer, 2000).

8. THE PUBLIC ATTITUDE

There is a need for introducing the products of biotechnology through the peer-reviewed press than through the general media, balanced presentation of the technology to the general public, and an understanding of the trade offs. The role of transgenics in reducing the load of pesticides in the environment needs to be considered seriously.

The first agricultural products of biotechnology have already reached the world markets. These products have received a frosty response in some parts of the world. Despite some sensational headlines in the press, people in North America have not reacted to foods containing ingredients developed through biotechnology. It is clear from consumer surveys that perceptions about biotech foods are strongly influenced by the type of information, confidence in government, and cultural preferences. Most of the people are cautiously optimistic about the benefits of biotechnology. They will accept the products if they see a benefit to themselves or to the society. Consumer responses to foods developed through biotechnology are basically the same as for any other food. Taste, nutrition, price, safety, and convenience are the major considerations. How seeds and food ingredients are produced is relevant only for a small group of concerned "organic" consumers. In general, consumers see considerable value in human genetic testing, the development of new medicines, and the use of biotechnology to develop insect-resistant crop plants. Consumers are less likely to accept the use of biotechnology with animals, and food products compared to crop plants. The most acceptable applications are those that offer a clear consumer benefit, as well as those perceived to be ethical and safe. However, public attitudes about agricultural biotechnology vary considerably among countries.

Public opinion is sharply divided on the benefits and uses of transgenic technology. There is a need to disseminate the information about potential uses and limitations of transgenic plants in crop production, and their effects on the environment. Transgenic maize with Bt genes has been banned in European Union due to presence of antibiotic (ampicillin) resistance gene. The end result is a serious setback to the public acceptability of transgenic crops. Since many of companies are involved in producing the synthetic insecticides, some of them may retard the process of using agricultural biotechnology, and may be partly responsible for adverse reaction of the public to the transgenic crops. There is a continued need for the governments and the NGOs to actively peruse this area of research and extension.

Environmentalists often focus on possible ecological impact from the use of biotechnology. In countries where consumers are more negative about biotechnology, the media coverage and activist opposition have been more pronounced and a discussion on the benefits of biotechnology has generally been ignored, while the potential risks have been emphasized.

Public perception is likely to have a great impact on innovation, introduction, and diffusion of products of biotechnology. And negative public perception is likely to keep the products of biotechnology from reaching the market place of its utilization. Public perception is influenced by a broad range of issues, including environmental safety, ethics, legal repercussions, economic gain, and socio-economic impact. Public opinion can be influenced by non-scientific considerations based on the impressions created by media and the pressure groups. The concerns about recombinant technology were raised in early 70s (Leopold, 1995). This even led to protests over setting up of biotechnology laboratories in developed countries. As a result, regulations and guidelines for handling DNA based technologies were formulated in USA and Europe. With increasing stakes for economic gains and potential risks, public debate over the need to develop regulatory mechanisms became all the more important. This also led to intense public debate among the scientists, particularly between molecular geneticists and the ecologists. As a result, the politicians, the industry, and environmental groups began to take a more active stance. Much of this debate is likely to play itself out in the regulatory arena, with the shift of focus from the laboratory to commercial applications. This role needs to be taken up by the international and national research organizations for agriculture, health, food, and environment.

Most of the scientific community and the public agree that the risks of genetic engineering are largely exaggerated, but there is a need for strict regulatory mechanism. Generally speaking, the farmers and the public believe that biotechnology will lead to increased food production and improved nutrition (Pilcher and Rice, 1998). Level of education, religion, socio-economic factors, pressure by the environmental groups, and governmental policy are likely to shape the public opinion about biotechnology. Scientific literacy, scientific proof of the non-sustenance of presumed risks, informal dissemination of information through the public media, clear standards, food levelling, reducing the extent of exaggerated expectations and coming out of the science fiction, making public to be part of the decision making process, and reliability of information are important to have a clear picture of the benefits and risks of biotechnology.

In the developing countries, there is a need for an environment, that is institutionally, socially, culturally, politically, and educationally favorable. Developing countries that are entering biotechnology also have a large proportion of population capable of making rational decisions. If large sections of the population are under poverty line and illiterate, the nongovernmetal organizations become the advocates of public opinion, which at times may be guided by several extraneous factors. Under such circumstances, international organizations such as FAO, WHO, UNIDO, and international agriculture research centers (IARCs) can be expected to play an important role in enhancing the public perception of the usefulness and the risks associated with the introduction of genetically modified organisms.

9. REGULATIONS FOR PRODUCTION AND RELEASE OF TRANSGENIC PLANTS

For genetically engineered organisms to gain acceptance, decisions that address the concerns associated with the application of biotechnology to agriculture must be science based. Regulatory agencies should assure credibility, and use a rational basis for decision-making. Science and legal processes are inextricably linked for regulations that evaluate biological products. Review of any particular product should be based on scientific criteria relevant to that product. Advances in biotechnology should determine as to what has actually happened at the molecular and biochemical levels to scrutinize the product safety, and the effect of product modification upon food safety. The approach to review is constantly evolving due to new types of products and the availability of scientific information. Genetically modified plants have been released in several countries. The regulations governing the use of transgenic plants vary considerably in different countries. Existing regulations have been applied to the production and release of genetically modified plants, but have not been adequate to address the potential environmental effects from the transgenic plants.

In the USA, health and safety aspects of genetic engineered organisms are covered by Environment Protection Agency (EPA), United States Department of Agriculture (USDA), and Food and Drug Administration (FDA) (Levin and Strauss, 1993). The guidelines related to the use of genetically modified organisms have also been issued under the Commission of the European Communities (CEC, 1990a,b). Latin American countries have a plant quarantine and regulation system to deal with plant introduction, and these may have to be altered to deal with genetically modified organisms (IAICA, 1991). The permission for release of modified organisms is given after the risk assessment by the individual countries. The risk assessment is carried out by the competent authority based on data supplied by the applicant. The competent authority has to give the decision within a specified period. The legislation in USA is product specific, while in the European Union, it is process specific. Different countries in the world are adopting approaches related to these models, depending upon the type of legislation already in place. In future, when the use of transgenic plants will become widespread, the need for harmonizing of the standards and procedures will become all the more important. Some initiatives have already been taken in this direction by OECD (1992). International organizations such as UNIDO, UNEP, FAO, and WHO have published a 'Voluntary code of conduct' for release of organisms into the environment (UNIDO, 1991). UNIDO has also supported an international biosafety network and advisory service, which helps the developing countries in setting up of national authorities that are qualified to handle the release of genetically modified organisms.

Regulations should be developed to implement statutes, establish procedures and criteria by which specific types of products are evaluated; including those produced using biotechnology such as vaccines, plant varieties, pesticides, animal products, and pharmaceuticals. It usually takes five years of field-testing to collect the data needed to pass through the regulatory system, while another two years may be needed to complete the review process. The available scientific information (particularly from the peer-reviewed journals) should be used by regulatory authority in evaluating the specific products. Scientists can be asked to identify the issues and recommend approaches to particular types of products. Information can also be requested on specific products. Internationally, the appropriate scientifically based standards, guidelines, and recommendations for transgenic products as they move into the marketplace are being developed by the representatives of national governments at working groups, and task forces devoted to these issues under the standard-setting bodies — *Codex Alimentarius*, the International Plant Protection Convention (FAO, 1986), and Office des International Epizootics. The use of the existing standard-setting bodies to address concerns about products of biotechnology focuses attention on harmonization of risk-assessment methodologies and evaluation of specific products.

9.1. Regulations for handling and release of genetically modified plants in India

Department of Biotechnology (DBT), Ministry of Science and Technology, Government of India, has formulated guidelines for the release of rDNA in 1990 under the Environmental Protection Act 1986

(Anonymous, 1998). These guidelines cover:

- genetically engineered organisms,
- genetic transformation of green plants and animals,
- rDNA technology in vaccine development, and
- large scale production and deliberate/accidental release of organisms, plants, animals, and products derived by rDNA technology.

After the signing of the Convention on Biodiversity in 1992, DBT revised its earlier guidelines to accommodate the safe handling of GMOs in research applications and technology transfer in 1994. This includes the large-scale production and deliberate release of GMOs. To implement these guidelines, the government notification directs the creation of various committees. Currently these guidelines are being implemented through a three-tier mechanism. These include:

- institutional biosafety committees (IBSC) to monitor the research activities at the institute level,
- review committee on genetic manipulation (RCGM) functioning in the DBT, and
- genetic engineering approval committee (GEAC) of the Ministry of Environment and Forests, that has the authority to permit large scale use of GMOs at commercial level.

The DBT has set up the rDNA Committee to prepare modified draft of guidelines from time to time on the basis of current scientific information, and from the experience gained locally and outside the country. In general, the biosafety guidelines deal with the definition of rDNA, classification of pathogenic microorganisms, containment facilities, biosafety levels, rDNA research activities, large-scale experiments, release of GMOs, import and shipment of rDNA and its products, and quality of biologicals produced by rDNA technology.

10. RISK ASSESSMENT

The focus of biosafety regulations need to be on safety, quality, and efficacy. The management, interpretation, and utilization of information are important components of risk assessment, and determine the effectiveness and reliability of biotechnology. Various approaches addressing the risks are concerned with establishing good standards of laboratory practice, efficiency and security of the containment facilities, and effects of modified organisms on the human health and the environment (Levin and Strauss, 1993). The risk is assessed in the form of access - as a measure of the probability that a modified organism (or the DNA inserted in it) will be able to enter the human body and survive there, and expression - anticipated or known level of expression of the inserted DNA. Risk also measures damage in the form of harm likely to be caused to a person by exposure to the modified organism.

Hazards posed by a modified organism may include tendency of a self-pollinated line to outcross because of self-sterility or any other factors. Or a plant may have a tendency to become a weed, yield toxic substances in the product or change in the toxin(s) produced by the plant. Any of these attributes may pose a risk to people consuming the product, working with it or to the environment. In case of toxin genes inserted into the plants, the range of the expressed toxin may be much wider than expected, with adverse consequences for the environment. Plants may also display a change in appearance, reaction to other biotic and abiotic stress factors or the end use characteristics. However, it is not easy to measure weediness, as such characteristics are not easily defined. The ecological consequences in most cases are only qualitative. Therefore, risk assessment for the genetically modified plants requires a detailed assessment of the modified plant in comparison to the plant from which it has been derived. The procedures adopted should take cognizance of the environment where the plant is to be released.

The containment of the transgenic plants from tissue culture, growth rooms, and, greenhouse are covered by good laboratory practice. Care need to be taken that pollen and seed from the laboratory produced plants do not escape outside the facilities. The plants should be labeled properly, and there should be no mixing between the transgenic plants. High level of quality control is needed over the DNA sequences, gene constructs, transgenic plants, and the experimental results. Growing plants in the greenhouse involves the same level of controls as in the laboratory. The greenhouse should be properly designed to keep out insects and pollen. The facilities should be run under the control of a biosafety committee, and the level of containment should depend on the type of transgenic plants. The greenhouses should have a controlled and filtered airflow system, control of water outlets, and sterilization. Autoclaving of plant and soil material coming out of the greenhouse is very important.

A comprehensive risk assessment is necessary once a plant has to be released for small-scale experiments, and commercial production (NAS, 1989). At this stage, the scientists concerned, the biosafety committee, and the national/international regulatory authorities should determine whether it is acceptable to release the specific transgenic plants, and if needed, restrictions to be imposed. Field containment should be in place to limit the possible environmental impact of the release experiment. This may include isolation from the sexually compatible species, prevention of flowering, use of male-sterile lines, and subsequent monitoring protocols. Data required for risk assessment includes:

- Organization and the people involved,
- DNA donor, the receiving species, and the transgenic plant,
- target environment and the conditions of release,
- transgenic plant x environment interaction, and
- control, monitoring and waste treatment.

10.1. General information

This includes the institution and the people involved in development, and the people/organizations to be responsible for field containment, monitoring, and waste treatment.

10.2. DNA donor and receiving species

It includes complete information about the donor and the receiving species. The receiving plant species forms the baseline with which the transgenic plant should be compared. Information on the donor species indicates the type of information needed about the transgene. Information is also needed about the vector used in the transformation, and antibiotic or herbicide resistance genes used as a marker. Finally, there should be a complete information about the transgenic plant, molecular data on the genes inserted, stability of expression, changes in allergenicity, toxicity, persistence in particular environmental conditions, and ability to invade new habitats. The changes in the transplant should be measured against the unmodified control genotype.

10.3. Conditions of release and the target environment

The risk to the target environment requires qualitative judgment, and should be based on a case-by-case study, depending on experience. Information about the purpose of the release, size, design, and agronomic requirements are important for risk assessment at the national and international level. Ecological information about the release site, survey of plant species growing in the target region, and the nature of pollen dissemination are important. The anticipated target and nontarget organisms with which the transgenic plant will interact need to be determined. Information should be recorded whether the transgenic plant would become a better or worse host. The risk to the environment also includes harmful effects on the beneficial nontarget organisms.

10.4. Interaction between the transgenic plant and the environment

It is important to describe the invasiveness of the transgenic plant in the wild habitat, ability to propagate sexually/asexually, possibility of transferring the transgene to the same or related species, or to microorganisms, and the consequences of gene transfer.

10.5. Control, monitoring, and waste treatment

Once the transgenic plants are released into the environment, there is a movement of pollen, seed, and the plants outside the immediate environment of release. It is important to monitor the transgenes in the environment after the release, and the efficiency with which it is possible to destroy the plant material, in case necessary. Efficient methods of detecting the transgenic plants and the transgene in nontarget species is necessary. It can be done by visual marker (*e.g.*, β-glucorinidase) or a selectable marker (*e.g.*, antibiotic resistance, or molecular analysis, *e.g.*, PCR and southern hybridization). If necessary, methods of destroying the plant material at the end of the experiment should be described.

Commercial releases follow once the results of experimental releases have been found to be satisfactory. The aim of risk analysis is to find out the changes in experimental protocol and the method by which transgenic plant may be confined in order to minimize risk to the environment and human health. The risk assessment should be carried out by the multidisciplinary biosafety committee with expertise in molecular biology, environmental science, entomology, pathology, and any other field as appropriate. At this stage, the governmental authorities, environmental groups, social activists, NGOs, and progressive farmers may be involved in order to make the process transparent, and assure the public that care is being taken to minimize risks to the environment and human health. The institute biosafety committee should have the responsibility for monitoring the site after release for risk assessment to enhance accountability. All the unexpected events need to be documented and reported. The national biosafety committee should look into the proposal, and recommend any additional information that may be necessary. The oversight responsibility of the national biosafety committee should also involve interviews with the field experimentation staff, and the release team/institution. Finally, the release proposal may be examined by the regional or international agencies as appropriate to ensure a proper and transparent release process.

It is difficult to conclude that release of a transgenic plant would not involve a risk to the environment or human health. And it is not possible to put a value on the degree of risk. The essential point is to determine how the transgene might alter the risk compared to the nontransgenic crop. There will certainly be questions that cannot be answered in entirety. Not much can be said about the nature and extent of gene flow. In situations where enough knowledge is not available, it is important to use the knowledge available through the conventional plant breeding to aid in the risk assessment. Plant breeding has been carried out for several thousand years, and many of the genes that are being inserted through the recombinant technology fall into the same classes as those manipulated by the conventional plant breeding.

Once the transgenic plants are released for commercial cultivation, measures such as prevention of flower production, and destroying all plant parts is not possible. Therefore, the risk assessment should take into account the pollen transfer between the nontransgenic crop and the wild relatives. There may also be possibilities for taking the transgenic plants into areas where the sexually compatible wild relatives of the crop are present in large numbers. Transgene instability may be another cause of concern when the transgenic crops are grown on a large scale. Strategies for introducing herbicide resistance into several crops or different toxins against the prevalent pests and diseases have to be properly devised. Similarly, options for containment after large-scale cultivation of a transgenic plant are limited. Therefore, risk assessment must take these factors into account, and consider all factors available from small-scale experiments. To overcome such problems, it may be useful to use tissue specific expression (target site for insect feeding or infection by the pathogen) or use of male-sterile lines to limit the dispersal of pollen as is the case with hybrids produced by conventional plant breeding. Another possibility is growing the transgenic crops where there are no wild relatives of the crop, and in areas free from nontransgenic crop. The license for cultivation may be cancelled if the results of growing a transgenic crop are unsatisfactory or there is a risk to the human health and the environment.

11. RISK MANAGEMENT

Many agricultural practices are in place for risk management. The risk of gene transfer by outcrossing from an herbicide resistant crop to a weed, *e.g.*, from canola to weedy mustard can be managed by spraying an herbicide with a different mode of action. Crop rotations can also be used to control such weeds. The risk of introducing a fertile hybrid between the transgenic plants and weedy relative can be managed by seeds grown under strict certification procedures to identify crop weed hybrids in the seed production plots. Gene transfer within the same species (such as the terminator gene) can be avoided by keeping a safe distance between the adjacent plots. Such information to avoid outcrossing is available for most of the cultivated crops. In areas where there is a greater chance of gene transfer, *e.g.*, in the center of origin of a crop plant, serious thought should be given before introducing a transgenic crop with certain genes. Varieties or crops that are likely to be carried to next crop season or contaminate the same crop next season can be replaced by crops varieties with a less or no carryover of seed to the next season.

12. CONCLUSIONS AND FUTURE PROSPECTS

The use of crop protection traits through transgenics will continue to expand in future, and gene pyramiding might become very common. This approach of controlling insects would offer the advantage of allowing some degree of selection for specificity effects, so that pests, but not the beneficial organisms are targeted. While several crops with commercial viability have been transformed in the developed world, very little has been done to use this technology to increase food production in the tropics. It is important to make this technology available to farmers, who cannot afford the high cost of seeds marketed by the private sector. International research centers, advanced research institutions in the developed world, and the national agricultural research systems and the private sector can play a major role in promoting biotechnology for food, feed, and fiber production in the developing world.

REFERENCES

Alyokhin AV and Ferro DN 1999. Modifications in dispersal and oviposition of Bt-resistant and Bt-susceptible Colorado potato beetles as a result of exposure to *Bacillus thuringiensis* subsp. *tenebrionis* Cry3A toxin. *Entomologia Experimentalis et Applicata* **90** : 93-101.

Anonymous 1998. *Workshop on Biosafety Issues Emanating from use of Genetically Modified Organisms (GMOs). Background Document.* Biotech Consortium India Limited. New Delhi, India pp 289.

Arpaia S, Gould F and Kennedy G 1997. Potential impact of *Coleomegilla maculata* predation on adaptation of *Leptinotarsa decemlineata* to Bt-transgenic potatoes. *Entomologia Experimentalis et Applicata,* **82** : 91-100.

Barton K, Whiteley H and Yang NS 1987. *Bacillus thuringiensis* δ-endotoxin in transgenic *Nicotiana tabacum* provides resistance to lepidopteran insects. *Plant Physiol.,* **85** : 1103-1109.

Bauer LS 1995. Resistance: a threat to the insecticidal crystal proteins of *Bacillus thuringiensis. Florida Entomol.,* **78** : 415-443.

Benedict JH, Sachs ES, Altman DW, Deaton DR, Kohel RJ, Ring DR and Berberich BA 1996. Field performance of cotton expressing Cry IA insecticidal crystal protein for resistance to *Heliothis virescens* and *Helicoverpa zea* (Lepidoptera: Noctuidae). *J. Econ. Entomol.,* **89** : 230-238.

Berberich SA, Ream JE, Jackson TL, Wood R, Stipanovic R, Harvey P, Patzer S and Fuchs RL 1996. The composition of insect-protected cottonseed is equivalent to that of conventional cottonseed. *J. Agri. Food Chem.,* **44** : 365-371.

Bergvinson D, Willcox MN and Hoisington D 1997. Efficacy and deployment of transgenic plants for stem borer management. *Insect Sci. Applicat.,* **17** : 157-167.

Birch ANE, Geoghegan IE, Majerus MEN, McNicol JW, Hackett CA, Gatehouse AMR and Gatehouse JA 1999. Tri-trophic interactions involving pest aphids, predatory 2-spot ladybirds and transgenic potatoes expressing snowdrop lectin for aphid resistance. *Mol. Breed.,* **5** : 75-83.

Bottino MB, Girard C, Jouanin L, Metayer M le, Picard Nizou AL, Sandoz G, Pham Delegue MH, Lerin J, Le Metayer M and Thomas G 1998. Effects of transgenic oilseed rape expressing proteinase inhibitors on pest and beneficial insects. In *Proceedings of the International Symposium on Brassicas, 23-27 Sept 1997.* Rennes, France. Acta Horticulturae no. 459. pp 235-239.

Boulter D 1995. Plant biotechnology. Facts and public perception. *Phytochemistry* **40** : 1-9.

Brake J and Vlachos D 1998. Evaluation of transgenic event 176 "Bt" corn in broiler chickens. *Poul. Sci.,* **77** : 648-653.

Calgene 1990. *Kan^r gene: Safety and Use in Genetically Engineered Plants. Request for Advisory Opinion.* California, USA: Calgene Inc.

Chakrabarti SK, Mandaokar AD, Ananda Kumar P and Sharma RP 1998. Synergistic effect of Cry 1Ac and Cry 1F δ-endotoxin of *Bacillus thuringiensis* on cotton bollworm, *Helicoverpa armigera. Curr. Sci.,* **75** : 663-664.

Chevre AM, Eberm F, Baranger A and Renard M 1997. Gene flow from transgenic crops. *Nature,* **389** : 924.

Commission of the European Communities (CEC). 1990a. *Council Directive of 23rd April 1990 on the Contained use of Genetically Modified Microorganisms.* Reference no 90/219/EEC. Official Journal L117, Volume 33, May 1990. Brussels, Belgium: Commission of the European Communities.

Commission of the European Communities (CEC). 1990b. *Council Directive of 23rd April 1990 on the Deliberate Release into the Environment of Genetically Modified Organisms.* Reference no 90/220/EEC. Official Journal L117, Volume 33, May 1990. Brussels, Belgium: Commission of the European Communities.

Cook RJ 2000. Science-based risk assessment for the approval and use of plants in agricultural and other environments. In *Agricultural Biotechnology and the Poor* (Eds Persley GJ and Lantin MM). Washington, USA: Consultative Group on International Agricultural Research. pp 123 - 130.

Crecchio C and Stotzky G 1998. Insecticidal activity and biodegradation of the toxin from *Bacillus thuringiensis* subsp. *kurstaki* bound to humic acids from soil. *Soil Biol. Biochem.*, **30** : 463-470.

Cui JJ and Xia JY 1997. The effect of Bt transgenic cotton on the feeding function of major predators. *China Cottons,* **24** : 19.

Cui JJ and Xia JY 1998. Effects of early seasonal strain of Bt transgenic cotton on population dynamics of main pests and their natural enemies. *Acta Gossypii Sinica,* **10** : 255-262.

Dale EC and Ow DW 1991. Gene transfer with subsequent removal of the selection gene from the host genome. *Proc. Natl. Acad. Sci. USA* **88** : 10558-10562.

Daniels NE 1981. *Migration of Greenbugs in the Texas Panhandle in Relation to their Biotypes.* Texas Agriculture Experiment Station Miscellaneous Publication, MP-1487.

DeWit PJGM, Joasten MHAJ, Lauge R, Roth R, Luderer R, Westerink N, Honee G, Kooman-Gersmann M, Laurent F, van der Horn RAL, de Jorg CF, Vossen P and Weide R 1998. Gene for gene interactions and the role of avirulence genes in pathogeniticity of race-specific resistance. In *7th International Congress of Plant Pathology, Edinburg, Scotland, 9-16 August 1978.* Abstracts, Vol. 1.

Dogan EB, Berry RE, Reed GL and Rossignol PA 1996. Biological parameters of convergent lady beetle (Coleoptera: Coccinellidae) feeding on aphids (Homoptera: Aphididae) on transgenic potato. *J. Econ. Entomol.,* **89** : 1105-1108.

Donegan KK, Schaller DL, Stone JK, Ganio LM, Reed G, Hamm PB and Seidler RJ 1996. Microbial populations, fungal species diversity and plant pathogen levels in field plots of potato plants expressing the *Bacillus thuringiensis* var. *tenebrionis* endotoxin. *Transgenic Res.,* **5** : 25-35.

Doolittle WF 1999. Phytogenic classification and the universal tree. *Science* **284** : 2124-2127.

Ebinuma H, Sugita K, Matsunaga E and Yamakado M 1997. Selection of marker-free transgenic plants using the isopentenyl transferase gene. *Proc. Natl. Acad. Sci. USA,* **94** : 2117-2121.

Fearing PL, Brown D, Vlachos D, Meghji M and Privalle L 1997. Quantitative analysis of CryIA(b) expression in Bt maize plants, tissues, and silage and stability of expression over successive generations. *Mol. Breed.,* **3** : 169-176.

Federici BA 1998. Broad-scale use of pest-killing plants to be true test. *California Agriculture* **52** : 14-20.

Fischhoff DA, Bowdish KS, Perlak FJ, Marrone PG, McCormick SM, Niedermeyer JG, Dean DA, Kusano-Kretzmer K, Mayer EJ, Rochester DE, Rogers SG and Fraley RT 1987. Insect tolerant tomato plants. *Bio/Technol.,* **5** : 807-812.

Fitt G, Mares CL and Llewellyn DJ 1994. Field evaluation and potential ecological impact of transgenic cottons (*Gossypium hirsutum*) in Australia. *Biocontrol Sci. Tech.,* **4** : 535-548.

Fitter A, Perrins J and Williamson M 1990. Weed probability challenged. *Bio/Technol.,* **8** : 473.

Food and Agriculture Organization (FAO). 1986. *International Code of Conduct on the Distribution and Use of Pesticides.* Rome, Italy: Food and Agriculture Organization.

Foster JE, Ohim HW, Patterson FL and Taylor PL 1991. Effectiveness of deploying single gene resistances in wheat for controlling the damage by Hessian fly (Diptera: Cecidomyiidae). *Environ. Entomol.,* **20** : 964-969.

Fraley R 1992. Field testing genetically engineered plants. In *Current Combinations in Molecular Biology: Improvement of Agriculturally Important Crops* (Eds, Fraley R, Frey NM and Schell J). New York, Cold Spring Harbor, USA: Cold Spring Harbor Press, pp 83-86.

Franck Oberaspach SL and Keller B 1997. Consequences of classical and biotechnological resistance breeding for food toxicology and allergenicity. *Plant Breed.,* **116** : 1-17.

Gebhard F and Smalla K 1998. Transformation of *Acinetobactor* sp. strain BD413 by transgenic sugarbeet DNA. *Appl. Environ. Microbiol.,* **64** : 1550-1554.

Gill SS, Cowles EA and Pietrantonio FV 1992. The mode of action of *Bacillus thuringiensis* endotoxins. *Annu. Rev. Entomol.,* **37** : 615-636.

Gillard MS, Flynn L and Rowell A 1999. Food scandal exposed. *The Gaurdian* 12.2.99 : 1.

Girard C, Picard Nizou AL, Grallien E, Zaccomer B, Jouanin L and Pham Delegue MH 1998. Effects of proteinase inhibitor ingestion on survival, learning abilities and digestive proteinases of the honeybees. *Transgenic Res.,* **7** : 239-246.

Giroux S, Cot JC, Vincent C, Martel P and Coderre D 1994. Bacteriological insecticide M-one effects on predation efficiency and mortality of adult *Coleomegilla maculata lengi* (Coleoptera: Coccinellidae). *J. Econ. Entomol.,* **87** : 39-43.

Gould F and Anderson A 1991. Effects of *Bacillus thuringiensis* and HD-73 delta-endotoxin on growth, behavior, and fitness of susceptible and toxin-adapted strains of *Heliothis virescens* (Lepidoptera: Noctuidae). *Environ. Entomol.,* **20** : 30-38.

Gould F 1998. Sustainability of transgenic insecticidal cultivars: integrating pest genetics and ecology. *Ann. Rev. Entomol.,* **43** : 701-726.

Greene JK, Turnipseed SG and Sullivan MJ 1997. Treatment thresholds for stink bugs in transgenic B.t. cotton. In *Proceedings, Beltwide Cotton Conference, 6-10 January 1997.* New Orleans, LA, USA. Volume 2. Memphis, USA: National Cotton Council. pp 895-898

Gregorius HR and Steiner W 1993. Gene transfer in plants as a potential agent of introgression. In *: Transgenic Organisms* (Eds, Workman K and Tommie J). Basel, Switzerland: Birkhauser Verlag pp 83-107.

Griffiths W 1998. Will genetically modified crops replace agrochemicals in modern agriculture? *Pesticide Outlook,* **9** : 6-8.

Harris A 1991. Comparison of costs and returns associated with *Heliothis* resistant Bt cotton to non-resistant varieties. In *: Proceedings, Beltwide Cotton Conference, 1991.* Memphis, Tannesse, USA: National Cotton Council, Pages 249-297.

Hilbeck A, Baumgartner M, Fried PM and Bigler F 1998. Effects of transgenic *Bacillus thuringiensis* corn-fed prey on mortality and development time of immature *Chrysoperla carnea* (Neuroptera: Chrysopidae). *Environ. Entomol.,* **27** : 480-487.

Hilder VA and Boulter D 1999. Genetic engineering of crop plants for insect resistance – a critical review. *Crop Protect.,* **18** : 199-191.

Hoffmann MP, Zalom FG, Wilson LT, Smilanick JM, Malyj LD, Kiser J, Hilder VA and Barnes WM 1992. Field evaluation of transgenic tobacco containing genes encoding *Bacillus thuringiensis* delta-endotoxin or cowpea trypsin inhibitor: efficacy against *Helicoverpa zea* (Lepidoptera: Noctuidae). *J. Econ. Entomol.,* **85** : 2516-2522.

Hofte H and Whiteley HR 1989. Insecticidal crystal proteins of *Bacillus thuringiensis. Microbiol. Review* **53** : 242-255.

Inter-American Institute for Cooperation on Agriculture (IAICA). 1991. *Guidelines for Release into the Environment of Genetically Modified Organisms.* San Jose, Costa Rica: Inter American Institute for Cooperation in Agriculture.

Johnson MT 1997. Interaction of resistant plants and wasp parasitoids of tobacco budworm (Lepidoptera: Noctuidae). *Environ. Entomol.,* **26** : 207-214.

Johnson MT and Gould R 1992. Interaction of genetically engineered host plant resistance and natural enemies of *Heliothis virescens* (Lepidoptera: Noctuidae) in tobacco. *Environ. Entomol.,* **21** : 586-597.

Johnson MT, Gould F and Kennedy GG 1997. Effects of natural enemies on relative fitness of *Heliothis virescens* genotypes adapted and not adapted to resistant host plants. *Entomologia Experimentalis et Applicata,* **82** : 219-230.

Jouanin L, Girard C, Bonade Bottino M, Metayer M le, Nizou ALP, Lerin J, Delegue MHP le and Metayer M 1998. Impact of transgenic oilseed rape expressing proteinase inhibitors on coleopteran pests and honey bees. *Cahiers Agricultures,* **7** : 531-536.

Kaiser J 1996. Pests overwhelm Bt cotton crop. *Nature,* **273** : 423.

Kuiper HA and Noteborn HJM 1994. Food safety assessment of transgenic insect-resistant Bt tomatoes. Food safety evaluation. In *Proceedings of an OECD-sponsored Workshop, 12-15 September 1994, Oxford, UK.* Paris, France: Organisation for Economic Cooperation and Development (OECD) Pages 50-57.

Lang BA, Moellenbeck DJ, Isenhour DJ and Wall SJ 1996. Evaluating resistance to CryIA(b) in European corn borer (Lepidoptera: Pyralidae) with artificial diet. *Resistant Pest Management* **8** : 29-31.

Lehrer SB 2000. Potential health risks of genetically modified organisms: How can allergens be assessed and minimized. In *Agricultural Biotechnology and the Poor* (Eds Persley GJ and Lantin MM). Consultative Group on International Agricultural Research. Washington DC, USA, pp 149-155.

Leopold M 1993. The commercialization of biotechnology, the shifting frontier. In *Biotechnology R & D Trends. Science Policy for Development* (Ed. Tzotzos GT). The New York Academy of Sciences, New York, USA, pp 214-231.

Levin SA 1988. Safety standards for the release of genetically engineered organisms. *TIBTECH,* **6** : 547-549.

Levin M and Strauss HS 1993. Overview of risk assessment and regulation of environmental biotechnology. In *Risk Assessment in Genetic Engineering* (Eds Levin M and Strauss HS). McGraw Hill Inc., New York, USA, pp 1-17.

Llewellyn D and Fitt G 1996. Pollen dispersal from two field trials of transgenic cotton in the Namoi Valley, Australia. *Mol. Breed.,* **2** : 157-166.

Lozzia GC, Furlanis C, Manachini B and Rigamonti IE 1998. Effects of Bt corn on *Rhopalosiphum padi* L. (Rhynchota: Aphididae) and on its predator *Chrysoperla carnea* Stephen (Neuroptera: Chrysopidae). *Bollettino di Zoologia Agrariae di Bachicoltura,* **30** : 153-164.

Luginbill P Jr and Knipling EF 1969. Suppression of wheat stem fly with resistant wheat. *USDA/ARS Production Research Report,* **107** : 1-9.

MacIntosh SC, Stone TB, Jokerst RS and Fuchs RL 1991. Binding of *Bacillus thuringiensis* proteins to a laboratory selected strain of *Heliothis virescens. Proc. Natl. Acad. Sci. USA,* **188** : 8930-8933.

Malone LA, Giacon HA, Burgess EPJ, Maxwell JZ, Christeller JT and Laing WA 1995. Toxicity of trypsin endopeptidase inhibitors to honey bees (Hymenoptera: Apidae). *J. Econ. Entomol.,* **88** : 46-50.

Mascarenhas RN, Boethel DJ, Leonard BR, Boyd ML and Clemens CG 1998. Resistance monitoring to *Bacillus thuringiensis* insecticides for soybean loopers (Lepidoptera: Noctuidae) collected from soybean and transgenic Bt-cotton. *J. Econ. Entomol.,* **91** : 1044-1050.

McLaren JS 1998. The success of transgenic crops in the USA. *Pesticide Outlook* **9** : 36-41.

Meenakshisundaram KS and Gujar GT 1998. Proteolysis of *Bacillus thuringiensis* subspecies *kurstaki* endotoxin with midgut proteases of some important lepidopterous species. *Indian J. Exp. Biol.,* **36** : 593-598.

Miller HI and Flamm EL 1993. Biotechnology and food regulation. *Curr. Opini. Biotechnol.,* **4** : 265-268.

Moberg WK 1990. Understanding and combating agrochemical resistance. In *Managing Resistance to Agriochemicals* (Eds Green MB, LeBaron HM and Moberg WK). ACS Symposium Series, Washington, USA pp 3-16.

National Academy of Sciences (NAS). 1987. *Introduction of Recombinant DNA-Engineered Organisms into the Environment: Key Issues*, National Academy of Sciences, Washington DC, USA p 24.

NAS 1989. *Field Testing Genetically Modified Organisms.* National Academy of Sciences, Washington DC, USA. p 170.

Noteborn HPJM, Bienenmann Ploum ME, Alink GM, Zolla L, Reynaerts A, Pensa M, Kuiper HA and Fenwick GR 1996. Safety assessment of the *Bacillus thuringiensis* insecticidal crystal protein CRYIA(b) expressed in transgenic tomatoes. In *Agri Food Quality: An Interdisciplinary Approach* (Eds Hedley C, Richards RL and Khokhar S). Special Publication No. 179. Royal Society of Chemistry, Cambridge, UK pp 23-26.

Organization for Economic Cooperation and Development (OECD). 1992. *Report of OECD Workshop on the Monitoring of Organisms Introduced into the Environment.* Organization for Economic Cooperation and Development, Paris, France.

Orr DB and Landis DA 1997. Oviposition of European corn borer (Lepidoptera: Pyralidae) and impact of natural enemy populations in transgenic versus isogenic corn. *J. Econ. Entomol.,* **90** : 905-909.

Parker RD and Huffman RL 1997. Evaluation of insecticides for boll weevil control and impact on non-target arthropods on non-transgenic and transgenic B.t. cotton cultivars. In *: Proceedings, Beltwide Cotton Conference, 6-10 January 1997, New Orleans, LA, USA.* Volume 2. National Cotton Council. Memphis, USA pp 1216-1221.

Pham Delegue MH and Jouanin L 1997. The honey bee and transgenic plants. *Revue Francaise d'Apiculture,* **574** : 250-251.

Picard Nizou AL, Grison R, Olsen L, Pioche C, Arnold G and Pham-Delegue MH 1997. Impact on proteins used in plant genetic engineering toxicity and behavioral study in the honeybee. *J. Econ. Entomol.,* **90** : 1710-1716.

Pilcher CD and Rice ME 1998. Management of European corn borer (Lepidoptera: Crambidae) and corn rootworms (Coleoptera: Chrysomelidae) with transgenic corn: a survey of farmer perceptions. *Amer. Entomologist,* **44** : 36-44.

Porter DH, Burd JD, Shufran KA, Webster JA and Teetes GL 1997. Greenbug (Homoptera: Aphididae) biotypes: selected by resistant cultivars or preadapted opportunists. *J. Econ. Entomol.,* **90** : 1055-1065.

Potrykus I 1990. Gene transfer to plants: assessment and perspectives. *Physiol. Plant.,* **79** : 125-134.

Pusztai A, Ewen SWB, Grant G, Brown DS, Stewart JC, Peumans WJ, Damme EJM and Bardocz S 1993. Antinutritive effects of wheat-germ agglutinin and other N-acetylglucosamine-specific lectins. *British J. Nutrit.,* **70** : 313-321.

Pusztai A, Grant G, Brown DJ, Stewart JC, Bardocz S, Ewen SWB, Gatehouse AMR and Hilder V 1992. Nutritional evaluation of the trypsin (EC 3.4.21.4) inhibitor from cowpea (*Vigna unguiculata* Walp.). *British J. Nutrit.,* **68** : 783-791.

Pusztai A, Koninkx J, Hendriks H, Kok W, Hulscher S, Damme EJM van Peumans WJ, Grant G and Bardocz S 1996. Effect of the insecticidal *Galanthus nivalis* agglutinin on metabolism and the activities of brush border enzymes in the rat small intestine. *J. Nutrit. and Biochem. Sci.,* **7** : 677-682.

Raybould AF and Gray AJC 1993. Genetically modified crops and hybridization with wild relatives: a UK perspective. *J. Appl. Ecol.,* **30** : 119-219.

Riddick EW and Barbosa P 1998. Impact of Cry3A-intoxicated *Leptinotarsa decemlineata* (Coleoptera: Chrysomelidae) and pollen on consumption, development, and fecundity *of Coleomegilla maculata (*Coleoptera: Coccinellidae*). Ann. Entomol. Soc. Amer.,* **91** : 303-307.

Riddick EW, Dively G and Barbosa P 1998. Effect of a seed-mix deployment of Cry3A-transgenic and nontransgenic potato on the abundance of *Lebia grandis (*Coleoptera: Carabidae) and *Coleomegilla maculata* (Coleoptera: Coccinellidae). *Ann. Entomol. Soc. Amer.,* **91** : 647-653.

Riggin Bucci TM and Gould F 1997. Impact of intraplot mixtures of toxic and nontoxic plants on population dynamics of diamondback moth (Lepidoptera: Plutellidae) and its natural enemies. *J. Econ. Entomol.,* **90** : 241-251.

Sachs ES, Benedict JH, Stelly DM, Taylor JF, Altman DW, Berberich SA and Davis SK 1998. Expression and segregation of genes encoding CryIA insecticidal proteins in cotton. *Crop Sci.,* **38** : 1-11.

Serageldin I 1999. Biotechnology and food security in the 21st Century. *Science,* **285** : 387-389.

Serratos JA, Willcox MC, Castillo-Gonzalez F (eds.). 1997. *Gene Flow Among Maize Landraces, Improved Maize Varieties, and Teosinte: Implications for Transgenic Maize.* International Wheat and Maize Research Institute. Mexico, DF, Mexico p 122.

Sharma HC 1993. Host plant resistance to insects in sorghum and its role in integrated pest management. *Crop Protect.,* **12** : 11-34.

Sharma HC, Ananda Kumar P, Seetharama N, Hari Prasad KV and Singh BU 1999. Role of transgenic plants in pest management in sorghum. In: *Symposium on Tissue Culture and Genetic Transformation of Sorghum, 23-28 Feb 1999.* International Crops Research Institute for the Semi-Arid Tropics (ICRISAT) Patancheru, Andhra Pradesh, India.

Sharma HC and Ortiz R 2000. Transgenics, pest management, and the environment. *Curr. Sci.,* **79** : 421-437.

Sharma HC, Sharma KK, Seetharama N and Ortiz R 2000. Prospects for using transgenic resistance to insects in crop improvement. *Electronic Journal of Biotechnology* Vol 3, no 2. http://www.ejb.org/content/vol 3/issue 2/full/20.

Sims SR and Holden LR 1996. Insect bioassay for determining soil degradation of *Bacillus thuringiensis* subsp. *kurstaki* CryIA(b) protein in corn tissue. *Environ. Entomol.,* **25** : 659-664.

Sims SR 1995. *Bacillus thuringiensis* var. *kurstaki* (CryIA (c) protein expressed in transgenic cotton: effects on beneficial and other non-target insects. *Southwestern Entomologist,* **20** : 493-500.

Sims SR and Berberich SA 1996. *Bacillus thuringiensis* Cry IA protein levels in raw and processed seed of transgenic cotton: determination using insect bioassay and ELISA. *J. Econ. Entomol.,* **89** : 247-251.

Tabashnik BE 1994. Evolution of resistance to *Bacillus thuringiensis. Annu. Rev. Entomol.,* **39** : 47-79.

Tabashnik BE, Finson N and Johanson MW 1991. Managing resistance to *Bacillus thuringiensis*: lessons from the diamondback moth (Lepidoptera: Plutellidae). *J. Econ. Entomol.,* **84** : 49-55.

Tang JD, Collins HL, Roush RT, Metz TD, Earle ED and Shelton AM 1999. Survival, weight gain, and oviposition of resistant and susceptible *Plutella xylostella* (Lepidoptera: Plutellidae) on broccoli expressing Cry1Ac toxin of *Bacillus thuringiensis. J. Econ. Entomol.,* **92** : 47-55.

Thomas DJI, Morgan JAW, Whipps JM and Saunders JR 1997. Plasmid transfer by the insect pathogen *Bacillus thuringiensis* in the environment. Pages 261-265 *in Microbial Insecticides: Novelty or Necessity? Proceedings of a Symposium held at the University of Warwick, 16-18 April 1997.* British Crop Protection Council. Coventry, Farnham, UK.

Tiedje JM, Colwell RK, Grossman WL, Hodson RE, Lenski RE, Mack RN and Regal PJ 1989. The planned introduction of genetically modified organisms: ecological considerations and recommendations. *Ecology,* **70** : 298-315.

Tzotzos T (ed.). 1995. *Genetically Modified Microorganisms: A Guide to Biosafety.* Commonwealth Agricultural Bureau, International. Wallingford, Oxon, UK.

United Nations Industrial Development Organization (UNIDO). 1991. *Voluntary Code of Conduct for the Release of Organisms into the Environment.* July 1991. United Nations Industrial Development Organization. Vienna, Austria.

Vaeck M, Reynaerts A, Hofte H, Jansens S, DeBeuckleer M, Dean C, Zabean M, Van Montagu M and Leemans J 1987. Transgenic plants protected from insect attack. *Nature,* **327** : 33-37.

Wang CY and Xia JY 1997. Differences of population dynamics of bollworms and of population dynamics of major natural enemies between Bt transgenic cotton and conventional cotton. *China Cottons,* **24** : 13-15.

Williamson M, Perrins J and Fitter A 1990. Releasing genetically engineered plants. Present proposal and possible hazards. *Trends in Ecology and Evolution,* **5** : 417-419.

Wood EA Jr 1971. Designation and reaction of three biotypes of the greenbug cultured on resistant and susceptible cultivars of sorghum. *J. Econ. Entomol.*, **64** : 183-185.

Wu JY, He XL, Shu C, Chen S, Fu CX and Huang JQ 1997. Influence of waterlogging on the bollworm resistance of Bt cotton. *Jiangsu J. Agricult. Sci.*, **13** : 231-233.

Wyngaarden JB 1990. The future role of biotechnology in society. In *: Advances in Biotechnology.* Swedish Council for Forestry and Agricultural Research. Stockholm, Sweden, p 260.

Yoder JI and Goldsbrough AP 1994. Transformation systems for generating marker-free transgenic plants. *Biotechnology,* **12** : 263-267.

Chapter 17

INTELLECTUAL PROPERTY RIGHTS AND INDIAN AGRICULTURE

K Hariprasanna, AK Singh★ and Gyan P Mishra

Division of Genetics, Indian Agricultural Research Institute, New Delhi - 110 012 India

Summary

The subject of Intellectual Property Rights (IPR) is the most debated issue under Trade Related Intellectual Property Rights (TRIPs), which is one of the major areas covered in the Uruguay Round of GATT. The IPR can be granted through patents, trade secrets, copyrights, trademarks, geographical indications etc. However, the existing patent laws of many countries including India did not provide for IPR on living organisms including plant varieties. Implications of TRIPs and Plant Breeders' Rights (PBR) will compel the countries to formulate policy issues on production and distribution of seeds, forming patents or legislation, farmers' rights and contribution of communities in varietal protection. All these aspects will have significant impact on diversification of Indian agriculture and also on the ongoing crop improvement programmes. Implications on seed saving by farmers and traditional seed exchange among farmers or in other words the farmers' privilege have to be debated critically. India has taken the lead by passing the Geographical Indication Bill and the Protection of Plant Varieties and Farmers' Rights Bill. Patenting of plant genetic resources associated with food and agriculture may limit or minimize their use and exchange, likely resulting in a narrowing of the genetic base even in gene-rich countries. India has introduced Biological Diversity Bill in order to provide proper legislation regarding ownership and benefit sharing

★Corresponding author

with respect to plant genetic resources, and also to reward the local communities and indigenous knowledge. Some of the strategies India has to follow to negate the ongoing patent related problems are documenting its bio-wealth and traditional knowledge, modification of existing patent laws, strengthening working efficiency of patenting authority and spreading patent awareness.

Keywords : Intellectual property rights, plant variety protection, farmers' rights, patents

1. INTRODUCTION

"Necessity is the mother of invention" this adage underlies the fact that in general human endeavor is directed at fulfilling some need of human and monetary profit is one of the motive behind man's relentless toil, inventiveness and ingenuity. Societies and governments have devised various ways to reward their inventor so that they can work with greater zeal and devotion to develop newer more useful inventions.

After the Second World War, many countries got together to work on ways and means to promote international trade and as a result, the General Agreement on Trade and Tariff (GATT) was signed by 23 countries in 1948. India was one of the founder members of GATT. After eight rounds of Multilateral Trade Negotiations, the GATT was finally concluded and signed by over 120 nations in December 1993. India is among 64 countries that had ratified this multilateral agreement by June 1994. GATT has been replaced by World Trade Organization (WTO) since 1st January 1995. The major areas covered during negotiation of the historical eighth round *i.e.*, Uruguay Round are - agriculture, multi-fibre agreement, market access, TRIPs and services and dispute settlement body (DSB). The Intellectual Property Rights (IPR) remains the most debated issue under Trade Related Intellectual Property Rights (TRIPs) (Rai *et al.*, 1998).

Intellectual property rights, like other categories of rights or law, are historically situated (Granovetter, 1985). They are conceived, constructed and interpreted "in context". Their emergence - as applied to a particular type of innovative activity - is related to economic, technological and political factors - and cannot be seen simply as the unfolding of disembodied legal logic (Fowler, 1994). This article aims at providing information on various aspects of IPR and its impact on Indian Agriculture.

2. INTELLECTUAL PROPERTY RIGHTS (IPR)

As mentioned earlier the IPR is the most debated issue under TRIPs. The term 'property' is often found associated with physical objects only, such as household goods or land, for which ownership and associated rights are guaranteed and protected by law prevalent in a country. This property is described as tangible. Intellectual property, on the other hand, is intangible and includes patent, trade secrets, copyrights, trademarks, geographical indications, industrial designs and layout designs of integrated circuits (Anonymous, 1993). The right to protect this property prohibits others from making, copying, using or selling the proprietary subject matter. Under biotechnology, one of the most important examples of intellectual property is the processes and products like transgenic plants, animals and microorganisms, which result from the development and use of genetic engineering techniques. It is the characterization of these research that results into intellectual properties, which has encouraged industries to allocate more and more labour, research and development units and funding to facilitate the production of commercially marketable items.

2.1. History and definition

The term 'intellectual property' has a European origin. Protection of intellectual property is not an innovation of the modern industrial society. Earlier records reveal that even primitive communities recognized the right of a person over his intellectual creation. Trademarks were recognized by the Zhou dynasty in China about 3000 years ago. It has been recorded in history that the first patent law was established in about 15th century AD, in the city of Venice, known for its excellent artisans and craftsmen. The artisans, the goldsmiths and the jewelers, were outraged that their apprentices soon became competitors who often exceeded their masters in skill and excellence. A patent law was enforced in 1474 on the junior apprentices to ensure that they could open their own workshop only after 14-21 years of serving in their masters' workshop. Although, no clear picture emerges from history, it is quite possible that patent laws then depended on the 'might' of the monarch. Legal redressal for conflicts arising then out of intellectual property issue was, perhaps, quick and decisive (Granovetter, 1985).

The idea of protection, especially in relation to copyright began to emerge with the invention of printing press, which made it possible for literary work to be duplicated by mechanical process instead of being copied by hand.

The term IPR is now widely used by legal systems all over the world. The TRIPs Agreement does not define in detail the term "intellectual property right". IPR are usually defined as rights granted by a state authority for certain products of intellectual effort and ingenuity. Most often the term IPR is used as a collective name for rights such as those dealt with by the TRIPs Agreement. From an analysis of the other IPR covered by TRIPs one can further draw the conclusion that IPR are rights either to exclude others from certain commercially interesting acts in relation to the described subject matter and/or rights to obtain a remuneration in respect of those acts (Leskien and Flitner, 1998). The scope of intellectual property and the relevant legal rights have in recent years kept pace with industrial, scientific and technological developments.

2.2. Different forms of IPR

The various forms of protecting "Intellectual Property" include:

2.2.1. Patent

A patent is a monopoly right granted to a person who has: a) invented a new and useful article, b) improved an existing article, or c) has developed a new process of making an article. A patentable invention should have novelty, inventiveness and commercial utility. If invention lacks any of these, the patent can not be granted. Although, the objective of seeking patent on an invention is primarily driven with commercial interest of the patent holder, there are several advantages of granting patent, which include: a) encouragement for innovative research/invention, b) inducing an inventor to disclose his invention instead of keeping it as a trade secret, c) offering rewards for efforts made (in terms of both time and money spent) on invention, and d) inducing capitalist for making investment for commercialization of a patented invention.

The patent application for an invention can be filed with the patent office in the respective country along with required fee and specifications to be provided within 12 months from the date of filing the patent. Any person interested may, within four months from the date of advertisement give notice of opposition to grant of a patent. The applicant in turn is given an opportunity to reply to the opponent. Both parties are enabled to support their cases by evidence.

By grant of a patent, the State provides to the patentee a statutory provision for an exclusive right, which can lead to the creation of monopoly. Monopolies affect a wide variety of interests of the patentee, the licensee, the manufacturer, the consumer and the State. Monopolies can also lead to restrictive practices, cartel formations including output restrictions, price hijacking, limited availability and under-utilization of the products. And when two countries, a developed and a developing country, are involved these practices result in imports and/or outflow of royalties and hinder indigenous R&D. Thus there is incompatibility between the patent system and the developmental process of developing countries. While patents protect the interest of an individual (or an organization), the patent and other industrial laws of a country are meant to provide a correct balance between all these concerns and interests.

2.2.2. Trademark

A trademark is a visual symbol in the form of a word, slogan or a label applied to articles of commerce with a view to indicate to the purchaser, that those are the goods manufactured or dealt in by a particular person as distinguished from similar goods manufactured or dealt in by other persons. A trademark helps in identifying a product and its origin; guarantees in unchanged quality and at the same time advertise the product.

2.2.3. Copyright

Copyrights are concerned with the right of protecting of physical material existing in the field of literature, art, informatics, databases, computer software and integrated circuits. Copyrights subsist only in original work *i.e.,* the work must be an expression of inventive original thought. There is no copyright on ideas, which are not put in black and white.

2.2.4. Geographical appellation

The term geographical appellation refers to the goods originating in particular country or a territory with their quality, or other characteristics, attributable to their geographical origin. In order to have geographical appellation or geographical indication (GI) on a product, it is mandatory that GI is first protected within the country of origin otherwise it can not be protected outside. In case, the GI is not protected, it becomes a generic name allowing others also to use the same name/device for the product of similar nature. India, was not having GI bill, which itself was an impediment in having GI on speciality products like Basmati rice,

Alphonso mango and Darjeeling tea etc. With an Act in place, now suitable protection for the above said commodities can be granted.

2.3. IPR applied to plants

The first intellectual property rights law in the United States to cover biological materials explicitly was the Plant Patent Act of 1930 (Anonymous, 1930). This act provided patent protection for asexually reproduced varieties of domesticated plants (*e.g.* apple, pear, rose, etc.). Since that time, patent or patent-like protection has been expanded through legislation and court decisions to include sexually reproduced plant varieties, microorganisms, genes and gene complexes. Trade secrets, contracts, and use of the tort theory of conversion have also been used for the protection of plant germplasm (Innen and Jondle, 1989).

The wide geographic spread of seeds and planting materials since the earliest days of agriculture, (some 12,000 years ago) and the ease with which seeds could be obtained and moved about, rendered exclusive ownership and control almost impossible at the species or genotype level (Fowler, 1997). As breeding programs and commercial seed activities came to encompass more and more crops in the middle decades of 20th century, the need for IPR protection for a broader range of crops (grains, vegetables, etc.) was felt by emerging commercial interests. After several unsuccessful attempts in the 1960s, the seed industry persuaded U.S. Congress to pass legislation, the Plant Variety Protection Act, in 1970. This act requires that an applicant should demonstrate that the protected variety is distinct, uniform (genetically), and stable. It does not require disclosure of an inventive step (Anonymous, 1970).

Until recently it was the variety itself that was the primary unit of economic value. With the advent of the new biotechnological tools, the focus of value began to expand and shift to genes. Once unthinkable, genes can now be transferred across species "boundaries". The expression of firefly genes transferred to tobacco plants is one such particularly evocative example of the power of recent scientific advances. Less dramatic transfers are becoming routine activity. The application of these new tools to plant breeding has brought new economic opportunities and fresh challenges to the existing IPR regimes (Fowler, 1997).

3. PLANT BREEDERS' RIGHTS

The Plant Breeders' Rights (PBR) are exclusive rights granted to the breeder of a new plant variety to exploit the protected variety for

1991 Act provides for the eventual protection in all UPOV member States of all plant genera and species.

Secondly, under the 1978 Act, the breeder's protection enabled him only to control marketing of the reproductive material of his variety and production of such material for the purpose of marketing. A number of difficulties arose with this formulation of the breeder's right. It had the advantage for farmers that the production of seed on their farms for sowing on their farms fell outside the scope of protection but it had the effect also that a person could buy one fruit tree and use it, after propagation, to plant a vast orchard with no obligation to the breeder. The modern technique of tissue culture multiplies the potential for this kind of misuse of the breeder's variety. The 1991 Act accordingly extends the breeder's protection to production and reproduction of his variety but permits member States on a discretionary basis to allow any traditional form of saving seed on the farm, which they wish to retain, particularly in case of small and marginal farmers.

Thirdly, under the 1978 Act, a variety can be taken to a country which does not provide protection for new plant varieties and used there to produce an end product, say, cut flowers, which is exported back to a country where the breeder's variety is protected. The breeder receives no remuneration from the exploitation of his variety in this way. The 1991 Act extends the breeder's protection in very limited circumstances to the harvested material of his variety so as to enable him to seek some reward from the exploitation of his variety in the kind of circumstance described above.

Fourthly, under the 1978 Act, a protected variety can be modified in a very limited respect, *e.g.* by reselection, mutation, backcross breeding or transgenesis and provided, that the modified variety is clearly distinguishable from the protected variety, it can be separately protected without any obligation to the breeder of the protected variety. The 1991 Act provides that varieties that are "essentially derived" from a protected variety in this way can still be protected but cannot be marketed without the permission of the breeder of the protected variety from which they are derived. Varieties are "essentially derived" for this purpose only when they are virtually entirely reconstructed upon the basis of the protected variety from which they are derived. This provision is designed for discouraging parasitical breeding approaches. An overview of some key provisions of the UPOV Acts of 1978 and 1991 is given in Table 2 (Ghijsen, 1998).

Table 2. Overview of some key provisions of the UPOV Acts of 1978 and 1991 (Ghijsen, 1998).

Key provisions	UPOV 1978	UPOV 1991
Breeders' exemption	Included	Included
Principle of essential derivation	Not included	Included
Scope of protection	Only traded material	All materials + harvested product + end product (optional)
Farmers' privilege	Farm saved seed not under scope of PBR	Included
No. of species to be protected	Minimal 24	All
Duration of protection	15-18 years	20-25 years
Double protection (*e.g.* PBR and patent)	Not possible	Possible

However, under *sui generis* system, breeder exemption has been retained by the national PVP systems allowing the breeders to use a protected variety as a parent in breeding programs. An interesting situation in this context, which may lead to serious conflicts, is described hereunder:

Consider a transgenic variety of cotton, which carry a patented gene, for example, '*Bt*'. Under breeder exemption one could use this '*Bt*' cotton variety as a donor parent to transfer the '*Bt*' gene through backcross method of breeding to a commercial variety that is susceptible to boll worm. The newly developed variety will not fall under the category of 'essentially derived variety', as it is not virtually reconstructed entirely upon the basis of protected variety, but upon the recipient variety. Thus the consent of breeder of donor variety would not be essential for the commercialization of this variety. However, if one looks at the problem from different angle, the entire merit of newly developed variety is solely because of single gene added from the donor. One might argue that in this case the transferred gene *per se* is protected and therefore it doesn't come under the purview of breeder exemption. But this argument would hold well only in those countries where gene patents are allowed. Therefore one may like to reconsider breeder exemption *vis-à-vis* use of transgenics in breeding program under the *sui generis* system.

4.5. Gene protection technologies

The implementation of Plant Breeders' Rights or other forms of protecting plant varieties seems to have operational difficulty in developing countries where large number of farmers with very small holdings are engaged in agriculture. Realizing this difficulty some of the multinational companies have devised several foolproof technologies using genetic engineering to control seed germination/growth and development processes. Such technologies include terminator, verminator and traitor etc. These technologies completely prevent the use of farm saved seed for planting next crop. This might have serious consequences in countries like India where majority of the farmers depends on farm saved seeds in the case of self-pollinated crops like rice, wheat, pulses and many oilseeds. Worldwide opposition of such technologies by farming communities and NGOs has resulted into their temporary withdrawal.

5. FARMERS' RIGHTS

The concept of "Farmers' Rights" has been the subject of intense debate at FAO for a decade. "Farmers' Rights" have been defined as "rights arising from the past, present and future contributions of farmers in conserving, improving and making available plant genetic resources, particularly those in the centers of origin/diversity" (Anonymous, 1989). The definition highlights a crucial distinction between these and other resources, or raw materials.

Plant genetic resources as represented in traditional "farmer-varieties" of agricultural crops are an already-improved raw material. There is an "intellectual" component to the resource. The resource consists of thousands of genetically distinct varieties of wheat, rice, tomatoes, etc., which owe their existence in large part to the people who have tailored them through selection over the years. This work of selection on variability generated through crossing and maintenance of elite lines is the result of innovative and creative approaches with goals in mind. It needs to be remembered that farmwomen and men have not only created several thousand races of food and cash crops, they have also identified valuable genes and traits in these crops and maintained them over generations.

Communities have not only developed complex systems of pest management and biological control, they have identified and managed a series of genes conferring valuable traits for commercial and domestic needs. The gene(s) for traits as diverse as disease resistance, high salt tolerance, resistance to water logging and drought tolerance have been maintained in the repertoire of communities. Along with these commercial traits, characteristics like cooking time, taste, digestibility and milling are recognized and maintained. Women who have been the traditional custodians of the seed and responsible for its selection, are the repositories of this knowledge and in the true sense owners of this complex seed technology and know-how (Sahai, 1996). It is this genetic diversity which is now being captured and used in modern plant breeding programs.

The traditional farmer-varieties although impressive and useful, usually cannot meet the UPOV criteria of distinctness, uniformity and stability. In particular, they are rarely uniform (Leskien and Flitner, 1997). A peasant farmer depends on the stability and dependability of production offered by genetic heterogeneity because uniformity in the field leads to vulnerability to pest and disease attacks. Thus, while the intellectual property of modern plant breeders might be protected legally, the breeding work of farmers is not and cannot be through the same statutes or similar approaches (Fowler, 1994).

The perceived imbalance in this situation led to political controversy and finally to the agreement on "Farmers' Rights" at FAO, as noted above. The definition for these rights however, falls short of something that can be operationalized or enforced. "Farmers' Rights" could be seen, or constructed, as a form of IPR, which might be extended either to individuals or to communities (Swaminathan, 1996). Such legislation is under consideration in many countries.

5.1. *Sui Generis* legislation and farmers' rights

It has been suggested that the *sui generis* system provided for in Art. 27.3(b) TRIPs could be implemented for Farmers' Rights. Given that the *sui generis* system has to provide for an IPR, this approach might not conform with the TRIPs Agreement if one accepts that Farmers' Rights are a concept rather aiming at counter balancing IPR than creating any additional intellectual property right (Louwaars, 1998).

This does not, however, prevent the inclusion of provisions that implement Farmers' Rights and/or traditional resource rights into a *sui generis* system. Such an incorporation of Farmers' Rights based, for example, on the Indian proposals for Community Intellectual Property Rights and the Plant Variety Recognition and Rights Model Act, could complement the *sui generis* system by compensating those who have been conserving plant genetic resources (PGR) for the past centuries and thereby have contributed until now to the development of plant varieties protectable under the *sui generis* system foreseen in the TRIPs Agreement (Leskien and Flitner, 1997).

5.2. Plant variety protection and farmers' rights

Farmers have a fundamental right to land. Prime farmland should be protected for social and environmental reasons. No agricultural land should be acquired without the farmers' consent, and no agricultural land should be made available to MNCs. According to the Indian Constitution, every citizen has the freedom to practice his/her occupation. Seed being the most important input of agricultural production, and therefore part of the occupation of farmers, the Constitution ensures the right of farmers to produce, reproduce, modify and sell seed.

Legislation which will simultaneously deal with breeders' and farmers' rights needs to be introduced in all countries. This can be titled the "Plant Variety Protection and Farmers' Rights Act". This act should clearly distinguish between the rights of Farmer-cultivators and Farmer-conservers (Swaminathan, 1998). Keeping in view the above, a Joint Committee was set up by Government of India, which drafted "The Protection of Plant Varieties and Farmers' Rights Bill" in 1999 and later redrafted in 2000 (Anonymous, 1999). The bill has been passed by the Parliament, which is formulated to provide for the establishment of an effective system for protection of plant varieties, the rights of farmers and plant breeders and to encourage the development of new varieties of plants. Under the bill a) it is considered necessary to recognize and protect the rights of the farmers in respect of their contribution made at any time in conserving, improving and making available plant genetic resources for the development of new plant varieties, b) for accelerated agricultural development in the country, it is necessary to protect PBR to stimulate investment for research and development, both in the public and private sector, for the development of new plant variety, c) such protection will facilitate the growth of the seed industry in the country which will ensure the availability of high quality seeds and planting material to the farmers, and d) to give effect to the aforesaid objectives, it is necessary to undertake measures for the protection of the rights of farmers and plant breeders.

The best way for the local communities to obtain remuneration for their work done in the present and the future is the breeding of improved varieties by local farmer-breeders. This also minimizes the chance that farmers have to buy seeds of foreign variety that have been improved from farmers' original landraces (Ghijsen, 1998). An approach gaining emphasis in the recent years is the concept of Participatory Plant Breeding. This has been defined in its broadest sense, ranging from plant breeder-controlled decentralized breeding to various degrees of farmer involvement in the breeding or improvement process. Participatory approaches draw upon the comparative advantages of both conventional plant breeding ("formal") and farmers' plant breeding ("informal"). Professional plant breeders carry out germplasm enhancement in much the same way they would if they were working alone but the final selection and variety testing is done by farmers (Participatory variety selection), or by breeders and farmers working together (Participatory plant breeding). Participatory approaches can be used at various stages in the breeding cycle and hence requires cooperation between breeders and farmers in partnership (Joshi and Witcombe, 1998).

For successful implementation of participatory approaches a number of constraints in the institutional systems would need to be overcome. The issue of incentives and rewards to plant breeders (PBR) is one such serious concern, as farmers' involvement has to be duly considered. So necessary amendments in the existing regulations may have to be brought about. Support to national agricultural research systems may be needed to stimulate the use of local landraces, to strengthen the role of farmers in realizing more demand-driven breeding programs and to cover risks involved in their participation apart from addressing the question of Farmers' rights.

6. IPR AND PLANT GENETIC RESOURCES

Today, one of the most vibrant and volatile areas in the field of intellectual property rights concerns the status of plant genetic resources associated with food and agriculture. These resources come in a variety of forms from gene and gene complex to a finished crop variety; from a peasant-selected and maintained landrace to the inbred line used in a private sector plant breeding program. Arguably, both the farmer-variety and the modern, scientist-produced cultivar contain a measure of creative, intellectual input. Beyond debate is the fact that plant genetic resources constitute the biological foundation of agriculture. They are clearly among the most valuable of all resources as the raw material for the evolution and development of the world's food crops. The questions of ownership, control and benefit sharing in relation to plant genetic resources are inextricably linked with the more practical questions of access to, and development and use of these resources. A mixture of policies and laws can be employed to address such issues (Fowler, 1997).

Plant breeding involves the rearrangement and manipulation of genes. Some 6 million samples of agricultural crops (typically in the form of seed) are conserved in cold storage in more than 1300 "gene banks" around the world (Anonymous, 1998). Gene bank collections contain much of the intra- as well as inter-species genetic diversity of the world's agricultural crops - most of it coming from developing countries where virtually all major crops were domesticated and initially developed (Zeven and Zhukovsky, 1975). These genetic resources - the legacy of 12,000 years of agriculture and eons of plant evolution before that - are the raw material used in plant breeding programs. Without it, plant breeders would not be able to fashion new varieties resistant to pest and diseases, or adapted to new farming or climatic conditions.

Major efforts are underway to catalog and document these resources and to construct maps of the genomes of a number of crops. Such work will allow for more efficient deployment and use of specific genes and gene complexes. Moreover, as genes and their functions are identified, the prospect of inserting a particular characteristic in one species by using genes from another becomes more and more feasible, and common.

The desire to gain legal ownership and control over genes has led to numerous and widely publicized court cases, and to the effective expansion of coverage of patent statutes. It has also led to the intensification of conflict over ownership and benefit-sharing associated with biological diversity in general, and more specifically with the thousands of distinct varieties of domesticated crops and the genes they contain which are conserved *ex situ* or still found in the fields and gardens of traditional farmers, particularly in developing countries (Anonymous, 1996).

6.1. FAO undertaking on plant genetic resources for food and agriculture (PGRFA)

FAO undertaking on Plant Genetic Resources for Food and Agriculture (PGRFA) formulated in 1980's aimed at addressing broad range subjects including collection, conservation and utilization of Plant Genetic Resources for food and agriculture, access, benefit sharing and farmers' rights. As per FAO undertaking, the plant genetic resources are the 'common heritage of mankind' meant for unrestricted use by anyone anywhere in the world for the benefit of human race.

6.2. Convention on biological diversity (CBD)

The Biodiversity convention was signed by over 150 countries at Rio de Janeiro in June 1992. The convention is a landmark document, which highlights concerns regarding saving, protecting and sharing of genetic wealth of planet earth and also preventing unfair exploitation of the rich genetic wealth and traditional knowledge of the developing countries by the developed world (Glowka, 1997). The convention on Biodiversity came into force on 29th December 1993 and has by now been ratified by 170 countries. It is the only International legal instrument that addresses biological diversity in a comprehensive way. The most important of the provisions of the UNCED-CBD and one which has received the greatest attention, relates to the control of and access to plant genetic resources. The Convention lays down that the plant genetic resources are the 'national heritage' of the country in whose territory they are found and has placed them under the sovereign control of individual nation (Principle of ownership). In addition, CBD lays down two other conditions of great importance to germplasm owning countries. These are that of Prior Informed Consent (PIC) and Material and Information Transfer Agreements with respect to the transfer of genetic resources from owner countries to countries/companies/individuals that want to use these resources (Anonymous, 1994). Thus, access of other countries to plant genetic resources will have to be negotiated on mutually agreed terms. The Convention also calls for the results of research and development, and the benefits arising from the commercial and other utilization of plant genetic resources, to be shared in a fair and equitable way with the country which provided these resources in the first place.

It is the most obvious example of the continuation of the centuries-long struggle over ownership and control of biological materials. The Convention "reaffirms" that "States have sovereign rights over their own biological resources" (Anonymous, 1992). Under terms of the Convention, access to these resources is granted by the "country of origin" on the basis of mutually agreed terms. Article 2 states that the country of origin of genetic resources "means the country which possesses those genetic resources in *in-situ* conditions." And *in-situ* conditions "means conditions where genetic resources exist within ecosystems and natural habitats, and, in the case of domesticated or cultivated species, in the surroundings where they have developed their distinctive properties."

6.3. FAO undertaking on PGRFA and CBD : The conflict

The ideology of plant genetic resources being considered as "Common Heritage" of mankind as per FAO undertaking and "National Sovereignty" principle as per CBD are in conflict with each other. The conflict is particularly with regard to sharing/ownership of the PGRFA collected prior to the arming into force of CBD (*i.e.*, the bulk of material currently under gene banks of various CG Centers). Whether the sovereign right of the country of origin on such PGR will still be applicable or these will fall under the sovereign right of the holder is not yet clear. If, latter is the case, the gene rich seed poor centers will be at great loss.

6.4. Conservation of plant genetic resources *vs.* PVP

A PVP system can prevent the piracy of landraces, but their *in situ* maintenance and the sharing of benefits by farmers or communities should be regulated outside the main PVP system, to avoid complicated regulations. Farmers' rights as well as provisions for PIC in the CBD have a tendency to limit the access to germplasm. This may have an adverse effect on plant breeding as a whole and create the opposite result of their original aims (Ghijsen, 1998).

On the other hand the experience of UPOV member countries has shown that PVP increases the number of breeders and, consequently, widens the spectrum of improved varieties available to farmers, with a potential increase in genetic variability. The fact that new variety offer substantial advantages to farmers does mean that farmers may choose to stop growing their existing varieties or landraces in favour of new varieties, whether or not such plants are protected by PBR. Ways must be found to make important new varieties available to farmers generally, whilst encouraging the continued use by some farmers of their existing varieties or landraces, through some sort of incentives, so as to prevent genetic erosion (Greengrass, 1996).

7. PATENTS ON INDIAN PLANTS: AN UNETHICAL ATTEMPT

In India, turmeric, neem, black jeera, mustard and amla were used from ancient times for traditional curing besides as edible foods. Under globalization, it is very difficult for India to seek its right as American companies have got either these plant varieties patented or are in the way of getting it. The U.S. companies are claiming that through researches they have developed these plant varieties and thus these companies have monopoly on the trade of such plant materials. The U.S. companies further argue that the qualities of turmeric, neem, karela and amla were developed in USA and therefore, they have the property rights on these plant varieties.

7.1. Patent on Neem

In 1990, W. R. Grace and Company (USA) procured the property right of neem products through a U.S. patent (No. 5124349) (Subbaram and Thyagarajan, 1998). The company invented a chemical treatment for stabilizing the azadirachtin, thereby increasing its shelf life and making it possible for it to be transported worldwide. Although the use of neem oil by the Indian farmers was mentioned in the patent application, there can be no provision for sharing the resultant huge commercial benefits from the sales of the insecticide with the Indian farmers under the present regime. Though the patent would not prevent the Indian farmers from using neem oil locally as a pesticide, as long as it was produced on the farm or purchased in crude form from neighbors or local market, any Indian entrepreneur would be prohibited from developing and marketing a similar commercial invention (Ghate *et al.*, 1999).

In India, quite a good number of researches have been done in research institutes, universities, Malarial Research Centre, Tata Energy Research Institute and Khadi and Gramodyog Commission. Presently, these organizations are still carrying out their research work in this field. Based on the researches brought out by Indian Research Centers, many medicines, cosmetics and pesticides from neem were successfully launched in the markets. Keeping the medicinal and industrial value of neem, some developed countries like USA and Japan are running behind, to get the products patented in their names (Singh *et al.*, 1998).

In 1993, M/s Neem Campaign, a non-government organization (NGO) filed a petition against the property right of neem product obtained by W. R. Grace and Company. M/s Neem Campaign in its challenging petition provided the facts that Indians knew the use of neem oil against fungal infection since century back and various products of neem were used as medicines for curing against wound, skin diseases, etc. On the basis of facts given by the Neem Campaign, the property right of neem which was once granted to the U. S. company was declared illegal and cancelled.

7.2. Patent on Turmeric

In March 1995, two Indian born scientists, namely, Suman K. Das and H. P. Kohli working at Medical Centre of University of Mississippi got patent right of turmeric products to be used as healing medicine against wounds from the Patent Office of USA (No. 5401504) (Subbaram and Thyagarajan, 1998). The patent was granted on the grounds that use of turmeric in the powder form for wound healing was not known in the U. S. The patenting of turmeric was challengeable because Indians had known about the qualities of turmeric since ancient times and they had used it since then.

In 1996, the patenting of turmeric was challenged by C.S.I.R. (India) on the basis that product patent was normally given only when its product is invented by the scientists/institutes of the particular country who is applying for getting patent right. In case of turmeric, Indian people were familiar with the use of turmeric from centuries, therefore, it was unjustified to grant property right to a country other than India. C.S.I.R. (India) provided numerous evidences including 32 documents through which it was proved that turmeric was used as indigenous medicine in ancient literature. The use of turmeric has been mentioned in the ancient Indian literature as for healing wounds. As C.S.I.R. could present published evidence in an appeal in the U. S. court that such usage was known in India, and hence not novel, the patent was revoked consequently (Singh *et al.*, 1998).

7.3. Patent on Basmati Rice

The RiceTec Inc. (USA) applied for patenting of Basmati rice on 8th July 1994 in the Patent Office of USA, and this office provided property right on Basmati rice to RiceTec by the granting of a U. S. patent titled "Basmati rice lines and grains" (No. 5663484) on 2nd September 1997 (TIFAC, 1998). After having this right of Basmati rice now RiceTec Inc. is authorized for sale of rice in the international market by the trade names having prefix or suffix Basmati or similar name. As a matter of fact RiceTec has already launched several brands such as American Basmati, Kasmati, Jasmati and Texamati, which may confuse the consumers and give a setback to Indian export. This patent right may deprive India, the leading Basmati rice grower, of exporting its Basmati rice to USA, European Union and Middle East countries. Although there is a great demand of Indian Basmati in the international market and India has been exporting Basmati rice nearly worth rupees 2500 crores annually, due to property right earned by RiceTec, India may face great legal problem to trade Basmati rice in world market.

One of the claims of the patent granted to RiceTec provides coverage of all rice varieties with Basmati quality, which are in the height range of 80 cm to 140 cm. Such broad range claims are likely to have negative impact on future growth and development of Basmati improvement. By granting the Basmati patent to RiceTec, the U. S. patent office had essentially deprived Pakistan, India or anyone else of their prior use-rights to all the genetic traits and genes that give rise to the essential characteristics of Basmati and other similar aromatic fine grain rice, and so denied them the right to sell such grain, in North, Central or South America or Caribbean Islands (Benjavan, 1999).

An inter-ministerial committee was organized by Indian Government to protest against Basmati patent. The expert committee thoroughly examined with the help of legal experts each of the 22 claims made by RiceTec Inc. and based on strong evidences revoked the claim numbers 15-17 of the patent pertaining to Basmati grains and won the case in its favour. The remaining claims (which have far-reaching consequences on research on Basmati breeding) have also been set aside by U.S. patent office recently, except claims 8, 9, 11, 12 and 13, which relate to three specific rice lines developed by RiceTec.

8. GATT TEXT ON AGRICULTURE

India, along with 129 other countries, is now a signatory to GATT with TRIPs as one of its components, which has several provisions more favourable for developed countries than for developing ones (Dhar and Chaturvedi, 1998). Its Article 27 compels member countries to protect through patents, innovations in all fields including food, health and other biotechnology related fields. Member countries may however exclude plants, animals and essentially biological processes for reproduction from patenting.

Agriculture is one of the several areas, which are excluded from the ambit of India's present patent law. The new GATT, however, explicitly provides for the introduction of IPR in agriculture. Section 27.3 of the draft text on TRIPs provides that India would have to accept the system of "patents or an effective *sui generis* system or a combination thereof" for plant varieties. The *sui generis* system in fact refers to the PBR system, the globally recognized framework of which is provided by the UPOV. The UPOV convention, adopted in 1961, lays down the legal rights of the plant breeders. Unlike the industrial patent system, UPOV provides for an exemption of breeders: they are free to use a protected variety to develop a new variety. Farmers, *de facto*, are exempted too, as they were in practice allowed to save their own seed. For these reasons, the system of PBR has been less stringent than the patent system. However, in March 1991 amendments to the UPOV Convention have made the PBR system closer to that of the industrial patents. The amendments, among others, aim at strengthening of the monopoly of the plant breeder, by restricting the farmers' and breeders' exemptions. Section 15.2 of the revised UPOV Convention provides that farmers can be permitted to propagate any protected variety on their own farm, but this act of seed saving has to be "subject to the legitimate interests of the breeder". These "legitimate interests" are the royalty that the breeder would claim for the use of the protected varieties. This implies that the farmer can no longer use the seeds from a protected variety for further propagation without compensating the breeder.

9. IMPLICATIONS FOR INDIAN AGRICULTURE

Indian farmers fear that accepting the new GATT will have an immediate impact on seed prices. This may impede most farmers, those who are low-income earners, in purchasing improved seeds because of the higher prices. It is however unclear, why the public sector would stop the sale of propagating material at affordable prices.

The uncertainties that farmers in countries like India face as a result of the new GATT have another critical dimension. In India, exchange of new seeds between farmers would certainly be affected once India adopts a PBR system similar to the UPOV 1991 Act. As stated earlier, this traditional seed exchange among farmers, has been the major cause for the rapid dissemination of new improved varieties in remote areas, and has been considered to be the principal reason behind achieving self-reliance in Indian agriculture. The Indian government has argued that this farmers' privilege for seed saving will not be affected since "limited exchange of seeds according to prevailing traditional systems" can be retained under PBR. Commercial exchange of protected seeds among farmers, however, will be prohibited under PBR. Moreover, the fact that only around 20-25 per cent of India's annual seed requirement is met by formal private and public agencies, makes it doubtful whether the inter-farmer seed sales, which make up the remaining, can be seen as "limited seed exchange".

One may has to reconsider the fact that the 1991 Act of UPOV permits saving seed on the farm for sowing the next crop, in case of small and marginal farmers. The criterion followed for classifying the farm holdings in developed countries may include all the farm holdings in India under the scope of small and marginal holdings. So, the subject of whether withdrawal of farmers' exemption will really have any impact on seed saving in our country has to be debated.

10. STRATEGIES TO BE FOLLOWED

India is well known for a number of items like turmeric, neem, pepper, Darjeeling tea, Basmati rice and host of traditional medicinal plants. The MNCs are making their efforts to bring about new strains of similar quality and also specific purpose medicines and chemicals and patent them. The patent on products from neem for medicinal and insecticidal purposes, turmeric for medicine and Basmati rice, etc. are the major cases under MNCs control. Although the patents on Basmati rice, neem and turmeric have been cancelled, many traditional remedies and recipes are becoming patented products with the rights for commercial production. U.S. holds many patents on the plants extensively grown in India such as 'pomegranate' (as anti-viral agents), 'mustard' and 'soapnut' (as fire retardant), bitter gourd (treatment of tumours and HIV infections), amla (anti-viral activity), pepper (nutritional use) while India holds few patents on them. Thus legally India's biological wealth is open for global exploitation (Rai *et al.*, 1998). Some of the strategies that can be adopted to safeguard from such exploitation are listed bellow.

10.1. Modification in patent laws

After signing WTO Accord, changing, from Process patent to Product patent is inevitable to India to continue as a global trade partner under the WTO regime. There should be provisions regarding conservation, sustainable use and benefit sharing. Geographical indication law provides the legal means for interested party to prevent use of 'native' names in designation or presentation of a 'Good'. Unfortunately, India was not having any specific law or geographical indications for long. Presently, the Geographical Indication Bill has been passed by the Indian Parliament and got approval of President of India.

10.2. Documentation of bio-wealth and heritage

India is a repository of world bio-wealth and rich cultural heritage and also center of diversities in the world. Unfortunately, very little of our bio-wealth are made known in terms of appropriate catalogues. If India catalogues all such information on scientific basis, then many of our patent disputes either would not occur at all or can be solved with least problem. Efforts are being made by Ministry of Environment and Forest and Spices Board in this direction, but similar attempts must be made in other fields also.

10.3. Improvement in the working efficiency of patent office

In India, at present there are only four patent offices with head quarters in Calcutta and only one Patent Information Centre in Bangalore, which are insufficient to be fully used by whole country. Also in India, it takes about seven years to get the patent while in USA it hardly takes two years. To speed up the process, improvement in staffing anu most efficient working condition is needed. Establishment of a 'National Patent Office' is also under the serious consideration for easy patent registration.

In this context, after signing the Paris Convention for Protection of Industrial Property and Patent Cooperation Treaty (PCT) on 11th August 1998, India has become able to get international cooperation for overcoming patent disputes and making more efficient patent system.

10.4. Patent literacy

Patent awareness is relatively at lower ebb in India as compared to the developed nations. Entrepreneurs, scientists, technologists and related persons must be given good awareness through mass media exposures, periodic seminars and symposia. Patents are published in Indian Gazettes. The prerequisites and procedure for filing the patent application etc. should be clear to the person(s)/institutions, otherwise these can be denied even for a useful and commercially attractive invention.

11. CONCLUSIONS AND FUTURE PROSPECTS

Though India had initially not been in favour of inclusion of TRIPs in the scope of the Uruguay Round, it ultimately decided to go along with the rest of the world community. India, which does not presently recognize product patents in the field of drugs, food products and chemicals has been allowed a transition period of ten years for establishing a product patent regime for such items. The requirements for India to put an end to the on going patent related problems are documenting its bio-wealth and traditional knowledge, bringing out necessary changes in existing patent laws particularly on product patents, strengthening working efficiency of patenting authority and finally popularizing patent literacy and other related aspects. The expert committee on patents must understand the TRIPs related issues in the true national perspective and suitably act as 'think tanks' to the government and general public. India must rethink on its exclusive 'defensive strategy' to fight patent problems. As India is one of the hotspots for bio-diversity, proper legislation regarding ownership and benefit sharing with respect to plant genetic resources should be enacted. Besides protecting plant genetic resources, the agents and/or communities responsible for maintenance of bio-wealth through generations should be suitably rewarded.

India has taken the initiative by drafting the Protection of Plant Varieties and Farmers' Rights Bill, which has been passed by the Parliament. Realising that there is no comprehensive legislation dealing with bio-diversity, Government of India has introduced the Biological Diversity Bill 2000 into Parliament, which is currently being examined by a committee of MPs. Further, India should approach the WTO to treat patent violations and spurious patenting using pirated genetic materials as punishable offences. India has already succeeded in revoking patents granted for neem, turmeric and Basmati rice either completely or partially. Effective strategies have to be drawn to prevent the recurrence of any such unfavourable incidences in future, so as to augment the growth and development of science in our country.

REFERENCES

Anonymous (1930). Townsend-Parnell Plant Patent Act, 46 Stat. 376 (May 23, 1930, Ch. 312, § 1).

Anonymous (1970). 227 U.S.P.Q. (BNA) 443.

Anonymous (1989). FAO - Conference Resolution 5/89.

Anonymous (1992). Preamble to the Convention on Biological Diversity. U.N. Environment Programme.

Anonymous (1993). TRIPs Agreement, Art. 1.2.

Anonymous (1994). International Code of Conduct for Plant Germplasm Collecting and Transfer. FAO, Rome.

Anonymous (1996). Access to plant genetic resources and the equitable sharing of benefits : A contribution to the debate on systems for the exchange of germplasm. In : Issues in Genetic Resources, No. 4, IPGRI, Rome.

Anonymous (1998). The State of the World's Plant Genetic Resources for Food and Agriculture. FAO, Rome. p. 510.

Anonymous (1999). Bill No. 123 of 1999, Government of India.

Benjavan R (1999) Intellectual property rights and plant genetic resource. *TDRI* (Thailand Development Research Institute) *Quarterly Review*, **14**(2) : 7-11.

Chopra VL (2000) Intellectual property right issues and plant breeding. **In**: *Plant Breeding: Theory and Practice*, 2nd edn. Chopra, V. L. (ed.) Oxford & IBH Publishing Co. Pvt. Ltd, New Delhi. pp. 461-472.

Dhar B and Chaturvedi S (1998) *Implications of the Regime of Intellectual Property Protection for Bio-diversity: A Developing Country Perspective.* Research and Information System for the Non-Aligned and Other Developing Countries (RISNOPC), New Delhi.

Fowler C (1994). Unnatural Selection: Technology, Politics and Plant Evolution. (International Studies on Global Change, Vol. 6). Gordon and Breach Science Publishers, USA. p. 321.

Fowler C (1997). By policy or law?- The challenge of determining the status and future of agro-biodiversity. *J. TECH. L. & POL'Y*, **3**(1). [http://journal.law.ufl.edu/~techlaw/3-1/fowler.html]

Ghate U, Gadgil M and Rao PRS (1999) Intellectual property rights on biological resources: Benefiting from bio-diversity and peoples knowledge. *Curr. Sci.,* **77**(11) : 1418-1425.

Ghijsen H (1998). Plant variety protection in a developing and demanding world. *Biotechnology and Development Monitor*, **36** : 2-5.

Glowka L (1997) *A Guide to Designing Legal Frameworks to Determine Access to Genetic Resources.* IUCN, The World Consevation Union, Gland.

Granovetter M (1985). Economic Action and Social Structure, *Am. J. of Soc.*, **50** : 21-28.

Greengrass B (1996) What is plant variety protection? Lecture delivered in *National Seminar on the Nature of and Rationale for the Protection of Plant Variety Under the UPOV Convention,* September 12, 1996. New Delhi.

Ihnen J and Jondle R (1989). Protecting Plant Germplasm: Alternatives to Patent and Plant Variety Protection. ASA-special publication (American Society of Agronomy, USA), **52** : 123-143.

Joshi A and Witcombe JR (1998) Farmer participatory approaches for varietal improvement. **In**: *Seeds of Choice*. Witcombe, J. R. *et al.* (eds.) Oxford & IBH Publishing Co. Pvt. Ltd, New Delhi and Calcutta. pp. 171-189.

Leskien D and Flitner M (1997). Intellectual Property Rights and Plant Genetic Resources: Options for a *sui generis* system. **In**: *Issues in Genetic Resources*, No. **6,** IPGRI, Rome. p. 77.

Leskien D and Flitner M (1998). The TRIPs agreement and intellectual property rights for plant varieties. **In**: *Signposts to Sui Generis Rights*, No. **4**. GRAIN, Philippines.

Louwaars NP (1998) *Sui generis* rights: From opposing to complementary approaches. *Biotechnology and Development Monitor*, **36** : 13-16.

Rai VK, Singh AK and Singh SK (1998) Intellectual property rights in Indian context: A review. Souvenir, *National Seminar on Intellectual Property Rights,* December 12-13, 1998. BHU, Varanasi, India. p. 29-38.

Sahai S (1996). Genetic resources, intellectual property rights and rights of indigenous communities. Madras Agric. J., **83**(7) : 419-421.

Singh S, Singh AK and Singh R (1998) Enforced patenting of Indian plants and agricultural commodities: A challenging issue for future. Souvenir, *National Seminar on Intellectual I perty Rights,* December 12-13, 1998. BHU, Varanasi, India. p. 52-56

Subbaram NR and Thyagarajan G (1998) Committee on Science and Technology in Developing Countries (COSTED), Chennai, India.

Swaminathan MS (ed.) (1996) *Agro-Biodiversity and Farmers' Rights.* Konark Publishers Pvt. Ltd, New Delhi. p. 280

Swaminathan MS (1998) Farmers' rights and plant genetic resources. *Biotechnology and Development Monitor*, **36** : 6-9

Technology Information Forecasting and Assessment Council (TIFAC). (1998) *Intellectual Property Bulletin*, **4** : 1-3

Zeven AC and Zhukovsky PM (1975) *Dictionary of Cultivated Plants and their Centres of Diversity*. Centre for Agric. Pub. And Document, Wageningen. p. 219.

Chapter 18

BIOINFORMATICS STUDIES IN PLANT SYSTEMS

Shipra Agrawal and BN Mishra*

Department of Biotechnology, Institute of Engineering and Technology, Sitapur Road, Lucknow - 226 021, Uttar Pradesh, India

Summary

Plant bioinformatics has recently emerged as an important area of research for designing gene manipulation in several agronomically and commercially valuable plants. Gene annotation through computational analysis in plant systems is continuously growing with the developments in several large-scale plant genome sequencing projects. The complete sequencing of the model plant system, Arabidopsis thalliana followed by computational analysis of its genome sequence revealed that approximately 26,000 genes make up a basic set of genes for a fully functional flowering plant. Likewise, genome sequencing of various widely grown crops, medicinal and aromatic plants are now underway. Plants like barley, canola, corn, wheat, cotton, lettuce, loblolly pine, peach, potato, poplar, rice, sorghum, soybean, sunflower, tomato, mentha and cymbopogon are being sequenced and voluminous sequence data has been accumulated. Recognition of rice genome as a precise model system for cereal crops led to the establishment of International Rice Genome Sequencing Project (IRGSP). The IRGSP group is expected to produce the details of complete nucleotide sequence of rice genome that may further be harnessed to perform gene annotation, in silico genomics, functional genomics and comparative genomics in cereal crops. Furthermore, such gene and genome sequence analysis are inevitable to bring the breakthrough in the

*Corresponding author : E-mail : profbmishra@rediffmail.com

field of plant sciences. Besides, chapter also covers various databases and softwares available on World Wide Web (WWW) for the analysis of nucleotide and protein sequences.

Keywords : Database, DNA chips, genome sequencing, *in silico* tools, rice genome, web designing.

1. INTRODUCTION

The advent of bioinformatics followed by its implication in studying genomics has been proved to have a significant impact on biotechnology industries involving human health and agricultural production. The various developed procedures of bioinformatics are being implied to microbial, animal and plant genomics in order to predict/characterize genes, their regulatory network and protein products. Besides animals and microbes, the genome of several plants including crops, cereals, medicinal and aromatic plant systems are being sequenced and the sequences are annotated to establish practical methods in genomics and proteomics. The developments in plant bioinformatics has been tightly linked with the International Genome Sequencing Projects. Broadly, the bioinformatics in plants covers: database management and analysis of gene sequences including regulatory sequences; the regulation of gene expression; analysis and recognition of genomic sequences; gene structure prediction; modeling of transcriptional and translational control; and large scale genome analysis. Rapid advancements in this area, particularly during the last five years have enabled scientists to create plant varieties of desired characters. In addition, the collection, integration, analysis and management of this ever growing research data in plant molecular biology, biochemistry, physiology, taxonomy and other biological streams are becoming challenging and require standardization/development of new procedures/methods /softwares in bioinformatics for systemizing data mining, accurate genome analysis and gene annotation. As a result, research in bioinformatics is undergoing sweeping changes and the field is evolving rapidly.

The recent progress in research and product developments in the area of plant bioinformatics has been found confined to crop informatics, gene finding, comparative genome informatics, host-pathogen interaction, gene function discovery including identification of genes for salt, drought/ water and pest resistance, starch biosynthesizing and nitrogen fixing genes and chip informatics. The challenges of genome sequencing of important plants are also being paid attention.

Currently, the DNA chips and Micro array methods are considered to bring the breakthrough in the field of bioinformatics. Oligonucleotide based DNA arrays are used for monitoring gene expression and detection of DNA/gene polymorphisms. Each array contains thousands of different oligonucleotides synthesized *in situ* on a glass substrate using light directed solid phase combinatorial chemistry, which are used in parallel DNA hybridization analysis, yielding sequence expression and genotype information. Pairs of perfect and single base mismatch probes to each target gene are synthesized on adjacent areas, which serve as control to identify and subtract non specific hybridization during DNA hybridization. Usefulness of several application products presently available for expression analysis (genetic Hu6800, Mu6800, Ye6100), mutation detection (Gene Chip CYP450 assay, p53) and mapping were described. The bioinformatic methods based on DNA chips and microarrays are also being implemented in the molecular computing of plant systems (Sasaki and Burr, 2000).

Complete sequencing of *Arabidopsis thalliana* (a model plant system), genome analysis and gene annotation introduced a new concept for crop revolution through gene annotation of crops and other plants of importance. Requirement of specific and precise gene annotation also encouraged starting of sequencing projects for plant systems (Sato *et al.*, 1997; Meinke *et al.*, 1998; Marra *et al.*, 1999; Motoaki *et al.*, 2001).

The rice has been recognized as a model cereal crop plant and an International Rice Genome Sequencing Project (IRGSP) was established in 1998 involving various nations as participating centers for sequencing specific parts of specific chromosome. The above project is supposed to accomplish by the end of 2004 (Sasaki and Burr, 2000). The complete rice genome sequence will serve instrumental in cracking the codes of rice and other cereal crops. The genome sequencing of other crops i.e. sorghum, wheat, barley, rye and maize is also advancing. All the above mentioned sequencing projects have targeted on development of reliable methods for accurate gene annotation of the concerned plant. The study of comparative genomics of above plants is also required to answer the basic questions of biology of crop plants such as pattern of phyllogeny, orthologous and paralogous genes as well as molecular phyllogeny of different enzymes and proteins (Sakata *et al.*, 2000 a&b; Sasaki and Burr, 2000; Yuan *et al.*, 2001; Sakata *et al.*, 2002).

In agricultural industry, increasing crop production and quality principally rely on the identification of new plant traits that can be incorporated into new plant varieties by breeding programs. *In silico* genomics is aiding plant biotech industries in accelerating the crop improvement by rapid identification and manipulation of new traits. Further, the techniques of high throughput sequencing and expression and functional analysis that facilitated the discovery of new drug targets in pharmaceutical industry are playing a major role.

The developing methods in plant bioinformatics further show great prospects for bringing breakthrough in the area of crop cultivation. The tools will identify genes involved in seed structure, composition and germination, genes influencing flavour and fragrance, limiting fertility (male sterility) and genes determining yield performance under specific environmental conditions. The outcome of such results will add to the milling quality of wheat, to improve the malting performance of barley, to generate novel grain flavours, to produce grain more resistant to pre-harvest sprouting and the environmental stresses i.e. heat, cold and drought. Further, identification of genes responsible to create resistance against biotic and abiotic stresses will be identified.

The genome structures and their nucleotide arrangements are highly variable from plant to plant; they have drastic variation even at interspecies and variety levels. Such variations indicate that bioinformatic tools developed for genome analysis and gene annotation of specific plant / crop plant will not be appropriate for the other systems. Whereas, most of the *in silico* tools and protocols of plant genomics and proteomics are based on the genome structure of model plant systems. Hence, the annotations performed with such tools for different plant systems cannot be accurate. In addition, the bioinformatics related predictions are accurate only in those plants, which have been worked out widely from breeding, genetics, molecular biology, functional genomics, proteomics, biochemistry and physiology point of views and whose specific and suitable databases are available.

Precisely, unavailability of suitable annotation tools including softwares and databases are leading towards misleading predictions of genes, regulatory network and expression patterns of the organisms/plants.

Lab-automation tools for high throughput sequencing, better algorithms for assembly and gene prediction and full length cDNA sequence for gene identification are some of the essential requirements for accurate genome and gene analysis in plant bioinformatics.

2. TOOLS FOR BIOCOMPUTING

2.1. Databases

The genomics revolution has been the democratization of genome projects. These projects fall into diverse categories: EST sequences and expression data collection, large-scale mapping, limited and complete genome sequencing, as well as compilations of mutants and phenotypes. The progress in such projects has resulted in the creation of large, associated datasets of diverse types that are often scattered in space, format, and time. This has made comparative analyses difficult. Henceforth, the creation of integrated data methodologies and approaches for compiling significant informations as discrete datasets has been proved of great importance in analyzing and curating the related data in order to come with concrete scientific calculations.

The concerned datasets are related with plant genome data to facilitate the efficient on line gene annotation. The datasets are modular, extensible, inter-operable data structures and interfaces that can be queried using manual and automated methods.

2.1.1. Components of datasets

The types of data in a dataset can be divided broadly into two categories: static data and dynamic data. Static data represents data that do not change frequently and reflect primary types of information. Dynamic data are defined as data that undergo frequent changes and updates and generally reflect descriptive information and secondary (analyzed or comparative) types of information. Although it is difficult to state explicitly which data (and to what level those data) need to be archived and for how long, standardization/selection of database during biocomputing is very essential. The data should be portable across databases over time with minimal effort required to upgrade it to new formats. A meeting of representative curators and developers from all plant databases should be held to develop a collective list of standards for archiving.

Data archives are available at three levels: public archives, specialized databases, and private/lab databases. Public data archives generally contain all the primary/ static data. Specialized databases contain some aspect of the static, primary data and some aspects of the dynamic data. Important functions of the specialized databases include the storage of the more dynamic and complex data types that are kept most current and accurate, serving the need of particular research community and the development and establishment of standards for data exchange. Lab/ individual databases are databases that manage data in individual groups as they are produced. Public data archives and specialized databases allow public access at all times.

The public archives stick to some minimal guidelines for standardization of data and that the data be made easily accessible *via* both individual queries and bulk queries. Funding agencies must take the lead in establishing as what type of resources will be available, not just in the short term, but also in the long term for specialized databases. Specialized databases such as species-specific databases do have important roles in plant biology but their quality and compatibility is essential with public archives and with other species-specific databases.

2.1.2. Concept of an ideal database for mapping data analysis

As mentioned earlier, comparative genetic mapping in plants is of great importance in studying genomics of plants. A database required for comparative mapping analysis with accuracy should have following characteristics.

1. All of the raw segregation data for individual mapping populations should be available to the research community, especially those data that were used for the development of comparative maps. All newly generated segregation data should be deposited in the species-specific databases maintained for different organisms.

2. The evidence for proposed orthologous relationships must be defined and supported in comparative mapping research papers and databases.

3. Sequences of all RFLP, SSR, STS and SNP markers used in a project to construct genetic or physical maps should be deposited in GenBank at the time the map is made public. Accession numbers must be provided so that all of these markers can be traced back to a publicly available sequence. For groups using AFLP-like markers or others where no sequence is easily available, there should be a parallel requirement for public availability of all probes, primers and protocols.

4. Funding should be made available to encourage all workers to sequence historic, current and future DNA markers that are used

in comparative maps. Quality controls for these data should be especially stringent and deposition should be accompanied by a description of the quality controls undertaken.

5. When comparative maps are constructed using physical mapping data, the underlying information used to construct the physical maps (e.g., FPC fingerprints, BAC end sequences, BAC hybridization, etc.) should be made publicly available in a project database.

6. The raw data for all genetic and physical mapping, and the software used for their analysis, must be made available. All different analyses that are overlaid to produce a given map comparison must be described and the individual analysis components should be made available.

2.1.3. Role of data analysis in establishing experiments for functional genomics

The MIAFGE working group recommended standards for making a dataset for the description of functional genomics experiments. These standards encompass the minimum information about methodologies, parameters, and the design of functional genomics experiments that should be documented to enable the interpretation, verification and comparison of functional genomics data. "Functional genomics" is defined as any genome-scale phenotypic analysis, including microarray experiments. Such phenotypic analyses could be applied at a variety of levels from the molecular and cellular to the individual or population or ecosystem levels.

2.1.4. Databases for crop plants

The various plant bioinformatics networks and genomic centers have constructed their own databses compiling different plants for their characteristics and other related informations such as genes, gene markers, QTL, cDNAs, ESTs, ORFs, physical maps, genetic maps, germplasm informations including phenotypic details. The followings are the major centres, which have specific databases.

UK CropNet, ARS-USDA, ASPP-Plant Gene Register, GrainGenes, MaizeDB, IGGI, RGP, RiceGenes, Soybase Soybean Home Page, TAIR, TIGR, ICIS, SINGER, GEM, GRIN, Hybrid Corn, Gramene. The description of contents of databases and their websites are given in the table 1.

2.1.5. Comprehensive databases for crop plants

Details of some of the comprehensive databases are given below:

2.1.5a UK CropNet

UK CropNet was established in 1996 and was formed in the collaboration of 6 centers, each centre took over the task of genome analysis of a crop plant. They are as follows: *Arabidopsis* (University of Nottingham), Barley and Potato (Scottish Crop Research Institute), Brassicas (John Innes Centre), Cereals (John Innes Centre), Comparative Genome Analysis (John Innes Centre) and Forrage grasses (Institute of Grassland and Environmental Research). The above UK CropNet groups have contributed eight UK databases using the ACEDB (http://www.sanger.ac.uk/Software/Acedb/) database management system. The server of UK CropNet (http://synteny.nott.ac.uk/) gives access to databases AGR (Arabidopsis Genome Resorce), BarleyDB, BrassicaDB, CerealsDB, CompDB (Comparative mapping), Forage Grasses (FoggDB), MilletGenes and SpudBase.

2.1.5b TIGR

The Institute of Genomic Research (TIGR) provides a wide range of genome related informations on various plants including crops and cereals. The TIGR gene indices (http://www.tigr.org/tdb/tgi.shtml) are a collection of species specific databases in terms of gene indices. The TIGR gene indices for a genome is created by using a highly refined and well tested protocol. The protocol is used for organizing (EST sequences are retrived from dbEST and trimmed to remove vector, poly A/T tails and adapter sequences), clustering and assembling ESTs and gene sequences to produce high fidelity consensus sequences for the represented genes and eliminating misclustered and low quality gene sequences.

The details of ESTs of some crop plants available on TIGR gene indices are indicated as follows:

Plant	TIGR Gene Index
Glycine max	GmGI 3.0
Arabidopsis thalliana	AtGI 4.0
Lycopersicon esculentum	LGI 5.0
Zea mays	ZmGI 3.0
Oryza sativa	OsGI 4.0
Sorghum bicolor	SbGI 1.0
Triticum aestivum	TaGI 1.0
Solanum tuberosum	StGI 1.0
Medicago trancatula	MtGI 1.0

Table 1. Databases compiling specific details of plants, genes, nucleotide sequences, proteins, ESTs, markers, maps etc.

Database	Database contents	Web site
AGR	Physical maps of chromosomes IV and V, Recombinant Inbred maps, DNA sequences (all publically available *Arabidopsis* DNA sequences including 64 MB of genome sequence against major databases) Sequence contigs and germplasm resources.	http://synteny.nott.ac.uk/
Barley DB	Nordic Barley database (59 trait studies and over 1100 germplasm accessions), EMBL and Swissport sequences (1000 DNA and over 200 protein sequences for *Hordeum* species), bibliographic data and 25 barley linkage maps from other UK institutions. Also has 5 barley linkage maps, sequence data, primer data, trait study data, micro-malting data.	http://synteny.nott.ac.uk/
Brassica DB maps	Contains genetic maps developed for *Brassica rapa* developed in collaboration with BBSRC and AAFC Saskatoon Research Centre. It also has above 2202 *B. napus* DNA sequences from EMBL (1709 ESTs and 493 genomic sequences coding for different 276 gene products), results of BLASTX analyses (32,167 proteins showing significant homology)	http://synteny.nott.ac.uk/
CerealsDB	2400 RFLP probes (insert sizes, copy numbers, chromosomal locations and polymorphism data with additional data on DNA end sequences and PCR amplification conditions for the wheat anchor set), scanned and annotated images (DNA hybridization patterns), genetic maps (wheat, rye, *Aeglops umbellulata*) and the corresponding segregation data, the JIC wheat germplasm and pedigree database, over 40, 000 cereal sequences from EMBL and a catalogue of probes.	http://synteny.nott.ac.uk/
CompDB	Comparative mapping links between various monocots (Including barley, foxtail millet, rice, wheat, maize, perl millet, sugar cane, sorghum and oat) and dicots (arabidopsis and brassicas), conserved chromosomal segments of various monocots (relative to rice) and bibliographic data.	http://synteny.nott.ac.uk/

Table 1. Continued

Database	Database contents	Web site
FoggDB	Genetic maps, loci, sequences, genetic maps,loci, sequencing, associated data (phenotypes), trait scores, QTLs and images (phenotypes, disease symptoms, recombinant chromosomes for introgression mapping) and also possessing useful traits.	http://synteny.nott.ac.uk/
MilletGenes	Mapping data (11 perl millet crosses and a finger millet cross) with Downy mildew QTLs for all maps, RFLP and STS polymorphism data for all perl millet probes, scanned autoradiographs (for RFLP probes) and gels (for sequence tagged site markers), probe insert sizes, copy numbers and locations for perl millet and foxtail millet probes, DNA end sequences, primer sequences and PCR amplification conditions for a subset of perl millet probes and a set of JIC AFLP profiles.	http://synteny.nott.ac.uk/
SpudBase	5 potato linkage maps, EMBL and SwisPort sequences (over 800 DNA and 200 protein sequences from *Solanum tuberosum*) and binliographic data.	http://synteny.nott.ac.uk/
AgDB	Agricultural related databases and datasets.	http://www.agdb
AtDB	An *Arabidopsis thalliana* database. It provides database search, has links with Arabidopsis genomic resources, *Arabidopsis* Genome Initiative	http://www.AtDb
GrainGenes	Databases on wheat (Tritiaceae) and other grains: Comparative genetic maps; Location of Triticeae ESTs on the rice genome sequence; the databses also has molecular and phenotypic informations on wheat, barley, oat, rye and sugarcane.	http://www.GrainGenes.org
Tritiaceae Repeat Sequence Database	Over 300 user contributed repetitive sequences are stored as FASTA and BLAST files.	http://wheat.pw.usda.gov/

Table 1. Continued

Database	Database contents	Web site
HarvEST database	Consists data from the U.S. wheat EST project and U.S. barley EST project	http://www.HarvEST
RiceGD	Rice genome database	http://www.RiceGd
RiceBlastDB	Contains details about pathogen causing rice blast disease	http://www.RiceBlastDB
RiceGenes	Informations on rice genes,ESTs, molecular marker, Physical map, genetic map etc.	http://www.RiceGenes
MaizeDB	Maize genetic database	http://www.MaizeDB
SoyBase	Database on soyabean genes	http://www.SoyBase
New UK BLAST databases	BLAST searching against 110, 000 ESTs sequences	
CropSeqDB	Database serves two main purposes: as repository of sequences from UK crop species and these sequences have been linked to AGR database through BLAST analysis	http://www.CropSeqDB
Brassica DB	It contains 400 BBSRC funded microsatellite markers and maps using these SSR markers. It also has for nucleotide and protein sequences for *B. napus*, *B. oleracea* and *B. rapa*.	http://www.BrassicaDB
Gramene databases	Databases for proteins, RiceGenes, Sequence databases, genetic maps and physical maps of maize, rice, sorghum and other members of grass family.	http://www.gramenc.org base

Table 1. Continued

S.No.	Database	Database contents	Web site
	MAtDB	MIPS *Arabidopsis thalliana* database compiles information for each individual gene of *Arbidopsis*. It has links with several other database revealing informations on mutants, proteins, DNA sequences, clones and gene predictions.	http://www.mips.gsf.de/ proj/thal/db/index.html
	MENDEL database	MENDEL plant gene nomenclature database, MENDEL – ESTS : Sequence of plants; MENDEL GFDb : Plant and organelle protein sequence families.	http://www.MENDEL database
	RGenes	Database on plant disease resistance genes	http://ars-genome.cornrll. edu/rice/
	SolGenes	DBCAT: databases for gene of solanaceous members	http://ars-genome.cornell. edu/solgenes/welcome.html
	DictyDB	AceDB database for *Dictyostellium discoideum*	http://ars-genome.cornrll. edu/cgi-bin/WebAce/ webace?db= dictydb
	Yeast and *Arabidopsis* database	Gene expression database	http://cmgm.stanford.edu/ pbrown/explore/index. html http://www.inra.fr/ Versailles/BIOCEL/Filters
31.	General and *Arabidopsis* plasma membrane	Proteome database	http://expasy.hcuge.ch/ ch2/2d-index.html http://sphinx.rug.ac.be: 8080

2.1.5c GRAMENE

Gramene is a comprehensive collection of databases and tools for gene and protein annotation for members of cereal family. It has following databases:

- Oryzabase (http://shingen.nig.ac.jp.rice/oryzabase) used for genetic resource stock information, gene dictionary, chromosome maps, mutant images, and fundamental knowledge of rice science).
- International Rice Institute (http://www.cgair.org/irri)
- UK CropnetDB (http://ukcrop.net/db.html)
- Maize database (Maize DB) (http://Zmdb.iastate.edu/)
- Maize Genome database (ZmDB) (http://www.agron.missouri.edu/)
- North American Barley Genome Project (http://css.orst.edu/barleynabgmp/nabgmp.html)
- Plant Arrays: A Virtual Library, INRA (http://www.Univ-montap2.fr/plant_arrays/index.html)
- Sorghum Genomics Project (http://sorghumgenome.tamu.edu/)
- TIGR Gene Indices (http://tigr.org/tdb/tgi.shtml)
- Grass Genera of the World (http://nmnh.www.si.edu/botany/projects/grasspag.html)

The Gramene (http://www.gramene.org) has details about rice genome sequencing projects, its density maps, physical map, annotation status, besides regular progress in sequencing project and other details. It also has databases on following cereals (monocots): barley, maize, sorghum, oat, sugarcane, wheat and other monocots.

2.1.6. Databases on rice genome and genetic resources

There are more than 20 databases and websites available online for rice genomics. A brief review of such databases are as follows:

2.1.6a RiceGenes

The RiceGenes established in 1993 is one of the earliest databases and it was collectively developed by USDA-ARS Center for

Bioinformatics and Cornell University (http://www.genome.cornell.edu/rice/). This database has details on comparative genomic maps, molecular markers, morphological markers, and QTL and germplasm data and has been updated with microsatellite marker data and microsatellite maps of the 12 rice chromosomes. The updated site has been renamed as http://www.gramene.org that can be switched from one web resource of one crop to another and serve as portal to interconnect such sites.

2.1.6b SHIGEN

National Institute of Genetics in Mishima, Japan established a rice genome database which has a huge biological collection of germplasm databases called SHIGEN (Shared Information of Genetic Resources) (http://www.shigen.nig.ac.jp).

The database consists a wide range of information about rice including classical genetic maps with phenotypic genes, molecular maps, an integrated map with RFLP markers and phenotypic markers for cross-referencing of marker positions in each map. The database also features details of variety information, information on rice classification, morphology, rice cultivation, mutant collection and has a dictionary of all identified genes in rice genome with corresponding references.

2.1.6c MAFF

The MAFF databank consists details of sequence analysis of all rice cDNA clones generated from RGP website of the Japanese Ministry of Agriculture, Forestry and Fisheries (MAFF) (http://gene.affrc.go.jp). The Korea Rice Genome Database having compilation of EST database, genetic maps and chloroplastic genome sequence has been established and updated by Korea Rice Genome Program of Myongji University, Korea (http://bioserver.myongji.ac.kr). An integrated database for rice genome namely Integrated rice genome explorer (INE) was developed by RGP (Sakata *et al.*, 2000a).

A comprehensive database constructed by National Institute of Genetics (Mishima, Japan) has been compiled on about genetic stock of 11,000 mutants developed by different Japanese Institutes and Universities. The database shows details of marker genes, mutant lines, autotetraploid lines, trisomics, and reciprocal translocation mutant lines (http://www.shigen.nig.ac.jp).

All the contributing centers of IRGSP show progress of their sequencing results and have their independent databases. The more specialized databases on rice are: Rice Gene Index (http://www.tigr.org/tdb/ogi) and Rice Repeat Database (http://www.tigr.org/tdb/rice/blastsearch.shtml). These databases are integrated parts of TIGR' s web site on rice genome (http://www.tigr.org/tdbrice). Rice Gene Index consists gene indices as well as 51,000 ESTs (Expressed Sequence Tags) and Ets (Expressed Transcript Sequences) from international sequencing projects of rice. Whereas the Rice Repeat Database contains systematic detailing of repeats and mobile elements 9 DNA transposons, retroelements etc. of 50 % of the rice genome. The rice genes and repeat sequences are searched using BLASTN and these sequences can be downloaded via a FTP link. A database namely Rice blast DB was developed by the RiceGEne groups gives informations about fungus (*Magnaporthe grisea*) causing rice blast (http://www.genome.cornell.edu/riceblast).

2.2. Softwares

2.2.1. *In silico* genomics of plants

As mentioned above that higher plants have been targeted for their complete nucleotide sequences and the nucleotide sequences are desired to annotate using suitable and integrated databases. The plant gene annotation/ in silico genomics is desired to have predictions about presence of type of genes, number, position, splice junctions, ORFs, cDNA, exons and cis-regulatory gene network.

Genomes of various plant systems of agronomical/medicinal/ herbal are analyzed and annotated to have prior informations/prediction about their genes of significance. Such predictions using different bioinformatic approaches help a molecular biologist / plant biotechnologist for planning a precise strategy to develop a plant variety of desired characters. The bioinformatics in cereals is generally desired for better grain quality with high vitamin, protein contents, high starch biosynthesizing potential, pest/ insect resistance, water resistance, high nitrogen fixation capability etc. Whereas, the gene annotation for pulses/legumes is generally desired for high yielding varieties, expression of quality proteins at desired level etc. Horticultural plants are also targeted for annotation of genes responsible for better size and quality of fruits and flowers.

The gene identification is performed through various available gene finding softwares based on similarity with already explored genes in functional genomics. In another words, an unidentified nucleotide is compared with identified genes from several gene sequence databases in order to attain the information. Gene function can be defined and elucidated from several views such as cellular function, biochemical function, developmental function (e. g. role in pattern formation) or adaptive function (the contribution of the gene product to the fitness of the organism). Furthermore, the genome of *Arabidopsis thalliana* (model system for higher plants) has been completely sequenced and several annotation tools have been developed considering the arrangement of genomic sequences on all its 5 chromosomes. The softwares and annotation results for *A. thalliana* are serving a base for other plant systems. The general tools being used for gene annotation of plant systems have been given in the table 2.

Since gene predictions and sequence annotations are supposed to be very specific for a plant system, a need for generating widely accepted, robust and continuously updated sequence analysis methods integrated with coherent and efficient prediction system is required to perform computer assisted annotation of genomic sequences. In the process, two other plant systems namely rice (*Oryza sativa*) and alfalfa (*Medicago sativa*) has been identified as the model for cereal crops and model system for pulses respectively (Yuan *et al.*, 2001).

Rice holds the smallest genome size, stable transformability, diploid nature and a wide range of molecular and genetic data resources, which make it suitable as a model plant system for cereals. Considering the importance of rice and the concept of developing it as the model species for members in the grass family, an International Rice Genome Sequencing Project was started in 1998. The project took over the target of producing the complete rice genome as well as to develop databases and annotation tools for accurate prediction of genes, regulatory gene network, ORFs, ESTs, cDNA, splice sites etc in rice and other related crop systems (Sasaki and Burr, 2000).

The *in silico* genomics in plants are generally performed on the following on line tools:

2.2.1a AGAD at TIGR (http://www.tigr.org.tdb/ath1/htmls/ath1.htmt)

It can be used for gene or clone name searching, graphic display of clones and gene structures or browsing genes by role categories. BLAST sequence search (http://198.76.161.137/euk-blast/index.cgi?projects=ath1) pulls out sequence matches for a query sequence with results linked to annotation. Its FTP site (ftp://ftp.tigr.org/pub/data/a_thalliana/ath1)

Table 2. Programmes /softwares /tools for plant gene annotation

Annotation tools for plant genomics	Web sites
NCBI EST database	http://www.ncbi.nih.gov/dbEST/index.html
The Institute of Genomic Research	http://www.tiger.org
Arabidopsis database	http://genome-www.stanford.edu/Arabidopsis
Arabidopsis ESTs	http://www.cbc.med.umn.edu/Research Projects/Arbidopsis
Arabidopsis genome sequence	http://www.genome-stanford.edu/ Arabidopsis/AGI/AGI links.html
Intron prediction programs	GRAIL http://compbio.ornl.gov/tools/index.shtml; NetPlantGene http://www.cbs.dtu.dk/services/NetGene2/Nottingham
Arabidopsis stock centrs	http://nasc.nott.ac.uk http://www.ohio.aims.cps.msu.edu/aims
In silico digestion of BACs	NIP (software) http://www.mrc-Imb.cam.ac.uk
NetStart	http://www.cbs.dtu.dk/services/NetStart/
NetPlantGene	http://www.cbs.dtu.dk/NetPlantGene.html
NetGene2	http://www.cbs.dtu.dk/services/NetGene2/

Table 2. Continued

Annotation tools for plant genomics	Web sites
SplicePredictor	http://germlin1.zool.iastate.edu/cgi-bin/sp.cgi
FGENESP	http://genomic.sanger.ac.uk/gf/gf.html
MZEF	http://sciclio.cshl.org/genefinder/
GRAIL v1.3	http://compbio.ornl.gov/Grail-1.3/
GeneMark/GeneMark.hmm	http://genemark.biology.gatech.edu/GeneMark/hum.cgi
GENSCAN	http://CCR-081.mit.edu/GENSCAN.html
RiceGASS	http://www.RiceGAAS.dna.affrc.go.jp
Genscan and RiceHMM	http://www.RiceHMM and http://CCR-081.mit.edu/GENSCAN.html
Printrepeats and RepeatMasker	http://ww.Printrepeats http://www.RepeatMasker
TIGER Orthologous Gene Alignments	http://www.tigr.org./tdb/toga/toga.shtml
Orthologous genes in rice	http://www.tigr.org/tdb/e2k1/osa1/

contains genome sequence of plants and annotation files, including predicted cDNA, CDS, and protein sequences.

2.2.1b MATB at MIPS (http://mips.gsf.de/proj/thal/db/index html)

It can be used for gene or clone name or keyword searching, graphic display of clones and gene structures, browsing plant genes by functional categories, cross-genome comparisons. Its BLAST and pattern-matching box (http://mips.gsf.de/proj/thal/db/search frame.html) performs sequence searches on ESSA-generated sequences are also possible. Its FTP box (ftp://ftmips.gsf.de/cress/) contains clone sequences, tiling paths, standard ORF names, and annotation.

2.2.1c Entrez at NCBI (http://www.ncbi.nlm.nih.gov/entrez/query.fcgi?db=Nucleotide)

The site allows searching of nucleotide, protein, PubMed, and other data (use a search word followed by a plant name (e.g. arabidopsis) to retrieve Arabidopsis data. Its BLAST (http://www.ncbi.nlm.nih.gov/BLAST) can be used for plant gene sequence identification using GenBank sequences alignment. The Genome Section (http://www.ncbi.nlm.nih.gov/PMGfs?Genomes/3702.html) has genomic sequences and other related data on desired plants/organism. Its FTP (ftp://ncbi.nlm.nih.gov) can be used for seeing graphical view of gene annotation (http://sequence-www.stanford.edu/ara/annotation.html).

2.2.1d MapViewer (http://arabidopsis.org/servlets/mapper)

At TAIR (http://arabidopsis.org.home.html) displays ORFs at zoom levels of 200x and higher, and allows wildcard and alias searching on clone names, ORF names, genes and markers. BLAST, FASTA and PATTERN MATCH sequence analysis tools can be used with a complete dataset of AGI clone sequences. Its FTP site contains files of the genome sequence and predicted cDNA, CDS, and protein sequences made available by TIGR and MIPS.

2.2.2. *In silico* proteomics of plants

Advanced databases are essential for gaining insight into protein structure and function from the voluminous, heterogeneous and distributed molecular data. The *in silico* proteomics of plant is performed to know the following aspects (i) identification of proteins by homology searching in a specific protein database; (ii) secondary structure determination of proteins by using programmes available such as softwares for multiple protein sequence alignment (e.g. Clustal W) and machine learning methods. The structure determination is performed through global alignment (alignment of the whole protein against databases of already identified proteins) and local alignment (structure/fold recognition of protein motif/domain of some specific protein. Similarity among motifs of different proteins also helps in establishing the fundamentals for protein phylogeny; (iii) identification of the family/ superfamily of the proteins or to establish a molecular phylogeny of the newly identified protein with the existing protein families in a protein databank. Protein family classification is now well recognized as an effective means for large-scale functional characterization of genes. The proteins are classified in terms of precision, specificity and sensitivity. The protein sequence classification is done using sequence alignment softwares i.e. PSI-BLAST (position specific pairwise alignment) and FASTA and another method is based on Hidden Markov Models (HMM). The HMM (softwares i.e. SAM and HMMer) method uses a machine learning which employs probabilistic graphical models to describe protein motifs of protein superfamilies (identification of homology).

2.2.3. Tools for protein annotation

2.2.3a iProClass

It is an integrated resource that provides comprehensive family relationships at both global (whole protein) and local (domain and motif/ site) levels, as well as structural/functional classifications and features of proteins. The PIR superfamily/family organization allows complete and non-overlapping clustering of all proteins. The iProClass consists of more than 210,000 nonredundant PIR and Swiss-Prot proteins organized with more than 29,000 superfamilies, 100,000 families, 2,600 domains, 1,300 motifs, 280 post-translational modification sites and links to over 30 databases of protein families, structures, functions, genes, genomes, literature, and taxonomy. Protein and superfamily summary reports provide rich annotations, including membership information with length, taxonomy, keyword statistics, comprehensive enzyme and PDB cross-references and graphical feature display. The database facilitates classification-driven annotation for protein sequences and complete genomes and supports proteomics research. The iProClass is implemented in the Oracle 8i object-relational system and is available for sequence/text search and report retrieval at http://pir.georgetown.edu/iproclass.

2.2.2b HMM at TIGR (http://www..tigr.org- http://198.76.161.137/web-hmm/)

It searches protein sequence against a database of protein families formulated on the principles of hidden Markov chain models.

2.2.2c Stanford Gene Technology Centre (http://lugagefast.stanford.edu)

It has DAtA (http://lugagefast.stanford.edu/group/arabprotein/index.html) annotation program, which can be searched by gene or protein name, motif, blast hits. Genscan is used to predict coding sequences, and eMotif program is used to locate the motifs.

2.2.2d MapViewer (http://arabidopsis.org/servlets/mapper) at TAIR

It displays ORFs at zoom levels. Its FTP site (http://tairpub:tairpub@ftp.arabidopsis.org/home/tair/sequences/blast_dataset) contains files of the genome sequence and predicted cDNA, CDS, and protein sequences made available by TIGR and MIPS. Linking of ORFs in MapViewer to TAIR database, more sequence analysis datasets including all Arabidopsis proteins, both cDNA and genomic sequences of each gene.

3. STATUS OF GENOMIC RESEARCH IN PLANTS

3.1. Genome Sequencing Projects

As mentioned above, *Arabidopsis* is a model plant system for studying any aspect of biology of higher plants (Zhang, 1998) including genomics, proteomics, genetics, biochemistry , physiology etc. Considering the importance, this was the first plant system targeted for complete genome sequencing. During the procedure, rice and alfalfa genomes were also considered as specific models for cereals and legumes respectively. The sequencing of rice genome is under advanced conditions. Besides, sequencing projects for several crop plants is also advancing and details of some of the important ones can be retrieved from websites. They are:

- Arabidopsis: The Arabidopsis Information Resource http://www.arabidopsisorg/
- Lucerne: Alfagenes alfalfa (*Medicacago sativa*) http://ars-genome.cornell.edu/cgi-bin/WebAce/webace?db=alfagenes

- Beans: BeanGenes: Phaseolus and Vigna
  http://beangenes.cws.ndsu.nodak.edu

- Brassica: CabbagePatch
  http://brassica.bbsrc.ac.uk/
  http://jic-bioinfo.bbsrc.ac.uk/BrassicaDB/

- Legumes: CoolGenes: cool season food legumes
  http://coolgenes.cahe.wsu.edu/

- Cotton : CottonDB : Gossypium hirsutum
  http://algodon.tamu.edu/htdocs-cotton/cottondb.html

- Cereal GrainGenes: wheat, barley, rye and relatives
  http://wheat.pw.usda.gov

- Maize: MaizeDB
  http://www.agron.missouri.edu/

- Perl millet: MilletGenes
  http://jiio5. Jic.bbsrc.ac.uk:8000/cgi-bin/webace?db=millet

- Rice: RiceGenes
  http://ars-genome.cornell.edu/rice/

- Korean Rice Genome Project
  http://bioserver.myongji.ac.kr/

- Rosaceae: RoseDB
  http://ars-genome.cornell.edu/rosedb/

- Solanaceae: SolGenes
  http://ars-genome.cornell.edu/solgenes/welcome.html

- Sorghum bicolor: SorghumDB
  http://algodon.tamu.edu/sorghumdb.html

- Soybeans: Soybase
  http://129.186.26.94/

- TreeGenes: forest tree
  http://dendrome.ucdavis.edu/index.html

3.2. GENE ANNOTATION PERFORMED IN PLANTS

3.2.1. Annotation of Arabidosis Genomic Sequences

The genome of *Arabidopsis thalliana* has been considered as the model genome for the higher plants. It was also the pioneering plant to have its complete genome sequenced in the year 2000. The genome of this model plant was sequenced by an international collaboration, The Arabidopsis Genome Initiative. The genome is packaged into 5 chromosomes and contains 19 million base pairs and its complete sequencing took 5 years. The chromosomes 1, 2, 3, 4 and 5 have been analyzed completely for the presence of genes located on them. Analysis of the sequence of chromosome 5 yields further insights into physiological, developmental and biochemical properties of the Arabidopsis and other plants.

Its mutational breeding, genome/ gene annotation and function genomics have led to characterization of various important genes and properties in different plant systems i. e. identification and analysis of expression pattern of disease resistant gene in wheat, ripening of tomatoes, doubling of yield of rape seed oil, description of growth and development of a plant i.e. characterization /prediction of gene/genes responsible for the procedure. The genomic sequences at chromosomes 1, 2, 3, 4 and 5 have been annotated to have 6,756, 4,211, 5,246, 3,958 and 5,985 genes respectively (http://www.tigr.org).

The chromosome 5 is 26,000,000 bp long and it has 5,985 genes coding for several functions in plants. The annotation of Ch-5 reveals that there are 31,226 exons, 25,352 introns. The predicted genes include many genes having significant similarity with genes from other organisms. This indicates usefulness of Arabidopsis sequence in application beyond crop plant improvement. Generally, annotation in *Arabidopsis* is performed by HMM based softwares. The gene prediction tools used were Genemark.hmm and Genescan 2 (Sato *et al.*, 1997; Zhang, 1998; Marra *et al.*, 1999; Motoaki *et al.*, 2001).

3.2.2. Annotation of Rice Genomic Sequences

The genome annotation is done to add the value to its sequenced parts. A total of 54, 510 genes have been predicted out of 2,025 BAC/ PAC clones of rice genome carrying a total size of 270,690,300 nucleotide base pairs (on 10 th march 2002; http://www.RiceGAAS.dna.affrc.go.jp).

The same site shows details of 42,764 predicted DNA motifs and 2,562 predicted LTR pairs. The sequences of annotated genes along with supportive evidences can directly be retrieved

The different softwares currently used for rice gene annotation are FGENESH, Genmark.hmm (rice), Genscan (Maize), Genscan+, Glimmer R, GeneSplicer and tRNAscan-SE (Sakata *et al.*, 2000b; 2002).

The GlimmerR is a special version of GlimmerM, which has been trained to identify rice genes; in this software a second order Markov chain has been combined with the score of splice sites, which computed by maximal dependence algorithm. GlimmerR has a new dynamic programming algorithm that constructs the gene models. The algorithm prunes the large number of exons-intron combinations by scoring the exons and introns of a putative gene using decision trees.

In a more recent development in the direction of software development, an automated annotation system namely, RiceGAAS.dna.affrc.go.jp (Sakata *et al.*, 2002) integrated with a rice genome database has been launched (http://www.RiceGAAS). This automated system has a total of 14 gene analysis programmes for complete rice gene annotation (http://www.tigr.org./tdb/e2k1/osa1/). These programmes are : BLASTN for homology search against protein database, Genscan and RiceHMM for gene domain prediction; MZEF for exon prediction; SplicePredictor for splice site prediction; Printrepeats and RepeatMasker for repetative sequence detection; tRNAscan for transfer RNA prediction; SOSUI for classification and secondary structure prediction of membrane protein; and PLACE –SignalScan for cis-element detection. RiceHMM is gene domain prediction program for rice genomic sequences based on hidden Markov model and integrated with a database of rice ESTs composed of 15,000 cDNAs (Sakata *et al.*, 2000a&b; 2002; Yuan *et al.*, 2001).

The website http://www.tigr.org./tdb/e2k1/osa1/ has all the tool compiled for automated annotation of all rice BACs/PACs, this site also shows details of computationally analyzed sequences and identified putative orthologous genes using TIGER Orthologous Gene Alignments (http://www.tigr.org./tdb/toga/toga.shtml). Another website http://www.tigr.org/tdb/e2k1/osa1/ provides informations and tool for identifying candidate orthologous in rice.

3.2.3. Gene Annotation of other plants

It is obvious that the genome sequencing/mapping projects for several crop plants have been started and there is considerable progress in various genomes; the developments in the sequencing different cereals, legumes, fruits, vegetables, trees and ornamental plant can be obtained at http://www.nal.uada.gov/pgdic/Map_proj/). The raw nucleotide sequences of different plants are annotated to characterize/predict the genes and related gene net work. The plants are: barley, bean,, corn cotton, fungi, lettuce, mushroom, pinus, rye, sugarcane,, sorghum, wheat, apple, pea, plum, lillium,peach, potatto, tomato, tobacco, wheat,cucumber, spruce, rice, rose, sorghum, carrot etc.

4. CONCLUSIONS AND FUTURE PROSPECTS

The accumulation of plant genome sequencing data has been considered as the great source of genetic informations; annotation of genomes to bring accurate prediction of genome and proteome analysis will help a plant biologist to improve the existing varieties. The crop improvement can be performed through followings:

- Identification of important genes through *in silico* genomics and expression analysis using suitable softwares and databases. In conjugation with design of transgenic plants, the targeted gene responsible for improving quantitative and qualitative traits in commercially important crops could be identified.
- Designing of agrochemicals based on analysis of the components of signal and transduction pathways of the plant to select targets. Furthermore, with the help of chemiinformatics, identification of potential compound having herbicide/ pesticide/ insecticide properties is required which is supposed to block the concerned signal and transduction pathways.
- Identification of genes having pest/ insect/water/salt/cold/heat resistance properties can be identified in various crop systems. Such identification will help in performing functional genomics of plants in order to have desired expression of the concerned genes.
- The powerful tool of bioinformatics will help in organizing the genomic resource datas and to extend ones ability to analyze these complex biological systems. The analyzed data may help in answering basic question of plant biology.
- Efficient and constructive utilization of biological repository of clones, cell lines organisms and seed will take place. Existing on

line repositories are in the process of linking with each other's databases.

5. WEB RESOURCES

Followings are the major internet resources for plant bioinformatics:

- Arabidopsis Genome Initiative
  http://genome.stanford.edu/Arabidopsis/AGI
- Arabidopsis Genome Data Analysis, Cold Spring Harbour Laboratory
  http://nucleus.cshl.org/protarab
- Plant Genome Information Centre, USDA
  http://www.nal.usda.gov/pgdic
- UK Crop Plant Bioinformatics Network
  http://synteny.nott.ac.uk/agr/agr.html
- The Institute for Genomic Research Database (TIGR)
  http://www.tigr.org
- Arabidopsis Genome Centre at the University of Pennsylvania
  http://genome.bio.upenn.edu/ATGCU.html
- The Genome Sequencing Centre of Washington University in St. Louis
  http://genome.wustl.edu/gsc
- Grain Genes Database
  http://wheat.pw.usda.gov
- Maize Genome Database
  http://www.agron.missouri.edu
- Arabidopsis Internal Coding Exon Finder
  http://clio.cshl.org/genefinder/ARAB/arab.html
- NetPlantGene V2.0 Web Prediction Server
  http://cbs.dtu.dk/NetPlantGene.html

REFERENCES

Marra M, Kucaba T, Sekhon M, Hillier L, Martienssen R, Chinwalla A, Crockett J, Fedele J, Grover H, Gund C, MacCombie W, KenMcDonald, McPherson J, Mudd N, Parnell L, Schein J, Seim R, Shelly P, Waterstone R and Wilson R

(1999). zA map for sequence analysis of the *Arabidopsis thalliana* genome. *Nature Gen.* **22** : 265-269.

Meinke DW, Cherry JM, Dean C, Rounsley SD and Koornneef M (1998). *Arabidopsis thalliana* : A model plant for genome analysis. *Science* **282** : 662-682.

Sakata K, Antonio BA, Mukai Y, Nagasaki H, Sakai Y, Makino K and Sasaki T (2000a). INE : a rice genome database with an integrated map view. *Nucleic Acids Res.,* **28** : 97-102.

Sakata K, Nagasaki H, Idonuma A, Watanabe W, Kise M and Sasaki T (2000b). RiceHMM:gene domain prediction program for rice genome sequence. *Abstracts of 4th Annual Conference on Computational Genomics.* 31.

Sakata K, Nagamura Y, Numa H, Antonio BA, Nagasaki H, Idonuma A, Watnabe W, Shimizu Y, Horiuchi I, Matsumoto T, Sakai T and Higo K (2002). RiceGASS: An automated annotation system and database for rice genome sequence. *Nucleic Acids Res.,* **30** : 98-102.

Sasaki T and Burr B (2000). International Rice Genome Sequencing Project: the effort to completely sequence the rice genome. *Curr. Opin. Plant. Biol.,* **3** : 138-141.

Sato S, Kotani H, Nakamura Y, Kaneko T, Asamizu E, Fukami M, Miyajima N and Tabata S (1997). Structural analysis *Arabidopsis thaliana* chromosome 5. I. Sequence feature of the 1.6 Mb regions covered by 20 twenty physically assigned P1 clones. *DNA Research,* **4** : 215-230.

Motoaki S, Mari N, Hiroshi A, Mie K, Kazuko YS, Carninci, Hayashizaki and Kazuos (2001). Monitoring the expression pattern of 1300 *Arabidopsis* genes under drought and cold stresses using a full-length cDNA microarray. *Plant Cell,* **13** : 61-72.

Yuan Q, Quackenbush J, Sultana R, Pertea M, Salzberg SL and Buell CR (2001). Rice Bioinformatics: Analysis of rice sequence data and leveraging the data to other plant species. *Plant Physiol.* **125** : 1166-1174.

Zhang M (1998) Identification of protein coding regions in *Arabidopsis thaliana* genome based on quadratic discriminant analysis. *Mol. Plant Biol.* **37** : 803-806.

WEBSITES FOR REFERENCES

http://bioserver.myongji.ac.kr
http://demeter.bio.bnl.gov
http://gene.affrc.go.jp
http://www.genome.clemenson.edu/where/nippon bac/index.html
http://www.rgp.dna.affrc.go.jp
http://www.genome.cornell.edu/rice/
http://www.genome.cornell.edu/riceblast)
http://www.gramene.org
http://www.RiceGAAS
http://www.shigen.nig.ac.jp
http://www.tigr.org/tdb/e2k1/osa
http://www.tigr.org/tdb/ogi
http://www.tigr.org/tdb/rice/blastsearch.shtml
http://www.tigr.org/tdbrice
http://wheat.pw.usda.gov/
http://synteny.nott.ac.uk/
http://www.GrainGenes.org
http://www.HarvEST

http://ars-genome.cornell.edu/solgenes/welcome.html
http://www.inra.fr/Versailles/BIOCEL/Filters
http://cmgm.stanford.edu/pbrown/explore/index.html
http://expasy.hcuge.ch/ch2/2d-index.html
http://sphinx.rug.ac.be:8080
http://www.arabidopsisorg/
http://ars-genome.cornell.edu/cgi-bin/WebAce/webace?db=alfagenes
http://beangenes.cws.ndsu.nodak.edu
http://brassica.bbsrc.ac.uk/
http://jic-bioinfo.bbsrc.ac.uk/BrassicaDB/
http://coolgenes.cahe.wsu.edu/
http://algodon.tamu.edu/htdocs-cotton/cottondb.html
http://wheat.pw.usda.gov
http://www.agron.missouri.edu/
http://jiio5. Jic.bbsrc.ac.uk:8000/cgi-bin/webace?db=millet
http://ars-genome.cornell.edu/rice/
http://bioserver.myongji.ac.kr/
http://ars-genome.cornell.edu/rosedb/
http://ars-genome.cornell.edu/solgenes/welcome.html
http://algodon.tamu.edu/sorghumdb.html
http://129.186.26.94/
http://dendrome.ucdavis.edu/index.html

AUTHOR INDEX

SUBJECT INDEX

$ = 650 (Sat)